T0091807

Science of Vision

K.N. Leibovic
Editor

Science of Vision

With 171 Figures in 291 Parts

Springer-Verlag
New York Berlin Heidelberg
London Paris Tokyo Hong Kong

K.N. Leibovic
Department of Biophysical Sciences
Department of Ophthalmology
State University of New York at Buffalo
Buffalo, New York 14214
USA

Library of Congress Cataloging-in-Publication Data
Science of vision / K.N. Leibovic, editor.
p. cm.
Includes bibliographical references.
1. Vision. 2. Visual perception. 3. Computer vision.
I. Leibovic, K. N., 1921–
QP475.S44 1990
612.8'4–dc20 90-9589
Printed on acid-free paper

Softcover reprint of the hardcover 1st edition 1990

Typeset by Publishers Service of Montana, Bozeman, Montana.

9 8 7 6 5 4 3 2 1

ISBN-13:978-1-4612-7998-3 e-ISBN-13:978-1-4612-3406-7
DOI: 10.1007/978-1-4612-3406-7

Preface

Over the past several years there have been important developments in our understanding of the biological bases of vision on the one hand and in the formulation of computational theories on the other hand. Contributions to this progress have come from a wide range of disciplines in the biological and physical sciences. Although the implementation of function in biological and man-made systems may be very different, there are principles applicable to a task which they share in common. The theme of this book is that Vision Science is an interdisciplinary subject. In this it is in the tradition of *Information Processing in the Nervous System* which I edited and which was published by Springer-Verlag in 1969.

A few years ago Malcolm Slaughter and I with the help of our colleagues formed an Interdisciplinary Group in Vision and I put together a Vision Course for graduate and postdoctoral students who came from anatomy, biophysics and biochemistry, computer science, electrical engineering, ophthalmology, physiology, and psychology. This book grew out of that course. We have been fortunate to have contributions from both inside and outside SUNY/Buffalo. Had it not been for this unique participation from a number of disciplines we may not have written yet another book on vision of which there are already enough to fill several libraries.

We make no attempt to be encyclopedic. But we should like to believe that the topics we have chosen are interesting and representative of current knowledge and research in the vision sciences. The contributors to this volume have all been involved in active research. Naturally, they have developed their individual views, which may not always be in the groove of the most repeated and prevailing dogma. This is another motive for writing yet another book on vision. In some places we have included previously unpublished material and we have put forward hypotheses which hopefully will encourage critical thinking.

It is always difficult to address a diverse audience. Different authors deal with this in different ways and this, inevitably, is reflected in the following chapters. Our aim has been to review the present status and take the reader to problems in current research including some as yet unresolved issues. We have tried to make the material accessible to the scientifically mature, nonspecialist reader. But the specialist may find a fresh perspective, while the scientist from another or related discipline will become aware of the progress being made. Hopefully, the reader will gain an appreciation, if he or she does not already have one, of the rich variety and depth of the science of vision.

Our coauthors range from the young scientists near the beginning of their careers who are helping to advance the exciting new research in vision, to those of us looking back on a life dedicated to pushing aside the curtains enfolding our ignorance. Among the latter it is a particular pleasure to acknowledge the collaboration of Robert Boynton who will be retiring in 1991. He contributed to Information Processing in the Nervous System as well as to this volume. With his present chapter he can look back on 40 years of work starting with adaptation and proceeding to color vision.

Finally, on a personal note, to the extent that my efforts as author and editor may be of any merit, they are dedicated to the memory of my parents Joseph Avigdor and Chassia Leibovic and of my son David Avigdor Leibovic.

K.N. Leibovic

Acknowledgment: The preparation of this manuscript was aided by a gift to K.N. Leibovic from the U.B. School of Medicine Class of '39.

Contents

Part 3: Theory and Computation

Contributors

ROBERT M. BOYNTON Department of Psychology, University of California, San Diego, La Jolla, California 92093, USA

JOHN R. COTTER Department of Anatomical Sciences, State University of New York at Buffalo, Buffalo, New York 14214, USA

SVEN J. DICKINSON Computer Vision Laboratory, Center for Automation Research, University of Maryland, College Park, Maryland 20742, USA

BRUCE M. DOW Department of Physiology, State University of New York at Buffalo, Buffalo, New York 14214, USA

THOMAS E. FRUMKES Department of Psychology, Queens College of CUNY, Flushing, New York 11367, USA

VICTOR KLYMENKO Research Division, Radiology, University of North Carolina at Chapel Hill, Chapel Hill, North Carolina 27514, USA

K.N. LEIBOVIC Department of Biophysical Sciences and Department of Ophthalmology, State University of New York at Buffalo, Buffalo, New York 14214, USA

WILLIAM MAGUIRE Department of Radiology, Long Island Jewish Medical Center, New Hyde Park, New York 11042, USA

ALEX P. PENTLAND Vision Sciences Group, The Media Laboratory, Massachusetts Institute of Technology, Cambridge, Massachusetts 02139, USA

AZRIEL ROSENFELD Computer Vision Laboratory, Center for Automation Research, University of Maryland, College Park, Maryland 20742, USA

PETER D. SCOTT Department of Electrical and Computer Engineering, State University of New York at Buffalo, Buffalo, New York 14214, USA

MALCOLM SLAUGHTER Department of Biophysical Sciences, State University of New York at Buffalo, Buffalo, New York 14214, USA

SUSAN B. UDIN Department of Physiology, State University of New York at Buffalo, Buffalo, New York 14214, USA

DEBORAH WALTERS Department of Computer Sciences, State University of New York at Buffalo, Buffalo, New York 14214, USA

NAOMI WEISSTEIN Department of Psychology, State University of New York at Buffalo, Buffalo, New York 14214, USA

RICHARD P. WILDES Department of Computer Sciences, State University of New York at Buffalo, Buffalo, New York 14214, USA

Overview

K.N. Leibovic

Vision is the paramount sense in man. Not only are there more fibers dedicated to transmitting to the central nervous system the output from the eyes than from any other peripheral sense organ, but our mental processes, including memory, rely heavily on vision. We remember a visual scene more readily than a tactile or odorous experience and our language is full of visual metonyms: "vide" the following: "I can see behind his facade" when we mean "I can figure out in my mind what is going on in his"; "I take a different view" when we mean "I have a different opinion"; "I shall look into this" when we mean "I shall investigate this"; "He is a man of foresight" when we mean "He can think ahead" and so on. If no well-formulated thoughts can exist without language, then these examples suggest that an understanding of vision is preliminary to an understanding of thought processes in the brain. Because vision is so important to man it is also important in operations serving man, such as robotics, image analysis of geological and weather surveys, and satellite observations of distant planets.

The purpose of this brief overview is to provide a contextual setting for the chapters that follow. This book is divided into three parts. The first deals with the biological bases of vision, the second with psychophysics, and the third with theory and computation.

We begin with a brief review of the visual pathways in Chapter 1, so as to put the structures which are subsequently discussed into perspective. Chapter 1 also introduces some basic characteristics of neural tissue and terminology frequently encountered in neuroanatomy and physiology. Over the past 20 years a surprisingly large number of primate cortical areas have been found to be involved in vision (Chapter 1, Fig. 1.8, see also Van Essen, 1979; Cowey, 1981; Van Essen and Maunsell, 1983) further attesting to the dominant role of vision in sensory information processing. These areas of cortex are richly interconnected and we are beginning to get glimpses of their functions.

In Chapter 2 we deal with photoreceptors which transform the physical signal into a neural response. The most publicized discovery in recent years has been the elucidation of the transduction cycle by which the absorbed photon energy is amplified some 10^5 times, converted to a membrane conductance change and transmitted to subsequent neurons. The hypothesis advanced by Hagins and his colleagues several years ago (Hagins, 1972) that calcium is the "internal transmitter" has not been confirmed, but it has been a major stimulus which ultimately led to our present understanding of the transduction biochemistry and the role of cGMP.

The properties of photoreceptors are complex and fascinating. They make it possible for us to adapt to an enormous range of light intensities. The mechanisms subserving this adaptation are at the frontiers of present research. In addition many psychophysical phenomena to which retinal and cortical networks make their contribution can be traced back to photoreceptors. This includes adaptation, flicker fusion, temporal summation, and color vision.

Chapter 2 focuses on the cellular processes in photoreceptors and while these specifically subserve the transduction of light to a neural signal, they are representative in a more general sense. Thus, membrane potentials, currents, and conductance changes are observed in all neurons, G proteins participate in many cellular reactions where a cascade involving cyclic nucleotides is activated, and Ca^{2+} plays a ubiquitous regulatory role in cells.

In Chapter 3 we turn from cellular processes to interactions between cells in the retinal network. Major advances in retinal research have occurred with the physiological and pharmacological identification of segregated channels for increasing and decreasing light signals, the ON and OFF channels, with their associated transmitters and receptors. Although the postsynaptic effect is either excitation or inhibition, there is a rich modulation of the state of the cell due to the control and interplay of different transmitters and receptors, and this is beginning to show us a rationale for the variety of transmitters and receptors which are found in the nervous system. The ON and OFF channels begin with the bipolar cells.

A type of cell, the interplexiform cell, has relatively recently been discovered in teleost fish (Dowling and Ehinger, 1978; Ehinger, 1983). It feeds back from the inner to the outer plexiform layer and uncouples horizontal cells in dim illumination. This reduces surround inhibition and increases the transmitted postsynaptic photoreceptor signal. Here we have an example of a network mechanism for adaptation.

Originally, ganglion cell responses were characterized as ON, OFF, and ON–OFF depending on whether they responded to light ON, OFF, or both. In addition to these types, large (α) and small (β) ganglion cell receptive fields have now been characterized as linear or nonlinear in their input summation, brisk or sluggish and transient or sustained in their responses. A correlation of these properties with local or global computations, form, or movement seems to be emerging. At least since the pioneering studies of Lettvin, Maturana, McCullough, and Pitts (1959) it has been an aspiration to correlate neuronal activity with behavior and perception. Studies of the retina have contributed much to progress in this respect. The retina can appropriately be described as an "approachable part of the brain" (Dowling, 1987).

What Lettvin and his colleagues achieved in the retina and tectum has its counterpart in the discoveries of Hubel and Wiesel in cortex: cells tuned to the orientation of bars and edges could be the neural basis for the perception of form.

Hubel and Wiesel also adduced evidence for the functional organization of cortex in columns perpendicular to the cortical surface. Since then the anatomy and physiology of the many visual areas of the cortex have been intensively studied. Of particular interest are the findings that cell types of different characteristics in the retina give rise to pathways that can be traced to the highest levels in cortex. In the parvocellular layers of the lateral geniculate nucleus (LGN) the cells are sensitive to color and their small receptive fields have good resolution, whereas the magnocellular layer cells have large receptive fields, poor resolution, and are not sensitive to color. In V1, form and color arise from parvocellular LGN input, whereas movement is represented separately and arises from the magnocellular LGN input. Form and color cells from V1 project to separate areas of V2 and movement cells project to MT. Cells tuned to stereoscopic disparity and direction are also found in V1 and they project to V2 and MT. V2 and V3 project to V4 and MT to MST. Even at these more distant stages there is evidence for functional specialization, e.g., for color in V4 and for movement in MST (Livingstone and Hubel, 1988). The picture that emerges from this is that there are separate streams of information flow in which perceptual variables are processed in stages. In cortex the rich interconnections occur between areas which all have the same or very similar gross structure, but different perceptual variables are transmitted in streams of labeled lines. The thesis of Dow in Chapter 4 is that while the retina is mapped topologically onto V1, each variable is also mapped continuously across and parallel to the surface of the cortex. This necessitates that all variables be represented in a cortical column, the address for a particular retinal location and for a particular set of values of the variables. These values must be broadly tuned across the cortical surface if there is to be a continuum of computed values.

It is not yet clear how integrated percepts arise within the cortical machinery for information processing. This problem is immanent in much of the best in neuroscience. Its threads run through Chapter 9 and it is discussed in Chapters 6 and 10. An interesting, recent proposal (Damasio, 1989) is that different subsets of sensory input, which are processed in different brain regions, project onto "convergence zones." The latter feed back onto the same brain regions and synchronize their activity. The percepts are then represented in this synchronized activity.

Though much remains speculative, a miracle of design and performance is already revealed to us. How is it all put together? How do the immature networks develop and get connected in such intricate yet beautifully functioning systems? This is the topic of Chapter 5. The biological design is contained in the DNA of the individual. But this does not specify each detail in development, which proceeds by trial and error based on the guidelines of the design. Neurons send out growth processes which find their targets through molecular recognition, including factors that are influenced by activity and usage. Some processes survive and establish the appropriate contacts; others are withdrawn or atrophy.

Molecules, cells, and networks are the content of the biology dealt with in Part 1. But the curiosity that draws many of us is to understand behavior and perception as they arise from the biological substratum. Psychophysics continues to offer the quantitative approaches to such understanding. This is the content of Part 2.

Chapter 6 begins with questions of time and space and how our ideas on these may be rooted in our perceptual apparatus. We are looking here at the broad picture before getting into more specialized concerns in the following chapters. Rather than reviewing the enormous literature in this field we have chosen to confront some notions on temporal simultaneity and succession and on our experience of the present, while with regard to space perception we consider some results on binocular vision which have implications for a perceptual coordinate system and for the representation of percepts in the neural machinery of the brain. Ideas of the world we live in and of life itself are shaped by our experience of space and time. Chapter 6 touches on some of this experience.

A remarkable property of the visual system is its ability to detect just a few photons in the dark as well as to function at light intensities many orders of magnitude brighter. Yet through all these changes and under lighting conditions of diverse spectral composition we can preserve a constancy in our perception of contrast and color. What are the characteristics of human color vision and of adaptation to light and dark which describe these phenomena? Chapter 7 deals with adaptation and Chapter 8 with color vision.

While aspects relating to photoreceptors are discussed in Chapter 2, the psychophysical results of Chapter 7 also encompass interactions in networks. It has become clear in recent years that such interactions already are present between rods and cones in the retina and continue right through to cortex.

Psychophysical studies show how the system, which is tuned to maximal sensitivity at low light intensities, changes to preserve perceived contrast at higher light intensities. The quantitative formulations of these characteristics are expressed in the square root law, which implies optimal detection, and the Weber law, which implies contrast constancy. Not only sensitivity, but response kinetics also change with adaptation: at low light levels there is a long integration period permitting the capture of as many photons as possible; at higher light levels, sensitivity is traded for speed of response (cf. Bloch's law).

While classical studies emphasized thresholds, more recent work has been concerned with the many other aspects of visual perception influenced by adaptation. These are considered in Chapter 7 and the psychophysics is related to neural function.

Color, the other aspect of perceived constancy, is treated in Chapter 8, which clarifies the basic parameters of human color vision. Perceived color cannot be understood simply in terms of wavelength mixtures. Color constancy is a case in point. Colors appear "the same" in sunlight as in artificial illumination. Together with the preservation of contrast, as already alluded to, the result is a stability of perceived form and color over a wide range of conditions. Thus, familiar things remain familiar. Yet there are subtle effects which show that color constancy is not perfect. The factors which influence color vision include light sources, surface reflectances, and the physiology of sensation.

Advances in psychophysics as well as physiology have shed new light on the trichromatic theory of Helmholtz and the opponent color theory of Hering (see also Chapter 4). Chapter 8 provides a background and appreciation of our present state of knowledge.

Part 2 concludes with a review of information processing channels in Chapter 9. These channels, explored through psychophysical techniques, echo the physiologically identified channels which have been mentioned earlier. Psychophysicists speak of analyzers when physiologists speak of target cells and their receptive fields. The one speaks of stimulus dimensions when the other speaks of tuning to a range of orientations or binocular disparities or velocities. This is not just a parochial terminology difference. The methodologies are different and the corresponding functional entities need not have the same anatomical substrate. But the questions treated in Chapter 9 are at the heart of the neural bases of perception.

Fourier techniques have been used extensively in deriving the results in Chapter 9. The application of these techniques to vision was stimulated by the research of Campbell and his associates using various gratings of different spatial frequencies (see, e.g., Campbell, 1968, 1974). We now speak of spatial as well as temporal frequency channels in the visual system. There have even been suggestions that the brain is a Fourier analyzer, and there have been models of perceptual and memory processes based on holography. As usual, early enthusiasm with new results encouraged some unwarranted claims. In spite of these, important insights have been achieved. This is clear from reading Chapter 9, which presents a wide ranging review of psychophysical phenomena and what they tell us about information processing channels.

In the past psychophysics has led physiology, such as in color vision, binocular vision, and light and dark adaptation. Discoveries in physiology in turn raise new questions for psychophysics. This is apparent from the chapters in Parts 1 and 2 and the case is stated particularly well by Boynton in connection with color. The analysis of form perception in three-dimensional space appears to be more complex than color as noted in Chapter 9. An example of the links between physiology and psychophysics concerning space perception is discussed in Chapter 6. It has to do with a perceptual coordinate system and with the neural representation of percepts. Psychophysicists discovered that binocular disparity was the relevant stimulus in reduced cue situations. Other cues, like accommodation or convergence of the eyes, apparently have minimal effects. Later, physiologists discovered cortical cells tuned to disparity, so that one could speak of a representation of the three-dimensional world in terms of the activities of such disparity sensitive cells (see Chapter 4). Apparently there are also cells tuned to radial direction with respect to the observer. Earlier, psychophysicists had derived a set of curves which appear equidistant (the "Vieth–Müller circles") and another set which appears to radiate from the observer (the "Hillebrand hyperbolae"). A theoretical analysis shows these two sets of curves to have the property that as the point of fixation moves over them, their retinal images remain invariant (Balslev and Leibovic, 1971). Thus, there are grounds, based on psychophysics, physiology, and theory, for postulating the presence of a perceptual, egocentric "polar coordinate system." Another result from the psychophysics of binocular space perception concerns the perceived frontoparallel lines or planes. A set of cells which is activated by such a frontoparallel plane in a reduced cue setting does not represent a frontoparallel plane when the fixation point changes (Leibovic et al., 1970, 1971). The activity of the same cells no longer represents the same percept (see Chapter 6). What does this tell us about percepts in general and what other than disparity cues are involved here? It is a challenge to psychophysicists and, ultimately, to physiologists as well.

Among theoretical problems such as these, there are some that require mathematical analysis and computational approaches. This is the content of Part 3.

We believe that biological structures are designed optimally. Can we see optimality principles at work in specific cases? In Chapter 10 the diameter and spacing of photoreceptors and the lengths of their outer segments are analyzed in this spirit. In Chapter 10 we also consider the convergence and divergence of fibers between different cell groups conveying signals to each other. This is a universal feature in nervous systems and it is different from present man-made systems. What are the properties of such a system? What principles of information processing apply to it? We can ask specific questions in this context, such as why some receptive fields are large and others small, and why some overlap extensively and others hardly at all. A convergence and divergence of signal flow can have implications for the way information is encoded and processed in the nervous system. This is relevant to notions of emergent percepts and the discussions in Chapters 4, 6, and 9.

Another set of problems concerns basic strategies that any visual system can adopt in extracting information or generating a set of descriptors of a visual scene. For example, how are three-dimensional percepts or features generated from two-dimensional representations either in binocular vision or in a video transmission from an interplanetary probe? Is there a useful set of primitive elements from which our world as we see it can be constructed? What would these be for three-dimensional objects? And finally, in machine vision, what is the present status of the technology and how does it relate to studies of biological vision?

In Chapter 11 we consider the philosophy motivating computational vision as it is based on "artificial intelligence" (AI), a subject defying precise definition. Chapter 11 contains a discussion of some fundamental approaches and illustrates their application with examples from stereoscopic imaging. It emphasizes the similarities and differences between biological and artificial systems.

As noted in Chapter 10, image reconstruction is not interesting per se. Rather, the interest lies in the encoding of an image in terms of a set of descriptors and the transformations and processing of the image which this makes possible. This kind of problem must be solved in biological as well as in man-made systems. It follows that there are parallels between these systems on a fundamental level.

In human vision we can recognize an object we know in the real world from a representation in a shaded two-dimensional picture, from a line drawing, from suitable blobs of color, from a conjunction of elementary components, and so forth. Computer scientists have taken at least some of their cues from psychophysics. As Walters points out in Chapter 12 all computer vision has biological significance in the sense that it is we who use it and communicate with it. It is therefore not surprising that some authors have derived primitives from psychophysical experiments and considerations; and it is interesting that the early use of line elements in image reconstruction drew reinforcement from the discovery of edge and bar detectors in the visual pathway and cortex.

Several different approaches are represented in Part 3 of this book.

In Chapter 14 Pentland adopts an evolutionary approach starting with simple filters to extract visual parameters, as presumably may have been the case in creatures at an early stage of development. Complexity can be achieved by combinations and extensions of simple filters and by applying various computations to their output. It turns out that image features can be captured to a surprising degree as attested, for example, by the success of reconstructing the image in a different pose or under some other transformation (see, e.g., Chapter 14, Fig. 14.2). While much work in computational vision, and AI in general, involves networks of the threshold elements originally created by McCullough and Pitts (1943), Pentland also uses Fourier analysis. In this he is closer to the spirit which motivates Weisstein and her colleagues in Chapter 9. Pentland shows how his methods can be used in psychophysically meaningful tasks such as recovering shape from shading and range from movement and from depth of focus.

Different approaches are taken by Walters in Chapter 12 and by Dickinson et al. in Chapter 13. In both cases, objects are considered as being composed of elementary units that yield perceptually significant image primitives. In Chapter 12 these primitives come from psychophysical data which have a bearing on recognizing line drawings. In Chapter 13 the primitives are Biederman's psychophysically based "geons" (Biederman, 1985).

Vision comes naturally to us and without conscious effort. Because of that it is more difficult to specify analogous paradigms in computer vision with the necessary precision. This is a question addressed by Wildes as well as by Walters and Dickinson et al. Is it more efficient, with respect to recognizing an object or a scene, to have two-dimensional, viewer-centered or three-dimensional, object-centered representations? Since there are infinitely many aspects which an object can present to a viewer, do we need a huge viewer memory for recognition? Both nature and computer science have developed solutions to minimize the size and maximize the efficiency of memories. One such solution is presented in Chapter 13. It is based on making the viewed aspect invariant to minor changes of shape. It emphasizes the qualitative rather than the quantitative character of the primitives.

Dickinson et al. use a hierarchy of forms composing the primitives of which, in turn, more complex objects are composed. Going from the bottom-up and top-down in the hierarchy they assess the efficiency of reconstructing primitives from the different lower level structures. They do this by evaluating the probabilities that a lower level structure belongs to a higher level structure in their scheme (see Chapter 13, Fig. 13.9).

Integrating viewer-centered and object-centered approaches, Dickinson et al. have developed a novel method for pattern recognition.

The combination of object- and viewer-centered models is also fundamental to a description of binocular vision, which takes us back to Wildes' topic in Chapter 11.

Undoubtedly, significant applications of computer science occur in machine vision, and we would be remiss if we did not mention this topic. Scott addresses it in Chapter 15. Machine vision is now used in areas such as quality control and robotics in industry, diagnostic imaging and reading machines in medicine, and in geophysical mapping and satellite communications. It is a rapidly growing industry which is bound to develop both inside and outside a biologically based environment. It is not easy to predict where these changes will lead us.

Some day, no doubt, at least some of our theories will be seen to be wrong or incomplete, as has happened in the past. A case in point is the nature of light itself on which our vision depends: is it corpuscular or wavelike as Newton and Huygens, respectively, believed? Newton was proved wrong in the case of refraction and Huygens' theory cannot explain why the wave does not propagate backward. And yet they are both right, in a sense, in terms of the wave–particle duality of matter and radiation and both their models do describe a restricted set of phenomena. The danger of being proved wrong cannot stop our theorizing. For if there is any rationale to our existence, it is to comprehend nature and to satisfy our curiosity in a responsible manner.

References

Balslev E, Leibovic KN (1971). Theoretical analysis of binocular space perception. J. Theoret. Biol. **31**:77–100.

Biederman I (1985). Human image understanding: Recent research and a theory. Comp. Vision, Graphics & Image Process. **32**:29–73.

Campbell FW (1968). The human eye as an optical filter. Proc. IEEE **56**:1009–1014.

Campbell FW (1974). The transmission of spatial information through the visual system. *The Neurosciences. Third Study Program.* MIT Press, Cambridge, MA.

Cowey A (1981). Why are there so many visual areas? In *The Organization of the Cerebral Cortex* (FO Schmitt, FG Worden, G Adelman, S Dennis, eds.). MIT Press, Cambridge, MA.

Damasio AR (1989). The brain binds entities and events by multiregional activation from convergence zones. Neural Computation **1**:123–132.

Dowling JE (1987). *The Retina. An Approachable Part of the Brain.* The Belknap Press of Harvard University Press, Cambridge, MA.

Dowling JE, Ehinger B (1978). The interplexiform cell system 1. Synapses of the dopaminergic neurons of the goldfish retina. Proc. R. Soc. London B **201**:7–26.

Ehinger B (1983). Functional role of dopamine in the retina. In *Progress in Retinal Research* (N Osborne, G Chader, eds.), Vol. 2, pp. 213–232, Pergamon, Oxford.

Hagins WA (1972). The visual process: Excitatory mechanisms in the primary receptor cells. Annu. Rev. Biophys. Bioeng. **1**:131–158.

Leibovic KN, Balslev E, Mathieson TA (1970). Binocular space and its representation by cellular assemblies in the brain. J. Theoret. Biol. **28**:513–529.

Leibovic KN, Balslev E, Mathieson TA (1971). Binocular vision and pattern recognition. Kybernetik **8**(1):14–23.

Lettvin JY, Maturana HR, McCullough WS, Pitts WH (1959). What the frog's eye tells the frog's brain. Proc. I.R.E. **47**:1940–1951.

Livingstone M, Hubel D (1988). Segregation of form, color, movement and depth: Anatomy, physiology and perception. Science **240**:740–749.

McCullough WS, Pitts W (1943). A logical calculus of the ideas immanent in neural nets. Bull. Math. Biophys. **5**:115–137.

Van Essen DC (1979). Visual areas of the mammalian cerebral cortex. Annu. Rev. Neurosci. **2**:227–263.

Van Essen DC, Maunsell JHR (1983). Hierarchical organization and functional streams in the visual cortex. TINS **6**(9):370–375.

Part 1
Biophysics and Physiology

1
The Visual Pathway: An Introduction to Structure and Organization

JOHN R. COTTER

Introduction

The purpose of this chapter is to introduce the reader to the gross anatomy, microanatomy, and cell biology of the brain and the terminology that is commonly used in studies of the visual system.* In addition, pathways from the eye to various parts of the brain, including the cortex are reviewed. A brief account of what is known about the function of some visual structures is also included.

Gross Anatomy of the Brain

There are six major subdivisions of the vertebrate brain. In mammals the cerebral hemisphere and cerebellum are enormous and they conceal most of the other parts of the brain. The latter are collectively referred to as the brainstem and they include the thalamus, midbrain, pons, and medulla oblongata. These can be seen totally by cutting away the cerebellum and cerebral hemisphere (Fig. 1.1) or from the midline by slicing the brain in half (Fig. 1.2).

The Nerve Cell

Originating from the cell body or soma are two processes: the axon and one or more stem dendrites (Fig. 1.3). Dendrites are highly branched and short, being restricted to the neuropil. They may be coarse in appearance, being garnished with spines that represent regions of synaptic contact. In contrast to the dendrite, the axon is long, very thin, smooth, and relatively unbranched along its entire length. Branches, if any occur, arise near the soma, where they form recurrent

---

*This chapter is based on histology and visual science lectures given over many years. In preparing this chapter, the author was greatly influenced by a number of excellent more detailed sources that the reader may wish to consult. In particular, see end-of-chapter references for Fawcett, Truex and Carpenter, Crosby et al., and Alberts et al.

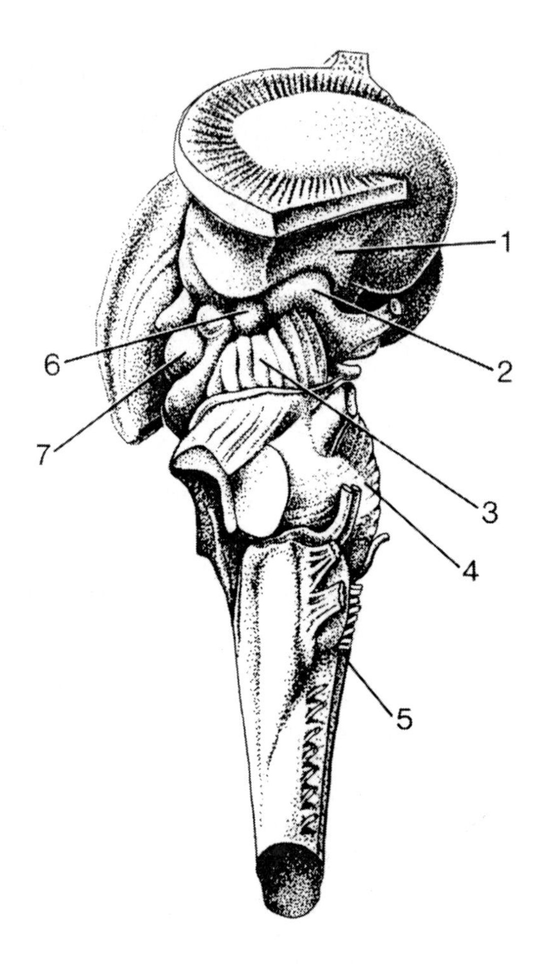

FIGURE 1.1. The brainstem: 1. thalamus, 2. optic tract, 3. midbrain, 4. pons, 5. medulla oblongata, 6. lateral geniculate body, 7. superior colliculus.

collaterals, or at the terminal end of the axon. Axons may also be covered with a myelin sheath.

The shapes of nerve cells vary widely and the form of the cell is greatly dependent on the degree of development and form of the dendrites (Fig. 1.4). Using morphological descriptions of the latter, neurons can be classified according to the orientation, length, and arborization of the stem dendrites.

Each division of the nerve cell subserves a specific function and in the case of the axon, dendrite, and synapse, unique properties of the cell membrane underlie the physiology of each cell part. The axon or fiber is involved in the conduction of information. This is accomplished by means of voltage-gated ion channels in the plasma membrane of the axon. In myelinated fibers, the channels are located at the nodes of Ranvier. At the end of the axon, the information is transmitted

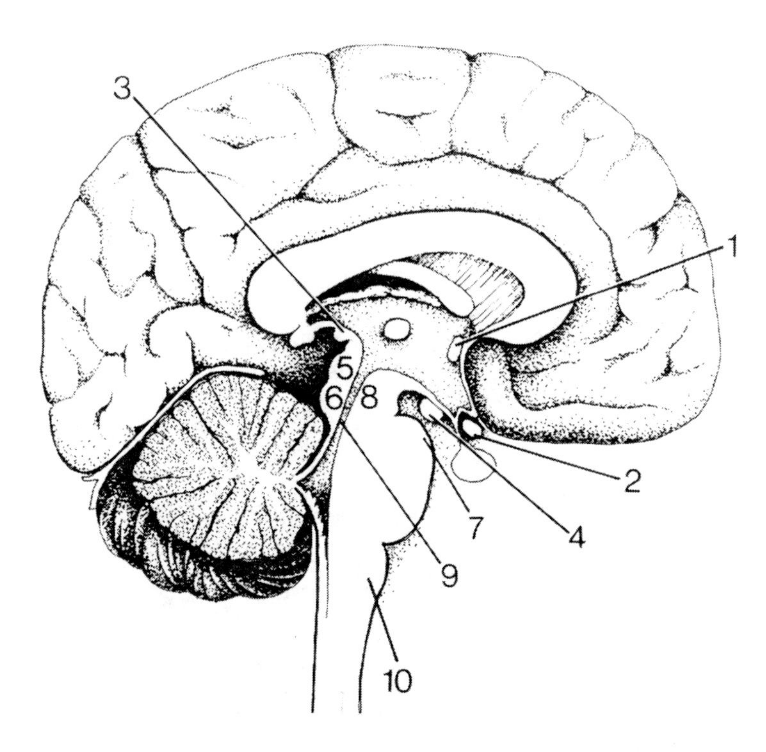

FIGURE 1.2. The brain sectioned in the midline. The thalamus is bounded by the anterior commissure (1), optic chiasm (2), posterior commissure (3), and the mammillary body (4). The midbrain is bounded by the posterior commissure (3), the mammillary body (4), the inferior colliculus (6), and the rostral tip of the pons (7). The superior (5) and inferior (6) colliculi form the roof or tectum of the midbrain. The tegmentum (8) lies beneath the cerebral aqueduct (9). The medulla oblongata (10) is located caudal to the pons.

from one cell to another by means of small morphologically and functionally specialized areas that are located on the cell surface. At such spots in the plasma membrane, electrical and chemical synapses are formed. Electrical synapses are gap junctions in which the cell membranes are closely opposed, thus permitting current flow between the cells. These synapses are far less numerous than chemical synapses and they operate by allowing ions to move through a channel formed in adjacent membrane surfaces. Transmission in the chemical synapses, on the other hand, is mediated by neurotransmitter materials such as acetylcholine, norepinephrine, epinephrine, and γ-aminobutyric acid (GABA). A number of amino acids and peptides are also utilized by neurons. These may be either transmitters or neuromodulators. The latter work in concert with neurotransmitters to enhance or diminish the effectiveness of the neurotransmitter.

Ultrastructural studies reveal that the chemical synapse has three morphological subdivisions: (1) a presynaptic component that is identified by the presence

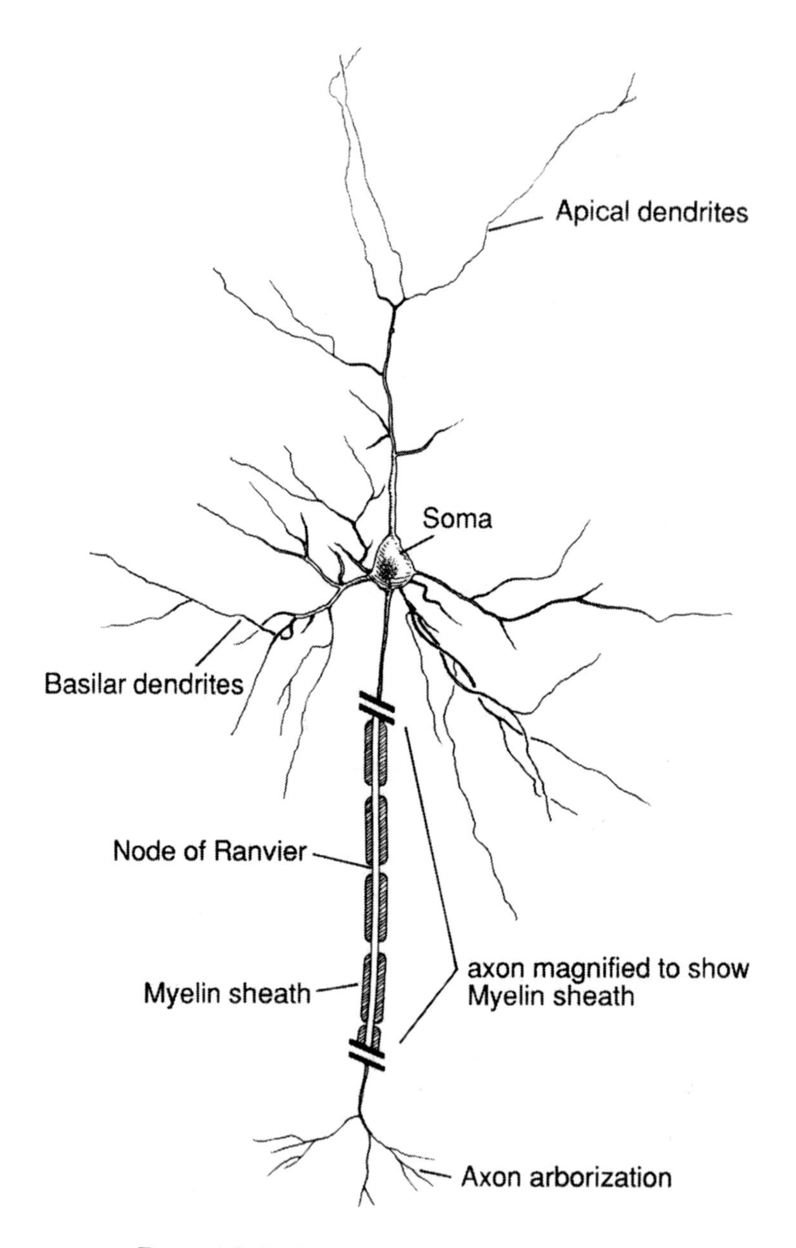

FIGURE 1.3. Typical nerve cell with major subdivisions.

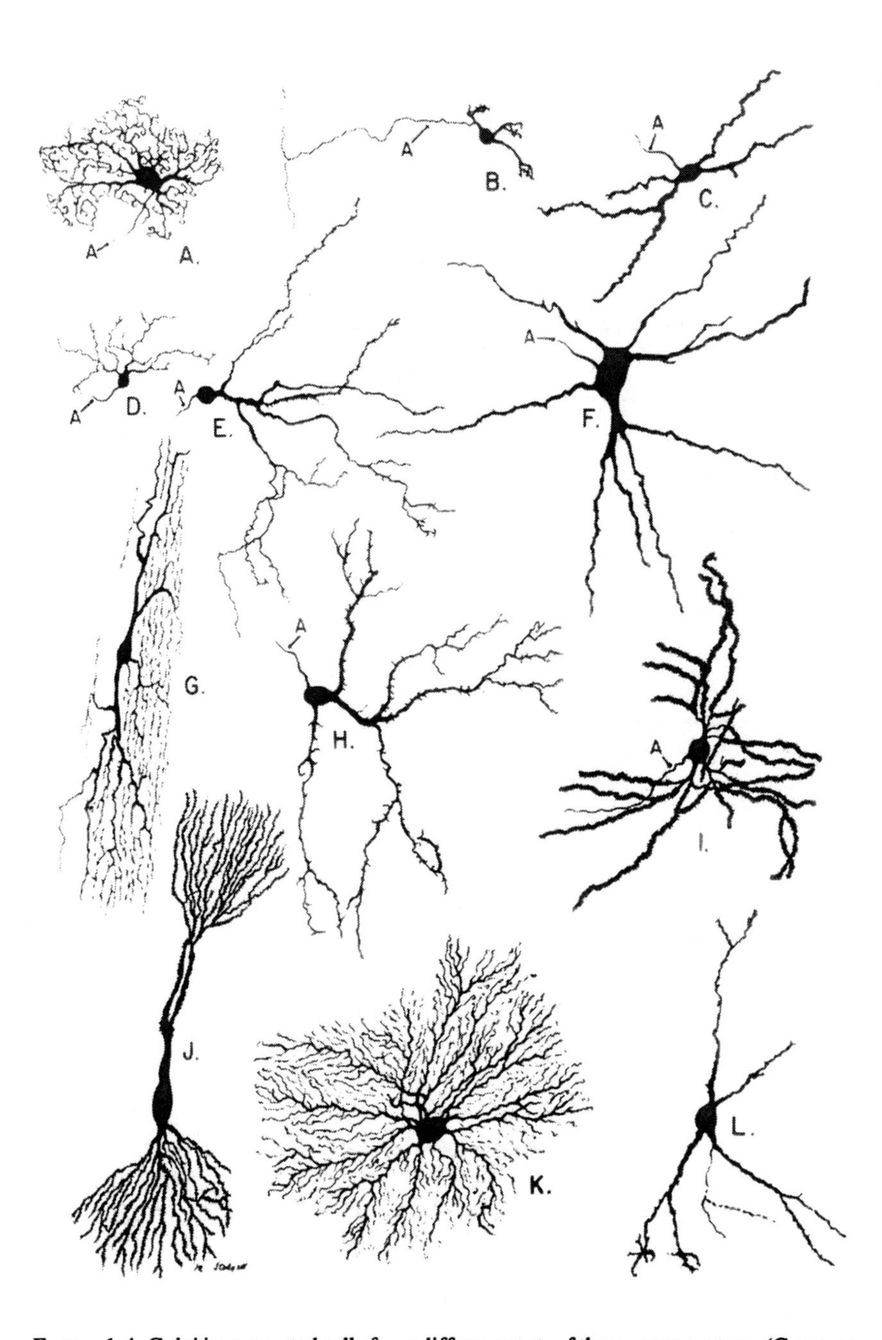

FIGURE 1.4. Golgi impregnated cells from different parts of the nervous system. (Courtesy of Dr. Clement Fox, Wayne State University, from Truex RC, Carpenter MB, *Human Neuroanatomy*, 6th ed., 1969. Reproduced with permission from the Williams & Wilkins Co., Baltimore, © 1969).

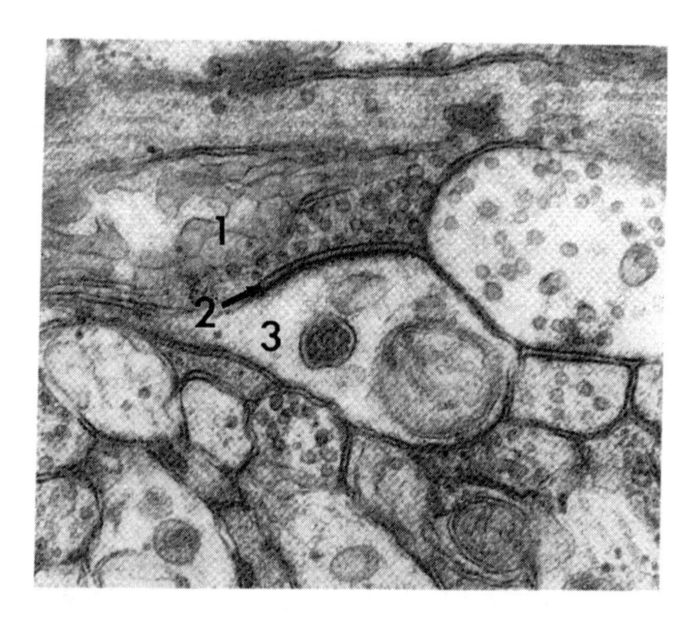

FIGURE 1.5. The three morphological subdivisions of a synapse. 1. Presynaptic component with synaptic vesicles. 2. Synaptic cleft. 3. Postsynaptic component with an electron-dense subsynaptic web underlying the postsynaptic membrane. The placement of synaptic vesicles polarizes the synapse so that information flows from the presynaptic side of the synapse. (Rat retina, inner plexiform layer, 150,000×.)

of synaptic vesicles; (2) a postsynaptic component that is often characterized by a plaque of electron-dense material; and (3) a synaptic cleft—a space of approximately 20–30 nm (Fig. 1.5).

Invasion of the presynaptic component by an action potential initiates a series of events that lead to the exocytosis of neurotransmitter material. On the presynaptic side, ionic and membrane mechanisms enable synaptic vesicles to move to the presynaptic membrane where they fuse with the membrane, thereby releasing neurotransmitter material into the intercellular cleft. On the postsynaptic side of the synapse, attachment of transmitter to receptor sites on the postsynaptic membrane activates ion channels thereby generating hyperpolarizing or depolarizing ion currents. In this way, information impinging on the cell is integrated by the postsynaptic structures.

Despite the physiological properties and structural complexity of the nerve cell, the interior of the cell contains the same fundamental subcellular structures that are found in all cells, irrespective of cell type. The concentration and distribution of many can be ascertained by judiciously selecting appropriate staining techniques. The application of transmission electron microscopy, however, reveals far more information about fine structure and the organelles themselves.

Neurons resemble secretory epithelial cells. Transmitters, like secretory products, are released by exocytosis and the organelles responsible for protein synthesis are especially prominent. This is not surprising since neurons are post-mitotic, very active, and being long lived, must continually replace senescent cellular constituents.

Ribonucleic acids of ribosomes and polyribosomes account for the basophilic nature, i.e., they stain with a basic dye such as cresyl violet, of the cell body and dendrites. At the ultrastructural level, ribosomes and polyribosomes are either attached to membranes to form the rough endoplasmic reticulum (Nissl bodies) or they are unbound and lie free in the cytoplasm. Both are involved in the macromolecular synthesis of proteins: in the case of unattached ribosomes, proteins are released directly into the cytoplasm. In the case of the rough endoplasmic reticulum, proteins synthesized by the structure are bound in membrane by the Golgi apparatus. The latter also alters the chemical composition of proteins produced by the rough endoplasmic reticulum and produces lysosomes that hydrolyze spent or aged organelles.

The smooth endoplasmic reticulum is a network of ribosome-free membranous tubules. A number of functions are attributed to the structure, including the production of lipid for membrane synthesis and axoplasmic transport of synthesized materials.

The interior of the nerve cell also contains cytoskeletal elements and inclusions. The cytoskeleton is composed of proteins that form microtubules and filaments. These are important in intracellular transport of materials and in providing structural support for the cell. Inclusions are storage products such as glycogen, a source of reserve energy, pigment granules, and the results of lysosome action (lipofuscin pigment).

Mitochondria are numerous and produce the energy needed to fuel biochemical reactions.

Organization of Neural Tissue in the Central Nervous System

The different parts of nerve cells can be thought of as being assembled in the central nervous system into nuclei, neuropil, and tracts. Nuclei consist of groups of cell bodies, and neuropil is formed by a dense concentration of dendrites and the terminal portions of axons. Although axons can terminate on any portion of the cell, most synapses are made on the dendritic tree. Together, neuropil and nuclei form the gray matter of the brain. White matter refers to the larger myelinated tracts of the brain and their appearance in fresh tissue. The tracts are formed by axons that may be either myelinated or nonmyelinated. The tracts form the pathways of the brain and in so doing create sensory and motor systems that interconnect nuclei. Nuclei, neuropil, and the tracts are closely associated with neuroglial cells (astrocytes, oligodendrocytes, and microglia). The oligodendrocytes form myelin.

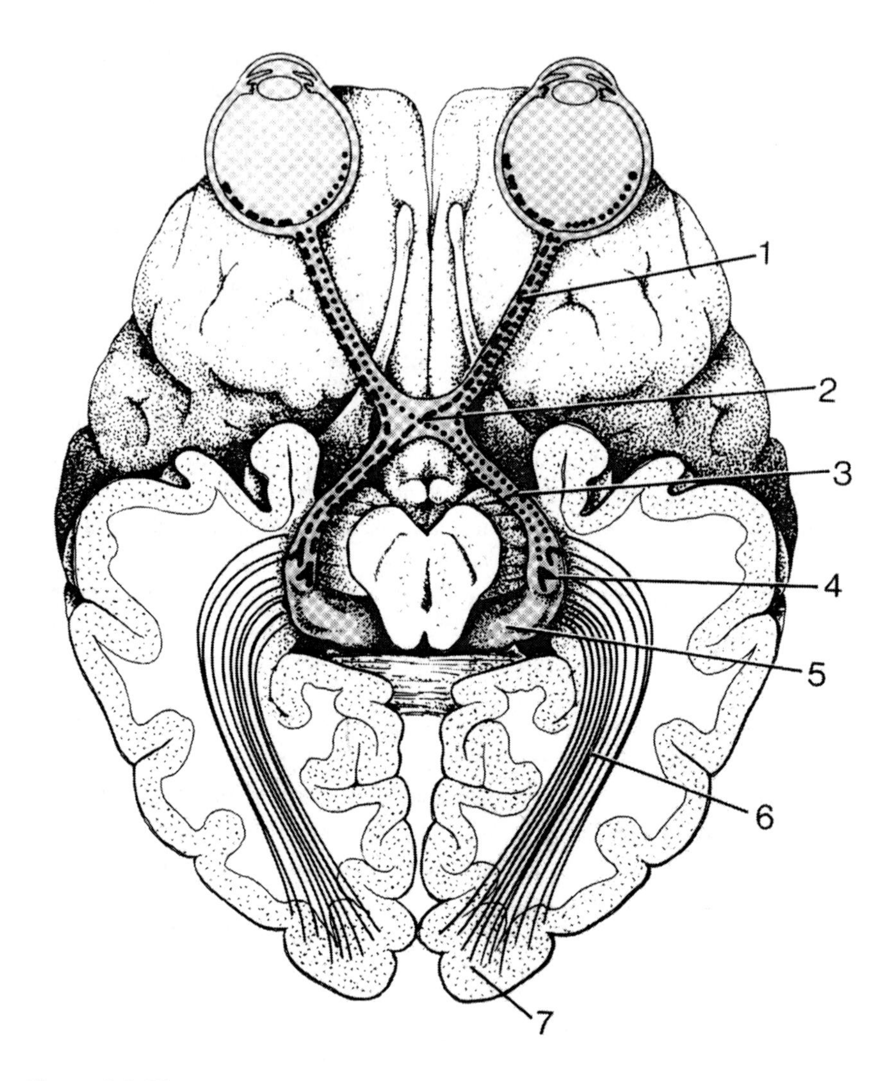

FIGURE 1.6. The eye is connected to the brain through the optic nerve (1). At the optic chiasm (2), fibers from the nasal retina cross to the opposite side of the brain and the optic nerves join to form the optic tract (3). The optic tract projects to the thalamus, pretectum, and midbrain. In this drawing fibers can be followed to the dorsolateral geniculate body (4). Fibers that originate in the geniculate body project to the visual cortex (7) to complete a retinogeniculocortical pathway (6). Brachium of superior colliculus (5).

The Optic Pathway

Studies of the primary optic pathway have been carried out in a large number of species. By and large the overall organization of the projection is very stable among mammals, i.e., the eye projects in an exact manner and always to specific regions of the brainstem. There are, however, differences so that details of the projection in one animal are not necessarily applicable to all species. Such differences – differences in the size of the pathway and development of nuclei in visual centers – represent variations on a theme that are due to the evolution of the visual system, the accentuation of specific sensory systems and ultimately the attainment of a specific ecological niche by individual species.

Visual stimuli falling on the retina activate the light-sensitive photoreceptors of the retina, which in turn initiate, through synapses with horizontal, bipolar, amacrine, interplexiform, and ganglion cells, information processing by cells in the inner retinal layers (see Chapters 2 and 3). The last cell in the circuit, the ganglion cell, relays the information to the visual centers of the brain.

The tract that connects the eye to the brain is called the optic nerve (Fig. 1.6). It is a collection of ganglion cell axons and it is referred to as a nerve because it is not embedded in the tissue of the brain and in fact grossly resembles cranial and peripheral nerves. Acting much like a cable of fibers, the arrangement of fibers in the system preserves visual field relationships of the viewer and relays them in an orderly manner to each station in the projection.

The number of optic nerve fibers varies according to species, there being approximately one million fibers in the human. At the optic chiasm, the optic nerves from each eye join and a portion of the fibers in the nerve cross, i.e., fibers from the nasal hemiretina cross to the opposite side (contralateral) of the brain. Fibers from the temporal hemiretina do not cross and project to the same side (ipsilateral) of the brain (Fig. 1.6). The number of crossing fibers varies with the species and is related to the position of the eye in the head, the more lateral the eyes, the more crossing fibers at the optic chiasm.

Fibers that cross are brought together with those that do not to collectively form the optic tract. This passes across the lateral surface of the brainstem and projects to the dorsal thalamus, a major region of termination for somatosensory, auditory, and visual pathways. At the level of the dorsolateral geniculate nucleus, the optic tract may bifurcate into a branch that continues along the surface and a medial branch that penetrates the tissue of the brainstem. Caudal to the dorsolateral geniculate nucleus, the branches reunite to form a surface structure, the brachium of the superior colliculus. At the level of the superior colliculus it divides to form a branch that is located along the lateral aspect of the colliculus and a branch that is located along its medial edge.

In addition to the major branches of the optic tract, there are smaller branches that exhibit great variation in both their size and number. These are the anterior (inferior) and posterior (superior, transpeduncular) accessory optic tracts.

Geniculostriate Pathway

The lateral geniculate nucleus of many animals consists of alternating cell body and fiber layers. Laminae are also found in other parts of the visual system, namely the superior colliculus and cerebral cortex. Generally, laminae represent areas in which fiber systems predominately terminate and originate. In the lateral

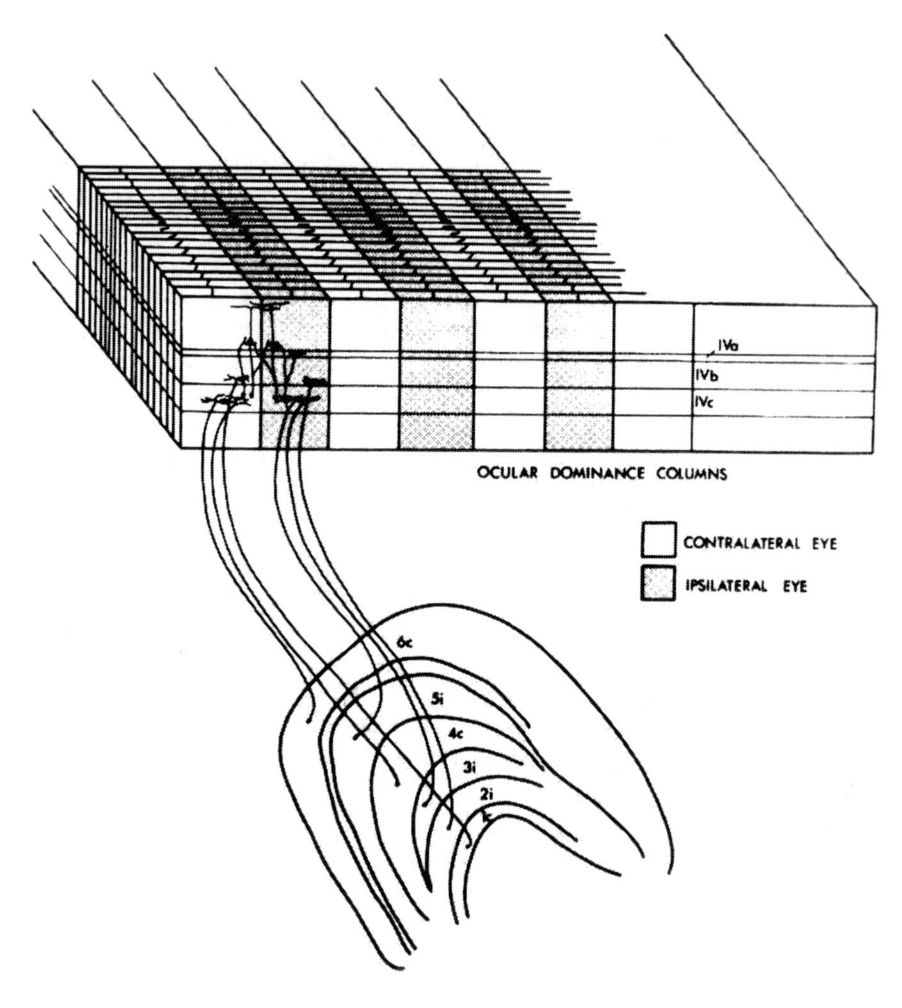

FIGURE 1.7. Projections from individual laminae of the lateral geniculate nucleus end in alternating ocular dominance columns of the primary visual cortex or area 17. (From Hubel DH, Wiesel TN, J. Comp. Neurol. **146**:421–450, 1972, reproduced with permission from Alan R. Liss, Inc.).

geniculate nucleus, the layering of somas is related to the projection from the nasal and temporal retina so that the projections from each eye are segregated from one another. Even in species that do not have obvious laminar patterns in the geniculate, there is a tendency to separate the terminals of crossed and uncrossed fibers. Although the functional significance of lamination in the dorsolateral geniculate is vague and the geniculate may not operate as a simple relay station, lamination in primates does appear to be a method of sorting different kinds of visual information, e.g., color.

The dorsolateral geniculate nucleus projects to the cerebral cortex through the geniculostriate pathway (Fig. 1.6). In this way one hemifield of visual space and

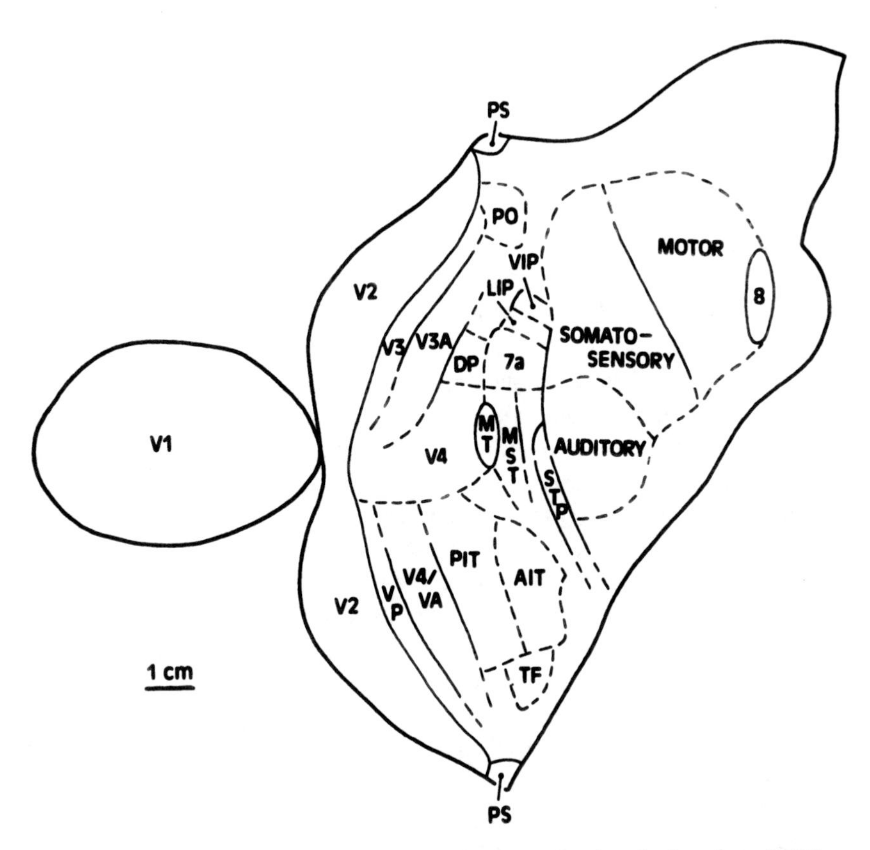

FIGURE 1.8. A map of the right cerebral hemisphere showing the location of different visual areas of the cortex. MT, middle temporal area; MST, medial superior temporal area; STP, superior temporal polysensory area; PO, parietal occipital area; VIP, ventral intraparietal area; VP, ventral posterior area; AIT, anterior inferotemporal area; DP, dorsal prelunate area; PIT, posterior inferotemporal area; PS, prostriate area; VA, ventral anterior area. (Reproduced, with permission, from the Annual Review of Neuroscience, Volume 10, © 1987 by Annual Reviews, Inc.)

information gathered through retinal processing of images by both eyes are brought to opposite sides of the brain. The projection is to area 17 of Brodmann, which is also referred to as the primary visual cortex (V1) or striate cortex. The projection terminates in layer IV. At this level projections from the laminae of the geniculate end in such a manner as to form ocular dominance columns (Fig. 1.7). Whereas cells in layer IV are mostly monocular, interlaminar connections combine input from the two eyes. This provides an anatomical basis for stereoscopic vision.

From area 17, connections are made with many other visual areas in the parietal, occipital, and temporal cortex (Fig. 1.8). Visual information concerning form, color, and movement is highly segregated in V1 and V2. Beyond these areas of cortex, the multitudes of visual areas in the cortex are thought to further segregate these aspects of visual perception, so much so that each discrete visual area may have a unique role to play in visual perception. Some areas to which functions have been assigned include V4, which is believed to have a role in color perception; the middle temporal visual area, MT, also known as V5, which may have a role in the detection of movement and depth perception; and the medial superior temporal area, MST, which may have a role in the detection of movement.

Projections to the Brainstem

The optic pathway begins in the retina and ends in three regions of the brainstem—the thalamus, midbrain, and pretectum. From these regions secondary projections are made to other areas of the brain. Although portions of the pathway—usually only the largest—can be demonstrated by gross dissection and examination of brain slices (Fig. 1.6), labelling the pathway and then cutting the brain into thin sections and tracing the pathway with a microscope, section by section, from beginning to end allows a much more detailed analysis.

In the course of the projection, the optic pathway terminates in a number of areas and nuclei. These are as follows: hypothalamus (suprachiasmatic nucleus), ventral thalamus (ventral lateral geniculate nucleus), dorsal thalamus (dorsolateral geniculate nucleus, pulvinar and lateral posterior nucleus), pretectum (olivary nucleus, posterior pretectal nucleus, nucleus of the optic tract), superior colliculus (superficial and intermediate gray layers), and accessory optic nuclei (medial, lateral, and dorsal accessory optic nuclei).

These are very specific cell groups that are defined by their position in the brainstem. To identify the various parts of the visual system, a thorough understanding of brain organization and terminology is required. Fortunately, atlases of the brains of many animals have been produced that identify by name and define by location the major pathways and nuclei of the brain. Some of these are done in stereotoxic coordinates so that small areas of the brain can be approached with great precision.

Assigning specific roles to each visual center has been difficult but some progress has been made. For example, the suprachiasmatic nucleus through its connection with the hypothalamus is believed to have a role to play in circadian rhythms, especially those related to the regulation of sleep. A number of functions including both reflexes and more discriminating roles have been attributed to other visual structures of the brain. In the case of the pretectum two clinically important eye reflexes, one involving lens thickness (accommodation) and the other the response of the pupil to light, are firmly established. These functions are subserved by connections of the pretectum with the oculomotor nuclei of the midbrain tegmentum. The superior colliculus is involved in integration of a variety of sensory stimuli and mechanisms of attention that are important in eye movements. The accessory optic nuclei may provide the means by which visual information reaches the floccular and nodular parts of the cerebellum; in so doing it has been hypothesized that the accessory optic system coordinates head and eye movements so that visual images do not blur as a result of head movement.

Although research has provided insight into how the different brain structures participate in vision, new concepts are likely to emerge as methods and approaches progress.

Acknowledgment. Original art work was done by John Nyquist.

References

Alberts B, Bray D, Lewis J, Raff M, Roberts K, Watson JD (1983). *Molecular Biology of the Cell.* Garland Publishing Inc., New York.

Berman AL (1968). *The Brain Stem of the Cat.* The University of Wisconsin Press, Madison.

Carpenter MB, Pierson, RJ (1973). Pretectal region and the pupillary light reflex. An anatomical analysis in monkey. J. Comp. Neurol. **149**:271–300.

Crosby EC, Humphrey T, Lauer EW (1962). *Correlative Anatomy of the Nervous System.* The Macmillan Co., New York.

Ebbesson SOE (1970). On the organization of central visual pathways in vertebrates. Brain Behav. Evol. **3**:178–194.

Fawcett DW (1986). *A Textbook of Histology.* W.B. Saunders Co., Philadelphia.

Garey LP, Powell TPS (1968). The projection of the retina in the cat. J. Anat. **102**: 189–222.

Hayhow WR, Webb C, Jervie A (1960). The accessory optic fiber system in the rat. J. Comp. Neurol. **115**:187–215.

Hayhow WR, Sefton A, Webb C (1962). Primary optic centers of the rat in relation to the terminal distribution of the crossed and uncrossed optic nerve fibers. J. Comp. Neurol. **118**:295–322.

Ito M, (1972). Neural design of the cerebellar motor central system. Brain Res. **40**:81–84.

Livingston M, Hubel D (1988). Segregation of form, color, movement, and depth: anatomy, physiology, and perception. Science **240**:740–749.

Maunsell JHR, Newsome WT (1987). Visual processing in monkey extrastriate cortex. Annu. Rev. Neurosci, **10**:363–401.

Paxinos G, Watson C (1982). *The Rat Brain in Stereotoxic Coordinates.* Academic Press, New York.

Schiller PH, Malpeli JG (1978). Functional specificity of lateral geniculate nucleus laminae of the rhesus monkey. J. Neurophysiol. **41**:788–797.

Simpson JI (1984). The accessory optic system. Annu. Rev. Neurosci. **7**:13–41.

Truax RC, Carpenter MC (1969). *Human Neuroanatomy.* Williams & Wilkins, Baltimore.

Turek FW (1985). Circadian neural rhythms in mammals. Annu. Rev. Physiol. **47**:49–64.

Van Essen DC, Maunsell JHR (1983). Heirarchial organization and functional streams in the visual cortex. TINS **6**:370–375.

Weiss L (1983). *Histology. Cell and Tissue Biology.* Elsevier Biomedical, Amsterdam.

Wurtz RH, Albano JE (1980). Visual-motor function of the primate superior colliculus. Annu. Rev. Neurosci. **3**:189–220.

2
Vertebrate Photoreceptors

K.N. LEIBOVIC

Introduction

Photoreceptors initiate the processes which mediate our perception of the visual world. They convert the light energy they absorb into a neural signal and they transmit it through intervening networks of neurons to the brain. In our retina there are about 120×10^6 rods and some 5×10^6 cones (see, e.g., Polyak, 1941), the former operating in dim illumination and the latter in bright light. Color vision and fine discrimination depend on cone vision. The light intensities which we encounter range from about a hundred photons incident on the cornea at the absolute visual threshold to more than 12 orders of magnitude above that in strong sunlight (see, e.g., Riggs, 1966). Such an enormous range could not be accommodated in one scaling without sacrificing the ability to discriminate intensity differences on which our form perception depends. Photoreceptors are designed to use an operating window spanning only three orders of magnitude of light intensity and to shift this window over a larger range of intensities through a process of adaptation to the ambient light background.

In this chapter we shall see how photoreceptors absorb light, how they transduce the photon energy into an electrochemical response which is transmitted to adjoining neurons, and how photoreceptors adapt to changing light intensities.

Structure of Rods and Cones

All vertebrate retinas have a similar gross structure and the same cell types, although the detailed cellular morphology, synaptic connections, chemical messengers, and abundance of different classes of cells may vary.

Figure 2.1a illustrates how light enters the retina through the ganglion cell layer and propagates to the rods and cones where it is absorbed by the photosensitive pigment. About 10% of the light incident on the cornea reaches the photoreceptors. The rest is lost in transmission through the ocular media, the blood vessels, and the neural cells of the retina. Figure 2.1b shows a rod and a cone isolated from the retina of the toad, *Bufo marinus*. Figure 2.1 also illustrates

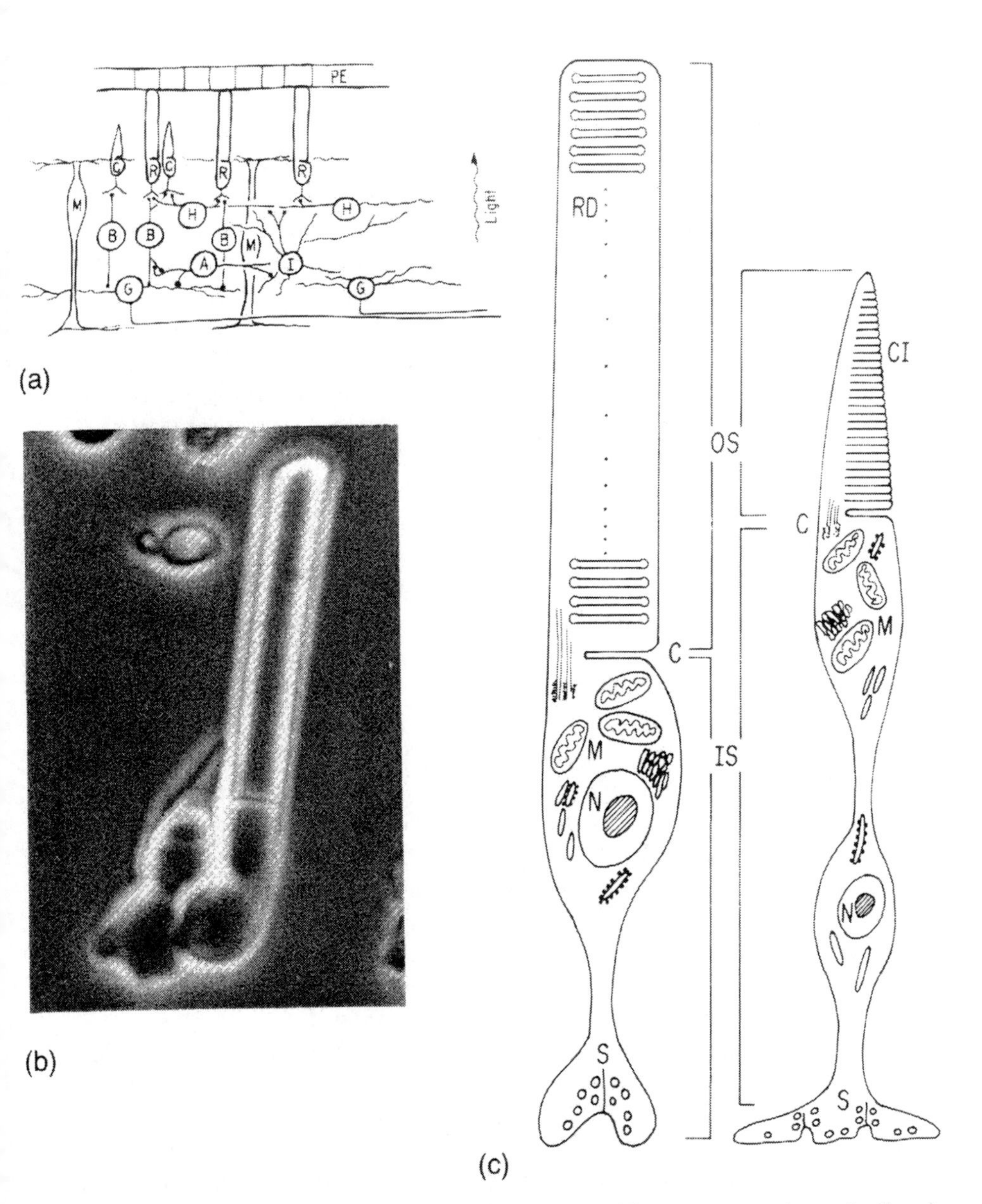

FIGURE 2.1. (a) A schematic diagram of the retinal elements. The wavy arrow shows the direction in which light travels through the retina. PE, pigment epithelium; R, rod; C, cone; H, horizontal cell; B, bipolar cell; A, amacrine cell; I, interplexiform cell; G, ganglion cell; M, Muller cell. (b) A rod and cone pair isolated from the retina of *Bufo marinus* after 1 day in KNL medium (Leibovic, 1986a). (c) Illustrations of a typical rod and cone. OS, outer segment; IS, inner segment; RD, rod discs; CI, cone invaginations; C, ciliary connection between inner and outer segments; M, mitochondria; N, nucleus; S, synaptic ending. The synaptic ending contains vesicles, some apposed to ribbon structures that are found in both rods and cones. The ciliary connection is a slender structure connecting inner and outer segments. In addition to the labeled organelles there are also rough and smooth endoplasmic reticulum, Golgi apparatus, and the other usual components of cells.

in more detail the structures of a typical rod and cone. The outer segments contain the photosensitive pigment molecules. These are embedded in the disc membranes of rods and in the invaginating cone plasma membrane. The inner segments contain densely packed mitochondria and the usual cellular machinery, including Golgi apparatus, endoplasmic reticulum, and nucleus. Inner and outer segments are linked by a narrow ciliary connection. At the synaptic ending are numerous vesicles, often arranged on two sides of a ribbon structure. As with cells in general the detailed morphology of rods and cones differs between different species and even between different regions of the same retina. In addition there are different kinds of rod and cone. There are three kinds of cones in the human retina mediating our color vision. They are known as "red," "green," and "blue" cones corresponding to the different regions of the visible spectrum in which each of these cells absorbs maximally. In the toad retina there are "red" and "green" rods having different absorption spectra.

Light Absorption

Figure 2.2 shows absorption curves for some rod and cone pigment molecules which capture the light. The best studied of these is the rhodopsin of rods. This contains a vitamin A chromophore, 11-*cis* retinal, attached to a protein, opsin, which is embedded in the disc membrane of rods. Light which is normally propagating down the long axis of a photoreceptor has its electric vector parallel to the disc surface. This is also the approximate orientation of the chromophore. The absorption probability is maximized at the vibrational frequency of the electric field vector which is optimal for the particular chromophore and protein combination. Hence, there is a peak in the absorption spectrum for each absorbing molecule at its optimum frequency or wavelength of the light. Although the chromophore is the same, the protein to which it is attached differs in the different rods and cones, and it is these structural differences which give rise to the different absorption spectra.

After absorbing a photon the 11-*cis* retinal goes through a series of conformational changes described in Figure 2.3 (Wald, 1968; Bonting, 1969; Abrahamson and Wiesenfeld, 1972; Kropf, 1972; Yoshizawa, 1972). The metarhodopsin II stage is reached within a millisecond and triggers a biochemical transduction cycle, which leads to membrane conductance changes. These are translated into membrane currents and potentials which modulate synaptic transmission to the second order neurons. At the end of the cycle in Figure 2.3 the chromophore separates from the opsin and makes a return trip to the pigment epithelium before the rhodopsin is regenerated and is once more fully able to absorb photons.

Based on measurements of optical density in the human eye it can be calculated that the probability of a rhodopsin molecule absorbing a 500 nm photon incident on a rod is about 2×10^{-4} (Pirenne, 1962; Leibovic and Kurtz, 1975). This probability is greatly increased by having some thousands of layers of rhodopsin in a rod outer segment interposed in the photon path. The result is that a human rod

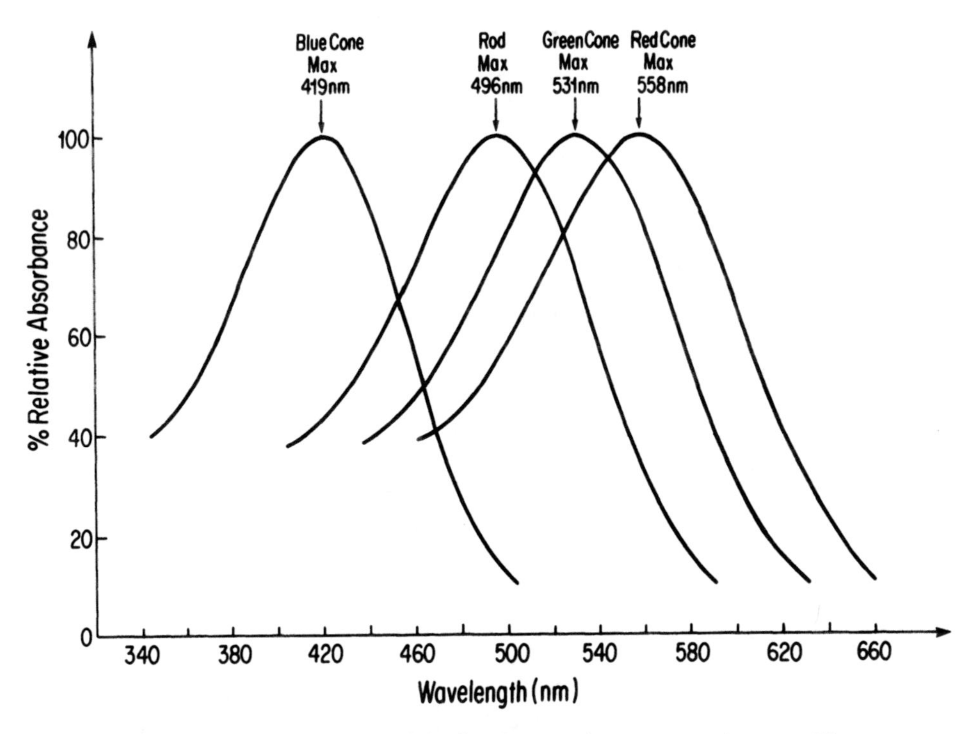

FIGURE 2.2. Relative absorbance of the four human photoreceptor pigments. The curve with a maximum at 496 nm is from the rod pigment; the other three are from the cone pigments: the long wavelength "red" cones, maximum at 558 nm; the medium wavelength "green" cones, maximum at 531 nm; and the short wavelength "blue" cones, maximum at 419 nm. (Curves according to Dartnall et al., 1983, following Barlow, 1982, © Academic Press, by permission.)

can absorb some 30% of the light to which it is exposed. The rhodopsin molecules are distributed over both sides of the discs, and a photon may be absorbed by any one molecule on any disc.

Since the discs are separate from the plasma membrane where the conductance changes occur, there must be a way to signal the absorption of a photon across this separation. But not only is this event transmitted, it is also amplified in the process more than 10^5 times. Thus, the 500 nm photon energy of about 4×10^{-12} erg leads to the displacement of some 10^7 electronic charges through a potential drop of 30–40 mV, an amplification of the order of 10^5.

A Digression on Membrane Potentials

In the steady state, in the dark, the inside of a rod or cone is about 40 mV more negative than the outside. The outer segment is quite permeable to Na^+ ions, and there is a steady Na^+ current flowing from outside the cell where the Na^+

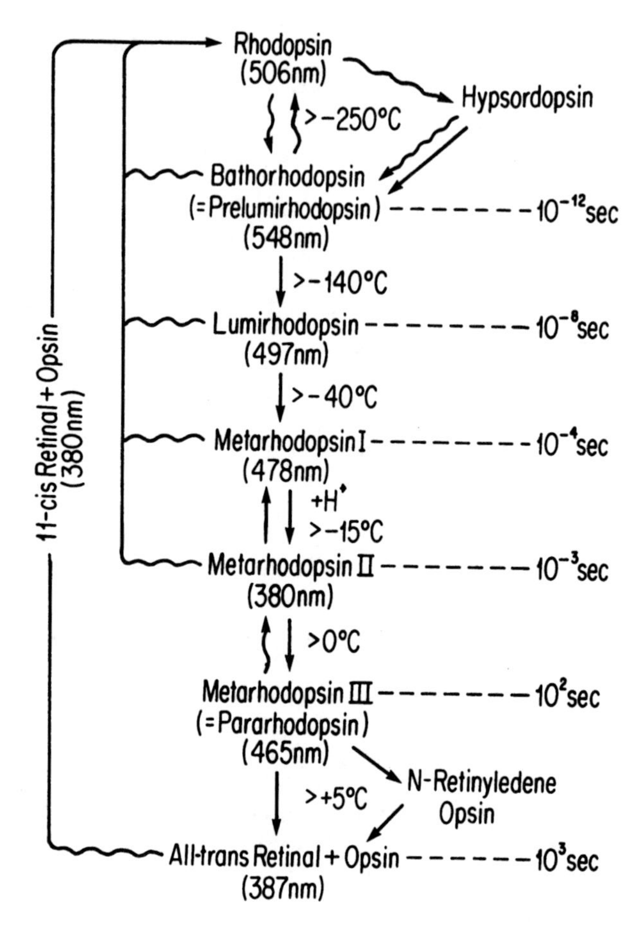

FIGURE 2.3. Rhodopsin and its intermediates, formed after light absorption. The wavy arrows denote transitions due to photon absorption. It will be noted that some intermediates are photoreversible. The straight arrows denote thermal transitions. The temperatures (°C) are those above which the intermediates make the transitions and below which they remain stable. The wavelengths (nm) are the maxima of the absorbance spectra. The times (sec) are of the orders of magnitude after photon absorption when the intermediates appear at room temperature (Shichi, 1983; modified from Yoshizawa and Horiuchi, 1973, © Springer-Verlag, by permission).

concentration is high through the outer segment and into the cell where the Na^+ concentration is low (Hagins et al., 1970; Korenbrot and Cone, 1972). K^+ concentrations are opposite to the Na^+ concentrations, being higher inside than outside, and K^+ flows out of the cell, mostly through the inner segment. To maintain a steady state a Na^+–K^+ pump in the inner segment pumps Na^+ out and K^+ into the cell (Frank and Goldsmith, 1967; Zuckerman, 1973; Stirling and Lee, 1980; Torre, 1982).

After a flash of light the outer segment Na^+ permeability is reduced (Tomita, 1965; Toyoda et al., 1969) and the membrane is hyperpolarized (Bortoff and

Norton, 1965; Baylor and Fuortes, 1970). A process involving membrane permeability changes is common in neurons where, as in cells in general, the ionic concentrations differ between the cytoplasm inside the cell and the extracellular fluid. As Nernst showed long ago:

$$V_m = \frac{RT}{zF} \ln \frac{[A]_e}{[A]_i} \tag{1}$$

where V_m is membrane potential, $[A]_e$ and $[A]_i$ are the external and internal ion concentrations, respectively, R is the gas constant, F is the Faraday constant, T is the absolute temperature, and z is the ionic charge. In the steady state, the diffusing tendency of ions from a higher to a lower concentration is balanced by an electrical potential opposing such movement. This is given by the Nernst equilibrium potential for that ion. The physiological distribution of Na^+ ions across the membrane is accompanied by a positive membrane potential, while the distribution of K^+ ions is accompanied by a negative potential. Of course, the membrane must be permeable to those ions for the diffusing tendencies and potentials to arise.

In the presence of several but not interacting ions to which the membrane is permeable, the potential assumes a more complicated form as, for example, in the Goldman–Hodgkin–Katz equation where

$$V_m = \frac{RT}{F} \ln \frac{P_{Na}[Na]_e + P_K[K]_e + P_{Cl}[Cl]_i}{P_{Na}[Na]_i + P_K[K]_i + P_{Cl}[Cl]_e} \tag{2}$$

with Na, K, Cl being sodium, potassium, chloride and P with the appropriate suffices being their permeabilities.

An assumption implicit in Eqs. (1) and (2) is that there is no net current flow across the membrane. When this is not the case different equations are required. Factors which limit the applicability of Eqs. (1) and (2) to photoreceptors involve not only the presence of steady currents, but also evidence that some ions interact in their movements through the membrane and the presence of active pumps such as the Na^+–K^+ pump in the inner segment and a Na^+–Ca^{2+} exchanger in the outer segment.

Equations (1) and (2) can be extended by including the contribution of a pump. For example, suppose that Na^+ is leaking into and K^+ out of the cell, and suppose a Na^+–K^+ pump maintains the ionic balance by pumping Na^+ out and K^+ into the cell. It has been found in a variety of tissues that in the steady state the pump normally exchanges three Na^+ for two K^+ ions. This results in a net outward current and thus the pump generates a potential which makes the inside of the cell more negative. When chloride is in electrochemical equilibrium across the membrane and there are no other ions involved, the membrane potential is given by (Mullins and Noda, 1963):

$$V_m = \frac{RT}{F} \ln \frac{r P_K[K]_e + P_{Na}[Na]_e}{r P_K[K]_i + P_{Na}[Na]_i} \tag{3}$$

where r is the ratio of Na^+:K^+ exchange of the pump, e.g., 1.5 when 3 Na^+ are exchanged for 2 K^+. For $r = 1$, i.e., a neutral pump which makes no contribution to the membrane potential, Eq. (3) reduces to (2) when Cl^- is in equilibrium.

In some tissues, such as nerve axons, the effect of the Na^+–K^+ pump on membrane potential may be negligible. But this is not the case in photoreceptors.

One can estimate the maximum contribution which a sodium pump can make to the membrane potential under some conditions. If V_P is the potential when the pump is active and V_L when it is inactive, the pump contribution is $V_P - V_L$. From Eq. (3), let $P_{Na}/P_K \equiv P$,[Na]/[K] $\equiv X$, and put $r = 1.5$ and $RT/F = 25.2$ mV. Then

$$V_P = 25.2 \ln \frac{1.5 + PX_e}{1.5 + PX_i}$$

where the suffices i and e stand for internal and external, respectively. Similarly, from Eq. (2) with $P_{Cl} = 0$ [or from Eq. (3) with $r = 1$]

$$V_L = 25.2 \ln \frac{1 + PX_e}{1 + PX_i}$$

Therefore

$$V_P - V_L = 25.2 \ln \frac{1.5 + PX_e}{1 + PX_e} \frac{1 + PX_i}{1.5 + PX_i}$$

We assume, for photoreceptors, that $[Na]_e = 120$ mM,$[K]_e = 2.5$ mM, and P is near 1 (Torre, 1982; Hodgkin et al., 1985). Then

$$V_P - V_L = 25.2 \ln \frac{1 + X_i}{1.5 + X_i}$$

approximately. This increases numerically with decreasing values of X_i, which generally are small in cells, including photoreceptors. It follows from the last equation that the pump contribution cannot exceed a hyperpolarization of about 10 mV. This is close to estimates of the actual value in dark adapted *Bufo* rods (Torre, 1982).

We can see that membrane potentials depend on a number of factors, including ion pumps, permeabilities, and concentrations.

Permeability in the rod outer segment is controlled by cGMP, which keeps the Na^+ channels open. The effect of light is to reduce the free cGMP and close these light-sensitive channels (Fesenko et al., 1985; Yau and Nakatani, 1985b; Nakatani and Yau, 1988b). As a result, in view of the relationships we have discussed, we would expect the cell to hyperpolarize or become more negative inside, and this is what happens.

We now return to the link between light and the membrane response.

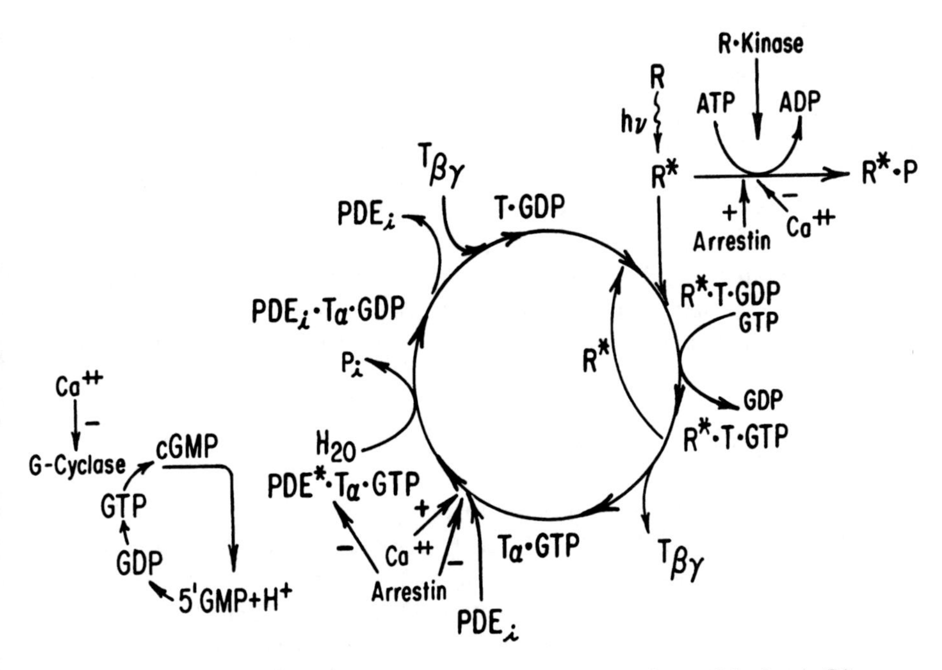

FIGURE 2.4. The transduction cycle is triggered by the active form of rhodopsin R*, corresponding to metarhodopsin II in Figure 2.3. See text for explanation of the steps in the cycle. (Modified from Stryer, 1986; © John Wiley & Sons, Inc., by permission.)

The Transduction Cycle

The absorption of a photon activates rhodopsin R to R* and sets in motion a biochemical cascade which involves transducin T (a G protein), phosphodiesterase (PDE), and guanisine 3′,5′-cyclic monophosphate (cGMP) (for reviews see Stryer, 1986; Pugh and Cobbs, 1986; Fung 1987). The biochemistry of this transduction is illustrated in Figure 2.4. In the inactive state a GDP is bound to transducin. On activation by R* the GDP is exchanged for a GTP and T splits into two parts. The active part T_α.GTP then activates PDE, which, in turn, hydrolyzes cGMP and renders it incapable of keeping the Na^+ channels open. Meanwhile R*, which has dissociated from T.GTP, recycles and activates more transducin. As a result some of the order of 10^2 PDEs are activated, each of which can hydrolyze some of the order of 10^3 cGMPs with an overall amplification of R* to cGMP hydrolysis of the order of 10^5.

The above processes initiate the response. To shut it off, R* is inactivated by the joint action of phosphorylation and a protein appropriately named arrestin. PDE is also inactivated when the T_α.GTP unit is hydrolyzed to T_α.GDP. At the same time the enzymatic activity of G-cyclase increases and regenerates new cGMP from GTP.

It has been found *in vitro* that Ca^{2+} inhibits rhodopsin kinase and G-cyclase and that it activates PDE. Since rhodopsin kinase phosphorylates R*, G-cyclase regenerates cGMP, and PDE hydrolyzes cGMP, the effect of Ca^{2+} opposes that of cGMP on the membrane. However, the extent to which these effects operate *in vivo* is as yet not fully determined.

Where do these steps of transduction occur? Rhodopsin is embedded in the disc membrane of rods but is free to diffuse in it. Transducin is at the disc surface but, when activated T_{α} (as well as the other part $T_{\beta}\alpha$) can leave the membrane and diffuse to neighboring discs, on its way activating many PDEs which may or may not be membrane bound. cGMP is found in the cytoplasm. The hydrolysis of cGMP in the cytoplasm creates many sinks which presumably draw off enough of these molecules from the plasma membrane to close the light-sensitive channels.

Ca^{2+} Feedback Control of Transduction

We have seen that there is a steady Na^+ current flowing into the cell through the outer segment in the dark. But Na^+, though the major, is not the only current carrier. The light-sensitive conductance will allow other ions to pass including Ca^{2+}, Mg^{2+}, K^+, and other monovalent cations (Nakatani and Yau, 1988a; Hodgkin et al., 1985; Stern et al., 1987). Under physiological conditions there is probably little K^+ flowing through these outer segment channels (Yau et al., 1981). On the other hand it has been shown that about 15% of the dark current consists of Ca^{2+} and about 5% of Mg^{2+} (Yau and Nakatani, 1985a; Nakatani and Yau, 1988a). The Ca^{2+} is removed by a Na^+–Ca^{2+} exchanger, which exchanges 1 Ca^{2+} for 3 Na^+ (Nakatani and Yau, 1989—or possibly 1 Ca^{2+} + 1 K^+ for 4 Na^+, Cervetto et al., 1989). When the outer segment permeability is reduced by a flash of light, the Ca^{2+} current is reduced together with the Na^+ current. But the reduction in Na^+–Ca^{2+} exchange lags the reduced current. This has some interesting consequences. As we saw earlier, increasing the internal Ca^{2+} tends to facilitate the processes involved in cGMP hydrolysis. By the same token, reduced Ca^{2+} accelerates recovery from the response. Since a flash of light leads to reduced cytosol Ca^{2+}, it not only initiates the response but also brings into play the means to shut it off again (Koch and Stryer, 1988). The change of internal Ca^{2+}, part of the output due to the light signal, feeds back, as it were, to terminate the output.

Control of OS Channel Conductance by cGMP and Ca^{2+}

When external Na^+ is reduced the resting potential becomes more negative and the light responses diminish (Capovilla et al., 1981, 1983; Hodgkin et al., 1984, 1985; Yau and Nakatani, 1984). This is to be expected. Since the membrane is permeable to Na^+, the resting potential responds to changes in Na^+ concentration. A smaller concentration difference reduces the contribution of Na^+ in depolarizing the membrane and it becomes more negative. Also, the outer segment current being carried mostly by Na^+, a reduction of the driving force due to the smaller Na^+ concentration difference will reduce the current. When the Na^+ conductance is now reduced by a flash of light the changes of membrane potential and current are smaller. In other words, the response amplitude is diminished. When external Na^+ is completely replaced by other monovalent cations such as K^+, Rb^+, or Cs^+, the current soon ceases to flow into the outer segment and there is no light response. This may seem puzzling at first, in view of the fact that these other ions can normally pass through the membrane. The answer can be sought once again in the effects of Ca^{2+}. As we know Ca^{2+} and Mg^{2+} leak into the cell together with Na^+. During this process Ca^{2+} and Mg^{2+} reduce the conductance of the light-sensitive channels (Stern et al., 1987). These channels have a high affinity for Ca^{2+} and Mg^{2+} and the consequent slow passage of these ions tends to block the channels. So, after Na^+ has been replaced, Ca^{2+} and Mg^{2+} continue to enter and reduce the conductance essentially to zero. It thus appears that there is a delicate control of the outer segment conductance. The flow of Na^+ prevents Ca^{2+} and Mg^{2+} blockage of the light-sensitive channels, while the Ca^{2+} and Mg^{2+} accompanying the Na^+ modulates channel conductance to a level appropriate for the Na^+ flux. When no Na^+ is flowing, the channels are blocked and exclude other ions.

It is possible to measure the flow of current through single channels using the patch clamp technique (see, e.g., Sakmann and Neher, 1983). Classically, a channel opens and closes on a millisecond time scale and allows a square pulse of current to pass. This is illustrated in Figure 2.5a. In photoreceptors such pulses are not observed, unless Ca^{2+} and Mg^{2+} are drastically reduced. Only a flickering current flows through the channels in the presence of Ca^{2+}, Mg^{2+}, and cGMP. But when the divalent cations are removed, large square wave like pulses appear. This is shown in the two recordings in Figure 2.5b. The noise has obviously increased in the second of these records. Therefore, an effect of the normal Ca^{2+} and Mg^{2+} entry through the OS channels is to reduce the channel conductance and the noise associated with the opening and closing of individual channels.

It has been known for some time that Ca^{2+} suppresses the dark current. In fact, based on such considerations Ca^{2+} had been proposed as a candidate internal transmitter (Hagins, 1972; Yoshikami and Hagins, 1973), even before the possible involvement of cGMP was appreciated. According to the Ca^{2+} hypothesis, there is a rise of internal Ca^{2+} after the onset of light, and this causes the Na^+ channels to close. But after a series of conflicting reports (see, e.g., Korenbrot, 1985 for a review) there is now substantial evidence that Ca^{2+} does not play this role and it has been shown that internal Ca^{2+} decreases after a flash of light (Yau and Nakatani, 1985a; McNaughton et al., 1986) contrary to the original Ca^{2+} hypothesis.

The electrophysiological demonstration that cGMP rather than Ca^{2+} is the internal transmitter came from patch clamp experiments (Fesenko et al., 1985). A patch of membrane was excised from a rod outer segment and the intracellular face was superfused with a solution containing cGMP. There was a dramatic increase of conductance, which depended monotonically on the cGMP concentration. From this dependence it can be calculated that the response, in the form of the current through the membrane patch, increases with the square or cube of cGMP concentration, suggesting that the cooperative binding of two or three

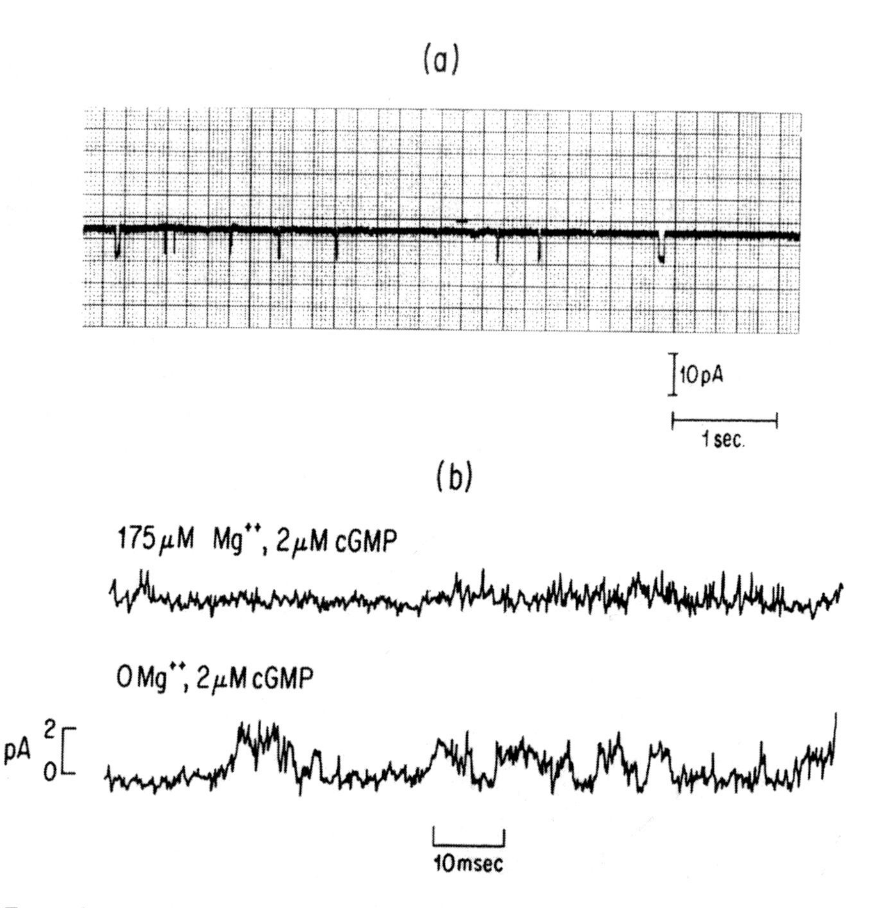

FIGURE 2.5. (a) Channel opening and closing gives rise to square current pulses as shown in this recording from chick embryo muscle cells (K. N. Leibovic and C. Lewis, unpublished). (b) Recordings from a patch of salamander rod show only a flickering current in the presence of Mg^{2+} (upper trace) and the appearance of current pulses, albeit somewhat irregular, in the absence of divalent cations (lower record). (From Haynes et al., Nature (London) **321**:66–70, 1986, Copyright © Macmillan Magazines Ltd., by permission.)

cGMP molecules is required to keep a channel open. Thus, it was demonstrated that cGMP acted directly on this conductance.

One may argue that a patch of membrane does not necessarily have the same properties as an intact outer segment. This was addressed by truncating a rod outer segment, dialyzing the interior with cGMP, and observing the effect (Yau and Nakatani, 1985b). The conclusion was the same. In the truncated outer segment it was also shown that the cGMP-activated conductance was suppressed by light (Nakatani and Yau, 1988b). However, for this to occur it was necessary to add GTP in the superfusing fluid. This is quite consistent with the transduction biochemistry we discussed earlier. For, in order to have a light response the photoactivated rhodopsin R* has to interact with transducin, which, in turn, must exchange its bound GDP for GTP to become active and promote the cycle.

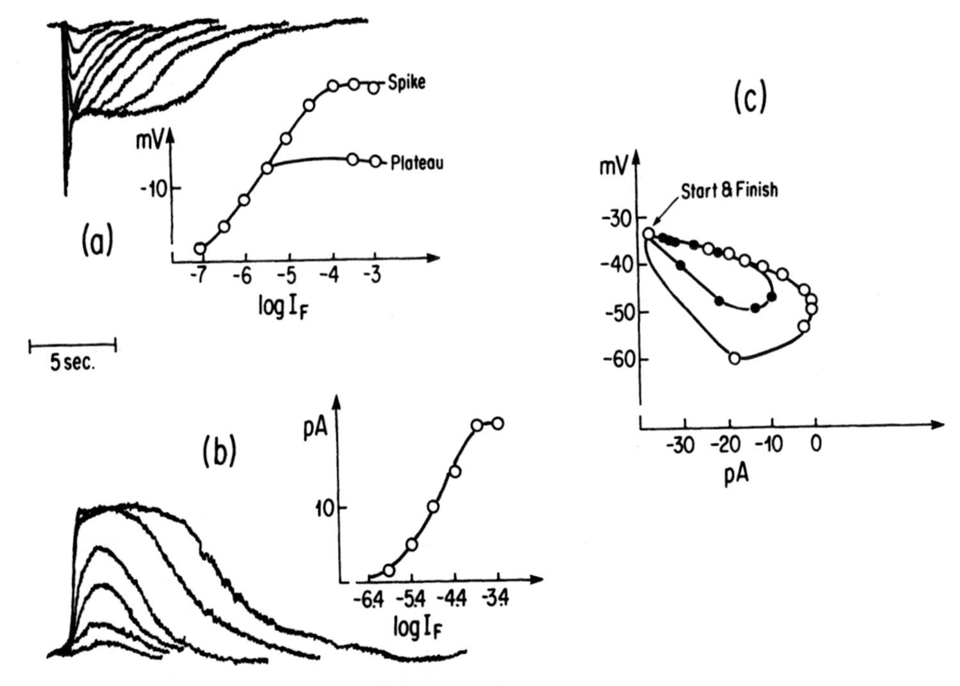

FIGURE 2.6. (a) Intracellularly recorded changes of membrane potential of a dark adapted rod of *Bufo marinus*. The responses to an increasing series of 100 msec flashes are superimposed in this record. The nominal flash intensities are −7, −6.5, −5.5, −5, −4.5, −4, −3.5, −3 in neutral density (ND) units of attenuation. Next to the recorded responses is a plot of peak amplitude versus flash intensity (Leibovic, unpublished). (b) Superimposed recordings of current responses to 100 msec flashes of increasing intensity from the outer segment of an isolated rod of *Bufo marinus*. The nominal flash intensities in ND units are −5.9, −5.4, −4.9, −4.4, −3.9, −3.7. Next to the responses is a plot of peak amplitude versus flash intensity (Leibovic, from unpublished data). (c) V–I trajectory from simultaneous recordings of voltage and current. The outer trajectory is for the largest, saturated, responses and the inner one is for the fifth largest, subsaturated, responses as derived from Figure 3A of Baylor and Nunn (1986).

In summary, although Ca^{2+} plays a modulating role, there is compelling evidence that the cGMP-activated conductance is the light-sensitive conductance of the outer segment.

Response Waveforms and Electrical Characteristics

The changes of membrane conductance which we have described are accompanied by changes of membrane potential and current which can be recorded in a variety of ways. The electrical responses of a rod to flashes of light of increasing intensity are shown in Figure 2.6. The membrane potentials in Figure 2.6a were obtained with an intracellular electrode which measures the potential between the tip of the electrode and the extracellular fluid. The inside of a photoreceptor is essentially isopotential (Bader et al., 1979; Attwell et al., 1982). Therefore, the potentials shown in Figure 2.6a are, in fact, the intracellular potentials at all points in the cell. The currents in Figure 2.6b were obtained with a suction electrode in which part of the cell, here the outer segment, was sucked into the electrode. The mouth of the electrode forms an effective seal around the cell membrane, and the current flowing across that part of the cell which is inside the electrode can thus be collected and measured.

Unlike the fixed amplitude nerve impulse, the hyperpolarizing membrane potentials increase with light intensity and at saturating intensities they develop a transient undershoot followed by a plateau. Like the potentials, the currents also saturate at high intensities, but unlike the potentials they do not exhibit a transient. During the saturated phase of the response all the light-sensitive channels are closed and a flash delivered at that time will evoke no additional change of either potential or current. The leading phase of the response waveforms reflects the activating processes of the transduction cycle, while the trailing phase represents the processes of inactivation and recovery of membrane conductance.

Potential and current have been recorded simultaneously in the same cell by sucking the outer segment into a suction electrode and inserting an intracellular electrode into the inner segment (Baylor and Nunn, 1986). An examination of such simultaneous records shows that the peak voltage response always precedes the peak current response. The maximum hyperpolarization occurs before the maximum reduction of inward current, and the potential starts to recover while the current is still diminishing. This is illustrated in Figure 2.6c.

These features of the response waveforms are shaped by the transduction biochemistry and the electrical properties of the cell. Having considered the former, we now turn to examine the latter, including the possible presence of voltage-sensitive conductances such as are found in nerve axons and almost all neural tissue. They are suggested by the divergence from ohmic behavior of potential and current which have been observed.

The relationship between current and voltage can be investigated with the voltage clamp technique. In this, for any current generated in the cell an equal and opposite current is delivered by the clamp circuit and thus the potential of the cell is held fixed at a desired value. If the membrane is ohmic the clamp current varies linearly with voltage. But if there are voltage-sensitive conductances which, moreover, may vary in time, the clamp current is nonlinear with voltage. Ohm's law $V = RI$ with R constant no longer holds, since R is now voltage dependent.

Figure 2.7a illustrates how the current changes when the potential of an isolated rod is stepped from its resting potential to hyperpolarizing or depolarizing levels in voltage clamp. An outward current depolarizes the membrane and an inward current hyperpolarizes it. When a voltage step is applied there is an instantaneous change in the current, followed by a slower one to a new, steady level. From such records one can construct the "instantaneous" and "steady state" current–voltage relationships. These are shown for a rod in Figure 2.7b and for a cone in Figure 2.7c. In the physiological range current varies linearly with voltage. But at depolarizing potentials the curves are strongly outward rectifying. In other words there is a strong deviation from linearity for the outward going currents. This is true for both the instantaneous and steady-state curves. No clamp current is required to keep the cell at its resting potential. Below it the clamp current must be negative and above it it must be positive. Thus, the current reverses direction at the resting potential as, indeed, is the case in Figure 2.7b and c.

As shown in Figure 2.7a, the steady-state currents are approached exponentially. This can be interpreted as the effect of the voltage step on the gate, as it were, of a voltage-sensitive channel, which opens or closes with an exponential time course depending on the voltage. It can be modeled by the following equations (Attwell and Wilson 1980, Attwell et al., 1982):

For a rod

$$I_A(V,t) = \bar{I}_A(V)A(V,t) \quad (4)$$

For a cone

$$I_B(V,t) = \bar{I}_B(V)B(V,t) \quad (5)$$

and

$$\frac{dX}{dt} = -\frac{X - X\infty}{\tau_x} \quad (6)$$

I_A and I_B are the currents in a rod or cone, respectively, and are functions of voltage and time; $\bar{I}_A$ and $\bar{I}_B$ are the corresponding steady state values, which are functions of voltage only; A and B are the voltage and time dependent functions which describe the exponential current changes after a voltage step. They are functions of voltage and time and obey Eq. (6) in which X stands for either A or B and X_∞ is the value of X as time $t \rightarrow \infty$; τ_x is the time constant of the exponential change and is a function of voltage only. When the parameters are estimated from the data, the theory agrees reasonably well with experiment.

Unlike I_A and I_B we shall see that the outer segment current depends very little on voltage. The implication therefore is that these currents originate in the inner segment.

The conclusion that outer segment currents are essentially independent of voltage in the physiological range (Bader et al., 1979; Baylor and Nunn, 1986) is based on work such as illustrated in Figure 2.8 from an experiment in which the outer segment current was recorded by a suction electrode, while the cell was voltage clamped by an intracellular electrode in the inner segment. Curves 1 to 3 show the relation between current and voltage in the dark (curve 1), in the presence of a half saturating background (curve 2) and a saturating background (curve 3). The curves were obtained by stepping the voltage to different values from the resting potential and measuring the resulting current. Curves 1 and 2 have a small slope between −80 and −20 mV, indicating that the outer segment conductance is fairly insensitive to voltage in the physiological range. The slope of curve 3 is essentially zero. No current flows through the cell regardless of

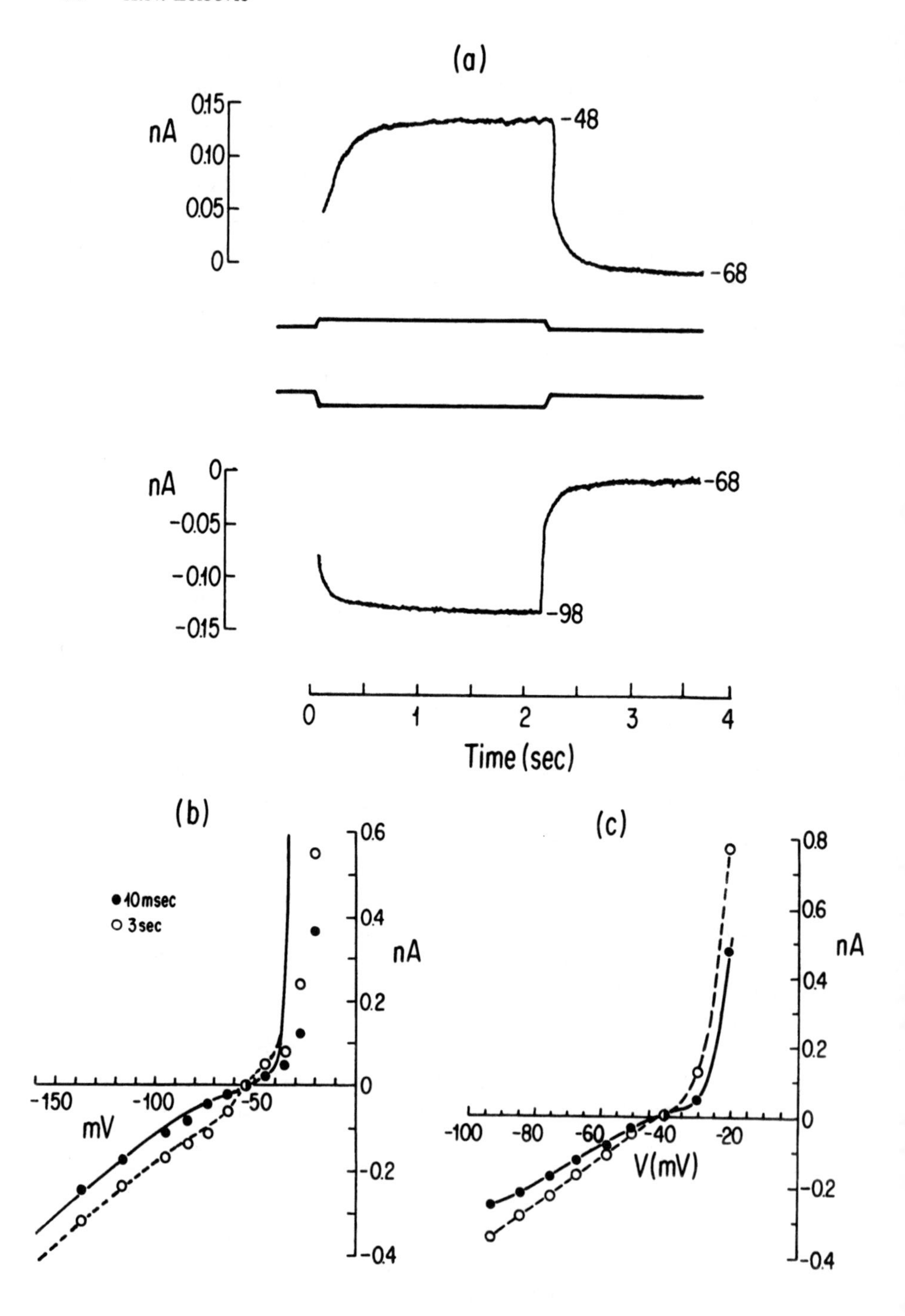
(a)
nA
0.15
0.10
0.05
0
-48
-68
nA
0
-0.05
-0.10
-0.15
-68
-98
0
1
2
3
4
Time (sec)
(b)
● 10 msec
○ 3 sec
0.6
0.4
nA
0.2
0
-0.2
-0.4
-150
-100
-50
mV
(c)
0.8
0.6
nA
0.4
0.2
0
-0.2
-0.4
-100
-80
-60
-40
-20
V(mV)

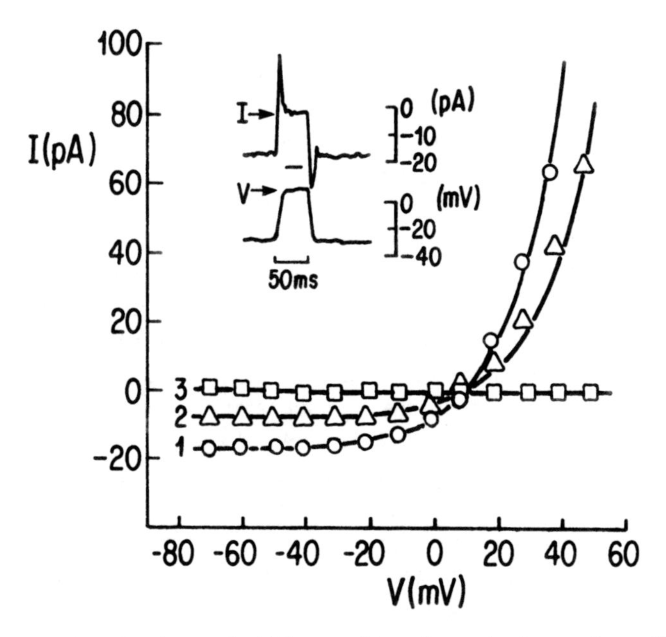

FIGURE 2.8. I–V relation for a rod which was voltage clamped using two intracellular electrodes in the inner segment while the current was recorded with a suction electrode in the outer segment. Clamp pulses of 50 msec were used, as shown in the inset, to step the membrane to different potentials and the currents were measured after the initial transient. (Baylor and Nunn, 1986 © The Physiological Soc. Hon. Treasurer, with permission.)

◄

FIGURE 2.7. (a) The membrane current of an isolated rod with a resting potential of −68 mV when it was depolarized to −48 mV (upper trace) and hyperpolarized to −98 mV (lower trace) by a voltage clamp step. (After Attwell and Wilson, 1980, © The Physiological Soc. Hon. Treasurer, with permission.) The traces in the center mark the voltage steps, up for depolarization and down for hyperpolarization. The initial rapid change of clamp current is unresolved on these records and the traces start at the beginning of the slow change. (b) I–V relations (●, instantaneous; ○, steady state) for an isolated rod obtained from voltage clamp records such as are shown in (a). (Attwell and Wilson, 1980, © The Physiological Soc. Hon. Treasurer, with permission.) (c) Same as in (b) for an isolated cone. (Attwell et al., 1982, © The Physiological Soc. Hon. Treasurer, with permission.) In accordance with the usual convention, current flowing out of the cell is positive and into the cell is negative.

voltage. At potentials more positive than −20 mV the conductance increases greatly. We have strong outward rectification as we have already seen in Figure 2.7. Figure 2.8 also shows that the reversal potential for the outer segment current is between 5 and 10 mV. According to our previous discussion and Eqs. (1) and (2) we know that this potential is under the influence of the ionic permeabilities and the concentrations on the two sides of the membrane. For example, K^+ by itself would be expected to contribute a negative potential and Na^+ by itself a positive potential. The reversal potential is between the presumed Na^+ and K^+ equilibrium potentials, each of which separately would be given by Eq. (1). This, however, is an oversimplified picture. Na^+ and K^+ are the principal ions. But, in view of what we know of the effects of other ions, of ionic interactions and the ion pumps, we can expect that the reversal potential is a complex function of several parameters, which remain to be explored.

It should be noted that Figures 2.7 and 2.8 refer to different experimental conditions. In Figure 2.8 the outer segment current is measured directly. In Figure 2.7 the current is the clamp current injected into the cell to keep the voltage constant. The reversal potential in Figure 2.7 is the resting membrane potential at which the clamp current is zero, although there is at that point a considerable current flowing into the cell through the outer segment. In Figure 2.8 the reversal potential is that at which the outer segment current reverses from inward to outward.

Photoreceptor Adaptation

Our ability to adapt to an enormous range of light intensities from the dark of night to bright sunlight is made possible, in the first instance, by our photoreceptors. There are, to be sure, adaptive changes in the retinal network and in higher centers, the latter of which can be demonstrated through binocular interactions. But it is our rod and cone vision which enables us to cover the physiological range of intensities (Boynton and Whitten, 1970; Leibovic, 1971; Dowling and Ripps, 1972; Normann and Werblin, 1974; Grabowski and Pak, 1975; Kleinschmidt and Dowling, 1975; Fain, 1976; Pepperberg et al., 1978; Shapley and Enroth-Cugell, 1984; Dowling, 1987; Leibovic et al., 1987a). The psychophysics of adaptation is treated in Chapter 7. Here we are concerned primarily with photoreceptors.

At a given level of adaptation a rod or cone can operate, typically, over a thousandfold range of light intensities. As the level of adaptation changes this operating window is shifted up or down the scale. Our rod vision operates over five orders of magnitude above the absolute dark threshold. It overlaps cone vision at the upper end of its range and then the latter takes over for the next six or seven orders of magnitude (Riggs, 1966).

The adaptive ranges of rods and cones appear especially remarkable in view of the fact that we can distinguish very small intensity differences. To get a better understanding of this, suppose a certain quantity of energy, such as might be represented by a packet of photons, is to be transduced into a detectable signal, such as a membrane conductance change. To amplify the photon energy into the signal energy we need a module which itself stores the appropriate energy for this process. There is noise associated with the operation of the module. As more energy packets are to be transduced, more modules must be added, each contributing its intrinsic noise to the system. This is so even in the absence of any input, for the module must always stand ready to respond. Thus, a point is reached at which the signal resulting from one energy packet entering a module is drowned out by the noise from the assembly of modules. In other words, there is a limit to the allowable input range which depends on the minimum energy packet to be detected. Given such a system, two signals differing by this minimum energy packet will always be detected, provided the intrinsic noise from a module remains constant.

The notion of modules is consistent with the structure and function of photoreceptors. In both rods and cones the light-absorbing molecules are arranged in layered fashion along the axis of the outer segment. Photon absorption is a local event and so is the spread of excitation due to such absorption (Cornwall et al., 1983), which may extend to less than 1 μm along the outer segment of a rod (Matthews, 1986). This implies that local mechanisms are involved in transduction, analogous to modular operation.

The above arguments favor a restricted operating range to achieve good discrimination of intensity differences. Then to extend the range we would look for a means to shift the operating range over the scale of intensities. This is precisely the way photoreceptors work. The detection of contrast is optimized and thus the perception of form and space is well subserved over a large range of intensities.

There is also a temporal aspect of photoreceptor adaptation. As the background of light gets brighter, so photoreceptor response get faster. Cones are faster than rods, for example, in terms of response latency and their ability to follow a flickering light. But the responses of each type of photoreceptor also speed up as the background intensity increases. This is illustrated in Figure 2.9 in which we compare the response waveforms of a dark-adapted rod with those in the presence of two backgrounds, one brighter than the other.

There is a functional significance to this as there is to threshold elevation accompanying the shift of the operating range. Speed of responsiveness is obviously important for speedy reactions. But at low levels of illumination, stimulus summation is required for detection. Thus, the responses are slow in dim light, allowing the absorption of more photons. But in bright light, when detection is easy, the responses are initiated and terminated more quickly.

We now consider in more detail the adaptive responses in terms of threshold, amplitude, and some temporal characteristics.

The smallest detectable or threshold response when recording membrane potential (Leibovic et al., 1987a) or current (Baylor et al., 1979b) from a dark

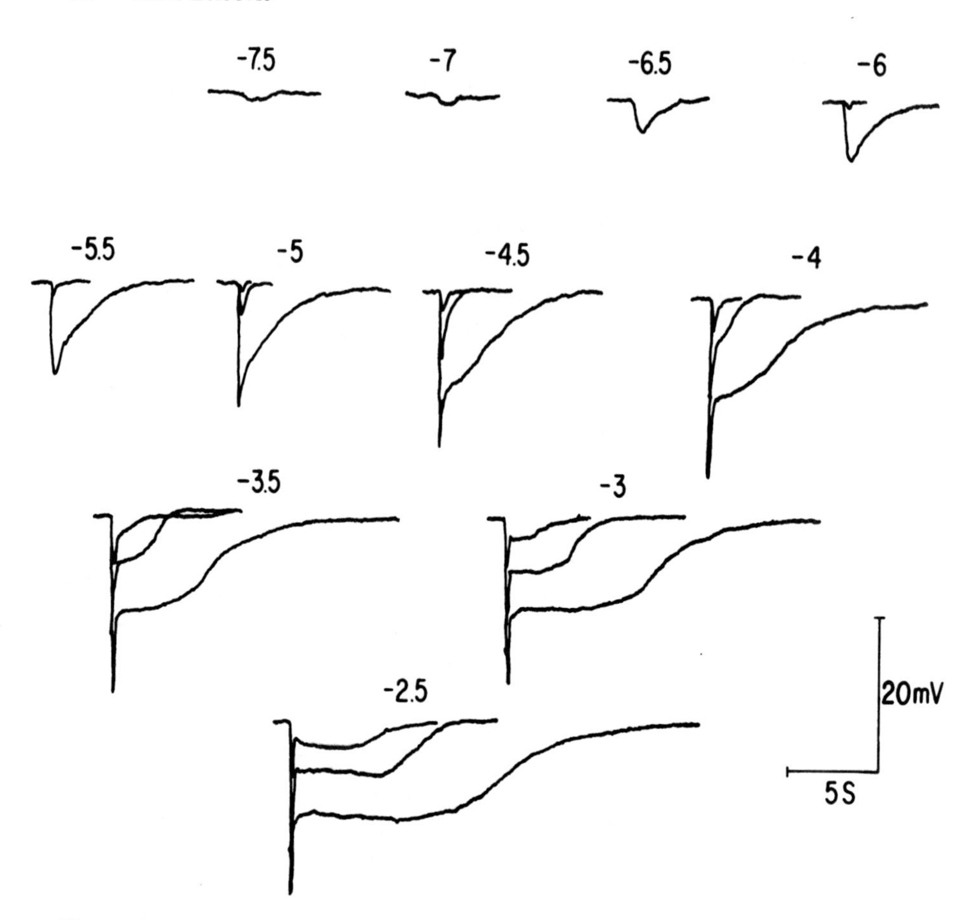

FIGURE 2.9. Response duration is shortened as the background intensity increases. This is shown for a series of rod responses to 100 msec flashes of increasing intensity. The intensity is given in terms of the neutral density (ND) of attenuation of the light beam. A dim light corresponds to a numerically large ND, a bright light to a numerically small ND. Each set of superimposed records shows the responses to the same flash intensity. Flash ND above the records. The largest of each of the three superimposed responses is from the dark adapted cell; and the other two were obtained in the presence of two backgrounds (ND 5.5 and 4.5, respectively). In addition to shortening, the responses also become smaller with increasing intensity of the background. (Leibovic, 1983, © IEEE, with permission.)

adapted rod is the single photon response. In the presence of a background the threshold rises and a more intense flash is required to elicit a detectable response. When a background is first turned on the threshold rises instantly to a high value but then relaxes and settles down to a lower one. When the background is turned off the threshold soon recovers again. This is illustrated in Figure 2.10a. It parallels our experience when illumination suddenly changes.

It is not only light but also the degree of bleaching which elevates thresholds. We have seen how, after absorbing a photon, a rhodopsin molecule undergoes a series of conformational changes and finally loses the chromophore. It is then no longer able to absorb another photon unless it regenerates, as indeed it does in the presence of the pigment epithelium at the back of the eye. A dark adapted retina looks red, but after intense light exposure it becomes white. This is the bleached state when the chromophores have detached from the opsin. Under moderate lighting conditions only a very small fraction of rhodopsin is bleached. For example, in comfortable reading light that fraction may be of the order of 10^{-4}, but when looking at a white surface in bright sunlight it may approach 10^{-1} and exposure to an even brighter tungsten filament lamp can bleach almost all the visual pigment. When the retina is isolated from the pigment epithelium and placed in a Ringer solution, rhodopsin does not regenerate and it is possible to study the steady-state effects of bleaching (Weinstein et al., 1967). Figure 2.10b and c show that the steady state threshold of toad rods depends on background and bleaching, respectively. In Figure 2.10b the threshold rises very little at first; then it rises in proportion to the square root of the background as the abscissa goes from −7 to −5.75; after that the rise is proportional to background and finally the threshold saturates. In Figure 2.10c the threshold is proportional to the percentage bleached up to some 15–20%, and thereafter it rises exponentially.

Threshold elevation is not the only effect of bleaching and backgrounds. There is also a reduction of the response amplitude. This is shown in Figure 2.11 for the saturated responses of toad rods. Threshold elevation and response compression are consequences of adaptation in rods and cones. Response compression is more marked in rods than in cones, as illustrated in Figure 2.12, and this is consistent with the suppression of rod vision at photopic light intensities. Figure 2.12 shows how photoreceptors shift their operating window over the adaptive range. But in different cells and different species the extent of threshold elevation and response compression differ, and the operating window may shrink more or less with increasing adaptation. In mudpuppy rods, for example, when a background light is turned on the threshold rises, but then partially recovers, and the operating window shrinks slowly as it moves over some five orders of magnitude of light intensity (Normann and Werblin, 1974). But in primate rods (Baylor et al., 1984b) and in human scotopic vision (cf. e.g., Barlow, 1972) the threshold shows little, if any, recovery in the presence of a background light (but see Tamura et al., 1989). It increases rapidly to saturation over only 2 to 3 orders of magnitude of light intensity and the operating window shifts little, if at all. These differences can be understood from the following observation. The recorded membrane potential and current are both reduced when a background light is turned on. In some cases these reductions exhibit a transient undershoot followed by a return to a steady state level. In other cases there may be no transient before the steady state. In the first case thresholds will recover as in Figure 2.10a; in the second case there may be little or no such recovery. Some authors, including Frumkes in

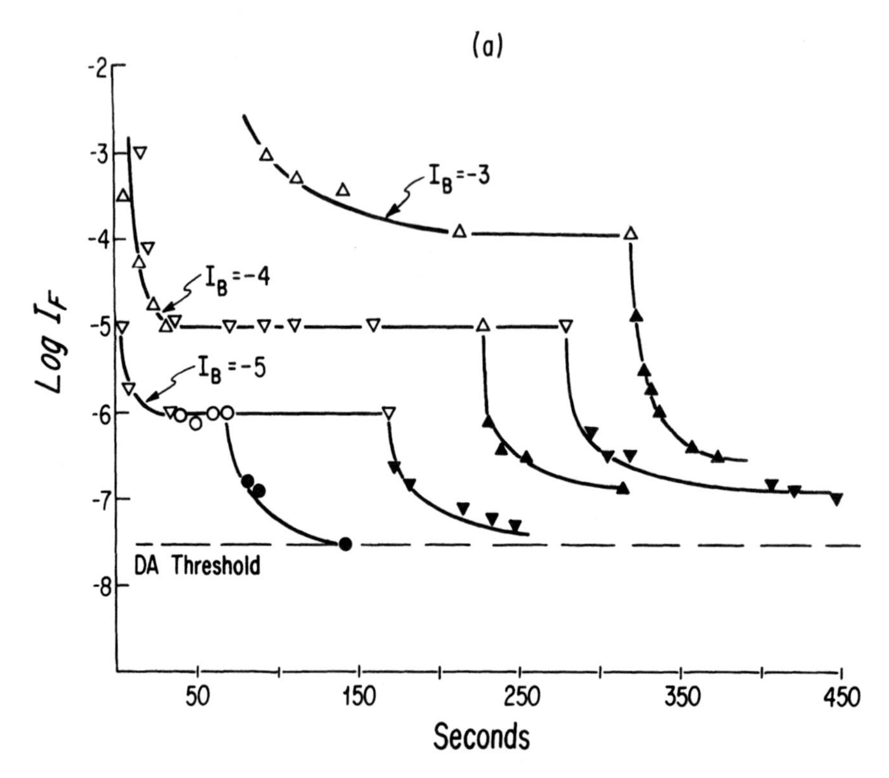

FIGURE 2.10. (a) The time course of the rod threshold at light on (open symbols) and light off (filled symbols) for three backgrounds (nominal ND −5, −4, −3). DA stands for dark adapted. I_F is flash intensity (in ND units). (Leibovic, 1986b, © Erlbaum Associates, with permission.) (b) Threshold elevation (Δ log I_{FT}) as a function of background log (I_B/I_{BO}). (Leibovic et al., 1987a, with permission of Oxford University Press, © the Society for Neuroscience.) (c) Same as (b) as a function of the percentage of bleached pigment B. Inset demonstrates linearity up to some 15–20%. The dashed curve is the threshold elevation due to the reduction of native rhodopsin by bleaching. (Leibovic et al., 1987a, © Oxford University Press, with permission of Oxford University Press, © the Society for Neuroscience.)

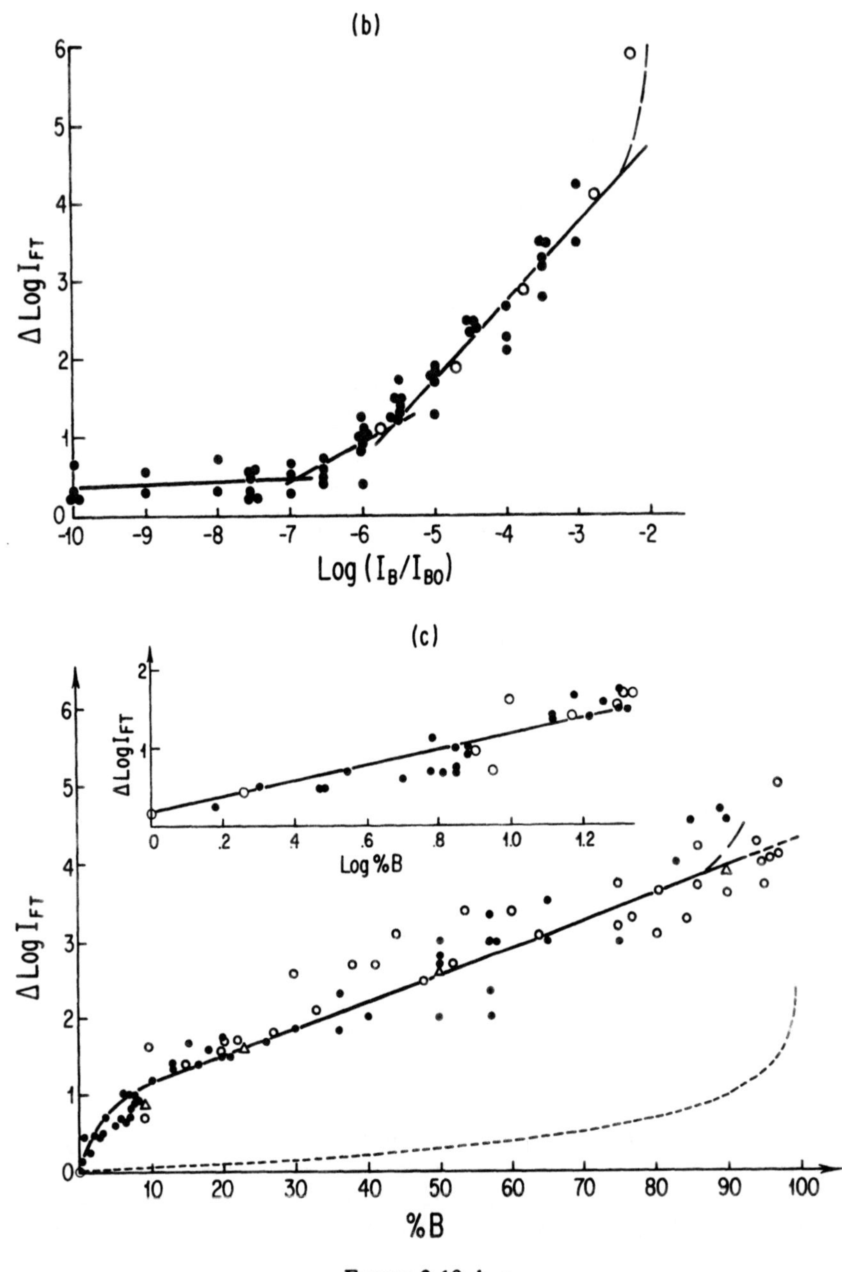

FIGURE 2.10. b, c

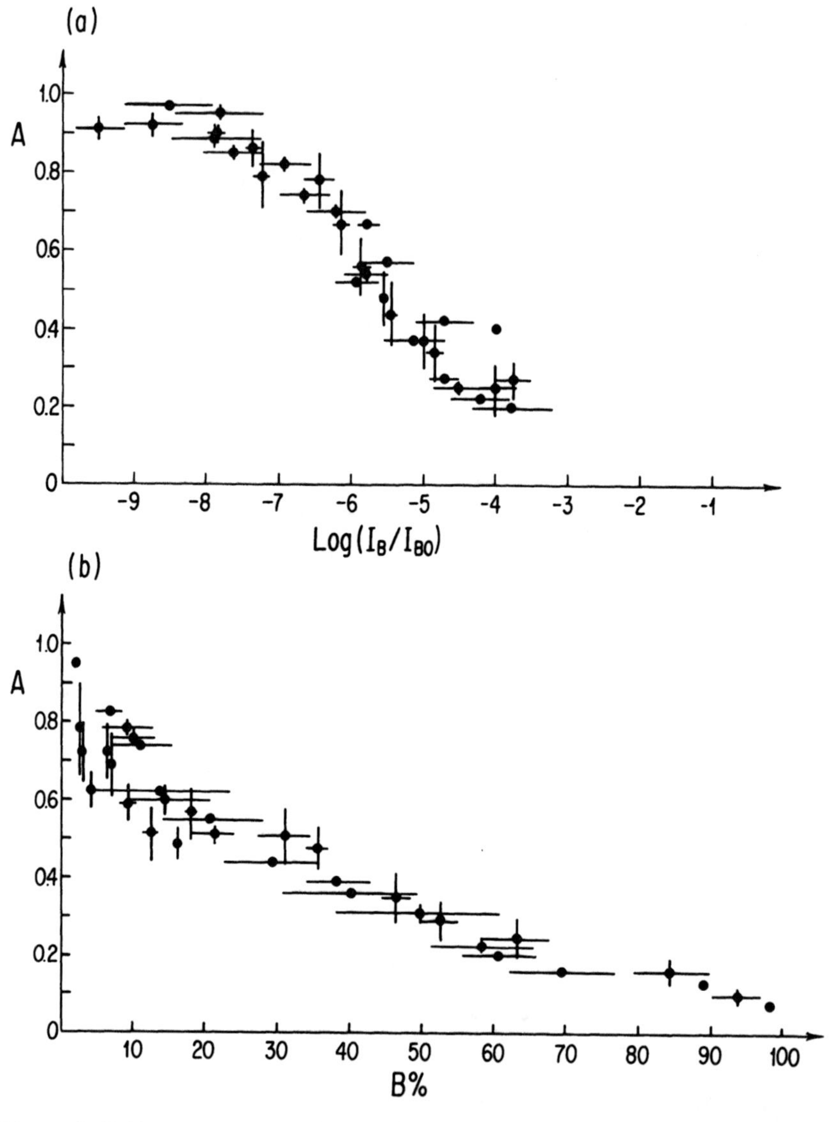

FIGURE 2.11. Response compression A is the reduced response, due either to background light or to bleaching, divided by the dark adapted response; (a) shows A vs. background, (b) vs. bleached fraction. Filled symbols are mean values; bars are standard deviations. (Leibovic et al., 1987a, with permission of Oxford University Press, © the Society for Neuroscience.)

Chapter 7, apply the term "adaptation" strictly to the first case. But threshold elevation and response compression as well as changes in the kinetics occur in both cases. Therefore we define photoreceptor adaptation to be the change in functional state as the operating range shifts in response to different levels of illumination and pigment bleaching.

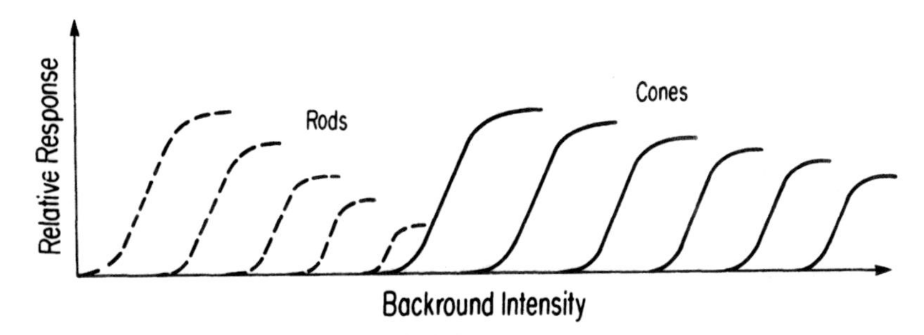

FIGURE 2.12. Each curve, spanning about a 1000-fold range of light intensities, represents a peak response vs. flash intensity plot for a rod (dashed lines) or cone (full lines). These operating curves shift along the intensity axis as the background intensity increases. Thresholds rise and response amplitudes diminish as illustrated.

As an example of temporal adaptation we consider the responses to a periodic stimulus such as a flickering light. When a train of flashes or a sinusoidally modulated light is projected onto a photoreceptor, the membrane potential responds as in Figure 2.13a. There is an initial hyperpolarization, denoted $(a + b)$, from the resting potential. Then the potential settles into an oscillation, of amplitude $2c$, which faithfully follows the periodicity of the stimulus. But, as the stimulus frequency increases, c decreases and tends to zero. At this point we have the "critical flicker fusion frequency" (CFFF). This is illustrated in Figure 2.13b for the dark adapted state (lower curve) and in the presence of a background (upper curve). μ is defined as the "modulation" analogously as in psychophysical studies:

$$\mu = \frac{c}{a + b} \tag{7}$$

It is noteworthy that in the presence of the background CFFF has increased, and the cell responds to higher frequencies. Also, μ is larger although $(a+b)$ gets smaller due to response compression. Thus, the output of the cell in terms of its response to stimulus changes is enhanced. This is shown in Figure 2.13c, which plots the gain of the system as a function of the input. The input is the flash intensity I_F and the gain G is output/input, i.e.,

$$G = \frac{\mu}{I_F} \tag{8}$$

For a given frequency the gain is larger in the presence of the background. It can be seen that the gain depends on the input which means that the system is nonlinear. Over much of the range we have

$$\log G = -\log I_F + \log K \tag{9}$$

where K is a constant for a given frequency and for a given state of adaptation. From Eqs. (8) and (9) it follows that μ depends on frequency and adaptation but not on input, in agreement with Figure 2.13b.

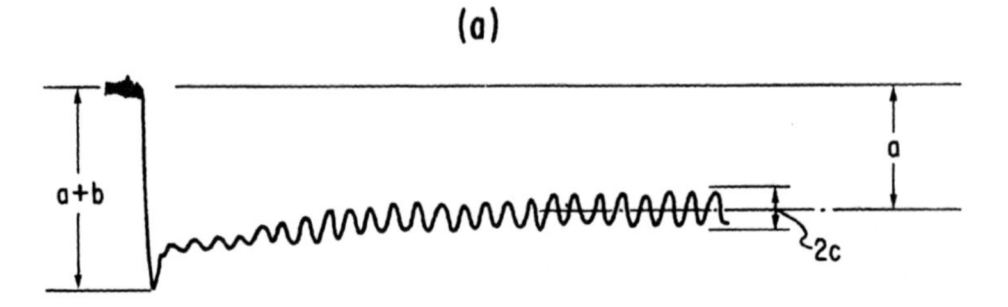

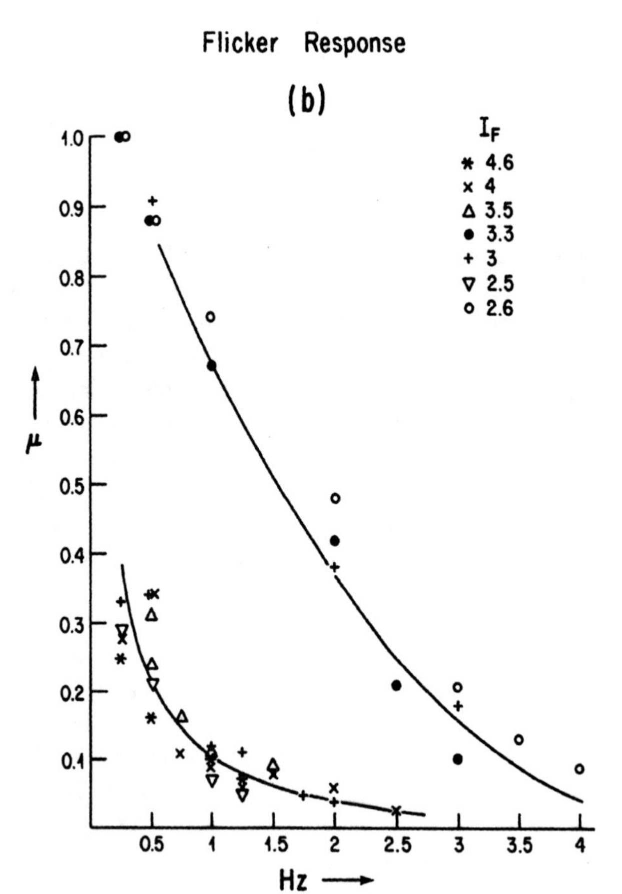

FIGURE 2.13. (a) Response of a toad rod to a periodic train of light flashes. After an initial hyperpolarization ($a+b$) the membrane potential settles to a lower mean value (a) around which it fluctuates with an amplitude (c) as it follows the flash train. (b) The modulation μ is a function of frequency and the state of adaptation but not of flash intensity I_F.

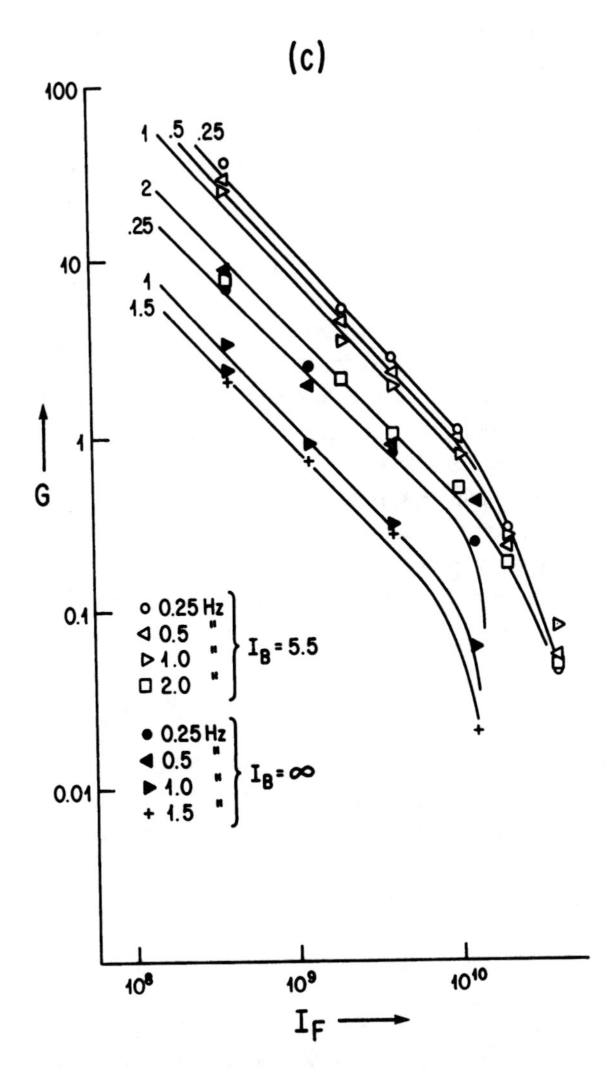

FIGURE 2.13. (c) The gain G is a function of the input I_F. It also depends on stimulus frequency and the state of adaptation. The filled symbols with $I_B = \infty$ are for the dark adapted state; the open symbols with $I_B = 5.5$ are for a background of neutral density attenuation 5.5 log units. The I_F scale reads in photons/cm²/sec; the G scale is multiplied by a factor of 10^{10}. Both scales are logarithmic (Leibovic and Kim, unpublished.)

In psychophysics it is well known that bleaching and backgrounds produce many equivalent effects (cf. Chapter 7). Is any such equivalence observed in photoreceptors? This apparently is the case with threshold elevation and response compression in *Bufo* rods. In Figure 2.14, the data of Figures 2.10 and 2.11 have been combined and rearranged to show that for each background there is a level of bleaching for which threshold elevation and response compression are the same. This raises the question of whether backgrounds and bleaching may be equivalent in all their consequences. The answer is negative. For example, the response kinetics are greatly accelerated in the presence of a background, but not in the presence of the threshold equivalent bleaching (Leibovic et al., 1987a). There are differences, including recovery of sensitivity after a flash, stimulus summation, and responsiveness to a flickering light where the equivalence with respect to threshold and response amplitude is not upheld (Leibovic et al., 1987b). This, as well as the different ways in which threshold and amplitude each depends on bleaching and background, respectively, in Figures 2.10 and 2.11, argue against a common mechanism for bleaching and background adaptation.

As in so many cellular processes, Ca^{2+} may also be involved in photoreceptor adaptation (Nakatani and Yau, 1988c; Matthews et al., 1988; Koch and Stryer, 1988). We have discussed the Ca^{2+} feedback control of transduction, and we have mentioned that in many photoreceptors there is a transient undershoot of the membrane potential followed by a return to a steady level, in response to a step of light. This transient can be abolished by blocking the Ca^{2+} feedback. Thresholds and response kinetics are altered in a manner consistent with a narrow definition of adaptation. Neither the sites of action of Ca^{2+} in producing these effects nor, more generally, the molecular bases of adaptation are known at this time. Ultimately, the adaptive changes we have discussed must be explainable in terms of the transduction biochemistry: Which of the steps in the cycle of Figure 2.4 determine thresholds and which determine response amplitude? How are these steps altered so as to elevate thresholds, reduce amplitudes, and change the response kinetics? We shall consider these questions further in the next section. Meanwhile we note some similarities of cellular and psychophysical behavior: The dependence of thresholds on time in Figure 2.10a and on backgrounds and bleaching in Figure 2.10b and c, as well as on flicker fusion illustrated in Figure 2.13 all have their counterparts in psychophysics. At the same time there are differences between photoreceptor and psychophysical adaptation. These have to do with postreceptor mechanisms and, to a lesser extent, with interactions due to coupling between the photoreceptors. Such similarities and differences will become more evident by a comparison of this chapter with the discussions in Chapter 7.

Some Current Problems

There are at least as many interesting and important problems in photoreceptor research as there are components and interactions, both molecular and cellular. In addition, there are questions of how photoreceptors fit into the larger scheme,

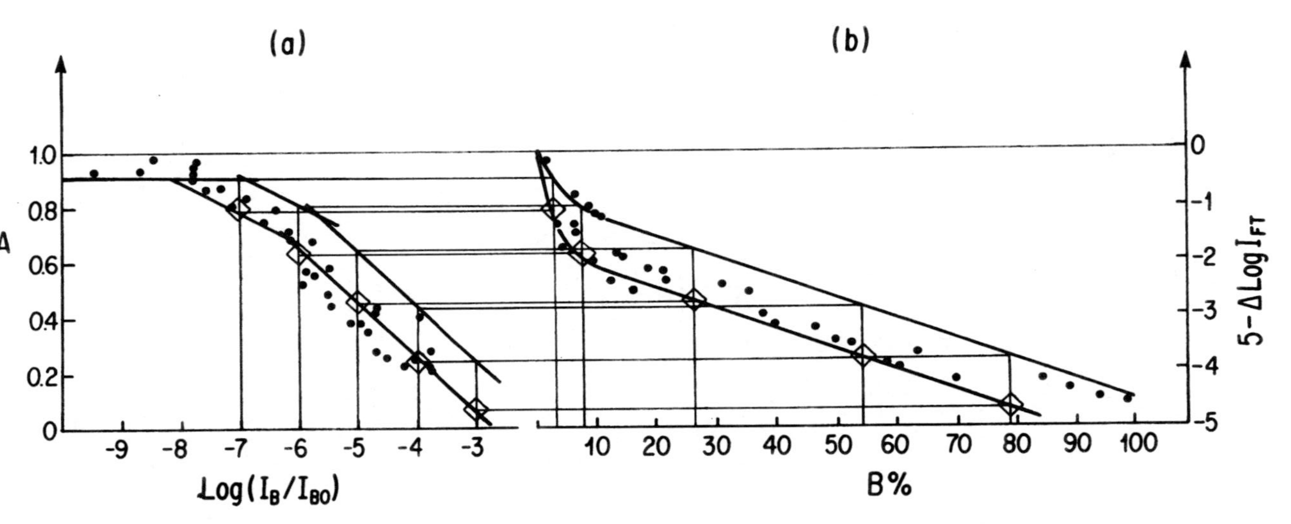

FIGURE 2.14. The filled circles in (a) and (b) are reproduced from Figure 2.11a and b, respectively. They are the response compressions for which the scale A is on the left. The upper curves in (a) and (b) are from Figure 2.10b and c, respectively. But they are shifted up 5 units and inverted by transforming to $(-\Delta \log I_{FT})$. The vertical scale for these curves is on the right. The lower curves passing through the diamond symbols are parallel to the upper curves. The vertical lines in (a) from −7 to −3 on the x axis are drawn to intersect the upper curve. From there are drawn horizontal lines to intersect the upper curve in (b) and from there vertical lines are drawn to the x axis. The backgrounds in (a) (viz. −7, −6, . . . , −3) are equivalent to the $\%B$ in (b) where the vertical lines intersect the x axis. It can now be seen that the lower curves through the diamonds reasonably fit the experimental points for A and thus when an (I_B,B) pair has the same threshold elevation it also has the same response compression.

the functions of the organism. Here we shall discuss just two areas related to our preceding discussions.

Transduction and Adaptation

In the transduction cycle we have described, the processes which activate the responses to light are better understood than those which inactivate them. We know that calcium and arrestin act as control elements. In addition there may be other factors. For example, some membrane proteins may be involved in PDE* inactivation (Lewis et al., 1987), and there may be several GTP binding sites on transducin which could speed its inactivation (Sitaramayya and Casadevall, 1987).

Whatever the details, we can imagine how thresholds and response amplitudes can be related to the biochemistry. To do this with regard to adaptation let us consider the effects of backgrounds and bleaching separately. Both raise the threshold to a flash of light and both reduce the response amplitude. But backgrounds also speed up the responses, while in the presence of a steady level of bleaching the shortening of the responses is not as pronounced. When a background is on, the transduction machinery is turning over steadily, but this is not the case when part or all of the pigment is bleached and no light is on.

Now, threshold elevation means that more photons are required to produce a measurable response to a flash. This could be due to less R being available. But since there is such an enormous amount of R in photoreceptors, this is not a serious limitation except when nearly all the pigment is bleached. It could be that the effectiveness of R* in activating T to T* is diminished, or that less PDE is activated to PDE* or that less cGMP is hydrolyzed. One might expect, a priori, that threshold elevation would occur at the beginning of the cycle. In that case, in the presence of a background, since the transduction cycle is turned on, arrestin is already present and R* is being phosphorylated one can imagine that R* inactivation might become more expeditious and the threshold would thus be raised. We cannot make the same argument in the case of bleaching. But a possibility might be that the presence of bleached pigment partly immobilizes the activation of the cycle following R* such as the exchange of GDP for GTP on transducin. Again, thresholds would be elevated.

Response compression suggests that fewer of the light-sensitive channels are open. Thus, a saturating flash which closes all the channels produces a smaller than the fully dark adapted response because fewer channels were open before the flash. This implies that there is less free cGMP. In the presence of a background this may come about through the continuous production of PDE* and, perhaps, compensatory changes in the activity of G-cyclase. In the presence of bleaching there might be some effect on G-cyclase and possibly PDE, traceable to the bleached pigment.

As regards response acceleration due to backgrounds it appears that the transduction cycle, once turned on, alters the reaction kinetics of both response initiation and turnoff.

These and other possibilities remain to be explored in detail. The adaptive control of the transduction cycle is the basis for adaptation in photoreceptors.

Interactions between Inner and Outer Segments

The inner segment contains the energy-producing mitochondria and the vegetative machinery of the cell. The outer segment is primarily concerned with transduction. Thus, there are many complex interchanges between these two differentiated regions of the cell. One of them, involving transduction and energy exchange, is the flow of a Na^+ current into the outer segment and its extrusion in the inner segment. This current can be considerable. For example, in *Bufo* rods in the dark it is about 25 pA or 16×10^7 positive charges per second (Baylor et al., 1979a). We estimate the internal Na^+ concentration to be 10 m*M* (Torre, 1982). The cytoplasmic volume of a 6-μm diameter, 80-μm-long rod, allowing for intracellular discs and organelles, is, say 10^{-9} ccs. Then the Na^+ current would replace the total cytoplasmic Na^+ content of the rod in about 44 sec. During the response of a rod to a saturating flash, the light-sensitive channels may remain closed for a number of seconds. Clearly, if the Na^+–K^+ pump in the inner segment were to continue working at the same rate as in the dark, and without any compensatory changes, the internal Na^+ concentration would be significantly altered. If the Na^+ and K^+ conductances of the inner segment were to change, mediated through such voltage-sensitive mechanisms as we have already discussed, this might counteract the pump currents and stabilize the concentrations. Whether this is so, or whether, by another mechanism, the Na^+–K^+ pump is responsive, either directly or indirectly, to outer segment conductance changes is unknown. There are some suggestions that it might be so. The possible interactions between inner and outer segments are not fully resolved, partly due to some uncertainties regarding voltage-sensitive conductances. We have considered these in an electrophysiological context. Another approach is pharmacological in conjunction with voltage clamping. For it is known that ion channels are affected by various drugs (see, e.g., Hille, 1984) and the latter have been used to identify various conductances in the inner segment (see Fain and Lisman, 1981; Owen, 1987 for reviews). In one line of work, isolated inner segments have been used (Bader et al., 1982). When dissociating a retina the photoreceptor outer segments are often lost. The inner segments may reseal themselves and remain viable for some time. An electrode can then be inserted, the inner segment can be voltage clamped, perfused with drugs and exposed to a variety of manipulations. Thus, Bader et al. (1982) found a current I_h which was blocked by extracellular Cs^+. It was activated by hyperpolarizations below −30 mV and it was interpreted as carried by Na^+ and K^+. Another current blocked by external tetraethylammonium (TEA) was activated by depolarizations above −70 mV and was interpreted as primarily a K^+ current. A current blocked by extracellular Co^{2+} was interpreted as a Ca^{2+} current. Intracellular Cs^+ was believed to affect a Ca^{2+} activated K^+ current and a Cl^- current.

The reader will have noticed that Attwell et al. (1980, 1982), whose work we discussed earlier, and Bader et al. (1982) provide different interpretations of inner segment currents.

In the work on inner segment currents and conductances it has been tacitly assumed that the Na^+–K^+ pump makes no contribution. It is known, however, that drugs such as TEA, which interfere with K^+ movement, and ions such as Cs^+ affect pump operation (see, e.g., Blaustein and Lieberman, 1984 for reviews). It is therefore interesting to see how responses are altered when the Na^+–K^+ pump is inactivated.

Figure 2.15 shows the changes of response waveform of vertebrate rods when the retina is superfused with Ringer solution containing ouabain (Leibovic, 1983). This substance binds very specifically to the Na^+–Ka^+ ATPase and inactivates the pump. It is evident from the figure that ouabain abolishes the transient spike as the response diminishes and before it disappears. This is different from the changes due to adaptation, where the amplitude also diminishes but where the general features of the response waveform are preserved. It suggests that the suppression of the pump by ouabain has preferentially affected the spike.

Significantly, the resting potential of the cell in Figure 2.15b hardly changed as the response and the pump ran down, and it remained essentially constant long after the response had disappeared. As we know from Eqs. (1)–(3) the membrane potential depends on the ion conductances and on the electrogenic pump. It is estimated that the latter contributes some 10 mV hyperpolarization to the dark membrane potential in *Bufo* rods (Torre, 1982). Therefore, when the Na^+–K^+ pump is shut off, and in the absence of other changes, the membrane should depolarize by about 10 mV. The fact that it does not do so suggests that there are simultaneous changes in ion conductances. For example, if the outer segment Na^+ conductance was to decrease it would tend to hyperpolarize the membrane and counteract the effect of the pump stoppage. But this is not enough. With the pump out of action the membrane potential will go to zero if the ion concentrations are allowed to equilibrate. Since the observed potential remains constant, it suggests that the conductances are affected in a manner to reduce or eliminate ion exchanges across the membrane.

The above observations are consistent with the proposition that the outer segment current and Na^+–K^+ pump are linked. This makes sense functionally. In particular, it would mean that the Na^+ current through the cell is under the coordinated control of outer segment conductance and inner segment pump.

An alternative view is that the Na^+–K^+ pump can be ignored in an analysis of the electrical responses of the cell. The response waveforms are then the result of conductance changes only. In particular the transient in the membrane potential at high flash intensities is due to voltage sensitive conductances. Indeed it is possible to simulate the transient spike by injecting a sufficiently strong, hyperpolarizing current step into the cell (Werblin, 1979). It has also been shown that Cs^+ abolishes the transient in the light response (Fain et al., 1978) and it has been suggested that the voltage sensitive conductances generating the transient are affected by Cs^+. This idea is supported by the following experiment (Baylor et al., 1984a): A rod

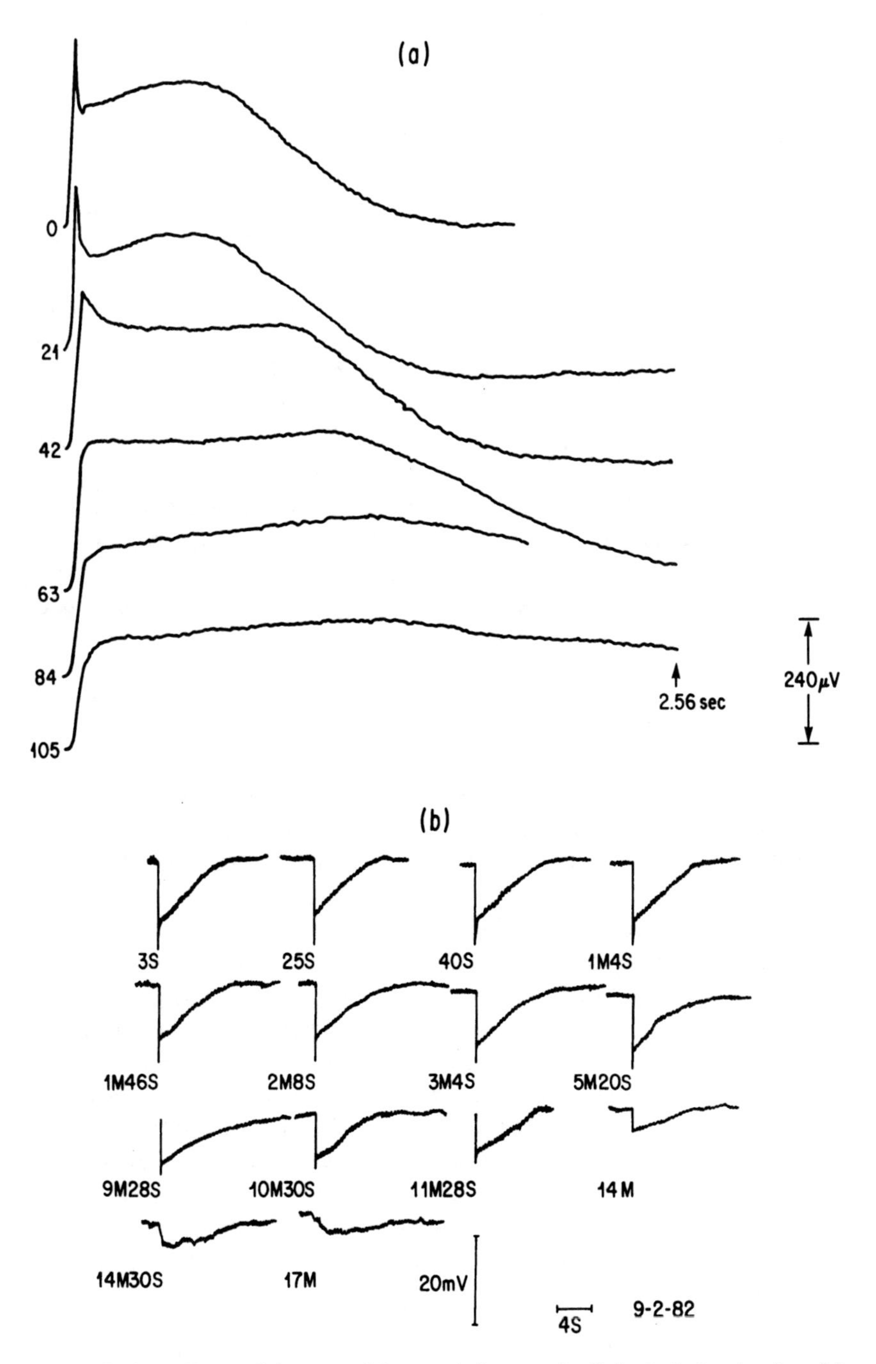

FIGURE 2.15. (a) Extracellular potentials recorded near rod cells in the isolated retina of the rat. A saturating flash of light was delivered after the start of ouabain perfusion at the times in seconds indicated at the beginning of each record. Note how the spike disappears while there is still a good sized response remaining. (b) Intracellular rod responses to a saturating flash in a toad rod before and during ouabain perfusion. Times after the start of perfusion are shown below each record. The spike disappears before the rest of the response as in (a). During the experiment and after the response was abolished the resting potential remained essentially constant. (Leibovic, 1983, Figs. 9, 10, © IEEE, by permission.)

is bathed in a solution containing Cs^+. A saturating light is projected onto the rod so that no further light response can be evoked by a superimposed flash. The cell membrane hyperpolarizes when the saturating light is applied, but it exhibits no transient in the presence of Cs^+. If a strong current step is now injected the membrane is further hyperpolarized and again there is no transient. But if the membrane is now depolarized to the resting potential in the dark and the same current step as before is injected the membrane hyperpolarizes and this time there is a transient spike. Since the light response has been eliminated in this experiment the result implies that the Cs^+ and voltage-sensitive conductances which underlie the spike are revealed as purely electrical effects, which depend on a hyperpolarizing step from the initial level of the membrane potential.

The different results we have presented here were obtained under different conditions. It will probably turn out that they reveal different aspects of the interactions, which make up the complex and very effective machinery subserving photoreceptor function. To pursue its detailed elucidation is the role of research.

Conclusion

In this chapter we have concentrated on cellular processes. Although these are considered in the context of photoreceptors, similar processes operate in numerous other cells. This applies to membrane potentials, the regulatory role of Ca^{2+}, the activation of biochemical cycles by G-proteins, and the intimate relationships between the electrical and biochemical states of the cell.

The physical stimulus for vision is light, and we have seen how light is absorbed in photoreceptors, how it initiates the cascade leading to the hydrolysis of cGMP, and how this produces a hyperpolarization which is transmitted to second order neurons. We have seen some of the delicate control mechanisms involved and we have probed the remarkable ability of photoreceptors to adapt to an enormous range of light intensities. Perhaps the greatest challenge facing photoreceptor research today is to elucidate the mechanisms of adaptation.

In subsequent chapters we shall discuss some theoretical problems concerning the distribution of photoreceptors on the retina, their spacing and structure. But first, in the next chapter, the attention shifts to the retinal network, the interactions of cells in ON and OFF pathways, and the transmitters and receptors mediating these interactions.

Acknowledgments. The comments of Drs. M.C. Cornwall, J.E. Dowling, T.E. Frumkes, G.J. Jones, M.M. Slaughter, F.S. Werblin, and K-W. Yau on this chapter were greatly appreciated.

References

Abrahamson EW, Wiesenfeld JR (1972). The structure, spectra and reactivity of visual pigments. In *Photochemistry of Vision. Handbook of Sensory Physiology,* Vol. 7/1. Springer-Verlag, New York.

Attwell D, Wilson M (1980). Behavior of the rod network in the tiger salamander retina mediated by membrane properties of individual rods. J. Physiol. **309**:287–315.

Attwell D, Werblin FS, Wilson M (1982). The properties of single cones isolated from the tiger salamander retina. J. Physiol. **328**:259–283.

Bader CR, MacLeish PR, Schwartz EA (1979). A voltage clamp study of the light response in solitary rods of the tiger salamander. J. Physiol **296**:1–26.

Bader CR, Bertrand D, Schwartz EA (1982). Voltage activated and calcium activated currents studied in solitary rod inner segments from the salamander retina. J. Physiol. **331**:253–284.

Barlow HB (1972). Dark and light adaptation: Psychophysics. In *Handbook of Sensory Physiology*, Vol. VII/4. Springer-Verlag, New York.

Barlow HB (1982). What causes trichromacy? A theoretical analysis using comb filtered spectra. Vision Res. **22**:635–643.

Baylor DA, Nunn BJ (1986). Electrical properties of the light sensitive conductance of rods of the salamander *Ambystoma tigrimum*. J. Physiol. **371**:115–145.

Baylor DA, Fuortes MGF (1970). Electrical responses of single cones in the retina of the turtle. J. Physiol. **207**:77–92.

Baylor DA, Lamb TD, Yau K-W (1979a). The membrane current of single rod outer segments. J. Physiol. **288**:589–611.

Baylor DA, Lamb TD, Yau K-W (1979b). Responses of retinal rods to single photons. J. Physiol. **288**:613–634.

Baylor DA, Matthews G, Nunn BJ (1984a). Location and function of voltage sensitive conductances in retinal rods of the salamander *Ambystoma tigrimum*. J. Physiol. **354**:203–223.

Baylor DA, Nunn BJ, Schnapf JL (1984b). The photocurrent, noise and spectral sensitivity of rods of the monkey *Macaca fascicularis*. J. Physiol. **357**:575–607.

Blaustein MP, Lieberman M (eds.) (1984). *Electrogenic Transport: Fundamental Principles and Physiological Implications*. Raven, New York.

Bonting SJ (1969). The mechanism of the visual process. Curr. Top. Bioenerg. **3**:351–415.

Bortoff A, Norton AL (1965). Simultaneous recording of photoreceptor potentials and the PIII component of the ERG. Vision Res. **5**:527–533.

Bortoff A, Norton AL (1967). An electrical model of the vertebrate photoreceptor cell. Vision Res. **7**:253–263.

Boynton RM, Whitten DM (1970). Visual adaptation in monkey cones: Recordings of late receptor potentials. Science **170**:1423–1426.

Capovilla M, Cervetto L, Pasino E, Torre V (1981). The sodium current underlying the responses of toad rods to light. J. Physiol. **317**:223–242.

Capovilla M, Caretta A, Cervetto L, Torre V (1983). Ionic movements through light sensitive channels of toad rods. J. Physiol. **343**:295–310.

Cervetto L, Lagnado L, Perry RJ, Robinson DW, McNaughton PA (1989). Extrusion of calcium from rod outer segments is driven by both sodium and potassium gradients. Nature (London) **337**:740–743.

Cornwall MC, Fein A, MacNichol EF Jr (1983). Spatial localization of bleaching adaptation in isolated vertebrate rod photoreceptors. Proc. Natl. Acad. Sci. U.S.A. **80**:2785–2788.

Dartnall HJA, Bowmaker JK, Mollon JD (1983). Microspectrophotometry of human photoreceptors. In *Color Vision* (JD Mollon, RT Sharpe, eds.), Academic Press, New York.

Dowling JE (1987). *The Retina: An Approachable Part of the Brain*. Harvard University Press, Cambridge, MA.

Dowling JE, Ripps H (1972). Adaptation in skate photoreceptors. J. Gen. Physiol. **60**:698–719.

Fain GL (1976). Sensitivity of toad rods: Dependence on wavelength and background illumination. J. Physiol. **261**:71–101.

Fain GL, Lisman JE (1981). Membrane conductances of photoreceptors. Prog. Biophys. Mol. Biol. **37**:91–147.

Fain GL, Quandt FN, Bastian BL, Gerschenfeld HM (1978). Contribution of a cesium sensitive conductance increase to the rod photoresponse. Nature (London) **272**:467–469.

Fesenko EE, Kolesnikov SS, Lyubarski AL (1985). Induction by cGMP of cationic conductance in plasma membrane of rod outer segment. Nature (London) **313**:310–313.

Frank RN, Goldsmith TH (1967). Effects of cardiac glycosides on electrical activity in the isolated retina of the frog. J. Gen. Physiol. **50**:1585–1606.

Fung B K-K (1987). Transducin: Structure, function and the role in phototransduction. In *Progress in Retinal Research*, Chap. 6. Pergamon Press, Oxford.

Grabowski SR, Pak WR (1975). Intracellular recordings of rod responses during dark adaptation. J. Physiol. **247**:363–391.

Hagins WA (1972). The visual process: Excitatory mechanisms in the primary receptor cells. Annu. Rev. Biophys. Bioeng. **1**:131–158.

Hagins WA, Penn RD, Yoshikami S (1970). Dark current and photocurrent in retinal rods. Biophys. J. **10**(5):380–412.

Haynes LW, Kay AR, Yau K-W (1986). Single cGMP activated channel activity in excized patches of rod outer segment membrane. Nature (London) **321**:66–70.

Hille B (1984). *Ionic Channels of Excitable Membranes*. Sinauer, Sunderland, MA.

Hodgkin AL, McNaughton PA, Nunn BJ, Yau K-W (1984). Effect of ions on retinal rods from *Bufo marinus*. J. Physiol. **350**:649–680.

Hodgkin AL, McNaughton PA, Nunn BJ (1985). The ionic selectivity and calcium dependence of the light sensitive pathway in toad rods. J. Physiol. **358**:447–468.

Kleinschmidt J, Dowling JE (1975). Intracellular recordings from *Gekko* photoreceptors during light and dark adaptation. J. Gen. Physiol. **66**:617–648.

Koch KW, Stryer L (1988). Highly cooperative feedback control of retinal rod guanylate cyclase by calcium ions. Nature (London) **334**:64–66.

Korenbrot JI (1985). Role of intracellular messengers in signal transduction in retinal rods. In *Progress in Retinal Research* (N Osborne, G Chader, eds.), Vol. 4. Pergamon, Oxford.

Korenbrot JI, Cone RA (1972). Dark ionic flux and the effects of light in isolated rod outer segments. J. Gen. Physiol. **60**:20–45.

Kropf A (1972). The structure and reactions of visual pigments. In *Physiology of Photoreceptor Organs. Handbook of Sensory Physiology*, Vol. 7/2. Springer-Verlag, New York.

Leibovic KN (1971). On the retinal basis of visual adaptation. Kybernetik **9**(3):96–111.

Leibovic KN (1983). Phototransduction in vertebrate rods: An example of the interaction of theory and experiment in neuroscience. IEEE Transact **SMC 13**(5):732–741.

Leibovic KN (1986a). A new method of nonenzymatic dissociation of the *Bufo* retina. J. Neurosci. Methods **15**:301–306.

Leibovic KN (1986b). Adaptation, brightness perception and their correlation with photoreceptor responses. Cognitive Science Soc., 8th Annual Conference. Erlbaum Associates.

Leibovic KN, Kurtz VT (1975). Flash photolysis of vertebrate photoreceptors. Biol. Cybern. **19**:83–94.

Leibovic KN, Dowling JED, Kim YY (1987a). Background and bleaching equivalence in steady state adaptation of vertebrate rods. J. Neurosci. **7**(4):1056–1063.

Leibovic KN, Kim YY, Pan Z-H (1987b). The relative contributions of gain control and response kinetics to photoreceptor adaptation. Invest. Ophthalmol. Vis. Sci. **28**(3):353.

Lewis JW, Schechter LE, Parker KP, Kliger DS, Dratz EA (1987). The retinal rod GTPase turnover rate increases with membrane concentration. Invest. Ophthalmol. Vis. Sci. Suppl. **28**(3):94.

Matthews G (1986). Spread of the light response along the rod outer segment. Invest. Ophthalmol. Vis. Sci. Suppl. **27**(3):300.

Matthews HR, Murphy RLW, Fain GL, Lamb TD (1988). Photoreceptor light adaptation is mediated by cytoplasmic calcium concentration. Nature (London) **334**:67–69.

McNaughton PA, Cervetto L, Nunn BJ (1986). Measurement of the itnracellular free calcium concentration in salamander rods. Nature (London) **322**:261–263.

Mullins LJ, Noda K (1963). The influence of sodium-free solutions on the membrane potential of frog muscle fibers. J. Gen. Physiol. **47**:117–139.

Nakatani K, Yau K-W (1988a). Calcium and magnesium fluxes across the plasma membrane of the toad rod outer segment. J. Physiol. **395**:695–729.

Nakatani K, Yau K-W (1988b). Guanisine 3′,5′-cyclic monophosphate activated conductance studied in a truncated rod outer segment of the toad. J. Physiol. **395**:731–753.

Nakatani K, Yau K-W (1988c). Calcium and light adaptation in retinal rods and cones. Nature (London) **334**:69–71.

Nakatani K, Yau K-W (1989). Sodium dependent calcium extrusion and sensitivity regulation in retinal cones of the salamander. J. Physiol. **409**:525–548.

Normann RA, Werblin FS (1974). Control of retinal sensitivity. I. Light and dark adaptation of vertebrate rods and cones. J. Gen. Physiol. **63**:37–61.

Owen, W.G. (1987): Ionic conductances in rod photoreceptors. Annu. Rev. Physiol. **49**: 743–764.

Pepperberg DR, Brown PK, Lurie M, Dowling JE (1978). Visual pigment and photoreceptor sensitivity in the isolated skate retina. J. Gen. Physiol. **71**:369–396.

Pirenne MH (1962). Spectral luminous efficiency of radiation. In *The Eye*, (H Davson, ed.), Vol. 2. Academic Press, New York.

Polyak S (1941). *The Retina*. Chicago University Press, Chicago.

Pugh EN, Cobbs WH (1986). Visual transduction in vertebrate rods and cones. Vision Res. **26**(10):1613–1643.

Riggs LA (1966). Light as a Stimulus for Vision, In *Vision and Visual Perception* (CH Graham, ed.), Chap. 1, p. 26. Wiley, New York.

Sakmann B, Neher E (eds.) (1983). *Single Channel Recording*. Plenum, New York.

Shapley R, Enroth-Cugell C (1984). Visual adaptation and retinal gain controls. In *Progress in Retinal Research*, Pergamon, Oxford.

Shichi H (1983). *Biochemistry of Vision*. Academic Press, New York.

Sitaramayya A, Casadevall C (1987). GTP may regulate GTPase activity of a G protein. Invest. Ophthalmol. Vis. Sci. Suppl. **28**(3).

Stern JH, Knutson H, MacLeish PR (1987). Divalent cations directly affect the conductance of excised patches of rod photoreceptor membrane. Science **236**:1674–1678.

Stirling CE, Lee A (1980). [^{2}H]Ouabain autoradiography of frog retina. J. Cell Biol. **85**:313–324.

Stryer L (1986). Cyclic GMP cascade of vision. Annu. Rev. Neurosc. **9**:87–119.

Tamura T, Nakatani K, Yau K-W (1989). Light adaptation in cat retinal rods. Science **245**:755–758.

Tomita T (1965). Electrophysiological study of the mechanism subserving color vision in the fish retina. Proc. Cold Spring Harbor Symp. Quant. Biol. **30**:559–566.

Torre V (1982). The contribution of the electrogenic sodium-potassium pump to the electrical activity of toad rods. J. Physiol. **333**:315–341.

Toyoda J, Nosaki H, Tomita T (1969). Light-induced changes in single photoreceptors of *Necturus* and *Gekko.* Vision Res. **9**:453–463.

Wald G (1968). Molecular basis of visual excitation. Science **162**:230–239.

Weinstein CW, Hobson RR, Dowling JE (1967). Light and dark adaptation in isolated rat retina. Nature (London) **215**:134–138.

Werblin FS (1979). Time and voltage dependent components of the rod response. J. Physiol. **294**:613–626.

Yau K-W, Nakatani K (1984). Cation selectivity of light sensitive conductance in retinal rods. Nature (London) **309**:352–354.

Yau K-W, Nakatani K (1985a). Light-induced reduction of cytoplasmic free calcium in retinal rod outer segments. Nature (London) **313**:579–583.

Yau K-W, Nakatani K (1985b). Light suppressible, cyclic GMP sensitive conductance in the plasma membrane of a truncated rod outer segment. Nature (London) **317**:252–255.

Yau K-W, McNaughton PA, Hodgkin AL (1981). Effect of ions on the light sensitive current of retinal rods. Nature (London) **292**:502–505.

Yoshikami S, Hagins WA (1973). Control of the dark current in vertebrate rods and cones. In *Biochemistry and Physiology of Visual Pigments* (H Langer, ed.) pp. 245–255. Springer-Verlag, New York.

Yoshizawa T (1972). The behavior of visual pigments at low temperatures. In *Photochemistry of Vision. Handbook of Sensory Physiology,* Vol. 7/1. Springer-Verlag, New York.

Yoshizawa T, Horiuchi S (1973). Studies on intermediates of visual pigments by absorption spectra at liquid helium temperature and circular dichroism at low temperatures. In *Biochemistry and Physiology of Visual Pigments* (H Langer, ed.), pp. 69–81. Springer-Verlag, New York.

Zuckerman R (1973). Ionic analysis of photoreceptor membrane currents. J. Physiol. **235**:333–354.

3
The Vertebrate Retina

MALCOLM SLAUGHTER

Introduction

The retina has often been described as a window to the brain because of its accessibility and suitability for scientific investigations. The expansion of brain research over the past decade has demonstrated that these two structures use similar mechanisms and this has spurred new interest in retinal research. The enormous advantage that retinal research enjoys within neuroscience is that we know the retina is responsible for transducing and encoding light signals and we can generate relevant stimuli to evaluate these functions. This is a rare attribute in complex neural tissue, shared by the visual cortex and lateral geniculate. The availability of relevant stimuli has often permitted visual neuroscience to be at the forefront of brain research. Twenty years ago the retina served as a model of cellular neurophysiology as the intricate patterns of intracellular responses and intercellular connections were discovered. Now, the retina is also a model of network systems and information processing, describing how neurons interact to codify and process sensory information. Questions about the role of multiple neurotransmitters, signal transfer at the synapse, and the integrative properties of membrane biophysics are being addressed. In this chapter I would like to give a brief review of retinal anatomy and physiology [see monographs by Rodieck (1973) and Dowling (1987) for detailed descriptions] and then concentrate on recent trends in retinal information processing.

The retina is a very thin (150–300 μm) layer of tissue that lies at the back of the eye. Light images are focused onto the retina through the combined refractive power of the cornea and lens and the aqueous and vitreous humor (fluid filled spaces that also act as lenses). The retina is oriented in the eye so that light must pass completely through it before the light stimulates photoreceptors. This arrangement dictates that the retina be both thin and transparent, but inevitably there is some distortion of the image reaching the photoreceptors. A compensatory design is found in the primate central fovea, the region of highest acuity. The retinal neurons are displaced laterally so that light rays stimulate photoreceptors directly without passing through the rest of the retina.

The eye is commonly likened to a camera, and there are indeed similarities between the iris around the pupil and the diaphragm of a camera, or between the adjustable focus of the lenses in both systems. But the analogy between the retina and a film strip belittles the extraordinary capabilities of this tissue. A film strip converts light energy into chemical energy, resulting in a precipitation reaction. The conversion of light energy into another energy form is termed transduction, a fundamental retinal function that is performed by the photoreceptor outer segments (see Chapter 2). But in addition to the transduction process, the retina also processes light signals so that by the time visual information is forwarded to the brain it carries much more than a point for point representation of an image. For example, in some species, information such as line orientation and direction of motion may be computed and encoded. Computational correlates of these retinal functions are discussed in Part 3 of this volume.

Retinal Circuitry

The neural retina, diagrammed in Figure 3.1, is a well-organized, laminated structure consisting of two plexiform (synaptic) layers sandwiched between three somatic (cell body) layers (Ramon y Cajal, 1973). Six types of cells form the building blocks for this circuitry (photoreceptor, bipolar, horizontal, amacrine, interplexiform, and ganglion cells) and they are distributed in an orderly array within these laminae. A prominent nonneuronal cell type within the neural retina is the Müller cell, a large columnar glial cell that extends from the somatic region of ganglion cells to the photoreceptors. Behind the neural retina, interdigitating with the photoreceptor outer segments, is the pigment epithelium. The cells of the pigment epithelium contain melanin, a black pigment that absorbs stray light and gives the pupil its dark appearance as one looks through the transparent vitreous, lens, and neural retina. The pigment epithelium has a number of functions, such as preventing light reflection, metabolic support for photoreceptors, and a role in adaptation.

The three somatic layers are the outer nuclear layer, the inner nuclear layer, and the ganglion cell layer. The outer nuclear layer contains the cell bodies of rods and cones. These photoreceptors send processes into the outer plexiform layer where they make contact with horizontal and bipolar cells and also contact other photoreceptors.

The inner nuclear layer contains the cell bodies of most classes of retinal neurons: bipolar, horizontal, amacrine, and interplexiform. The bipolar cell somas are near the center of this lamina. They send out two processes, one going to the outer plexiform layer where they receive input from photoreceptor terminals. The other process goes to the inner plexiform layer and synapses with amacrine and ganglion cells. The bipolar cells are the principal relay neurons, conveying afferent information between the plexiform layers. The majority of bipolar cells are thought to be excitatory, using glutamate as a neurotransmitter (Slaughter and Miller, 1983a). The horizontal cell bodies lie along the distal

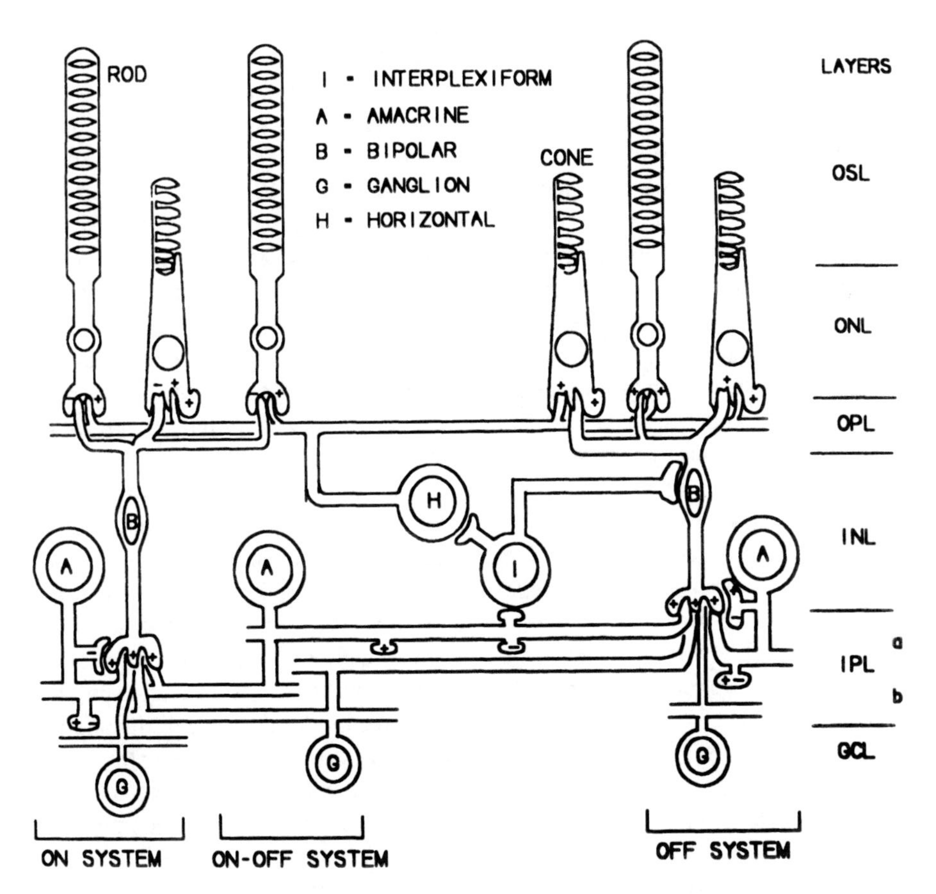

FIGURE 3.1. The principal anatomical components and synaptic connections in the vertebrate retina. The distal retina is at the top of the figure. The horizontal and bipolar cells connect with photoreceptors in the outer plexiform layer (OPL) of the distal retina, the amacrine and ganglion cells connect with bipolars in the inner plexiform layer (OPL) in the proximal retina. The ON bipolars end in sublamina b of the inner plexiform layer; the OFF bipolars end in sublamina a. Amacrine cells receive inputs and send outputs in the IPL; interplexiform cells receive inputs in IPL and send outputs to OPL. Outer segment layer (OSL) and outer nuclear layer (ONL) contain photoreceptors; inner nuclear layer (INL) contains cell bodies of bipolars, horizontal cells, amacrines, and interplexiform cells; the ganglion cell layer (GCL) contains ganglion cell somas. Details of the synaptic circuitry are included in the text.

border of the inner nuclear layer where they receive input from photoreceptors and send output to bipolar cells and photoreceptors. Horizontal cells are believed to be inhibitory, many of them release GABA (Lam et al., 1978; Ayoub and Lam, 1985), and therefore they are responsible for negative feedback control in the distal retina. The somas of amacrine and interplexiform cells lie at the proximal border of the inner nuclear layer. They both send processes into the inner plexiform layer where they receive input from bipolar cells and/or other amacrine cells. The interplexiform cells send processes, and most of their output, back to the outer retina where they synapse on horizontal cells and photoreceptors. The amacrine cells restrict their output to the inner plexiform layer, where they synapse on all available cell types: bipolar, interplexiform, ganglion, as well as other amacrine cells. Among the amacrine cells there is enormous diversity based on morphological and pharmacological criteria, indicating there are many subsets within this class of retinal neurons. The most populous subsets are inhibitory amacrines that contain GABA or glycine (Dick and Lowry, 1984), and excitatory amacrines containing acetylcholine (Masland and Mills, 1979).

The third and most proximal somatic lamina is the ganglion cell layer. Appropriately, it contains mainly ganglion cell somas but also some "displaced" amacrine cell bodies. The ganglion cells are the output cells of the retina. They receive input from amacrine and bipolar cells, converting these signals into a series of action potentials that are transmitted, via axonal projections, to the brain.

In summary, information flow begins with the photoreceptors as they transduce light to electrical energy and then send these signals through chemical synapses to bipolar and horizontal cells. The bipolar cells transmit this information across the two plexiform layers to excite amacrine and ganglion cells. The horizontal cells provide feedback and feedforward circuits in the outer plexiform layer while the amacrines provide both feedback and feedforward circuits in the inner plexiform layer. Most of this is negative feedback that dampens incoming signals. The interplexiform cell provides a feedback pathway between plexiform layers, taking signals from the inner plexiform layer and sending them back to the outer plexiform layer. Both the bipolar and interplexiform cells communicate between plexiform layers although the flow of information is opposite in the two cell types. The most direct information pathway across the retina is from photoreceptors to bipolars to ganglion cells and then to the brain. The other neurons intercede in this pathway to split or modify these signals. There are many refinements and some exceptions to this general model that will be described in ensuing sections, but overall it provides a basis for an introductory understanding of retinal function.

The number of photoreceptors is far greater than the number of bipolar cells, which in turn outnumber the ganglion cells. This forces a significant convergence, many photoreceptors contacting each bipolar cell, and many bipolars synapsing onto each ganglion cell. In man, there are approximately 120 million rods and 5 million cones, but only 1 million ganglion cells. This may appear to result in a loss of information as the signals merge. However, there is also a divergence of signals, since a single photoreceptor or bipolar cell may contact many neurons. This enables the integrative properties of the neurons to code for signals that were carried by sheer number in the presynaptic cells. Another compensatory system is the number of synapses formed. Although the ratio of bipolars to photoreceptors is small, the number of synapses is high so that again the bipolar cells can use these multiple inputs to avoid any loss of information, either through a temporal coding of these synaptic inputs or through nonlinear properties of the postsynaptic cells. One phenomenon attesting to the preservation of information is hyperacuity (Westheimer and McKee, 1977). This is the ability to resolve two closely apposed lines that are closer together than the separation between adjacent photoreceptors. Presumably this is accomplished by computational aspects of the network, based on a process of spatial comparison at localized sites in the retina. This may be based on interpolation of the points on each line and constraints that assume the interpolated points form a straight line.

The retina is a sensory tissue that sends information to the brain, but it also receives input from the brain. Retinal efferents, which have long been known in birds, have now been described in fish, amphibians, and reptiles, and suggested in mammals. They provide an extensive feedback loop to control retinal activity and suggest that higher brain functions can be integrated into responses even at this early level of sensory processing.

Cellular Responses and Interactions

The responses of retinal neurons are measured in electrical terms (see Fig. 3.2). All retinal neurons are normally negatively charged, meaning that the inside is at a more negative potential than the external environment. When a cell is depolarized, the inside of the cell becomes less negative or even positive with reference to the outside. The electrical state can be measured with a microelectrode that is placed inside the cell. Cells that are depolarized are termed excited, and excited cells can release neurotransmitters. The neurotransmitters affect other neurons, so that depolarization implies signal-transmitting capability. In contrast, when cells are hyperpolarized (the inside becomes more negative) then this chemical signaling capability is diminished. Contrary to our expectations, photoreceptors are excited in the dark and suppressed by light (Svaetichin, 1953; Tomita, 1963). Photoreceptors release glutamate, an excitatory neurotransmitter, at synapses with bipolar and horizontal cells (Murakami et al., 1972, 1975; Wu and Dowling, 1978; Slaughter and Miller, 1981, 1983b). Since photoreceptors are depolarized in darkness and release an excitatory transmitter, we would anticipate that bipolar and horizontal cells would also be depolarized in the dark. This is true for horizontal cells, but there are two kinds of responses in bipolar cells. One group follows the photoreceptors and is depolarized in the dark. These are the OFF bipolars. But another group of bipolars is suppressed by the photoreceptor transmitter. This inverting synapse means that these cells are excited in the light and are therefore termed ON bipolars. The net result of these interactions in the distal retina is that light signals are divided into two pathways, a signal to indicate when light is ON, carried by the ON bipolars, and a parallel pathway signaling when the light is off, carried by the OFF bipolars (Werblin and Dowling, 1969; Kaneko, 1970). It is unclear why there is this dual, ON–OFF system, since they seem to be redundant, although not exactly mirror images of each other. One possibility is that they increase the dynamic range of neuronal signaling. Transmitter release

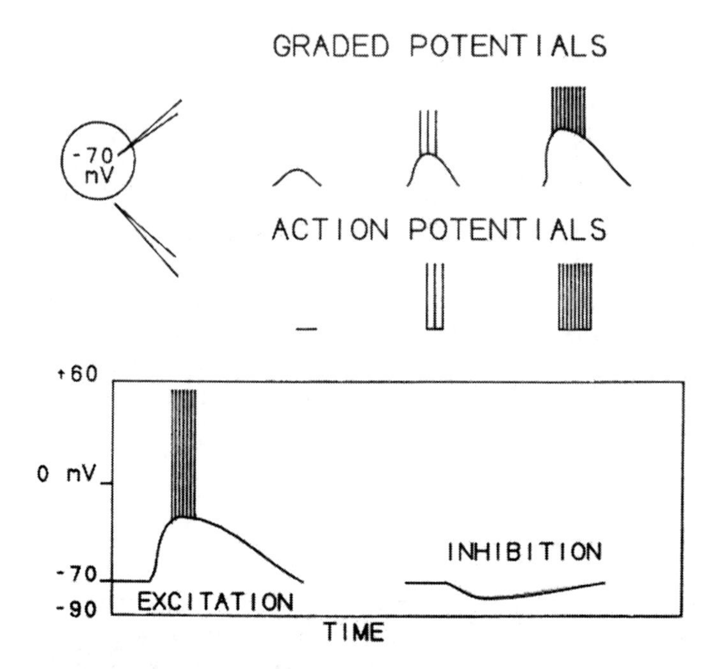

FIGURE 3.2. Basic neurophysiological principles. As shown in the upper left, a neuron normally rests at a negative internal potential, depicted here as −70 mV. This can be measured with an intracellular electrode inside the cell. Three sets of recordings obtained using intracellular (upper row) and extracellular (lower row) electrodes are shown at the upper right. Intracellular electrodes can measure both the slow, graded potentials and the action potentials, which ride on top of these slow potentials, while extracellular electrodes are used to monitor only spike activity. Small amplitude slow potentials do not reach threshold and produce no spike activity. Once the graded potentials exceed threshold they produce spikes that increase in frequency in proportion to the graded potential. Shown at the bottom, excitation refers to the depolarization of a cell and can be associated with spike activity. Inhibition is associated with a hyperpolarization (in most cases), moving the cell further from spike threshold.

is linked to membrane potential, but releasable stores of neurotransmitter are limited. If maximal transmitter release is fixed, then there is an inverse relationship between dynamic range and gain (or resolution) at the synapse. One means of overcoming this inherent biological limitation is to divide the dynamic range between two cells.

The mammalian rod system does not seem to use this dual pathway, being content to send signals with only one type of bipolar cell (Nelson, 1977; Dacheux and Raviola, 1986; Kolb and Nelson, 1983; Muller et al., 1988). But in this rod system the ON–OFF parallel processing is established in the inner retina, perhaps indicating the importance of this process. It is clear that the ON–OFF system is well conserved in all vertebrate families, from fish to mammals.

The bipolar cells send their dendrites and their signals to the amacrine and ganglion cells in the inner retina. There the synaptic connections are made primarily within the inner plexiform layer. In this layer, the ON and OFF bipolars send their processes to different sublaminae (Famiglietti and Kolb, 1976; Nelson et al., 1978). In general, the OFF bipolar synapses are made in the distal portion of this layer, termed sublamina a, while the ON bipolars send processes to the proximal portion, sublamina b. This anatomical segregation of ON and OFF signals is another manifestation of the emphasis placed on the separation of these two parallel channels.

The signals that reach the inner retina are transformed in many ways. As noted above, there is a very significant spatial convergence. The dendritic fields of many amacrine and ganglion cells are very large. The diameter of the outer segment of photoreceptors, which represents the photon capture field, is usually between 2 and 10 μm. The dendritic field of bipolar cells, which represents the extent of their synaptic sampling, is on the order of 50–200 μm. In contrast, the dendritic fields of some amacrine and ganglion cells are greater than 1000 μm in diameter.

In the distal retina, the light responses of photoreceptors, bipolars, and horizontal cells are relatively sustained. They generate slow wave potentials that are graded in amplitude with light intensity. Two very important physiological properties originate in the proximal retina: (1) the occurrence of action potentials (spikes) and (2) the appearance of transient responses (see Fig. 3.3). Both amacrine and ganglion cells can generate action potentials, which are nonlinear, all-or-none phenomena commonly found in neurons that transmit signals over long distances. Spikes are large-amplitude (approximately 100 mV), short-duration (a few milliseconds) responses that can be transmitted with high conduction velocities (10–100 m/sec). At the junction between bipolar cells and ganglion cells, the variable amplitude, graded potential signals of bipolar cells are converted to the fixed amplitude action potentials of ganglion cells. In relaying information across this pathway, the analog, graded potential amplitudes are translated into digital, spike frequencies. Larger amplitude slow potentials produce higher frequency spiking. Spiking is a common feature of ganglion cells, which is appropriate since all information leaving the retina must be relayed for relatively long distances. Amacrine cells also produce action potentials, although their spiking pattern is much more limited, often to only a single spike (Barnes and Werblin, 1986). For technical reasons, spikes are easier to monitor than slow potentials, and as a result we know much more about the responses of ganglion cells than amacrine or bipolar cells.

The responses of distal retinal neurons persist for the duration of the light signal. In the inner retina, many amacrine and ganglion cells follow this response pattern, producing sustained ON and sustained OFF responses (see Fig. 3.3). But in addition a class of transient neurons arises. In these cells, a brief response occurs at the onset or offset (or both) of a light stimulus. There are transient ON, transient OFF, and transient ON–OFF neurons. The latter response represents a convergence of signals from the ON and OFF pathways. It is still unclear what

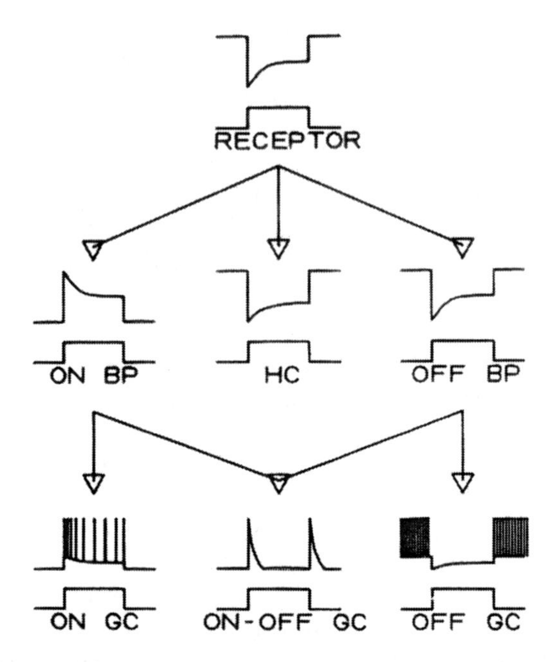

FIGURE 3.3. The responses of retinal neurons illustrated in Figure 3.1. Responses in the distal retina are graded potentials. Except for the ON bipolar, these neurons respond to light with a relatively sustained hyperpolarization. All photoreceptors respond to light (square wave signal) with a hyperpolarization. But at the first synapse, the ON bipolars are depolarized in light and the OFF bipolars are depolarized in darkness. In the inner retina, transient signals (ON–OFF cells) and spike activity originate. Amacrine and ganglion cell responses are often so similar that they are combined in this figure. The responses of interplexiform cells are not shown because of the limited data currently available on this cell type. Response amplitudes are not represented, but they generally range from 20 to 40 mV for the graded potentials and near 100 mV for the action potentials. By convention, an upward going response is a depolarization. Bipolar (BP); horizontal cell (HC); amacrine cell (AM); ganglion cell (GC).

mechanism is responsible for the generation of transient responses. One possibility is that a time-delayed negative feedback from amacrine cells shuts off a sustained signal, producing a phasic signal. Amacrine cells make two kinds of synaptic connections: serial, where one amacrine synapses on another, which in turn synapses on a third cell; and reciprocal, where an amacrine receives synaptic input from a bipolar, on which it synapses back. The reciprocal synapse could provide the negative feedback responsible for transient responses. A particularly attractive model from this perspective has been proposed (Maguire et al., 1989), suggesting that the feedback involves the turning off of a calcium current in some bipolars. This is appealing because it is known that calcium controls transmitter release and the modulation of a calcium current could make transmitter release to the inner retina very transient and still not significantly modify the bipolar cell response. Thus, the bipolar voltage response could be tonic, but its transmitter release could be phasic. Transient amacrine and ganglion cells would receive synaptic input from bipolar cells subject to this feedback; sustained amacrine and ganglion cells would get their input from bipolars lacking this feedback. Another possible explanation for the generation of transient responses is based on the observation that many bipolar cells are not purely tonic; there is an initial, phasic maximal response that then relaxes to a sustained response (see Fig. 3.3). If some amacrine and ganglion cells had high thresholds, set to respond only to the phasic component of the bipolar cell's response, this could also generate transient responses. Then transient and sustained amacrine and ganglion cells could sample the phasic and tonic response components of the bipolar cells.

Classification of Amacrine and Ganglion Cells

Amacrine and ganglion cells have been categorized in many ways, often reflecting their responses to particular stimulus parameters rather than a basic understanding of their true functions. By analogy with the bipolar cells, one system of classification is to divide ganglion cells into ON cells or OFF cells (Werblin and Dowling, 1969; Kaneko, 1970). Each class can be subdivided into sustained and transient signals. Thus, there are sustained cells encoding information about the steady state (is it light or dark?) and there are transient cells encoding the derivative of the light stimulus (is it lighter or darker than a moment ago?).

Another method of classifying ganglion cells is based on their spatial response properties (Enroth-Cugell and Robson, 1966). Some cells, termed linear, respond equally to stimuli throughout their receptive field. This is commonly measured by moving a grating pattern across the retina. Once the test pattern covers the receptive field, a linear cell shows no response to movement of the grating. As the grating moves, some regions of the receptive field will go from darkness to light, but other areas will go from light to darkness. If these regions are equal, the net response of a linear cell will remain unchanged. A nonlinear cell will respond to the movement of this grating, since some subsets of its receptive field respond differently than others. Sustained cells generally respond in a linear manner, while transient cells are nonlinear (Cleland et al., 1971). In the cat retina, these distinctions have been correlated with other properties of ganglion cells. For example, nonlinear cells, in addition to being transient, also have larger dendritic fields and their axons have faster conduction velocities than linear cells.

A third method of classification is based on information content. It is difficult to directly relate nonlinear spatial properties of ganglion cells with a particular feature of a visual image, but there are many response characteristics of ganglion cells that do directly correlate with the external image. For example, many ganglion cells are color coded. There are also a number of so-called trigger features of ganglion cells that relate to particular properties of the image. One very well-studied trigger feature is direction selectivity and another is orientation sensitivity. In these cases, we can deduce what components of the image are extracted by particular cells. In the frog retina, several studies have focused on what information is most relevant to ganglion cell output. This was initiated by the intriguing paper by Lettvin et al. (1959) entitled "What the frog's eye tells the frog's brain." Maturana et al. (1960) categorized ganglion cells into five functional groups: (1) sustained edge detectors, (2) convex edge detectors, (3) changing contrast detectors, (4) dimming detectors, and (5) dark detectors. This was refined by Lettvin et al. (1961), suggesting that the specific information sent from the retina to the brain fits into a few categories. However, in most cases it has not been possible to classify ganglion cells into a few functional types; it may not even be relevant since it is unlikely that most ganglion cells serve to carry only one type of information. There are not enough of them to permit such complete specialization (see Chapter 10). The opposite approach is to identify ganglion cells by particular distinguishing features. This naturally produces a larger number of categories. Anatomically, Kolb et al. (1981) described 23 groups of ganglion cells in the cat retina, while Caldwell and Daw (1978) found 16 physiological types of ganglion cells in the rabbit retina. Which classification is used may depend on the problem addressed: the ON and OFF classification is important in evaluating synaptic connections, the linear/nonlinear distinction relates to synaptic transfer to ganglion cells, while a more extensive classification scheme identifies specific information processing and encoding properties of the retina.

To summarize, although photoreceptors are excited only in darkness, they send opposite sign signals at their synapse with bipolar cells, resulting in the formation of ON and OFF channels. The amount of photoreceptor transmitter release is related to the logarithm of the light intensity, but the relationship is not linear and transmitter release reaches saturation while the photoreceptor response is still increasing. Signal amplification is greatest between photoreceptors and horizontal cells. The light response latency is less for the OFF than the ON bipolar. The horizontal cells generate lateral interactions in the distal retina, generating antagonistic center-surround properties. The ON and OFF signals carried to the inner retina undergo a number of additional transformations, resulting in responses that in some cases resemble bipolar responses (sustained and linear) while in other cases they are modified temporally (transient signals) and spatially (nonlinear responses). At the level of the amacrine and ganglion cells, some responses can be characterized by their integrative properties and their information content. The ganglion cells are the output neurons of the retina and they send axons to several brain regions. This forms the background for a description of several specialized systems in the retina that are of current research interest.

The Electroretinogram

The light responses of retinal neurons are manifested in a field potential, termed the electroretinogram (ERG). The ERG is of clinical importance because it can be measured easily and noninvasively by placing an electrode on the surface of the eye. The ERG is similar to other field potentials, such as the electrocardio-

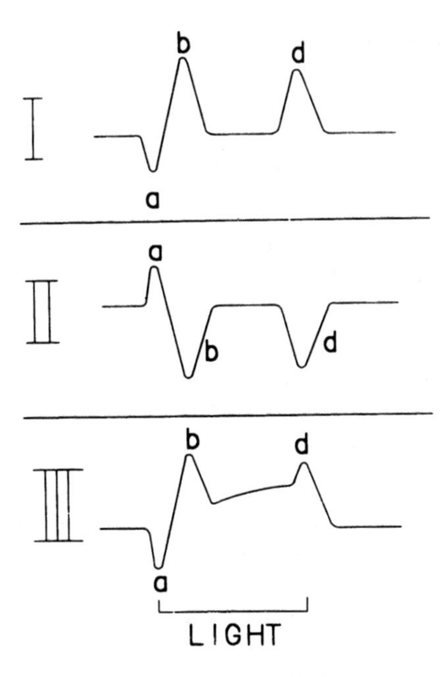

FIGURE 3.4. The major components of the ERG. The a-wave reflects photoreceptor activity, the b-wave reflects ON bipolar activity, while the d-wave probably reflects the activity of several cell types, including the OFF bipolar. (I) The field potentials that would be recorded at the cornea. (II) The same response if it was recorded within the retina. The difference in polarity represents the change in the position of the electrode relative to the current sources producing the various components of the ERG. (III) The response of a retina in which there is a large c-wave, between the b- and d-waves, which is produced by the pigment epithelial cells.

gram (heart) or the electroencephalogram (brain), in that it results from the synchronized electrical activity of large numbers of excitable cells. The ERG consists of several components as illustrated in Figure 3.4. At the onset of a light stimulus, the a-wave is the first response, a corneal negative potential that represents the activity of photoreceptors. The b-wave is the largest amplitude component and consequently the most studied. It is a corneal positive potential that follows the a-wave and is apparently generated by Müller cells when they are depolarized by the activity of the ON bipolars (Miller, 1973; Dick and Miller, 1978; Karwoski et al., 1985; Stockton and Slaughter, 1989). At light offset, a d-wave can be detected in cone-dominated retinas. In rod-dominated retinas a prominent c-wave, representing the activity of the pigment epithelium (Noell, 1954), follows the b-wave.

The relationship between retinal neurons and glia in the generation of extracellular currents has been extensively studied, offering insights into the interpretation of field potentials (Steinberg et al., 1970; Faber, 1969; Miller and Dowling, 1970; Newman, 1980). Although the Müller cells do not receive synaptic input from retinal neurons, they are affected by neuronal depolarizations.

Müller cells are permeable principally to potassium, so that as the potassium concentration around them changes, the potential of the Müller cell also changes. For example, when an ON bipolar cell depolarizes at light onset, sodium flows into the bipolar and potassium flows out. This potassium tends to depolarize Müller cells, producing a current along the length of this glial cell, resulting in a field potential across the retina that is detectable at the corneal surface.

A similar process occurs when the light is turned off, this time involving the OFF bipolars. This results in the production of the d-wave. However, the photoreceptors and the horizontal cells also depolarize, so the d-wave may represent responses of a number of neurons. The c-wave represents the responses of the pigment epithelium, which responds to photoreceptor depolarization in a process analogous to the relationship between Müller and bipolar cells.

The ERG can be used clinically to evaluate the physiology of the distal retina. However, amacrine and ganglion cells seem to make little contribution to the ERG (Stockton and Slaughter, 1989). This is unfortunate diagnostically, for example, in glaucoma where ganglion cell activity is impaired but the ERG appears to be normal. However, there are other retinal field potentials, generated by other patterns of light stimulation, which reflect the responses of inner retinal neurons.

The Distal Retina

Rod–Cone Pathways

Rods and cones serve very different functions. The former are involved in low light vision and are primarily designed for the efficient capture, summation, and transmittal of signals involving only a few quanta per rod per second. The cones function at much higher ambient light levels, convey color information, and have faster response properties. Paradoxically, although these two types of photoreceptors are distinctly different, their signals merge in all vertebrate retinas. It appears that the distinction between rods and cones is a requirement of the transduction process, but at subsequent synaptic stages the signals can converge and be carried through the same anatomical pathways.

The importance of this convergence is illustrated by comparing rod and cone circuitry in various vertebrates. Although the levels of separation differ, the end result is convergence in all known cases. In amphibian retinas, which are often a favorite for scientific studies, rods and cones synapse on the same cells. There are no distinct bipolars or horizontal cells that receive exclusive input from either photoreceptor, nor is there any known difference in the synaptic contacts. Thus, rod and cone signals merge at the first synapse in the distal retina. In the fish retina, using the goldfish as an example, there is some separation of rod and cone pathways. First, there is one type of horizontal cell that receives synaptic input only from rods (other horizontal cell types get mixed rod and cone input) (Saito et al., 1985). Second, rod and cone input to the bipolars differs. In the ON bipolar, rod input results in the closing of a channel, while cone input opens a channel (Saito et al., 1979, 1985; Nawy and Copenhage, 1987). But despite these mechanistic distinctions, both rods and cones converge on the same cell and produce a similar depolarization of the ON bipolar in the light. In mammals there is a more extensive separation. There are rod bipolars, which receive only rod input, and cone bipolars, which have cone input exclusively (Boycott and Kolb, 1973). However, even in mammals, this separation lasts for only two extra synapses. The rod bipolars do not send their signals to rod ganglion cells, but to rod amacrines instead. These rod amacrine cells then contact cone bipolars, so that the rod signal merges with the cone pathway (Kolb and Nelson, 1983). A notable exception to the mixing of rod and cone signals is the central fovea of the primate retina, a region responsible for high resolution vision, where rods are absent.

Actually, the rod and cone signals merge even earlier in the retina (Nelson, 1977). Rods and cones are connected by gap junctions (Kolb, 1977), which are anatomical bridges between adjacent cells that allow electrical signals to cross directly from one cell to another. The amount of current that passes between two cells is proportional to the number of active gap junctions between them. The existence of gap junctions between rods and cones indicates that signals in one photoreceptor system are carried to the other. Gap junctions act as high pass filters, so that the fast frequency components of the signal are preferentially transferred to neighboring photoreceptors. This filtering may allow the two signals to mix without losing the dual information content. For example, the light response of a photoreceptor may be sustained, but the input it gets from an adjacent photoreceptor would be filtered and therefore transient. A time comparison of the waveform could then be used to determine if the light response was direct or across a gap junction.

It has recently been suggested that gap junctions permit photoreceptors to signal high intensity light information that would be beyond the capability of a single photoreceptor (Attwell et al., 1987). This is because the dynamic range of the photoreceptor's electrical response to light is much greater than the dynamic range of its transmitter release. This means that the photoreceptor can code for a large range of light intensities but cannot directly communicate this information to bipolar cells. Coupling may compensate for this disparity, since large electrical signals can be carried across the gap junctions and cause neighboring photoreceptors to release transmitter. The combined activity permits synaptic communication of high intensity stimulation. This complex system also permits high gain at the photoreceptor synapse. Neurons release transmitter in proportion to their degree of excitation. If photoreceptors compact most of their transmitter release over a small voltage range, their response resolution increases and they can code for small changes in light intensity. The drawback is that they cannot at the same time signal a wide range of light intensities, but the gap junction system may alleviate this potential limitation. Thus, the system can obtain high gain and a large dynamic range. This is another means of tackling a recurring problem, which we previously suggested bipolars may handle by creating a reciprocal pair (ON and OFF).

Rod-cone coupling is dynamic and changes with the adaptational state of the retina (Yang and Wu, 1989). In the dark adapted retina, coupling is reduced. This tends to prevent the dilution of the rod signal, since coupling would allow leakage of current from the stimulated rod to other photoreceptors. When coupling is reduced the light generated currents produce a maximal voltage response in individual rods. A similar modulation of coupling with adaptational state is found in horizontal cells, described in the section on interplexiform cells.

Coupling of cells through gap junctions is a prominent feature throughout the retina. Horizontal cells are extensively coupled, there is coupling between some amacrine cells, and between amacrine and bipolar cells (as in the rod system described above), and not only are rods coupled to cones but different cones are coupled to each other.

In addition to direct gap junction connections between photoreceptors, there are also excitatory chemical synapses that connect cones of differing spectral sensitivities (Normann et al., 1984).

Returning to the topic of ON and OFF pathways, the mammalian rod system has a unique method for establishing these two important information channels. Again, it involves a mixing with the cone pathway. In our basic description of retinal circuitry we pointed out that the ON and OFF bipolars create two complementary pathways, one indicating when light goes on and the other when light goes off. In mammals, there appears to be only one type of rod bipolar. This was first noticed in the cat (Nelson, 1977), but has more recently been found in rabbit (Dacheux and Raviola, 1986) as well. Initially it was thought that this rod bipolar was an OFF bipolar, although recent evidence indicates that it is an ON bipolar (Dacheux and Raviola, 1986; Muller et al., 1988). Thus, the rod system, unlike the mammalian cone system, does not establish an ON/OFF system in the distal retina. But like the mixing of rod and cone systems, the establishment of ON and OFF pathways in the rod system is simply delayed by a couple of synapses (Kolb and Nelson, 1983; Famiglietti and Kolb, 1975). The rod ON bipolar contacts rod amacrines, which then establish sign conserving (ON) gap junction synapses with the cone ON bipolar and also sign-inverting (OFF) chemical synapses with the cone OFF bipolar. When a light is turned on, the rod ON bipolar is excited, it excites the rod amacrine, which in turn excites the cone ON bipolar. The net result is an ON response propagated to the inner retina. When the light is turned off, the rod bipolar is suppressed. This bipolar had been suppressing the cone OFF bipolar (through a sign-inverting glycinergic amacrine cell). In the dark, the cone OFF bipolar is released from inhibition, resulting in an OFF response. Thus, both ON and OFF depolarizing responses are generated and the ON–OFF rod pathways are established (Muller et al., 1988). This circuitous route to reestablish the ON and OFF pathways in the mammalian rod system suggests that this dual channel network is essential in vertebrate vision.

Rod and cone pathways interconnect at several levels. When both photoreceptor systems are active, in mesopic vision, the rod system has a suppressive effect on the cone system. Frumkes and colleagues (Goldberg et al., 1983; Frumkes and Eysteinnson, 1987) have shown that light stimulation which activates cones produces a larger response after rod responses have been eliminated by an adapting light. This was demonstrated by psychophysical testing and intracellular electrophysiology. At least part of this phenomenon appears to be due to a rod-generated, suppressive horizontal cell feedback onto cones.

Overall, there are many instances of a mixing of rod and cone signals. Although, intuitively, this would appear to be counterproductive, there are several instances where it has been found to be advantageous.

Center-Surround Mechanisms

Most sensory systems utilize a center-surround mechanism as a spatial filter to enhance contrast. In the retina, this system was first observed in ganglion cells (Kuffler, 1953), but it originates in the distal retina among bipolar cells, or perhaps even photoreceptors. The basic phenomenon is that a small spot of light will produce a larger response than a diffuse light of the same intensity. The small spot stimulates the receptive field center of photoreceptor and bipolar cells; the diffuse light stimulates not only these cells' receptive field centers but also neighboring cells. (Receptive field refers to the region of the retina that, when stimulated, will produce a response by the cell. Originally this was thought to represent the dendritic field of a neuron. However, cells can respond to stimuli outside their dendritic fields. This has resulted in the terminology of "receptive field center" to refer to direct effects and "receptive field surround" to refer to indirect effects.) The neighboring cells, which mediate the surround, are the horizontal cells. They produce a negative feedback into the receptive field center, which reduces the center response. The response to a small spot of light comes via the direct pathway from the photoreceptors; the response to a diffuse light is the summation of a center response and a surround response of the opposite polarity. Their interaction can be modeled as the sum of two Gaussians (Rodieck and Stone, 1965). The center response is stronger but generated over a smaller spatial extent; the surround is opposite in polarity, smaller in amplitude, but encompasses a larger area. The size of the surround is due not only to the larger dendritic field of horizontal cells, compared to bipolars, but also to gap junctions between horizontal cells.

This coupling is so extensive that the horizontal cells across the retina act as a functional unit, or syncytium. However, the gap junctions are dynamic structures and the electrical conductivity between horizontal cells can be reduced by neurotransmitters such as dopamine (Negishi and Drujan, 1979). Thus, the size and amplitude of the surround can be reduced by uncoupling horizontal cells and this may be important during dark adaptation (Mangel and Dowling, 1985, 1987; Witkovsky et al., 1988). In the light-adapted retina, the center-surround system works to suppress responses to stimuli that are not appreciably different from neighboring regions. This might be counterproductive under conditions of dim illumination, where resolution is sacrificed in order to maximize the detection of dim light signals. The uncoupling of horizontal cells eliminates the surround suppression, allowing effective summation over a larger number of photoreceptors.

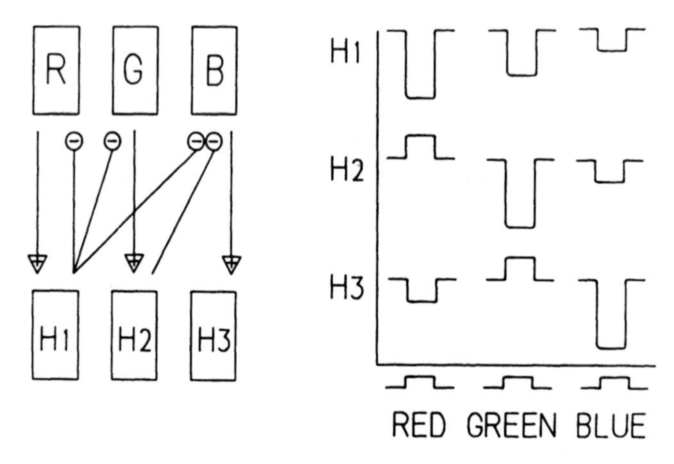

FIGURE 3.5. The predominant synaptic pathways that form the color responses in fish horizontal cells, as proposed by Stell and Lightfoot (1975). The H1 horizontal cell receives most of its input from the long wavelength sensitive cone, labeled "R" (red), and sends negative feedback to all three cones types. It hyperpolarizes to all wavelengths of light stimulation, although it is most sensitive to red light (illustrated on the top row on the right). The H2 horizontal cell receives most of its input from the middle wavelength sensitive cone ("G," green). Therefore in green light the "G" cone and the H2 horizontal cell are hyperpolarized. In red light, the "R" cone and the H1 horizontal cell are hyperpolarized. Since the H1 sends negative feedback to the "G" cone in the dark, red light reduces this feedback and thus depolarizes the "G" cone and the H2 horizontal cell. This is the biphasic response, a depolarization to long wavelengths and a hyperpolarization to medium and short wavelengths (second row at right). The feedback from the H2 to the short wavelength cone ("B," blue) produces a biphasic red/green feedback. Coupled with the direct blue, hyperpolarizing light response, the result is a triphasic color response. Physiological diagram is adapted from the results of Toyoda et al. (1982).

Color Vision

Color vision arises from the differences in the spectral sensitivities of the three cone types (see Chapter 2). The interaction between these cones underlies color perception (see Chapter 8). Two interesting examples of this type of interaction are found in the distal retina, one among horizontal cells and another in bipolar cells. Both of these examples come from experiments in the goldfish retina. As previously mentioned, without considering color, horizontal cells receive input from photoreceptors and light stimulation causes the horizontal cells to hyperpolarize. But there are color-coded horizontal cells that only hyperpolarize when the retina is stimulated with a particular wavelength of light and depolarize when another wavelength is used (Toyoda et al., 1982). These are called C-type (color), in contrast with L-type (luminosity) horizontal cells. Two types of C-type cells which have been described are: one that depolarizes to red light but hyperpolarizes to green light and a second that hyperpolarizes to both red and blue light, but depolarizes to green light.

The mechanism responsible for this color coding is the negative feedback synapse from horizontal cells to photoreceptors (Stell and Lightfoot, 1975; Stell et al., 1982; Murakami et al., 1982), illustrated in Figure 3.5. In the red/green dichromatic type it appears that red-sensitive horizontal cells (these are horizontal cells dominated by input from the red-sensitive cones) feed back onto green cones. When a red stimulus is used, the red cone and the red horizontal cell both hyperpolarize. The negative feedback from this red horizontal cell to the green cone is then reduced. This reduction in negative feedback makes the green cone depolarize (red light has much less direct effect on the green cone). When a green stimulus is used, the green cone is hyperpolarized directly by the light stimulus. Therefore, the green cone and its associated horizontal cell depolarize to red light and hyperpolarize to green light.

The trichromatic horizontal cell involves just one additional feedback pathway. The green horizontal cell provides a negative feedback to the blue-sensitive cone. This means that the dichromatic signals in the green horizontal cell are inverted in the blue cone; the blue cone hyperpolarizes to red light and depolarizes to green light. Blue light has a direct hyperpolarizing effect on the blue cone; the sum of all these inputs is a trichromatic response in the blue cone and its follower cell, the trichromatic horizontal cell.

Based on this model the circuitry required for color coding can be quite simple, accomplished using only color-specific pathways and negative feedback synapses. The result is the formation of cells that specify color (see also Chapter 4). Cones alone do not specify color, even though they are maximally sensitive to a particular wavelength of light. A single cone cannot distinguish between a small amount of light in the most sensitive part of its spectrum and a greater amount of light in a less sensitive part of its spectrum (see Chapter 2). This is the principle of univariance (Naka and Rushton, 1963). It is only the interaction between two or three cone types that permits a contrast between different regions of the color spectrum and forms the basis for color detection. The feedback from color-specific horizontal cells is an example of this interaction.

Another example can be found in color-coded bipolar cells. Although many bipolar cells have a center-surround antagonism that is based on luminance, in some bipolar cells the center and surround responses have different spectral sensitivities, leading to color coding (Kaneko, 1973; Yazulla, 1976; Kaneko and Tachibana, 1983). For example, there are "single opponent" bipolar cells that respond best to red light in their center, but to green light in their surround. Presumably the green surround is due to the spectral sensitivity of the negative feedback system through horizontal cells. Therefore, a diffuse red stimulus would preferentially generate the center response, but a diffuse green light would produce the surround response. A diffuse white light stimulus, which would stimulate both the red center and green surround responses (of opposite polarity), produces a negligible response. There are also single opponent bipolars with green centers and red surrounds, which would have the opposite color responses. A slightly more complicated arrangement is found in the "double opponent" bipolars, where both the center and the surround have antagonistic color properties.

For example, the bipolar may depolarize to red light in the center but hyperpolarize to green light in the center. In the surround the color responses are the reverse: the cell will depolarize to green but hyperpolarize to red light. Diffuse color stimuli would produce little if any response. In this case, the center response may involve negative feedback, and a pair of negative feedback synapses is required to produce the surround. It has been suggested (Daw, 1973; Dowling, 1987) that single opponent bipolar cells may be responsible for the phenomenon that after viewing a red light, a person will see green when looking at a white wall. Presumably, the red responses of the bipolar cells adapt, so that when a white stimulus is presented it stimulates the green system more than the red and the wall appears green. The double opponent system may account for the observation that a central patch of gray will appear green when it is surrounded by red.

Neurotransmitters

In the retina, as in other neuronal tissues, cells communicate with each other through synapses. There are two types of synapses, electrical and chemical. Electrical synapses, formed by gap junctions between cells, have already been discussed in reference to photoreceptors and horizontal cells. Chemical synapses, which represent the predominant means of communication, involve the release of neurotransmitter substances from presynaptic cells and their detection by specialized receptors on postsynaptic cells. Neurotransmitter release is (nonlinearly) proportional to the membrane potential of the presynaptic cell. When a neuron is depolarized (the inside becomes more positive) the neurotransmitter release increases. There is a threshold potential for transmitter release. Normally it is thought that a cell at rest (hyperpolarized) does not release neurotransmitter, although there are exceptions, such as retinal cholinergic neurons. There are two types of neurotransmitters: excitatory and inhibitory. Excitatory transmitters cause a depolarization in the postsynaptic cell, while inhibitory transmitters hyperpolarize the postsynaptic cell. The predominant neurotransmitters in the retina are the excitatory transmitters glutamate and acetylcholine and the inhibitory transmitters GABA and glycine [see Massey and Redburn (1987) for a review of retinal neurotransmitters].

Both photoreceptors and bipolar cells appear to use glutamate as their neurotransmitter (Murakami et al., 1972, 1975; Slaughter and Miller, 1983a). This is not surprising since excitatory amino acids, glutamate and/or aspartate, are ubiquitous in the central nervous system. The bipolar cells excite amacrine and ganglion cells, apparently acting on kainate-type receptors (Coleman et al., 1986) (a subclass of the glutamate receptors) in both cases. Some bipolar cells may release other neurotransmitters, such as serotonin (Marc et al., 1988) or glycine (Cohen and Sterling, 1986), but glutamate antagonists block light responses of amacrine and ganglion cells so effectively that the role of these other transmitters is unclear.

The Proximal Retina

Amacrine Cells

In the inner retina the bipolar cells synapse onto amacrine and ganglion cells. The amacrine, which means cell without an axon, may represent the most diverse cell type in the retina. Based on morphology and neurotransmitter content over 30 kinds have been identified. Based on neurotransmitter content alone, there are three cell types that constitute a majority of the amacrine cells: the glycinergic, GABAergic, and cholinergic amacrines. The former two are inhibitory and are found in a variety of morphological subtypes. GABAergic and glycinergic amacrine cells together make up almost two-thirds of all amacrine cells (Dick and Lowry, 1984), suggesting that a primary role of amacrine cells is feedback and feedforward inhibition. Although both GABA and glycine act through a similar mechanism, they seem to have different functions in the retina (Miller et al., 1977; Frumkes et al., 1981). It has been proposed that GABA preferentially acts on the ON pathway, while glycine acts on the OFF pathway (Ikeda and Sheardown, 1983). Alternatively, glycine may be responsible for fast, transient inhibition, while GABA controls more sustained inhibition (Belgum et al., 1984; Slaughter and Bai, 1989). Neither of these hypotheses completely defines the functions of these two transmitters, probably because they are contained in many different types of amacrines and perform many different functions. Based on the prevalence of inhibitory neurotransmitters, it was originally believed that the function of the amacrines was to provide lateral inhibition that would serve to modify excitatory inputs from bipolar cells. This would be the inner retinal equivalent of horizontal cell function. However, the cholinergic amacrines are excitatory (Masland and Ames, 1976) and this added another dimension to amacrine cell function. The cholinergic amacrines consist of two mirror image cells (conventional and displaced amacrines), apparently controlling transient OFF and ON excitation, respectively (Masland et al., 1984a,b; Tauchi and Masland, 1984; Massey and Redburn, 1985). The cholinergic amacrine has been linked with directionally selective ganglion cells (Ariel and Daw, 1982b; Ariel and Adolph, 1985), but its role is not limited to this trigger feature alone (Ariel and Daw, 1982a; Masland et al., 1984a,b). The discovery of catecholamine, indoleamine, and peptide-containing amacrine cells provides the substrate for further functional diversity. An additional permutation is that many amacrine cells contain two neurotransmitters that are capable of performing very different functions. For example, the excitatory cholinergic amacrine also contains GABA, a powerful inhibitor in the retina (Vaney and Young, 1988; Brecha et al., 1988). Presumably this enables one cell to either stimulate or suppress other neurons, depending on which transmitter is released. In another case, three transmitters have been localized to a single amacrine cell (Wu and Lam, 1988). The physiological importance of this plethora of amacrine cells is just beginning to be unraveled. One type of glycinergic amacrine is the rod amacrine cell described in the section on rod vision. The dopaminergic amacrine may also modulate rod vision. The cholinergic amacrine and at least some of the GABAergic amacrines are important in directional selectivity (see below).

Neurotransmitter diversity in amacrine cells is matched by a morphological variety, but it may be possible to divide amacrine cells into three anatomical categories (Masland, 1986). One type has a small, dense dendritic field and a high packing density (number of cell bodies per square millimeter of retina). An example is the glycinergic rod amacrine cell. Their constricted size and their synaptic connections suggest that these cells are involved in very localized information processing, probably related to spatial acuity. The second type has an extremely large, but sparse dendritic field that often traverses a large portion of the entire retina. Examples are the dopaminergic and serotonergic amacrines. This morphology suggests that these cells play no role in local processing, but instead serve to modulate the general state of the retina. For example, the dopaminergic amacrine may participate in controlling adaptation as the retina switches between dark adapted rod vision and light adapted cone vision. The third amacrine type seems to be a mixture of the other two groups; the cells have long dendrites and moderate packing density but a very high coverage factor (the number of dendrites within a given retinal location). The cholinergic amacrine cell is an example. The morphology of this group of amacrines cells seems contradictory since a large dendritic field implies diffuse information processing while a high coverage factor suggests the opposite. A possible resolution is that single dendrites are acting independently, receiving synaptic input and sending outputs at a local level. Thus, the functional unit is much smaller than a single cell and a large cell could still function as a local information processor (Miller and Bloomfield, 1983). Some models of directional selectivity (see below) require this type of local synaptic interaction.

Ganglion Cells

One of the most rewarding analyses of ganglion cells is based on the distinction between linear and nonlinear properties originating from the work of Enroth-Cugell and Robson (1966). Although, these properties may represent a general phenomenon in vertebrate retinas, it is best illustrated in the cat retina where correlations between ganglion cell anatomy (Boycott and Wassle, 1974) and physiology have been discovered. Sustained ganglion cells, termed X-type, respond with a graded response that is linear, both spatially and temporally. Anatomically, these cells, termed beta cells, have small dendritic fields and are densely packed. In contrast, a physiologically distinct ganglion cell class, termed Y cells, respond transiently and nonlinearly to light stimuli. Their axonal conduction velocity is higher than X cells. The Y cells match the anatomically identified alpha cell type, which is less densely spaced and has comparatively large dendritic fields. The alpha and beta ganglion cells represent about 50% of the cat's ganglion cell population. The remaining ganglion cells are termed W cells. Physiologically these cells have large receptive fields and slow axonal conduction velocities. Anatomically, these cells correlate with gamma and delta cells, which have large dendritic fields and small somas. The correlations between anatomy and physiology suggest that the beta cells may be involved in spatial resolution, while the alpha cells may be more integrative and generate trigger features such as directional selectivity. Stone (1983) has proposed three general ganglion cell functions: high-resolution pattern vision, movement vision, and ambient vision. He has suggested that they correlate with the X, Y, and W cells, respectively.

One view of amacrine and ganglion cells is that amacrine cells shape the bipolar inputs while ganglion cells integrate the amacrine and bipolar cell input to produce a coded signal of visual information that is transmitted to the brain. An opposing view is that much of the neural integration occurs before signals reach the ganglion cells. In the catfish retina this perspective has been extended to suggest that essentially all the integration is done by bipolar and amacrine cells and that the ganglion cells are passive followers of these inputs (Sakai and Naka, 1987a,b). In discussing integrative functions we will follow the conventional view, but similar results could result if processing occurred before the ganglion cells.

We have seen that luminance is coded in the photoreceptors, spot size analysis involves center-surround mechanisms generated by horizontal cell feedback, while color coding includes a further refinement on this feedback system with signal channeling along spectrally specific pathways. Inner retinal processing enables the extraction of other visual information from an image, such as orientation and direction selectivity. The latter has been a topic of particularly fertile research. The basic observation is that many ganglion cells will respond preferentially to movement in one direction. When a bar or a grating is moved in a particular ("preferred") direction across the ganglion cell's receptive field, the cell will respond vigorously, movement in the opposite ("null") direction will produce little or no response. Other directions of movement will produce intermediate responses. In many animals directional selectivity is a prominent feature of ganglion cells. It is present in 40% of turtle ganglion cells (Bowling, 1980) and 20% of rabbit ganglion cells (Oyster, 1968). Barlow and Levick (1965) demonstrated that segments of a ganglion cell receptive field could act as separate units for directional selectivity and they proposed a model, using delayed inhibition, to explain the generation of these responses. They proposed that synaptic excitation and inhibition are arranged so that movement in one direction evokes excitation first, followed by inhibition. The time delay prevents the inhibition from suppressing the excitation and a vigorous response results. However, movement in the opposite direction produces inhibition that coincides with excitation, the result being a weak response. This model has found support in pharmacological studies in rabbit showing that suppression of GABA inhibition eliminates directional selectivity, resulting in ganglion cells that respond to all directions of motion (Wyatt and Daw, 1976). In the rabbit, the cholinergic amacrine may also play an important role in directional selectivity since cholinergic blockers also disrupt this trigger feature (Ariel and Daw, 1982b).

There have been modifications of this model, for example, a low pass filter could replace a synaptic delay with the same effect (Poggio and Reichardt, 1976). Another refinement is to suggest that the inhibition is shunting (causing a decrease in membrane resistance without a membrane potential change) rather than the classical hyperpolarizing inhibition (Torre and Poggio, 1978).

DeVoe et al. (1985) have found that directional selectivity in turtle retina is not restricted to ganglion cells, but is also found in amacrine cells and bipolar cells. The spectral properties of directional selectivity in bipolars are distinctly different from those in amacrine and ganglion cells, suggesting that directional selectivity in the distal and proximal retina is formed independently. However, it does raise questions regarding the site of origin of directionally selective integration.

Centrifugal Pathways in the Retina

The retina is a sensory organ that sends information about the outside world to the brain. It is natural that there is a centripetal flow of information to accomplish this function. But there is also a reverse flow of information. In the most localized case these are the feedback synapses made by horizontal cells or amacrine cells. As already mentioned, these neurons often provide negative feedback to the neurons that synapse on them. But there are also feedback loops that are less localized and involve multiple synaptic levels. One example is the interplexiform cell, which sends information from the inner plexiform layer back to the outer plexiform layer, and another is efferent fibers that send information from various brain regions back to the retina.

The interplexiform cell body lies adjacent to amacrine cells in the inner nuclear layer and sends neurites to both the inner and outer plexiform layer. It receives input from bipolar and amacrine cells in the inner plexiform layer and synapses on all distal retinal neurons (Dowling and Ehinger, 1978). Its function is best understood in horizontal cells. In the fish retina the interplexiform cell contains dopamine and this neurotransmitter uncouples gap junctions between horizontal cells (Negishi and Drujan, 1979; Lasater and Dowling, 1985). This serves to shrink the receptive field of horizontal cells and thereby reduce negative feedback to photoreceptors. The result is that diffuse light stimuli produce a larger response in photoreceptors and bipolar cells than they would if feedback was present. Center-surround differentiation is sacrificed to increase the response-to-signal ratio. This increased sensitivity could be advantageous in very dim light and has led to the proposal that the interplexiform cell is involved in dark adaptation (Mangel and Dowling, 1985).

Centrifugal connections from various brain regions have been demonstrated by anatomical (Cowan and Powell, 1963; Springer, 1983; Zucker and Dowling, 1987; Schutte and Weiler, 1988; Uchiyama et al., 1988) and electrophysiological techniques (Yang et al., 1988; Tornqvist et al., 1988) but the functional importance of these connections is still speculative. One function, revealed by experiments in the fish retina, may be to regulate the interplexiform cell. Efferents from the olfactory bulb project to the retina and synapse on interplexiform cells, as well as other neurons. These efferents contain peptide neurotransmitters, LHRH and FMRFamide. When applied to the retina, these peptides have the same effect as dopamine, suggesting that the efferents stimulate the interplexiform cells. When fish are exposed to odorants the B-wave of the ERG is enhanced, indicating the olfactory input can modify retinal response properties (Weiss and Meyer, 1988).

Synaptic Receptors and Information Processing

There are many correlates in the retina. We have already mentioned correlations between cell structure and function (each physiological category of retinal neurons has a distinct morphology) and between anatomy and physiology (ON and OFF responses occur in distinct synaptic sublaminae in the inner plexiform layer). There appears to be a similar correlation between physiology and pharmacology. Specifically, particular synaptic receptors mediate particular information signals in the retina. This may serve to explain the plethora of transmitters and receptors.

If neurotransmitters simply depolarized or hyperpolarized cells, there would not be a need for so many types. Of course it is known that depolarization may involve many different mechanisms, with many different effects. For example, depolarization may be due to the opening of a sodium channel or a calcium channel, or the closing of a potassium or chloride channel. Each channel has a different population density and distribution, driving force, conductance, and opening and closing kinetics. This means that the depolarization produced by one channel may have a very different effect on a neuron than the depolarization produced by another channel. If neurotransmitters control this depolarization (voltage can be another means of control), then a large number of neurotransmitters are required to take advantage of this channel diversity.

Coupled with the large number of retinal neurotransmitters, an additional parameter increases synaptic coding capacity still further. Each neurotransmitter may activate more than one receptor channel. Thus, for many neurotransmitters there are multiple receptor subtypes. In the retina, where one type of cell may synapse on several postsynaptic neurons, this additional variable seems to be an important mechanism for information coding. Two examples of this phenomenon are the glutamate receptor subtypes in the distal retina and the GABA receptor subtypes in the proximal retina.

Glutamate is thought to be the photoreceptor neurotransmitter, since exogenous application of glutamate mimics the effect of the endogenous transmitter: OFF bipolars and horizontal cells are depolarized and ON bipolars are hyperpolarized. When analogs of glutamate are used experimentally, some only act on the ON bipolar while others act only on OFF bipolar and horizontal cells. Experiments indicate that there are three types of glutamate receptors and each class of second order neurons interprets the photoreceptor signal using a different receptor (Slaughter and Miller, 1981, 1983b, 1985a,b). In some respects this may be reasonable. All three cell types are presumably getting the same transmitter input from the photoreceptors, but they respond differently to this input and the three synaptic receptor subtypes are one mechanism for generating three different responses (Slaughter and Miller, 1985a,b).

The most dramatic example of this distinction between the three glutamate receptors is produced by APB (2-amino-4-phosphonobutyrate). This glutamate analog only binds to the ON bipolar's glutamate receptor (Slaughter and Miller, 1981). The result of APB treatment is that the ON bipolar is continuously hyperpolarized and is no longer responsive to light, while the OFF bipolar and horizontal cell continue to function. Since retinal excitation at light onset derives from the ON bipolar response, the retina no longer conveys stimulatory information when a light is turned on, but excitatory signals are still transmitted when the light is turned off (Figure 3.6). Behavioral studies in monkey aimed at clarifying the role of the ON and OFF pathways by using APB have not revealed dramatic losses in the animal's visually stimulated response. This seems to indicate that the OFF signal, plus compensatory brain mechanisms, compensate well for the loss of the ON channel (Schiller et al., 1986).

In the inner retina, approximately two-thirds of the amacrine cells contain GABA or glycine. Both of these inhibitory transmitters can open chloride channels. Often the chloride reversal potential of amacrine and ganglion cells is close to the resting membrane potential, so these channels do not hyperpolarize the cell very much. The major effect in these cases is not a voltage change, but a resistance change. If a large number of chloride channels are open, there is a large decrease in cell resistance. This reduces the light response (voltage) to excitatory input (current) according to Ohm's Law (the voltage is equal to the product of the resistance and the current). This is the classical effect of an inhibitory transmitter. But there is another GABA receptor in the retina, termed the GABA/B receptor (Bowery et al., 1980). The GABA/B receptor appears to close calcium channels on bipolar cells (Maguire et al., 1989) and open potassium channels on amacrine and ganglion cells (Slaughter and Bai, 1989). We have previously discussed the implications of calcium channel closings in the generation of transient responses. Whether GABA opens chloride or potassium channels has dramatic implications for the light responses of amacrine and ganglion cells. Activation of GABA/A receptors (chloride channels) suppresses all light responses, but activation of GABA/B receptors (potassium channels) suppresses only sustained responses. When GABA/B receptors are stimulated, transient signals predominate. In fact, cells that normally respond with sustained signals change their response characteristics and emit transient signals. This has interesting implications because sustained and transient signals can carry very different forms of information (see the section on ganglion cell responses). This implies that GABA/B receptors can alter the response properties of retinal neurons and thereby change the information content of these cells. There is evidence that indeed the signaling characteristics of cells do change when GABA/B receptors are stimulated. For example, retinal neurons that show no directional selectivity under normal conditions become directional when GABA/B receptors are stimulated (Pan and Slaughter, 1988). We have suggested that this may represent a mechanism for selective attention since the signal-carrying capacity of the retina

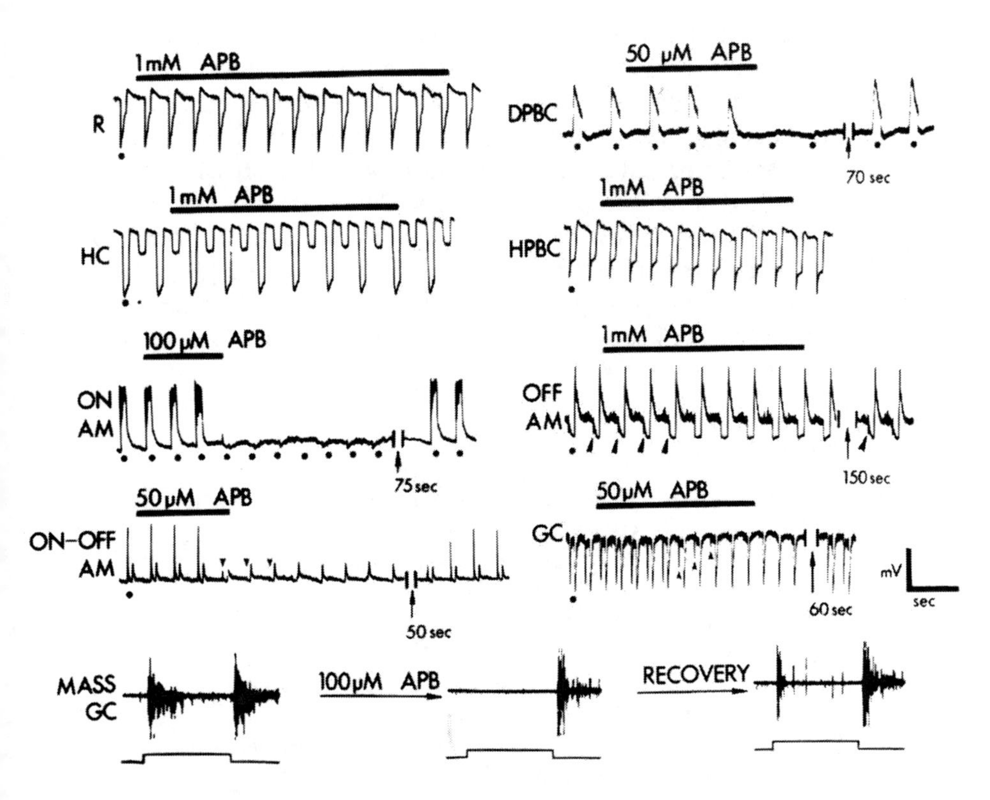

FIGURE 3.6. The effects of APB, a glutamate analog that acts selectively to suppress the light response of ON bipolar cells. The dark bars above the traces indicate when APB was applied while the dots or square pulses below each voltage trace mark the light pulses. In the distal retina note that the light responses of the photoreceptor (R), horizontal cell (HC), and the OFF bipolar (HPBC, also called a hyperpolarizing bipolar) are not diminished but the ON bipolar (DPBC, depolarizing bipolar) light response is eliminated. The ramification in the proximal retina is that ON responses are lost while OFF responses remain. Thus, the light responses of the ON amacrine (ON AM) are eliminated while the OFF amacrine (OFF AM) light responses persist. ON–OFF amacrine (AM) and ganglion cells (GC) lose their ON response components (arrows in fourth row). The bottom row shows that APB eliminates the ON component of mass ganglion cell activity, measured using an extracellular electrode that detects the spike activity of many ganglion cells. (From Slaughter and Miller, 1981. Copyright 1981 by the AAAS.)

can be modified under neurotransmitter control. There could be both a suppression of "unattended" information (sustained signals) and an enhancement of "attended" information (transient signals). The relevance of such a system is clear. If you have looked for a small animal after hearing a rustle in the bushes, it may be very difficult to detect that animal if it remains stationary. However, you are very quick to locate the animal if it moves. A selective attention mechanism designed to detect motion would be advantageous for this type of visual task.

In summary, the mysteries of how image processing is performed by the retina are just beginning to be unraveled. Many of the information coding strategies are not intuitively clear but reveal an intriguing design. The generality of these neural mechanisms makes them relevant to the central nervous system and are reflected in the descriptions of the visual cortex as well as in artificial intelligence systems described in other chapters of this book. Current research seems to have reunited two diverging forces: reductionist approaches at the membrane and molecular level with the more global questions of image processing and information processing. This is proving to be a very fertile area, but it is clear that the retina still has many secrets that may unfold with our changing approaches to visual science.

References

Ariel M, Adolph AR (1985). Neurotransmitter inputs to directionally selective turtle retinal ganglion cells. J. Neurophysiol. **54**:1123–1143.

Ariel M, Daw NW (1982a). Effects of cholinergic drugs on receptive field properties of rabbit retinal ganglion cells. J. Physiol. (London) **324**:135–160.

Ariel M, Daw NW (1982b). Pharmacological analysis of directionally sensitive rabbit retinal ganglion cells. J. Physiol. (London) **324**:161–185.

Attwell D, Borges S, Wu SM, Wilson M (1987). Signal clipping by the rod output synapse. Nature (London) **328**:522–524.

Ayoub GS, Lam DM-K (1985). The content and release of endogenous GABA in isolated horizontal cells of the goldfish retina. Vision Res. **25**:1187–1193.

Barlow HB, Levick WR (1965). The mechanism of directionally selective units in rabbit's retina. J. Physiol. (London) **178**:477–504.

Barnes S, Werblin F (1986). Gated currents generate single spike activity in amacrine cells of tiger salamander retina. Proc. Natl. Acad. Sci. U.S.A. **83**:1509–1512.

Belgum JH, Dvorak DR, McReynolds JS (1984). Strychnine blocks transient but not sustained inhibition in mudpuppy retinal ganglion cells. J. Physiol. (London) **354**:273–286.

Bowery NG, Hill DR, Hudson AL, Doble A, Middlemiss DN, Shaw J, Turnbull MJ (1980). (−)Baclofen decreases neurotransmitter release in the mammalian CNS by an action at a novel GABA receptor. Nature (London) **283**:92–94.

Bowling DB (1980). Light responses of ganglion cells in the retina of the turtle. J. Physiol. (London) **299**:173–196.

Boycott BB, Kolb H (1973). The connexions between bipolar cells and photoreceptors in the retina of the domestic cat. J. Comp. Neurol. **148**:91–114.

Boycott BB, Wassle H (1974). The morphological types of ganglion cells of the domestic cat's retina. J. Physiol. (London) **240**:397–419.

Brecha N, Johnson D, Peichl L, Wassle H (1988). Cholinergic amacrine cells of the rabbit retina contain glutamate decarboxylase and gamma-aminobutyrate immunoreactivity. Proc. Natl. Acad. Sci. U.S.A. **85**:6187–6191.

Caldwell JH, Daw NW (1978). New properties of rabbit retinal ganglion cells. J. Physiol. (London) **276**:257–276.

Cleland BG, Dubin MW, Levick WR (1971). Sustained and transient neurones in the cat's retina and lateral geniculate nucleus. J. Physiol. (London) **217**:473–496.

Cohen E, Sterling P (1986). Accumulation of (3H) glycine by cone bipolar neurons in the cat retina. J Comp Neurol **250**:1–7.

Coleman PA, Massey SC, Miller RF (1986). Kynurenic acid distinguishes kainate and quisqualate receptors in the vertebrate retina. Brain Res. **381**:172–175.

Cowan WM, Powell TPS (1963). Centrifugal fibers in the avian visual system. Proc. R. Soc. Lond. Ser. B **158**:232–252.

Dacheux RF, Raviola E (1986). The rod pathway in the rabbit retina: A depolarizing bipolar and amacrine cell. J. Neurosci. **6**:331–345.

Daw N (1973). Neurophysiology of color vision. J. Neurophysiol. **53**:571–611.

DeVoe RD, Guy RG, Criswell MH (1985). Directionally selective cells of the inner nuclear layer of the turtle retina. Invest. Ophthalmol. Vis. Sci. **26**, Suppl:311.

Dick E, Lowry OH (1984). Distribution of glycine, gamma aminobutyric acid, glutamate decarboxylase, and gamma aminobutyric acid transaminase in rabbit and mudpuppy retinas. J. Neurochem. **42**:1274–1280.

Dick E, Miller RF (1978). Light-evoked potassium activity in mudpuppy retina: Its relationship to the b-wave of the electroretinogram. Brain Res. **154**:388–394.

Dowling JE (1987). *The Retina an Approachable Part of the Brain*, p. 282. Harvard University Press, Cambridge, MA.

Dowling JE, Ehinger B (1978). The interplexiform cell system. I. Synapses of the dopaminergic neurons of the goldfish retina. Proc. R. Soc. Lond. Ser. B **201**:7–26.

Enroth-Cugell C, Robson JG (1966). The contrast sensitivity of retinal ganglion cells of the cat. J. Physiol. (London) **187**:517–552.

Faber DS (1969). Analysis of the slow transretinal potentials in response to light. Ph.D. Thesis, State University of New York, Buffalo.

Famiglietti EV Jr, Kolb H (1975). A bistratified amacrine cell and synaptic circuitry in the inner plexiform layer of the retina. Brain Res. **84**:293–300.

Famiglietti EV Jr, Kolb H (1976). Structural basis of 'ON'- and 'OFF'-center responses in retinal ganglion cells. Science **194**:193–195.

Frumkes TE, Eysteinnson T (1987). Suppressive rod-cone interaction in distal vertebrate retina: Intracellular records from *Xenopus* and *Necturus*. J. Neurophysiol. **57**:1361–1382.

Frumkes TE, Miller RF, Slaughter M, Dacheux RF (1981). Physiological and pharmacological basis of GABA and glycine action on neurons of mudpuppy retina III. Amacrine-mediated inhibitory influences on ganglion cell receptive-field organization: A Model. J. Neurophysiol. **45**:783–804.

Goldberg SH, Frumkes TE, Nygaard RW (1983). Inhibitory influence of unstimulated rods in the human retina: Evidence provided by examining cone flicker. Science **221**:180–182.

Ikeda H, Sheardown MJ (1983). Transmitters mediating inhibition of ganglion cells in the cat retina: Iontophoretic studies in vivo. Neuroscience **8**:837–853.

Kaneko A (1970). Physiological and morphological identification of horizontal, bipolar, and amacrine cells in goldfish retina. J. Physiol. (London) **207**:623–633.

Kaneko A (1973). Receptive field organization of bipolar and amacrine cells in the goldfish retina. J. Physiol. (London) **235**:133–153.

Kaneko A, Tachibana M (1983). Double color-opponent receptive fields of carp bipolar cells. Vision Res **23**:381–388.

Karwoski CJ, Newman EA, Shimazaki H, Proenza LM (1985). Light-evoked increases in extracellular K^+ in the plexiform layers of amphibian retinas. J. Gen. Physiol. **86**:189–213.

Kolb H (1977). The organization of the outer plexiform layer in the retina of the cat: Electron microscopic observations. J. Neurocytol. **6**:131–153.

Kolb H, Nelson R (1983). Rod pathways in the retina of the cat. Vision Res. **23**:301–312.

Kolb H, Nelson R, Mariani A (1981). Amacrine cells, bipolar cells and ganglion cells of the cat retina: A Golgi study. Vision Res. **21**:1081–1114.

Kuffler SW (1953). Discharge patterns and functional organization of mammalian retina. J. Neurophysiol. **16**:37–68.

Lam DM-K, Lasater EM, Naka K-I (1978). Gamma-aminobutyric acid: A neurotransmitter candidate for cone horizontal cells of the catfish retina. Proc. Natl. Acad. Sci. U.S.A. **75**:6310–6313.

Lasater EM, Dowling JE (1985). Dopamine decreases conductance of the electrical junctions between cultured horizontal cells. Proc. Natl. Acad. Sci. U.S.A. **82**:3025–3029.

Lettvin JW, Maturana HR, McCulloch WS, Pitts WH (1959). What the frog's eye tells the frog's brain. Proc. Inst. Rad. Eng. **47**:1940–1951.

Lettvin JY, Maturana HR, Pitts WH, McCulloch WS (1961). Two remarks on the visual system of the frog. In *Sensory Communication* (W. Rosenblith, ed.), pp. 757–776. MIT Press, Cambridge, MA.

Maguire G, Lukasiewicz P, Werblin F (1989). Amacrine cell interactions underlying the response to change in the tiger salamander retina. J. Neurosci. **9**:726–735.

Mangel SC, Dowling JE (1985). Responsiveness and receptive field size of carp horizontal cells are reduced by prolonged darkness and dopamine. Science **229**:1107–1109.

Mangel SC, Dowling JE (1987). The interplexiform-horizontal cell system of the fish retina: Effects of dopamine, light stimulation and time in the dark. Proc. R. Soc. Lond. Ser. B **231**:91–121.

Marc RE, Liu W-LS, Scholz K, Muller JF (1988). Serotonergic and serotonin-accumulating neurons in the goldfish retina. J. Neurosci. **8**(9):3427–3451.

Masland RH (1986). The functional architecture of the retina. Sci. Am. **255**:102–111.

Masland RH, Ames A III (1976). Responses to acetylcholine of ganglion cells in an isolated mammalian retina. J. Neurophysiol. **39**:1220–1235.

Masland RH, Mills JW (1979). Autoradiographic identification of acetylcholine in the rabbit retina. J. Cell Biol. **83**:159–178.

Masland RH, Mills JW, Cassidy C (1984a). The functions of acetylcholine in the rabbit retina. Proc. R. Soc. Lond. Ser. B **223**:121–139.

Masland RH, Mills JW, Hayden SA (1984b). Acetylcholine-synthesizing amacrine cells: Identification and selective staining by using radioautography and fluorescent markers. Proc. R. Soc. Lond. Ser. B **223**:79–100.

Massey SC, Redburn DA (1985). Light evoked release of acetylcholine in response to a single flash: Cholinergic amacrine cells receive ON and OFF input. Brain Res. **328**: 374–377.

Massey SC, Redburn DA (1987). Transmitter circuits in the vertebrate retina. Prog. Neurobiol. **28**:55–96.

Maturana HR, Lettvin JY, McCulloch WS, Pitts WH (1960). Anatomy and physiology of vision in the frog (*Rana pipiens*). J. Gen. Physiol. **43**(Sp12):129–171.

Miller RF (1973). Role of K^+ in generation of b-wave of electroretinogram. J. Neurophysiol. **36**:28–38.

Miller RF, Bloomfield SA (1983). Electroanatomy of a unique amacrine cell in the rabbit retina. Proc. Natl. Acad. Sci. U.S.A. **80**:3069–3073.

Miller RF, Dowling JE (1970). Intracellular responses of the Muller (glial) cells of mudpuppy retina: Their relationship to b-wave of the electroretinogram. J. Neurophysiol. **33**:323–341.

Miller RF, Dacheux RF, Frumkes TE (1977). Amacrine cells in *Necturus retina*: Evidence for independent gamma-aminobutyric acid and glycine-releasing neurons. Science **198**:748–750.

Muller F, Wassle H, Voigt T (1988). Pharmacological modulation of the rod pathway in the cat. J. Neurophysiol. **59**:1657–1672.

Murakami M, Ohtsu K, Ohtsuka T (1972). Effects of chemicals on receptors and horizontal cells in the retina. J. Physiol. (London) **227**:889–913.

Murakami M, Ohtsuka T, Shimazaki H (1975). Effects of aspartate and glutamate on the bipolar cells in the carp retina. Vision Res. **15**:456–458.

Murakami M, Shimoda Y, Nakatani K, Miyachi E, Watanabe S (1982). GABA-mediated negative feedback from horizontal cells to cones in carp retina. Jpn. J. Physiol. **32**:927–935.

Naka K-I, Rushton WAH (1963). S-potentials for colour units in the retina of fish (Cyrinidae). J. Physiol. (London) **185**:536–555.

Nawy S, Copenhagen DR (1987). Multiple classes of glutamate receptor on depolarizing bipolar cells in retina. Nature (London) **325**:56–58.

Negishi K, Drujan B (1979). Reciprocal changes in center and surrounding S-potentials of fish retina in response to dopamine. Neurochem. Res. **4**:313–318.

Nelson R (1977). Cat cones have rod input: A comparison of response properties of cones and horizontal cell bodies in the retina of the cat. J. Comp. Neurol. **172**:109–136.

Nelson R, Famiglietti EV Jr, Kolb H (1978). Intracellular staining reveals different levels of stratification for on- and off-center ganglion cells in the cat retina. J. Neurophysiol. **41**:472–483.

Newman EA (1980). Current source density analysis of the b-wave of frog retina. J. Neurophysiol. **43**:1355–1366.

Noel WK (1954). The origin of the electroretinogram. Am. J. Ophthalmol. **28**:78–90.

Normann RA, Perlman I, Kolb H, Jones J, Daly SJ (1984). Direct excitatory interactions between cones of different spectral types in the turtle retina. Science **224**:625–627.

Oyster CW (1968). The analysis of image motion by rabbit retina. J. Physiol. (London) **199**:613–635.

Pan Z-H, Slaughter MM (1988). A cellular model of selective attention. Soc. Neurosci. Abstr. **14**:987.

Poggio T, Reichardt W (1976). Visual control of orientation behaviour in the fly II. Q. Rev. Biophys. **9**:377–438.

Ramon y Cajal S (1973). La retine des vertebres. In *The Vertebrate Retina* (D Maguire, RW Rodieck, eds.), pp. 775–904. Freeman, San Francisco.

Rodieck RW (1973). *The Vertebrate Retina*. Freeman, San Francisco.

Rodieck RW, Stone J (1965). Response of cat retinal ganglion cells to moving visual patterns. J. Neurophysiol. **28**:819–832.

Saito T, Kondo H, Toyoda J (1979). Ionic mechanisms of two types of on-center bipolar cells in the carp retina. J. Gen. Physiol. **73**:73–90.

Saito T, Kujiraoka T, Yonaha T, Chino Y (1985). Reexamination of photoreceptor-bipolar connectivity patterns in carp retina: HRP-EM and Golgi-EM studies. J. Comp. Neurol. **236**:141–160.

Sakai HM, Naka K-I (1987a). Signal transmission in the catfish retina. IV. Transmission to ganglion cells. J. Neurophysiol. **58**:1307–1328.

Sakai HM, Naka K-I (1987b). Signal transmission in the catfish retina. V. Sensitivity and circuit. J. Neurophysiol. **58**:1329–1350.

Schiller PH, Sandell JH, Maunsell JHR (1986). Functions of the ON and OFF channels of the visual system. Nature (London) **322**:824–825.

Schutte M, Weiler R (1988). Mesencephalic innervation of the turtle retina by a single serotonin-containing neuron. Neurosci. Res. **91**:289–294.

Slaughter MM, Bai S-H (1989). Differential effects of baclofen on sustained and transient cells in the mudpuppy retina. J. Neurophysiol. **61**:374–381.

Slaughter MM, Miller RF (1981). 2-Amino-4-phosphonobutyric acid: A new pharmacological tool for retina research. Science **211**:182–185.

Slaughter MM, Miller RF (1983a). Bipolar cells in the mudpuppy retina use an excitatory amino acid neurotransmitter. Nature (London) **303**:537–538.

Slaughter MM, Miller RF (1983b). An excitatory amino acid antagonist blocks cone input to sign-conserving second-order retinal neurons. Science **219**:1230–1232.

Slaughter MM, Miller RF (1985a). Identification of a distinct synaptic glutamate receptor on horizontal cells in the mudpuppy retina. Nature (London) **314**:96–97.

Slaughter MM, Miller RF (1985b). The role of glutamate receptors in information processing in the distal retina. In *Neurocircuitry of the Retina, A Cajal Memorial* edited by (A Gallego, P. Gouras, eds.), pp. 51–65. Elsevier, New York.

Springer AD (1983). Centrifugal innervation of goldfish retina from ganglion cells of the nervus terminalis. J. Comp. Neurol. **214**:404–415.

Steinberg RH, Schmidt R, Brown KT (1970). Intracellular responses to light from cat pigment epithelium: Origin of the electroretinogram c-wave. Nature (London) **227**:728–730.

Stell WK, Lightfoot DO (1975). Color-specific interconnections of cones and horizontal cells in the retina of the goldfish. J. Comp. Neurol. **159**:473.

Stell WK, Kretz R, Lightfoot DO (1982). Horizontal cell connectivity in goldfish. In *The S-potential* edited by (BD Drujan, M. Laufer, eds.), pp. 51–75. Liss, New York.

Stockton RA, Slaughter MM (1989). B-Wave of the electroretinogram, A reflection of ON bipolar activity. J. Gen. Physiol. **93**:101–122.

Stone J (1983). *Parallel Processing in the Visual System*. Plenum, New York.

Svaetichin G (1953). The cone action potential. Acta Physiol. Scand. **29**:565–600.

Tauchi M, Masland RH (1984). The shape and arrangement of the cholinergic neurons in the rabbit retina. Proc. R. Soc. Lond. Ser. B **223**:101–119.

Tomita T (1963). Electrical activity in the vertebrate retina. J. Opt. Soc. Am. **53**:49–57.

Tornqvist K, Yang X-L, Dowling JE (1988). Modulation of cone horizontal cell activity in the teleost fish retina. III. Effects of prolonged darkness and dopamine on electrical coupling between horizontal cell. J. Neurosci. **8**:2279–2289.

Torre V, Poggio T (1978). A synaptic mechanism possibly underlying directional selectivity to motion. Proc. R. Soc. Lond. Ser. B **202**:409–416.

Toyoda J-I, Kujiraoka T, Fujimoto M (1982). The opponent color process and interaction of horizontal cells. In *The S-potential* (BD Drujan, M. Laufer eds.), pp. 151–160. Liss, New York.

Uchiyama H, Reh T, Stell WK (1988). Immunocytochemical and morphological evidence for a retinopetal projection in anuran amphibians. J. Comp. Neurol. **274**:48–59.

Vaney DL, Young HM (1988). GABA-like immunoreactivity in cholinergic amacrine cells of the rabbit retina. Brain Res. **438**:369–373.

Weiss O, Meyer DL (1988). Odor stimuli modulate retinal excitability in fish. Neurosci. Lett. **93**:209–213.

Werblin FS, Dowling JE (1969). Organization of the retina of the mudpuppy, *Necturus maculosus*. II. Intracellular recording. J. Neurophysiol. **32**:339–355.

Westheimer G, McKee SP (1977). Spatial configurations for visual hyperacuity. Vision Res. **17**:941–947.

Witkovsky P, Stone S, Besharse JC (1988). Dopamine modifies the balance of rod and cone inputs to horizontal cells of the *Xenopus* retina. Brain Res. **449**:332–336.
Wu SM, Dowling JE (1978). L-Aspartate: Evidence for a role in cone photoreceptor synaptic transmission in the carp retina. Proc. Natl. Acad. Sci. U.S.A. **75**:5205–5209.
Wu SM-S, Lam DM-K (1988). The coexistence of three neuroactive substances in amacrine cells of the chicken retina. Brain Res. **458**:195–198.
Wyatt HJ, Daw NW (1976). Specific effects of neurotransmitter antagonists on ganglion cells in rabbit retina. Science **191**:204–205.
Yang XL, Wu SM (1989). Modulation of rod-cone coupling by light. Science **244**:352–354.
Yan X-L, Tornqvist K, Dowling JE (1988). Modulation of cone horizontal cell activity in the teleost fish retina. II. Role of interplexiform cells and dopamine in regulating light responsiveness. J. Neurosci. **8**:2269–2279.
Yazulla S (1976). Cone input to bipolar cells in the turtle retina. Vision Res. **16**:737–744.
Zucker CL, Dowling JE (1987). Centrifugal fibres synapse on dopaminergic interplexiform cells in the teleost retina. Nature (London) **330**:166–168.

4 Nested Maps in Macaque Monkey Visual Cortex

BRUCE M. DOW

Introduction

The cerebral cortex is a flat sheet of neurons, roughly 2 mm thick in humans and other primates, which exhibits regional specialization largely on the basis of its inputs and outputs. The internal wiring pattern of the cerebral cortex appears to be remarkably similar from one region to another, though there are local differences in cell density and distribution (cytoarchitectonics) and fiber density and distribution (myeloarchitectonics). The primary visual cortex (Fig. 4.1), also known as striate cortex, area 17, area V1, is "visual" because its input comes from the retina, via the lateral geniculate nucleus of the thalamus. Distinct morphological features of the primary visual cortex in primates are the stripe of Gennari and the two bands of Baillarger, visible even in unstained tissue (see Brazier and Petsche, 1978). The primary visual cortex, in turn, sends output to a number of additional "visual" areas designated V2, V3, V3A, V4, and MT (or V5), and located in the tissue adjacent to area V1 (Fig. 4.2). These other visual areas lack the stripe of Gennari, but may have bands of Baillarger.

The essential defining features of the different visual areas, which currently total about 20 in primates (Van Essen, 1985), are primarily physiological rather than anatomical, though the precise physiological parameters that define each area remain somewhat fuzzy at present. Area MT appears clearly to be involved in the processing of motion information (Dubner and Zeki, 1971; Maunsell and Van Essen, 1983a; Erickson and Dow, 1989). Areas V1, V2 and V4 all seem to be involved in the processing of color and form, in addition to motion information (Zeki, 1983; Tanaka et al., 1986; Desimone and Schein, 1987; Yoshioka et al., 1988).

The functional organization of primate visual cortex exhibits both serial (hierarchical) and parallel features. Thus, the sequence V1–V2–V4 is largely serial. Receptive fields of single neurons (a neuron's "receptive field" being defined as the region of the visual field providing input to the neuron) become progressively larger as one proceeds from V1 to V2 to V4, and the modules or "columns" containing cells with similar functional properties seem to get progressively larger as

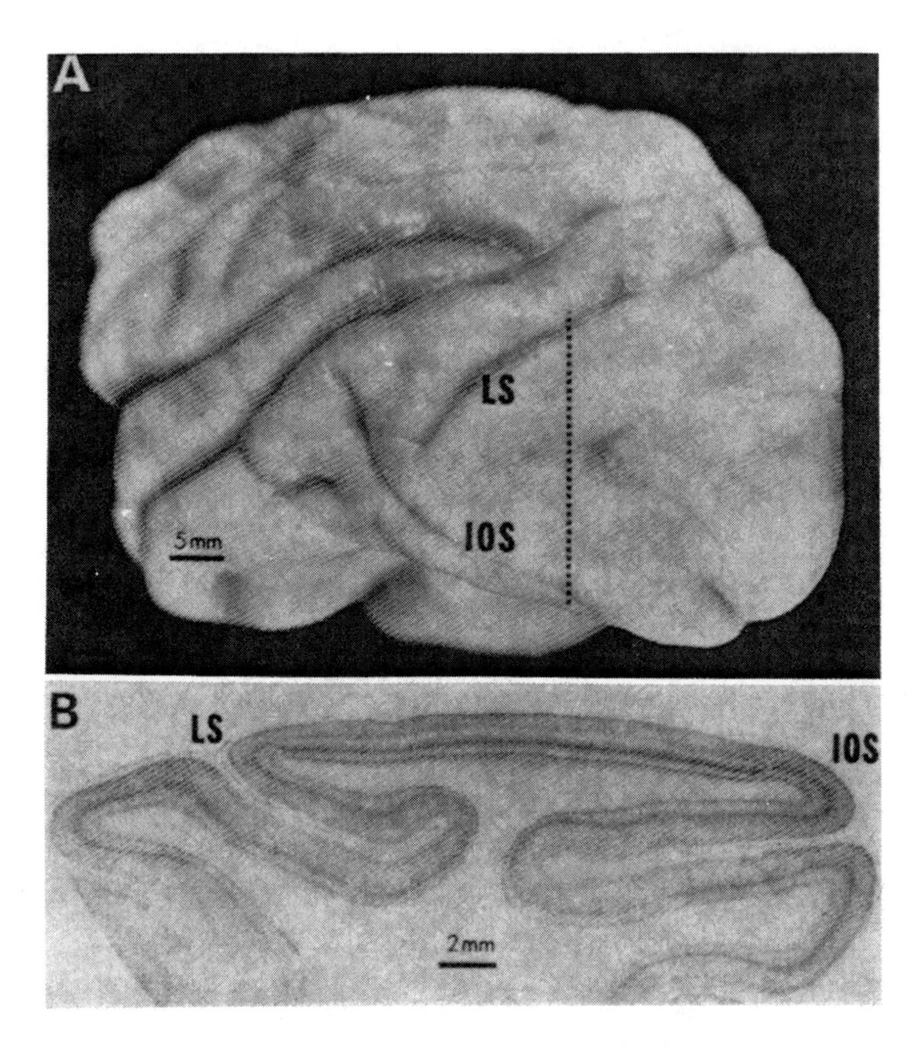

FIGURE 4.1. The primary visual cortex (striate cortex, area 17, V1) of a macaque monkey (Rhesus) as viewed from the surface (A) and in section (B). LS, lunate sulcus; IOS, inferior occipital sulcus. The dotted line in A indicates the plane of section for B. (From Dow et al., 1981.)

well (Yoshioka et al., 1990a,b). The sequence V1–V2–MT is also serial, but involves different functional modules in V1 and V2 than the sequence V1–V2–V4. This indication of parallel processing within V1 and V2 is of considerable interest and will be further discussed below.

The main goal of the present chapter is to describe the functional organization of the visual cortex in primates, including area V1 as well as several of the other well-characterized areas. Special emphasis is given here to the principle of

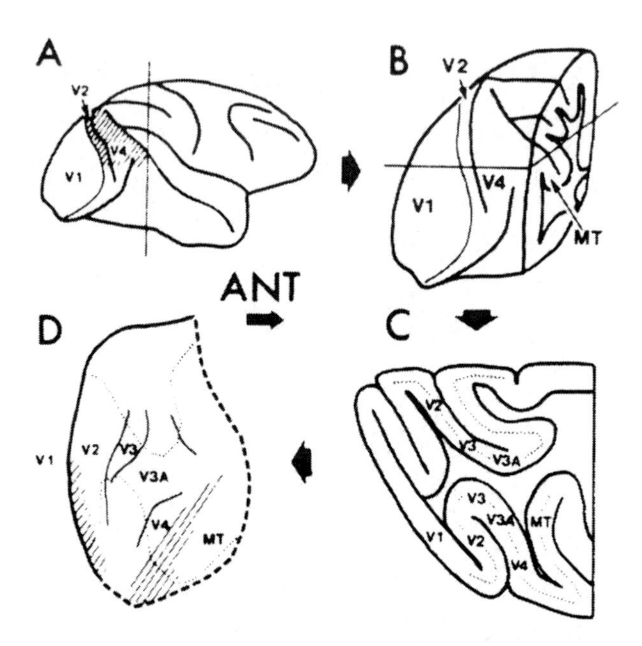

FIGURE 4.2. Locations of several visual areas in macaque monkey cortex as viewed from the surface and in section. Vertical line in A is the plane of section for B. Thin angled line in B is the plane of section for C. Dotted lines in C indicate the portion of extrastriate cortex (from this section) included in D, a flattened two-dimensional reconstruction of dorsal extrastriate occipital cortex. Shaded regions in D are portions of cortex visible from the lateral side of the hemisphere (see A). (Reproduced, with permission, from the Annual Review of Neuroscience, Volume 2, © 1979 by Annual Reviews Inc.)

"mapping." Evidence is presented for separate mappings of features such as retinal position (retinotopy), ocular dominance, line orientation, color, movement direction, and depth in space. Konishi (1986) and Knudsen et al. (1987) have made the important distinction between location-preserving maps, which simply reproduce body surfaces onto brain surfaces, and "computational" maps, which involve some kind of combinatorial transformation. In the auditory system, sound localization in space appears to be computed within the brain, on the basis of timing and intensity differences between the inputs received by the two ears. Knudsen et al. (1987) cite ocular dominance, orientation, and movement direction as examples of visual parameters that appear to be computationally mapped in the cortex. The present chapter proposes color, and depth in space, as additional examples (or potential examples) of computationally mapped visual features in monkey cortex. Of the various maps, only the retinotopic and ocular dominance maps are fully established in striate cortex. Although orientation is clearly mapped in V1, the mapping principle remains elusive. Evidence for mapping of the other three variables (color, direction of motion, and depth in space), while suggestive, is incomplete at the present time.

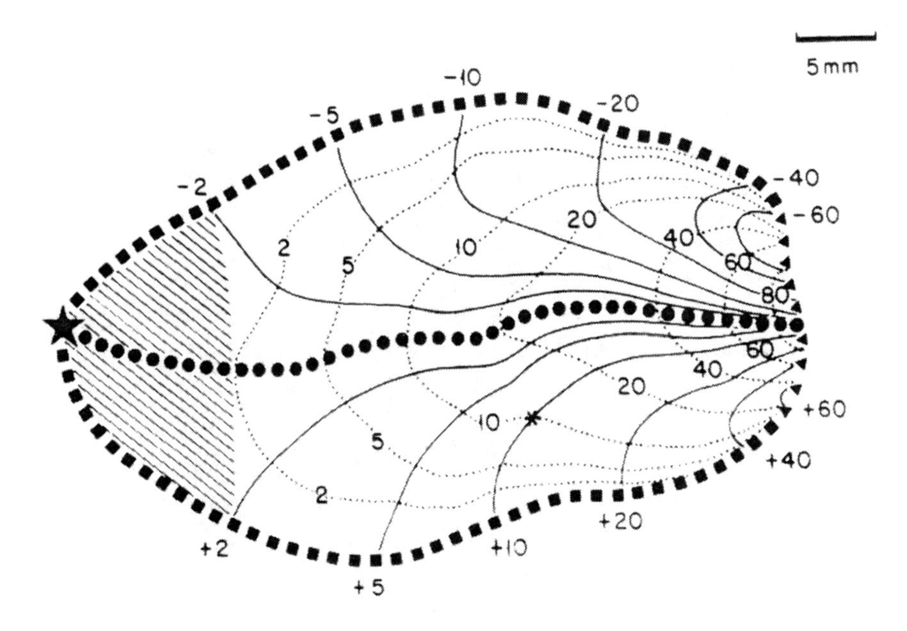

FIGURE 4.3. Schematic flattened retinotopic map of striate cortex in the *Cebus* monkey. Vertical meridian, black squares. Horizontal meridian, filled circles. Foveal center, star. Solid lines, elevations (degrees). Dotted lines, azimuths (degrees). (From Gatass et al., 1987.)

Retinotopy

All of the visual areas described above share the property of "retinotopy" (Gattass and Gross, 1981; Gattass et al., 1981, 1987, 1988). That is, each area contains an orderly map of the retina, or, more precisely, of one hemiretina of each eye. The topological details of the maps are different for each area, and some maps are better characterized than others. There is a general tendency for the map to be increasingly degraded as one proceeds from lower areas to higher areas. This is due to convergence and divergence of axons, resulting in larger and larger receptive fields of individual cells.

The retinotopic map of striate cortex in primates was first described by Talbot and Marshall (1941) more than 40 years ago. Subsequent studies (Daniel and Whitteridge, 1961; Allman and Kaas, 1971; Guld and Bertulis, 1976; Tootell et al., 1982; Van Essen et al., 1984; Dow et al., 1985; Gattass et al., 1987) have added to our understanding of specific details of the map. Perhaps the most striking feature of the striate retinotopic map is "magnification," with the foveal region (a mere speck on the retina) occupying nearly 25% of the total area of striate cortex, and the central 10 degrees occupying about 50% of the total area, in many primate species. A recent map of *Cebus* striate cortex (Gatass et al., 1987) is shown in Figure 4.3.

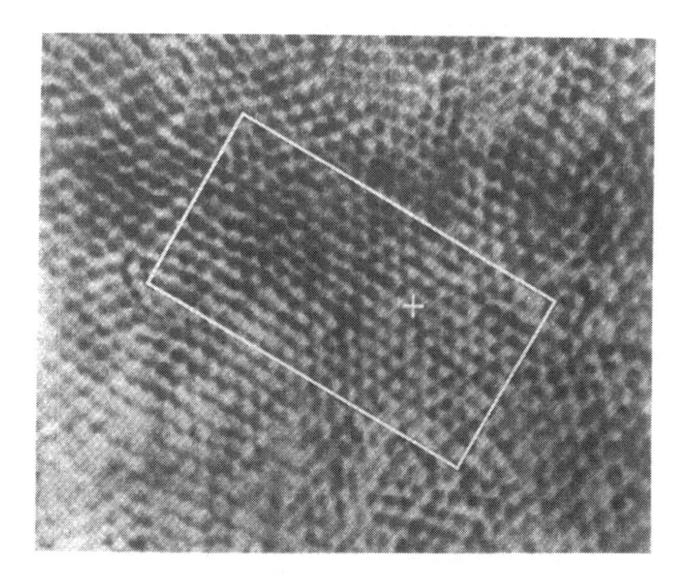

FIGURE 4.4. Photograph of the cone mosaic in the central fovea of a rhesus monkey (from Polyak, 1957) with the superimposed receptive field of a typical foveal striate cell drawn approximately to scale. The white cross, from Polyak's original figure, indicates the geographic center of the fovea. (From Dow et al., 1984, Copyright © 1984, John Wiley & Sons, Inc.)

What is the basis of magnification in the striate map? Certainly the density of cells in the retina plays a role, since the foveal cones, bipolar cells, and ganglion cells are smaller and more densely packed together than are their peripheral counterparts. Polyak (1957) proposed the existence of a "midget" system in the central retina of primates, which provided "private line" input from individual cones (through bipolar cells) to single ganglion cells. More recent work by Kolb, involving serial electron microscopic reconstructions of Golgi impregnated cells (Kolb, 1970; Kolb and DeKorver, 1988), tends to support Polyak's claim.

Polyak further suggested that the midget system might extend all the way to the cortex. This seems not to be the case, however, since more recent studies have shown that receptive fields of striate cells receiving foveal input are much larger than single cones (Dow et al., 1981). In fact, the mean size of a foveal striate cell receptive field, irrespective of eccentricity, is roughly 225 cones (Dow et al., 1984). This ratio of cones to average foveal striate receptive field is illustrated in Figure 4.4. Thus, there is a lot of convergence in the pathway from retina to striate cortex.

Furthermore, striate cortical magnification in the fovea is greater than would be predicted from either cone density or ganglion cell density (Malpeli and Baker, 1975; Myerson et al., 1977; Dow et al., 1981; see McIlwain, 1986). Some factor or factors other than retinal cell density must therefore be responsible for cortical magnification. We will return to this issue toward the end of the chapter.

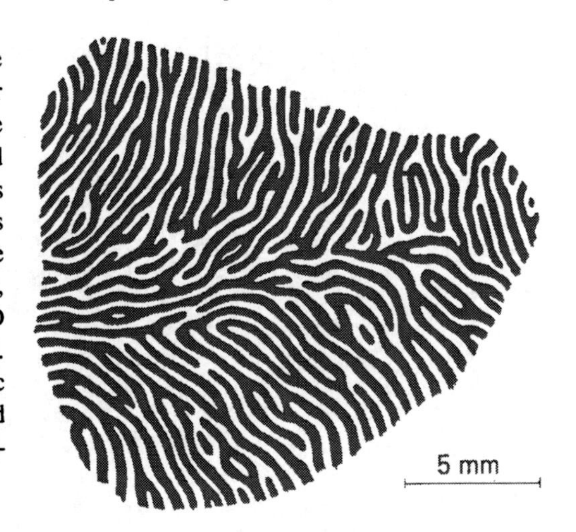

FIGURE 4.5. Ocular dominance stripes in layer 4 of the opercular region (central 7°) of macaque monkey striate cortex as viewed from the surface. One eye's stripes are black and the other's are white. Foveal center is to the right. (From LeVay et al. [1975], and from EYE, BRAIN, AND VISION. By David H. Hubel. Copyright © 1988 by Scientific American Library. Reprinted with permission by W.H. Freeman and Company.)

Ocular Dominance

Visual input from the two eyes is segregated in different layers within the lateral geniculate nucleus (LGN) in primates and most other mammals. Primate LGN consists of six layers, the outer four of which contain small cells and are called "parvocellular," and the inner two of which contain large cells and are called "magnocellular." The layers are numbered 6 to 1 from outside to inside. Layers 6, 4, and 1 receive input from the contralateral eye, and layers 5, 3, and 2 from the ipsilateral eye.

Hubel and Wiesel (1972) examined the pattern of LGN input to striate cortex by making localized lesions of layer 6 alone, or layers 6 and 5, of the monkey LGN. Such lesions result in degeneration of LGN axon terminals within the striate cortex. Careful examination of the terminal degeneration in serially reconstructed cortical sections revealed a series of alternating right and left eye stripes within layer 4 of striate cortex, where most LGN axons terminate. LeVay et al. (1975) later demonstrated the complete ocular dominance column "map" in the monkey's striate cortex, using a different staining technique, which did not require any LGN lesions. This map is shown in Figure 4.5.

The stripes are about 400 μm wide. It is not yet clear what physical forces determine this particular periodicity. It is clear, however, that the two eyes compete for their share of each repeating stripe segment, since occlusion of one eye during a "critical period" in postnatal development, which lasts typically from birth to about 1 year in macaque monkeys, results in expansion (Fig. 4.6) of the open eye's (white) stripes at the expense of the closed eye's (black) stripes (Hubel and Wiesel, 1977; Hubel et al., 1977; LeVay et al., 1980; Wiesel, 1982).

Under normal conditions the "critical period" is a time when binocular vision, including depth perception and stereoacuity (see below), is being established. The stripe width of 400 μm is thought to be determined by a combination of

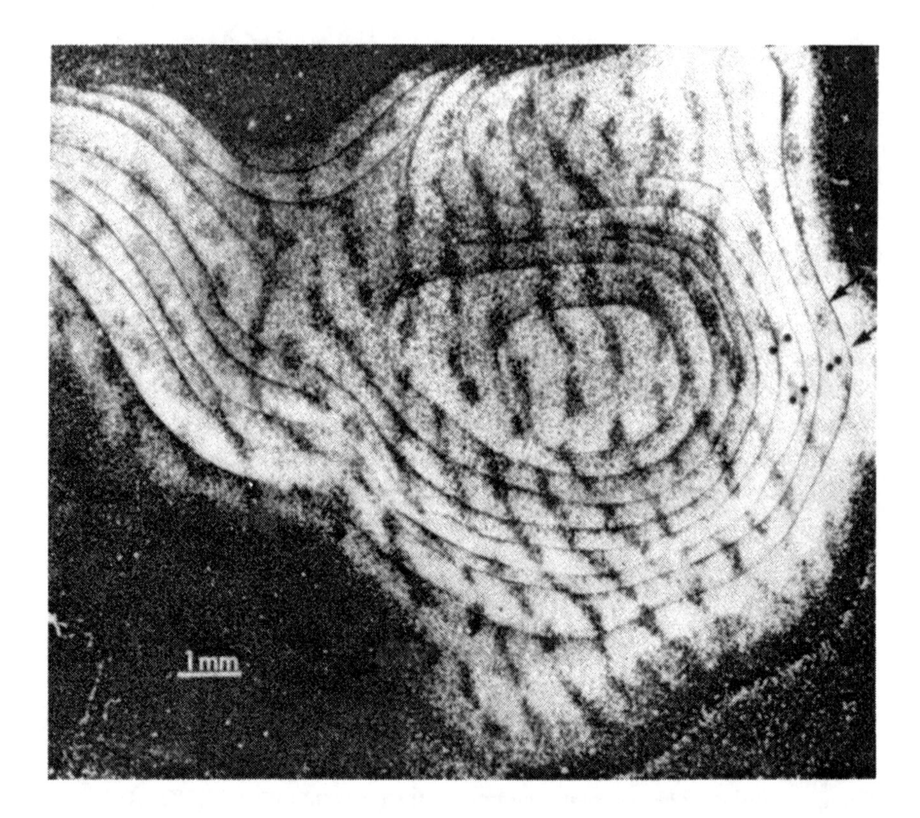

FIGURE 4.6. Ocular dominance stripes in the left occipital lobe of a macaque monkey whose right eye was sutured closed at 2 weeks. Left (normal) eye was injected with radioactively labeled amino acids 2 weeks before perfusion, at 18 months. White stripes (labeled regions) from normal eye. Black stripes (unlabeled regions) from sutured eye. The picture is a serial reconstruction from tangential sections. (From Hubel and Wiesel, 1977.)

prenatal and postnatal developmental mechanisms (see also Udin, Chapter 5). The postnatal process is likely to be a very dynamic one, involving coordinated movements of the two eyes. Any optical disparity between the eyes (i.e., strabismus or crossed-eyes, as well as astigmatism or myopia greater in one eye than the other) may, if left uncorrected during the critical period, interfere with the establishment of binocular vision. The critical period in humans may last up to 3 years (Banks et al., 1975).

Orientation

In the late 1950s and early 1960s Hubel and Wiesel (1959, 1962, 1965) made the important and completely unprecedented observation that cells in the primary visual cortex of the cat are sensitive to the orientation of lines. Hubel and Wiesel's basic experiment involved recording with extracellular metal microelectrodes from single cells in the visual cortex of animals in a state of anesthesia and muscle relaxation (to prevent movements of the eyes). The muscle relaxation necessitated the use of a respirator. The pupils were dilated and fixed, the eyes protected with contact lenses, plus additional correcting lenses for focusing on a screen placed at a convenient distance from the animal. By shining small spots and bars of light onto the screen, after some attention to the optics and the level of room illumination, they were able to locate the receptive fields of individual cells on the screen.

Hubel and Wiesel found three classes of orientation-selective cells in the cat's visual cortex, which they called "simple," "complex," and "hypercomplex." Simple cells had separate subregions within their receptive fields within which responses, in the form of extracellularly recorded action potentials, occurred at either light ON or light OFF. Some cells had two such subregions (one ON, one OFF), while others had three (an ON region flanked by two OFF regions or an OFF region flanked by two ON regions). The subregions tended to be elongated in the same axis as the cell's preferred orientation. Light bars elicited excitatory responses (increases in the rate of action potential production) from ON subregions and inhibitory responses (decreases in the rate of action potential production) from OFF subregions. Dark bars typically elicited the opposite effects (i.e., excitation from OFF regions, inhibition from ON regions). Complex and hypercomplex cells typically responded the same to dark and light bars, and did not have separate ON and OFF subregions; the preferred orientation was effective anywhere within the cell's receptive field, rather than having to be positioned at some particular location. Hypercomplex cells differed from complex cells in showing selectivity for the length as well as the orientation of lines. Some hypercomplex cells (called "higher order") seemed to respond to corners, or at least changes in curvature.

Hubel and Wiesel suggested that the three classes of orientation-selective cells (simple, complex, hypercomplex) represented the initial stages of form perception. Contours of visual forms were analyzed by extracting detailed information about their curvature in the form of tangents at each point. Simple cells extracted the information. Complex cells generalized for "orientation" without the need for a specific position. Hypercomplex cells indicated changes in orientation at regions of high curvature. Hubel and Wiesel proposed a hierarchical model according to which simple cells were combined to make complex cells and complex cells were combined to make hypercomplex cells. Using microelectrodes in a purely biological experiment, they had uncovered data that seemed to relate to visual perception, in the realm of psychology. An important bridge was being formed between biology and psychology. They received the Nobel Prize for this work in 1981 (see Hubel, 1982; Wiesel, 1982).

Examining the anatomical distribution of the cells, Hubel and Wiesel (1963) discovered a systematic arrangement of orientations in the tissue, such that an electrode directed vertically (i.e., orthogonal to the layers) would always encounter cells with the same orientation preference, while a horizontally (i.e.,

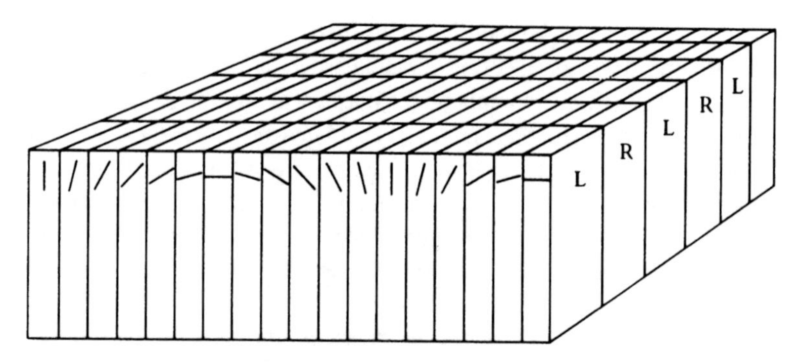

FIGURE 4.7. "Ice cube model" of orientation and ocular dominance columns in macaque monkey striate cortex. R and L indicate right and left ocular dominance stripes. Vertical dimension indicates cortical depth. Horizontal dimension represents cortical surface location. (From EYE, BRAIN, AND VISION. By David H. Hubel. Copyright © 1988 by Scientific American Library. Reprinted by permission by W.H. Freeman and Company.)

tangentially) directed electrode would encounter a regular "orientation drift." They proposed that orientation was organized in "columns" extending throughout the 2 mm of tissue depth.

They extended this work to monkeys a few years later (Hubel and Wiesel, 1968), where they reported evidence that the hierarchical model could explain the internal wiring of an individual orientation column, with nonoriented cells located in the input region (layer 4), simple cells located just above and below, complex cells just above and below simple cells, and hypercomplex cells in the upper- and lower-most layers.

In monkeys, where striate cortex includes a map of alternating ocular dominance stripes (see above), Hubel and Wiesel (1972, 1974a, 1977) also proposed that the orientation columns formed parallel slabs, about 18 in all (one for every 10°), running orthogonal to the ocular dominance stripes (Fig. 4.7). There is as yet no evidence (in macaque monkeys) to support the "parallel slab" or "ice cube" (Hubel, 1987) model, though it appears frequently in textbooks. Existing evidence from studies using voltage sensitive dyes (Blasdell and Salama, 1986) as well as the radioactively labeled glucose analogue 2-deoxyglucose (Hubel et al., 1978; Tootell et al., 1982, 1988a) suggests that isoorientation lines form complex whorls (rather than parallel slabs) that do not cross ocular dominance column borders at any fixed angle (Fig. 4.8). A notable problem with the ice cube model is its failure to explain "reversals" (Hubel and Wiesel, 1974a; Livingstone and Hubel, 1984) in the direction of orientation drift as a microelectrode advances horizontally or obliquely through the tissue (see below).

A possible clue toward the organization of orientation columns in striate cortex came in the early 1980s with the discovery of localized patches of tissue in layers 2 and 3, which contained a high concentration of the mitochondrial enzyme, cytochrome oxidase (Wong-Riley, 1979; Horton and Hubel, 1981; Hendrickson et al., 1981; Tootell et al., 1983; Horton, 1984; Carroll and Wong-Riley, 1984).

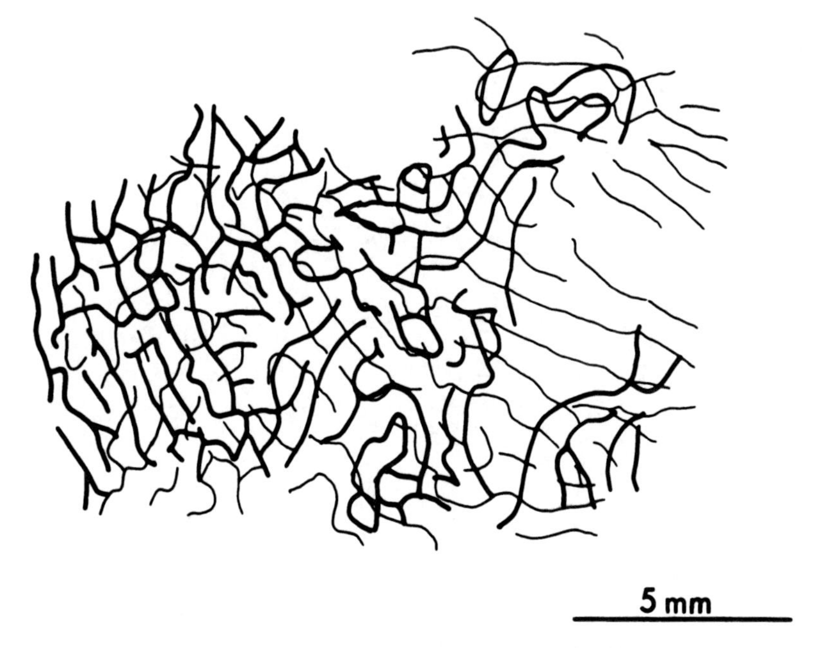

FIGURE 4.8. Orientation columns (thick lines) in relation to ocular dominance columns (thin lines) in a macaque monkey exposed to vertical stripes and injected with 2-deoxyglucose. (From Hubel et al., 1978.)

Horizontal sections through the tissue revealed that the patches form a regular array of polka dots (Fig. 4.13, far right). Cells in cytochrome oxidase (CO)-rich patches were subsequently found to lack orientation selectivity (Livingstone and Hubel, 1984).

These important findings have led to a new type of orientation column model, the "centric" model, of which four examples are shown in Figure 4.9 (Baxter and Dow, 1989). [The first centric model, proposed by Braitenberg and Braitenberg (1979), in fact preceded the discovery of CO patches.] The circles in Figure 4.9 represent CO zones, and are called "singularities," since orientation is not defined at these points. Singularities are called either "1" (squares) or "½" (triangles), depending on whether orientation changes by 360° or 180° as one makes a complete circuit around the center. They are called "plus" (open) or "minus" (filled), depending on whether orientation changes in a clockwise or counterclockwise fashion as one makes a complete clockwise circuit around the center.

Computer-simulated electrode tracks through the four centric models are shown in Figure 4.10 (Baxter and Dow, 1989). Note that "reversals" (see above) can be obtained with all four models, which makes them preferable to the ice cube model. Rates of orientation change and reversal rates per unit distance are quite different for the four models. Based on these differences (along with several

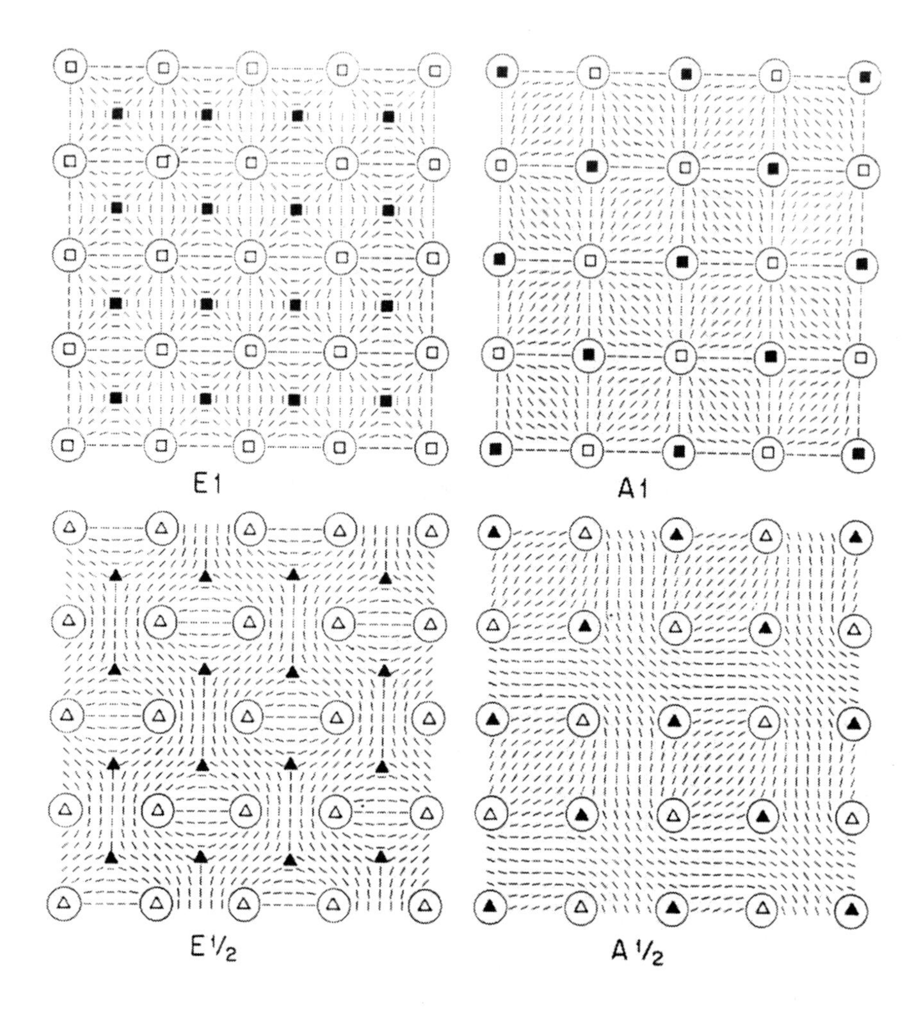

FIGURE 4.9. Four different centric models based on a square matrix. Each model represents a tangential section through layers 2 and 3 of striate cortex in a macaque monkey. The circles represent cytochrome oxidase "blobs." Each short line segment represents the preferred orientation at that particular location in the map. Open symbols are "clockwise" singularities in the orientation array; filled symbols are counterclockwise singularities. Squares are singularities of index "1" (see text). Triangles are singularities of index "½." (From Baxter and Dow, 1989.)

►

FIGURE 4.10. Computer simulations of horizontal electrode penetrations through the four models shown in Figure 4.9. Black ellipses represent cytochrome oxidase patches whose separation distances correspond to histological measurements. Two penetrations are shown through each model, one parallel to the long axis of the ellipses and the other oblique. Plots of orientation with electrode distance are shown above each model. In each case the parallel penetration yields a series of regularly occurring reversals, while the oblique penetration yields a nearly straight line. Orientation drift rate and reversal rate are least for the A ½ model (lower right) and greatest for the E1 model (upper left). (From Baxter and Dow, 1989.)

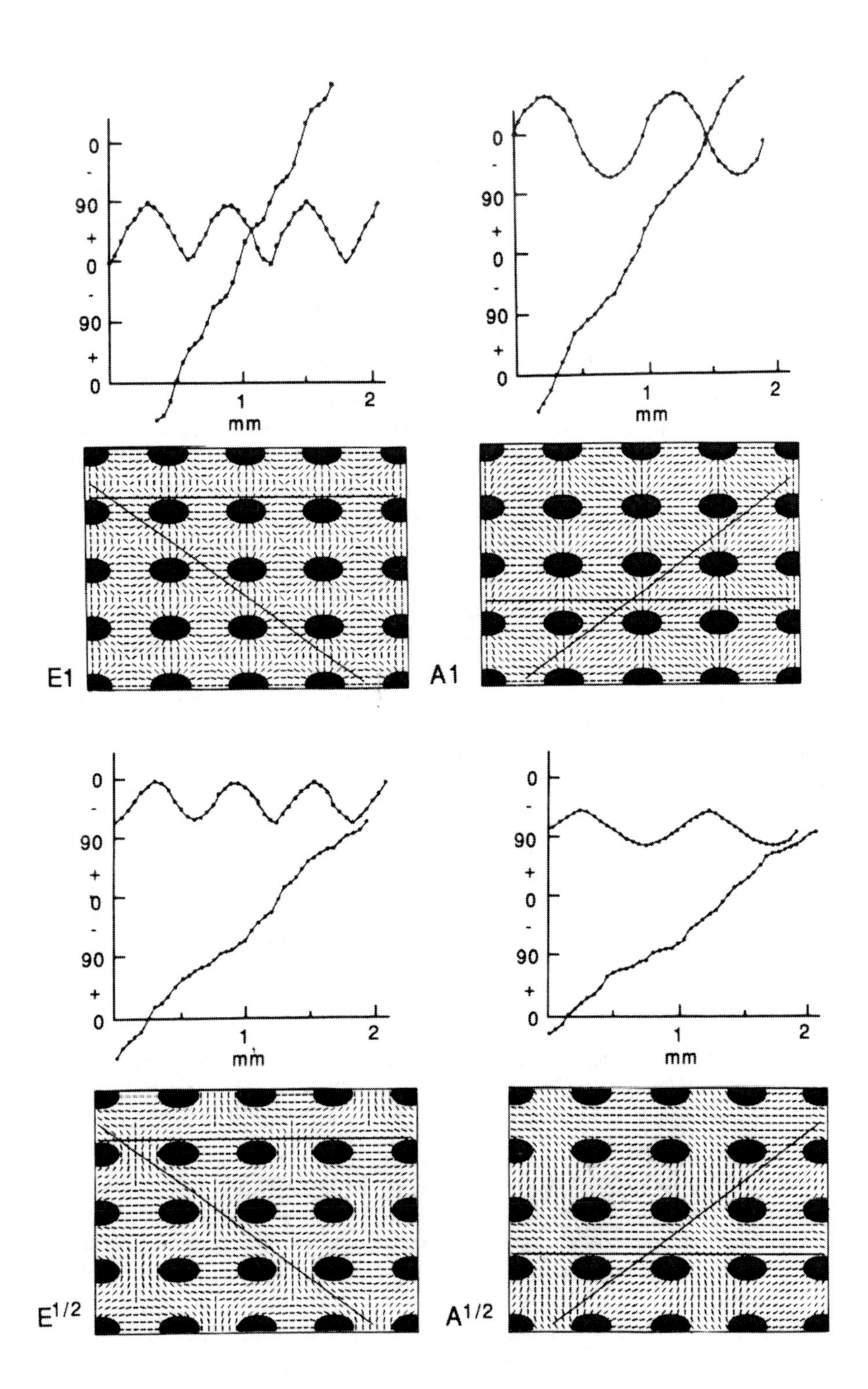
0
90
0
90
0
1
2
mm
E1
A1
E1/2
A1/2

others not illustrated here) it may be possible to select the best model. The problem remains unresolved at the present time.

Thus, the "parallel stripe" or "ice cube" orientation column model of Hubel and Wiesel is almost certainly incorrect. Another aspect of Hubel and Wiesel's orientation column model that has recently come into question is the issue of whether the columns extend through the entire cortical thickness. Several reports in both monkeys (Bauer et al., 1980, 1983; Kruger and Bach, 1982) and cats (Bauer, 1982, 1983; Lazareva et al., 1986; Bauer and Fischer, 1987; Bauer et al., 1989) have indicated that lower layer or "infragranular" cells and upper layer, "supragranular," cells participate in separate, often orthogonal, orientation column systems.

Murphy and Sillito (1986) were unable to find evidence for orientation "shifts" (i.e., major differences between upper and lower layer orientations) in electrode penetrations made orthogonal to the tissue layers in area 17 of the cat. In a recent report, Bauer's lab (Bauer et al., 1989) indicates finding single orientation "shifts" in 41% of such penetrations, with double shifts in 22%, and no shifts in 37% of penetrations. Examination of Murphy and Sillito's (1986) published data indicates the presence of some clear shifts, so the differences may relate to the actual number of such shifts, and possibly the level of anesthesia (Bauer and Fischer, 1987).

In the monkey, as mentioned above, the orientation columns are interrupted by zones in which the cells lack orientation selectivity: namely, the cytochrome oxidase patches. Both the cytochrome oxidase patches and the upper layer nonoriented zones identified on physiological grounds appear to occupy about one-third of the cortical surface area (Livingstone and Hubel, 1984; Dow and Vautin, 1987; Ts'o and Gilbert, 1988). Bauer et al. (1983) found an orientation shift in 70% of penetrations with oriented upper layer cells in monkey striate cortex, which would amount to about 47% of all penetrations, since only about two-thirds of all penetrations have oriented upper layer cells ($0.70 \times 2/3 = 0.47$).

A possible explanation of the orientation shift is that supragranular cells exhibit a "radial" orientation bias, such that their preferred orientation points to the center of gaze, while infragranular cells exhibit a "concentric" orientation bias, such that their preferred orientation runs tangent to a concentric circle around the center of gaze (Bauer and Dow, 1987, 1989). Such an arrangement in monkey recalls earlier reports by Leventhal and colleagues (Leventhal, 1983; Leventhal and Schall, 1983; Leventhal et al., 1984; Schall et al., 1986a,b) indicating a radial bias of retinal ganglion cells in cats and monkeys, and of visual cortical cells in cats.

The lower layer concentric orientation bias may in fact represent a radial bias for movement directions, since cortical cells typically respond to movement orthogonal to their preferred orientation (Hubel and Wiesel, 1962, 1968), and lower layer striate cells in both monkeys and cats tend to respond preferentially to movement (Dow, 1974; Hammond and MacKay, 1977; Hawken et al., 1988). The upper layer radial orientation bias is probably a true orientation bias (rather than a concentric direction bias), and may relate to binocular alignment of the eyes during fixation.

Thus, there are several indications that the parallel stripe ("ice cube") model (Fig. 4.7) of orientation architecture needs to be revised or replaced. Further studies using a combination of single cell recordings and either optical or histochemical labeling techniques (see above) will be required to resolve the issue.

Color

Color information is processed in striate cortex as well. The input derives ultimately from three classes of cone photoreceptors in the retina, whose light-absorption spectra as measured by Bowmaker and Dartnall (1980) are illustrated in Figure 2.2 of Chapter 2 in this volume. The first records of individual photoreceptor absorption were obtained in the early 1960s (Brown and Wald, 1964; Marks et al., 1964) by a technique known as microspectrophotometry, which involves the isolation of a single cone by manual dissection, followed by the passage of light through the outer segment of the isolated cone. The absorption spectrum thus derived represents the difference between the input and the output. What the subject "sees" is what the photoreceptor manages to extract from the light.

Cone responses have more recently been obtained electrophysiologically by means of a suction electrode (Baylor, 1987; Baylor et al., 1987). The results are consistent with the microspectrophotometry.

For purposes of comparison with cortical cell response properties in intact animals, it is necessary to correct the cone responses for light absorption by the lens, cornea, and macular pigment. The resulting curves agree closely with psychological functions obtained by Smith and Pokorny (1975) and Estevez (1979) (see Baylor et al., 1987).

The sum of the absorbencies of three cones provides a measure of our total visual sensitivity in daylight, referred to as "photopic" sensitivity (in contrast to "scotopic" sensitivity, mediated, at night, by rods). In fact psychophysical evidence suggests that photopic sensitivity is due largely to the long-wavelength-sensitive (L) and middle-wavelength-sensitive (M) cones, with only a small contribution from the short-wavelength-sensitive (S) cones (Tansley and Boynton, 1978; Eisner and McLeod, 1980; Boynton et al., 1985; Kaiser and Boynton, 1985). [The L, M, and S cone designations will be used here rather than R (red), G (green), and B (blue), to avoid confusion between types of cones and types of cortical color cells.]

Chromatic sensitivity, on the other hand, is thought to be due to subtraction of cone signals from one another. The most common type of subtraction involves the M and L cones, the predominant types in the primate retina. Subtraction of the form, +L-M, results in a cell that responds preferentially to red light, while subtraction of the form, +M-L, results in a cell that responds preferentially to green light. The other major type of subtraction – to generate yellow and blue – involves the S cones and either one or both of the other two cone classes. [There is some disagreement as to whether the L cones assist the M cones to produce yellow and blue, or whether the M cones act alone in opposing the S (and L) cones. Current evidence (see below) tends to support the latter arrangement, namely M/S (or M/SL) opponency.] Subtraction of the form, +S-M, results in a cell that responds preferentially to blue light, while subtraction of the form, +M-S, results in a cell that responds preferentially to yellow light. These are the four major types of color cells, sensitive to the "primary" Hering (1878) colors, red, yellow, green, and blue, that have been described in retina (DeMonasterio and Gouras, 1975), lateral geniculate nucleus (DeValois et al., 1966; Wiesel and Hubel, 1966; Dreher et al., 1976; Derrington et al., 1984; Lee et al., 1987), and striate cortex (Vautin and Dow, 1985).

Mean tuning curves of populations of red, yellow, green, and blue cells from foveal striate cortex are shown in Figure 4.11 (Vautin and Dow, 1985). Red and green tuning curves are shown as thick lines, yellow and blue tuning curves as thin lines. Dotted lines are extrapolations between spectral (left) and extraspectral (right) portions of the figure, with the extraspectral region represented as the percentage of 400 nm light in a mixture of 400 and 700 nm light. The figure is represented as a periodic function, the right and left margins continuous with one another. The shapes of the tuning curves in Figure 4.11 correspond fairly closely to the shapes of the tuning curves of psychophysical functions for the four colors red, yellow, green, and blue, obtained from human subjects (Werner and Wooten, 1979a,b). This correspondence suggests a process by which input from three classes of cones, with tuning curves plotted in wavelength units, is converted into the response patterns of four classes of primary color cells, with tuning curves plotted in units of psychophysical color perception.

The generation of such "primary color" cell responses, whether in striate cortex (Vautin and Dow, 1985) or in lateral geniculate nucleus (DeValois et al., 1966), is consistent with the Hering (1878) model of opponent color pairs (red/green, blue/yellow), just as the existence of three distinct types of cones is consistent with the Young–Helmholtz "trichromatic" model (Helmholtz, 1924) of color vision. There is thus no further need for these theories to be in conflict with one another. Both are in fact correct, though at different levels of visual processing.

Of crucial importance at the present time is to work out the details by which trichromacy is converted into opponency within the visual system. Preliminary work in this field by Vautin and Dow (1985) suggests that red and green cells are generated by means of roughly equal opponent inputs from L vs. M cones, while yellow and blue cells are generated by means of unequal opponent inputs from the M cones vs. the S and L cones (different proportions for yellow and blue).

In goldfish the opponency responsible for color vision appears to be generated within the retina itself, at the level of the horizontal and bipolar cells (Svaetichin, 1954; MacNichol and Svaetichin, 1958; Tomita, 1965; Kaneko, 1970; Stell and Lightfoot, 1975; Kaneko and Tachibana, 1981). In primates, however, there is an intervening step that may be necessary for high acuity form vision. The great majority of retinal ganglion cells in primates (there is essentially no information yet available on bipolar cells in primates) have concentric receptive fields with either an ON center and an OFF surround or an OFF center and an ON surround (DeMonasterio and Gouras, 1975). Chromatic ganglion cells, which are the most

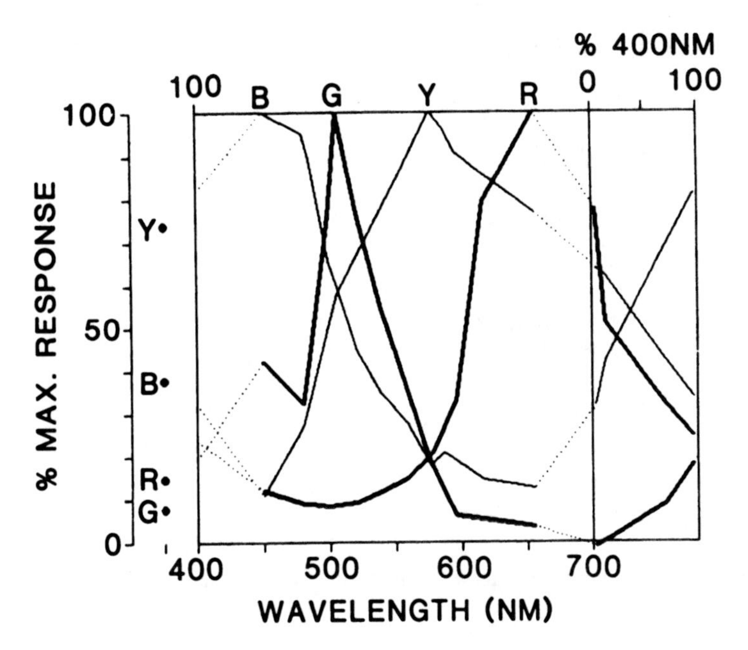

FIGURE 4.11. Normalized mean tuning curves for four classes of "primary" color cells in foveal striate cortex of the macaque. Right-most portion of the horizontal axis represents the extraspectral (purple) region, calibrated as the percentage of 400 nm light in a mixture of 400 and 700 nm light. Filled circles at the left indicate mean responses of the four cell classes to white light. (From Vautin and Dow, 1985.)

common type in the foveal region, receive their center input exclusively from a single class of cones (L, M, or S), and their surround input from a different cone class (or classes). Thus, there are L ON center cells with M OFF surrounds, M ON center cells with L OFF surrounds, M OFF center cells with L ON surrounds, and L OFF center cells with M ON surrounds (Fig. 4.12). In addition there are S ON center cells with either M or LM OFF surrounds. S OFF center cells are the least common of the six main types (Malpeli and Schiller, 1978; DeMonasterio, 1979; Zrenner and Gouras, 1981), though they do seem to exist, both in retina (DeMonasterio and Gouras, 1975) and in the lateral geniculate (Valberg et al., 1986).

The problem with this type of concentric color opponent receptive field organization is that the color opponency breaks down with small stimuli restricted to the receptive field center (Wiesel and Hubel, 1966). Such cells cannot, therefore, account for color vision when the colors are localized.

Cell properties within the lateral geniculate nucleus of primates are very similar to those exhibited by retinal ganglion cells. Chromatic responses are most characteristically seen in the parvocellular layers, and achromatic responses (LM ON or OFF center, with the opposite type of surround) in the magnocellular layers (Wiesel and Hubel, 1966; Derrington et al., 1984).

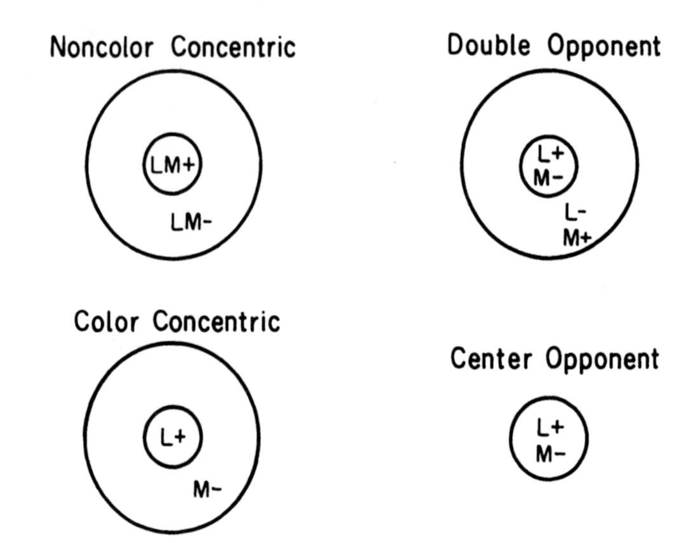

FIGURE 4.12. Cone inputs to nonoriented receptive fields in the macaque monkey visual system.

There is some tendency for the outer pair of parvocellular layers (6 and 5) to contain mostly ON center cells, and the inner pair of parvocellular layers (4 and 3) to contain mostly OFF center cells (Schiller and Malpeli, 1978; Michael, 1988), though the segregation is more complete in such carnivores as the mink (LeVay and McConnell, 1982) and ferret (Stryker and Zahs, 1983). There is also a tendency for S ON cells to be found in the inner pair of parvocellular layers, along with the L and M OFF center cells (Schiller and Malpeli, 1978).

Color cells were first reported in the striate cortex in the early 1960s (Motokawa et al., 1962; Lennox-Buchthal, 1962). Hubel and Wiesel (1968)

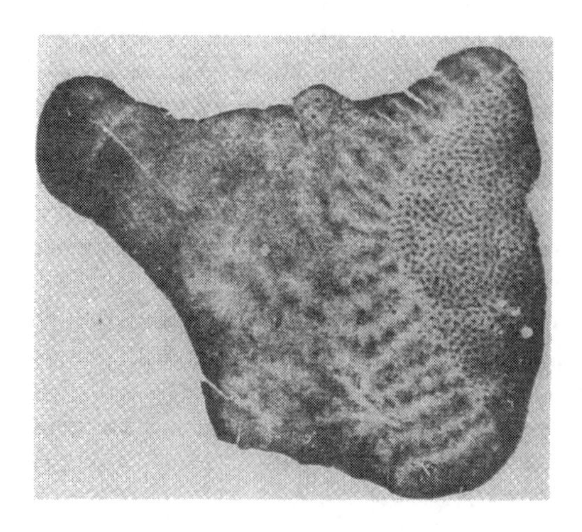

FIGURE 4.13. Cytochrome oxidase staining of a flat-mounted section from the lateral surface of squirrel monkey striate and prestriate cortex. Anterior is toward the left, dorsal toward the top. V1 is on the right (dots); V2 is on the left (black and white stripes). (From Tootell et al., 1983. Copyright 1983 by the AAAS.)

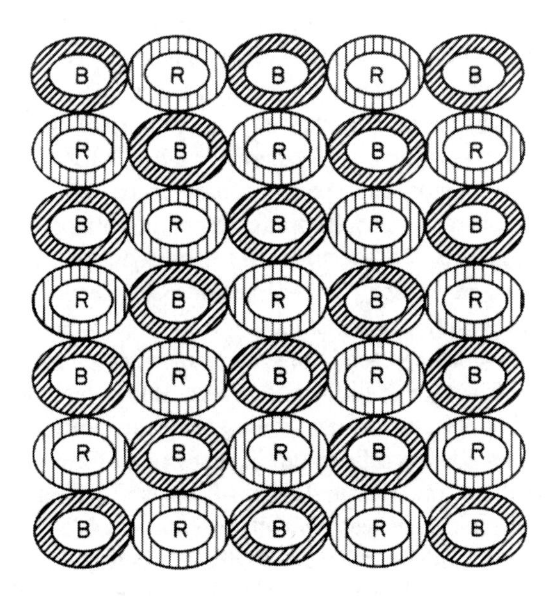

FIGURE 4.14. Model of the distribution of color cells in layer 4 of foveal striate cortex in macaque monkey as viewed from above. R and B indicate red and blue zones, respectively. Vertical and diagonal crosshatching indicate yellow (adjacent to red) and green (adjacent to blue) zones, respectively. Unlabeled regions indicate achromatic (white) zones. (From Dow and Vautin, 1987.)

found some clustering of color cells in color "columns." Dow (1974) found that cells in color columns tended to lack orientation selectivity, even in the upper layers of striate cortex. Michael (1981) reported separate red and green orientation columns, with achromatic orientation columns in between.

Of fundamental importance for color organization within primate striate cortex was the discovery in the early 1980s of a patchy distribution of cytochrome oxidase staining (see above) in the upper layers (2 and 3). The cytochrome oxidase (CO) patches or "puffs" or "blobs" (Fig. 4.13), as they are often called, are not evident with ordinary cell or fiber staining techniques. Cells within CO blobs lack orientation selectivity (Livingstone and Hubel, 1984), and about two-thirds of foveal striate blob cells show color selectivity, as compared with only one-third of upper layer cells outside of foveal striate blobs. There is evidence to suggest the existence of two different types of CO blobs (Dow and Vautin, 1987; Ts'o and Gilbert, 1988), one containing mostly red cells and the other mostly blue cells (Dow and Vautin, 1987; Tootell et al., 1988b). A checkerboard model (Fig. 4.14) has been proposed (Dow and Vautin, 1987), which would account for most if not all of the known data, including Michael's earlier report (1981) and a 2-deoxyglucose study by Crawford et al. (1982). The model proposes that red and blue cells are segregated in alternate blobs, with yellow, green, and achromatic cells in between.

Though there is evidence for distinct categories of red, yellow, green, and blue cells in striate cortex (Vautin and Dow, 1985), it is also apparent that color processing in striate cortex correlates with orientation selectivity in the form of a continuum, with midspectral cells showing the most orientation selectivity, and endspectral cells showing the least (Yoshioka et al., 1988).

A luminance correlation appears to be present as well. Midspectral cells require higher luminance levels for optimal responding than do endspectral cells (Yoshioka et al., 1990a,b). Endspectral cells respond quite well to stimuli that are equiluminant with a gray background, while midspectral cells fail to respond, or respond only very weakly, to such stimuli.

Figure 4.15 is a model representing the mapping of wavelength and luminance selectivity onto the surface of the striate cortex. Hue is represented as a plane with discrete foci of 400 and 700 nm, representing the locations of cytochrome oxidase "blobs" that contain cells exhibiting sensitivity for either short or long wavelengths (see Dow and Vautin, 1987). Midspectral wavelengths are represented in interblob regions, in keeping with the tendency of color cells with maximum sensitivity for midspectral wavelengths to show the highest degree of orientation selectivity (Yoshioka et al., 1990a,b). Other wavelengths are represented along a continuum between 400 and 500 nm, and between 700 and 550 nm, in keeping with the close correlation between hue preference and orientation selectivity (Yoshioka et al., 1990a,b). Luminance selectivity is represented as a third dimension, with a peak at 550 nm, in keeping with the close correlation between luminance threshold and hue selectivity, with midspectral cells showing high luminance thresholds and endspectral cells showing lower luminance thresholds (Yoshioka et al., 1990a,b).

There are several implications of this model. First, it suggests a mapping of the visible spectrum onto the cortical surface. The visible spectrum is clearly *not* mapped onto the retinal surface. Rather, information about the spectral distribution in any light stimulus is collected by three distinct cone channels (and one rod channel) and then relayed to central structures via the optic nerve. For the spectrum to be represented on the cortical surface requires a sophisticated arranging and recombining of the three cone channels. Such a mapping process is "computational" in the sense used by Knudsen et al. (1987) (see introduction). Vautin and Dow (1985) have suggested the ratios of cone inputs that would be needed to build cortical red, yellow, green, and blue cells, but there is no direct evidence at the present time as to the tangential distribution of the various types of parvocellular LGN axons as they synapse onto cortical cells. The model predicts that L and S ON center axons should project selectively to blob regions, in order to provide the inputs to short- and long-wavelength-sensitive cells located there. Vautin and Dow's (1985) data suggest that color cells with midspectral preferences should receive their excitatory input from the M ON center system of LGN and retina, and Dow and Vautin's (1987) model predicts that M ON center axons should project selectively to interblob regions.

There is not sufficient information at present in macaque monkeys to predict where the LGN OFF center systems (L, M, and S) should project. In mammals

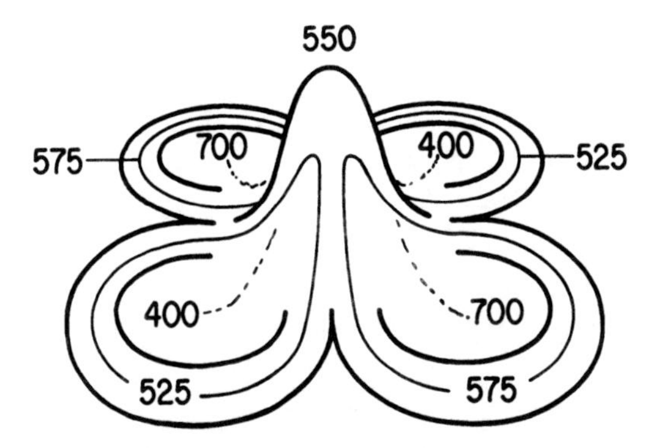

FIGURE 4.15. Three-dimensional version of the model shown in Figure 4.14, with wavelengths rather than colors; 400 nm and 700 nm are represented at the centers of alternating cytochrome oxidase "blobs," with 550 nm represented at the interblob center. The vertical axis represents luminance threshold, with blob cells having low thresholds, and responding well at equiluminance (in relation to a gray background), and interblob cells having high thresholds and requiring a substantial luminance contrast with the background.

with a high degree of segregation of ON and OFF systems in separate layers of the lateral geniculate nucleus (see above) there is evidence for separate ON and OFF columnar systems in primary visual cortex (McConnell and LeVay, 1984; LeVay et al., 1987; Zahs and Stryker, 1988).

If the model and the above line of reasoning are correct, then an interesting contrast can be made between the auditory and visual system mapping principles. In the auditory system sound pitch or frequency is "mapped" onto the cochlea (von Bekesy, 1960). Location of sound in space must be "computed" on the basis of differences in timing and loudness of the signals arriving at the two ears (Knudsen et al., 1987). In the visual system, location in space is mapped directly onto the retina. Light frequency (or its reciprocal, wavelength), after being broken down in the retina, must then be reconstituted in the cortex as a "computational" map. A recent study by Silverman et al. (1989) suggests that spatial frequency may also be mapped systematically across the striate cortex, with low spatial frequencies located near cytochrome oxidase blobs, and high spatial frequencies located at interblob centers.

The close association of luminance threshold and preferred hue implied by the model in Figure 4.15 is reminiscent of the locations of "focal" colors in three-dimensional color space, as shown in Figure 4.16, from the work of Berlin and Kay (1969). Note here that the white locus is at the highest brightness level, with yellow and orange slightly lower, and red and blue at about the middle of the brightness range. (See also Chapter 8 for a more complete discussion.) Figure 4.15 does not indicate where dark colors like black, brown, and purple (see Fig. 4.16) are located in the striate cortex, and is thus not a complete representation of three-dimensional color space, but its close association of luminance and hue seems to be a step in the direction of "surface colors" (which include a relative

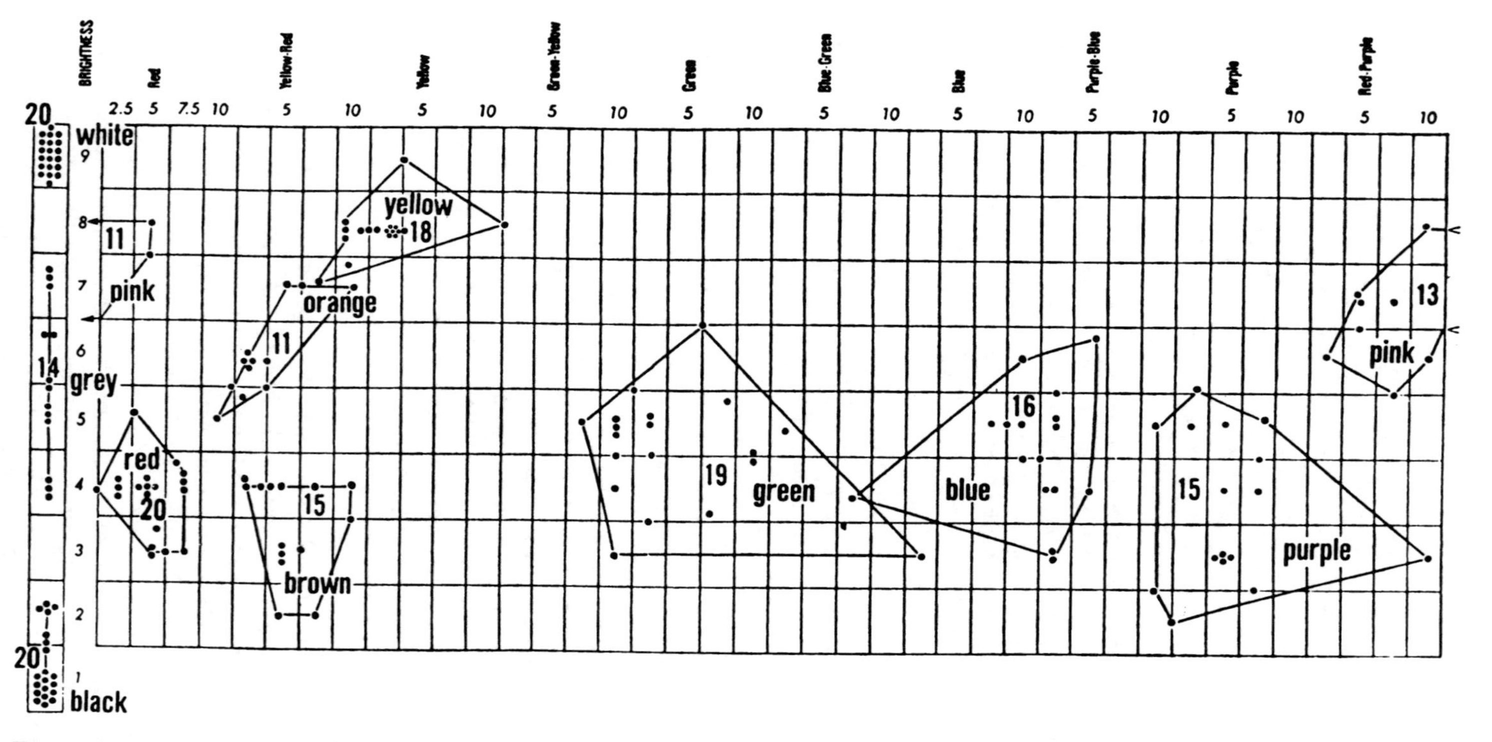

FIGURE 4.16. Normalized foci of basic color terms in 20 languages. Numerals at the top refer to Munsell color (hue) notations. Numerals in the brightness column at the left refer to Munsell brightness (value) notations. All colors (except achromatic ones) are at maximum Munsell saturation (chroma). Enclosed areas indicate the boundaries of named colors. Numerals within the boundaries indicate the number of languages that include that color category. The right and left ends of the chromatic portion of the chart are continuous with each other (and could be closed to form a cylinder, with the brightness axis in the center). (From Berlin and Kay, 1969.)

lightness parameter) and away from "aperture colors" (typically specified only by wavelength composition). Thus, if the model is correct, it suggests that the representation of surface colors is already underway in striate cortex.

Color information has been accumulating on areas V2 and V4 in recent years. Baizer et al. (1977) reported that V2 color cells lacked either orientation or direction selectivity, with orientation and direction cells likewise uniquely selective for their particular parameter. More recently it has been found that area V2 in squirrel monkeys and macaques shows its own distinct cytochrome oxidase staining pattern, consisting of a series of alternating light and dark stripes that run perpendicular to the border with area V1 (see Fig. 4.13). The pattern is most evident in squirrel monkeys whose largely lissencephalic (flat) cortical surface allows optimal visualization of areas V1 and V2. In macaques area V2 is mostly buried in the lunate sulcus and is consequently rather difficult to see. The human cortex is even more convoluted, though cytochrome oxidase staining of human cortex reveals V1 blobs (Horton and Hedley-Whyte, 1984).

Close examination of the V2 stripes in both squirrel monkeys and macaques reveals two distinct types of dark stripes, one thick and one thin, which alternate in a very regular pattern (see Fig. 4.13, to the left of the dots). The thin and thick CO stripes are more difficult to distinguish in macaque monkeys with ordinary staining procedures, but examination of serial sections reveals them quite clearly (Yoshioka et al., 1990a,b). They are also readily visible with the use of an antigen–antibody reaction involving "CAT-301," an antigen found in cat spinal cord, which turns out to be present in the thick but not the thin CO stripes of area V2 (Hendry et al., 1988) (see below for more information about this antibody).

Three groups (DeYoe and VanEssen, 1985; Shipp and Zeki, 1985; Hubel and Livingstone, 1987) have recently reported that cells in thin CO stripes of macaque V2 lack orientation selectivity in comparison to cells in interstripes or thick stripes. Many cells in CO thin stripes are reportedly color selective as well, which was to be expected, after Livingstone and Hubel (1984) found that the V1 blobs projected selectively to the V2 thin stripes, while the V1 interblobs projected selectively to the V2 interstripes. V2 thick stripes reportedly receive their input from layer 4B of striate cortex (Livingstone and Hubel, 1987a).

Yoshioka et al. (1990a,b) have further discovered that color cells in V2 thin stripes are typically selective for endspectral rather than midspectral colors, which is consistent with Dow and Vautin's (1987) report (confirmed and extended by Yoshioka et al., 1990a,b) that V1 blob cells are also selective for endspectral rather than midspectral colors.

As in area V1 (see above), the midspectral color cells in area V2 have distinctly higher luminance thresholds than the endspectral color cells in area V2 (Yoshioka et al., 1990a,b). The V2 color and orientation data of Yoshioka et al. (1990a,b) are summarized in Figure 4.17, which represents the CO stripes as crosshatched vertical bars, alternately thick and thin. The segregation of short- and long-wavelength-sensitive cells in different CO blobs of area V1 suggests a similar segregation of short- and long-wavelength regions within the thin stripes of area

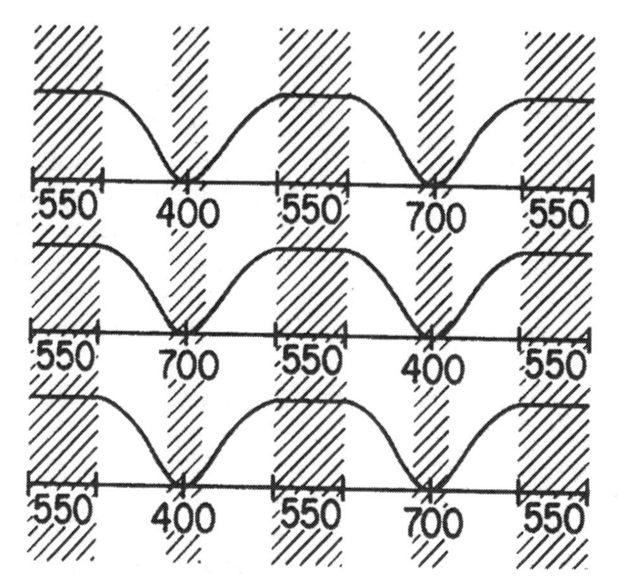

FIGURE 4.17. The model of Figure 4.15 as modified for area V2. Hatched areas represent cytochrome oxidase thin and thick stripes. White areas represent interstripes. The spectrum is represented in lines running perpendicular to the stripes, with the right (700 nm) and left (400 nm) ends located alternately within each thin stripe. The midspectral luminance threshold peak at 550 nm from area V1 has been broadened to a midspectral plateau in area V2, to make room for the largely achromatic thick stripe regions, within which the cells have been reported to show movement selectivity (Hubel and Livingstone, 1987). As indicated in Figure 4.23 the input to the thick stripes of area V2 comes from layer 4B of area V1 (not included in the model of Figure 4.15).

V2. One could postulate alternating red and blue thin stripes, but there is no evidence for this, and it would make the repeat distance for either red or blue unacceptably large. (The separation between alternate pairs of thin stripes is 4–6 mm.) If red and blue zones alternate within each stripe, as the model in Figure 4.17 suggests, and if midspectral colors are represented in the interstripes and thick stripes, as the data of Yoshioka et al. (1990a,b) indicate, then the model suggests one arrangement according to which the spectrum could be laid out on the V2 surface, with alternating right-to-left and left-to-right sequences. There are still other possible configurations, such as blue zones in one thin stripe lined up with blue zones in both neighboring thin stripes, and red zones aligned with red zones, but the configuration proposed in Figure 4.17 is the most parsimonious, and allows the shortest repeat distances.

The luminance thresholds of the cells with various wavelength preferences are represented in Figure 4.17 as topological lines with peaks at the interstripe–thick stripe junctions, analogous to the topological representation of the V1 interblob regions in Figure 4.15. The essential difference between the V1 map and the V2 map is that the thick stripes in V2 are interposed in the center of each interstripe region, thus forcing the luminance peaks of V1 to be broadened into plateaus.

Clearly the models presented in Figures 4.15 and 4.17 go beyond the existing evidence, but they are readily testable with established physiological and histochemical techniques.

Area V4 is likely to be involved in color processing as well, though not exclusively so, as had been initially claimed (Zeki, 1978, 1983). More likely, area V4 may represent the next stage in form and color processing after V2 (Tanaka et al., 1986; Desimone and Schein, 1987). Little is known at the present time about the internal organization of area V4, especially in relation to mapping. There is a crude retinotopic map (Maguire and Baizer, 1984), and there is preliminary evidence for columnar organization of color processing (T. Yoshioka, B.M. Dow, and R.G Vautin, unpublished).

Movement

In their earliest studies of visual cortex in both cats and monkeys, Hubel and Wiesel (1959, 1962) indicated that movement sensitivity was a common response property of single cells. Selectivity for direction of movement (Fig. 4.18), or "direction selectivity," which seems first to occur at the level of retinal ganglion cells in rabbits (Barlow et al., 1964; Barlow and Levick, 1965) and cats (Stone and Fukuda, 1974; Cleland and Levick, 1974), is not evident in primate retina. However, direction selectivity is present in primate striate cortex (Hubel and Wiesel, 1968; Wurtz, 1969), and is especially apparent in cells of layer 4B (Dow, 1974). Layer 4B is, in turn, the main source of input to area MT (Lund et al., 1975; Tigges et al., 1981), where most of the cells show direction selectivity (Dubner and Zeki, 1971; Maunsell and Van Essen, 1983a). Input to layer 4B comes from layer 4C alpha (Lund, 1988), the main terminus of the magnocellular axons from the LGN (Hubel and Wiesel, 1972).

Thus, there appears to be a separate magnocellular pathway for the processing of motion-related information (Livingstone and Hubel, 1987b). Layer 4B also projects to area V2 (see Martin, 1988, for a review), which contains a periodic pattern of cytochrome oxidase stripes, including thick stripes, thin stripes, and interstripes (see above). Thick stripes in area V2 receive their input from layer 4B of area V1 (Livingstone and Hubel, 1987a), and project to area MT (DeYoe and Van Essen, 1985; Shipp and Zeki, 1985). Cells in these stripes tend to be direction selective (Hubel and Livingstone, 1987).

The "magnocellular" pathway can be conveniently visualized through the use of an antigen–antibody reaction involving the CAT-301 antigen (see above). For some as yet unknown reason the magnocellular pathway contains high levels of this antigen. Processing of tissue slices with antibody to CAT-301 results in labeling of the magnocellular layers of the LGN, layers 4B and 5 of the striate cortex, the thick stripes of area V2, and all of area MT (Hendry et al., 1988). The only

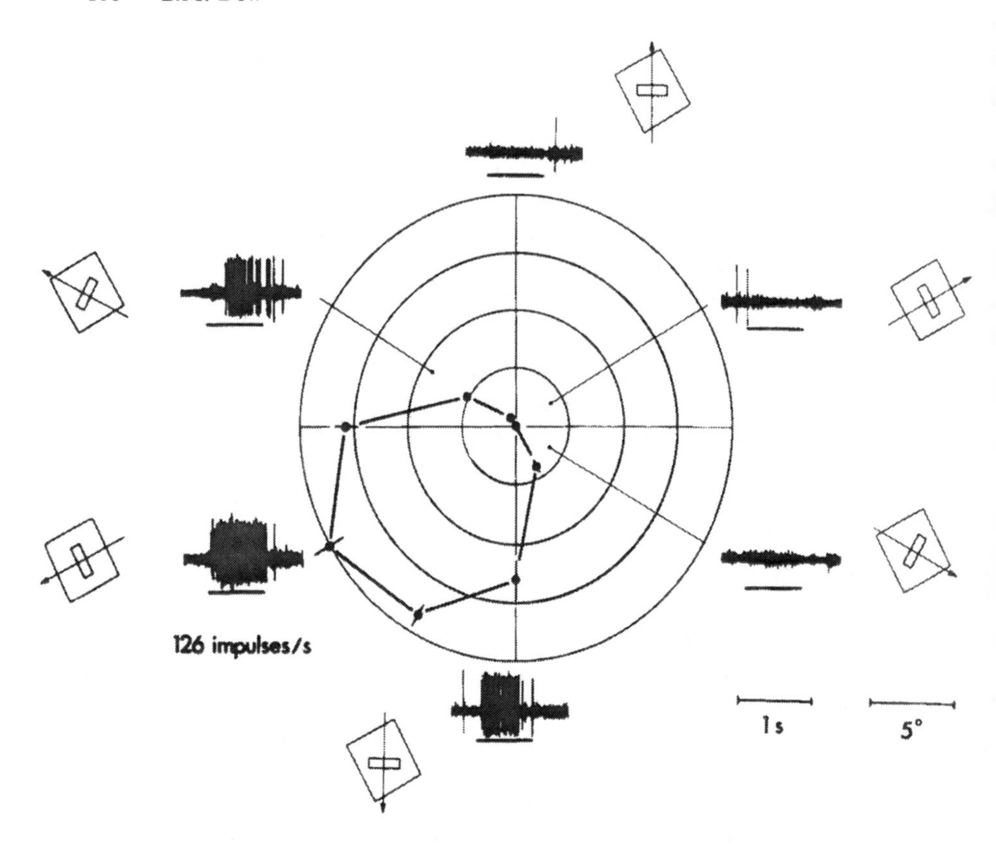

FIGURE 4.18. Direction selectivity of a single cell recorded from the middle temporal (MT) visual area of a macaque monkey. Oscilloscope records of extracellularly recorded responses are shown for individual presentations of the six indicated directions of motion. Bars below each trace mark the time the stimulus was on. The size of the stimulus and its direction of motion relative to the receptive-field outline are indicated alongside each trace. The polar plot is the average rate of firing during stimulus presentation for five repetitions of 12 directions of motion. Bars indicate the standard error of the mean of each point. (From Maunsell and Van Essen, 1983a.)

known component of the magnocellular pathway not to show label under these conditions is layer 4C alpha of striate cortex.

There is a tendency in all visual areas for cells with strong direction selectivity to be achromatic (Dow, 1974; Baizer et al., 1977; Zeki, 1978; Hubel and Livingstone, 1987). This is true in area V2, where color information is carried mostly in the thin dark stripes and interstripes (DeYoe and Van Essen, 1985; Shipp and Zeki, 1985; Hubel and Livingstone, 1987; Yoshioka et al., 1990a,b), while movement information is carried mostly in the thick dark stripes (DeYoe and Van Essen, 1985; Shipp and Zeki, 1985; Hubel and Livingstone, 1987). Direction selectivity is uncommon in area V4 (Zeki, 1978), to which the thin dark stripes project (DeYoe and Van Essen, 1985; Shipp and Zeki, 1985), while color selec-

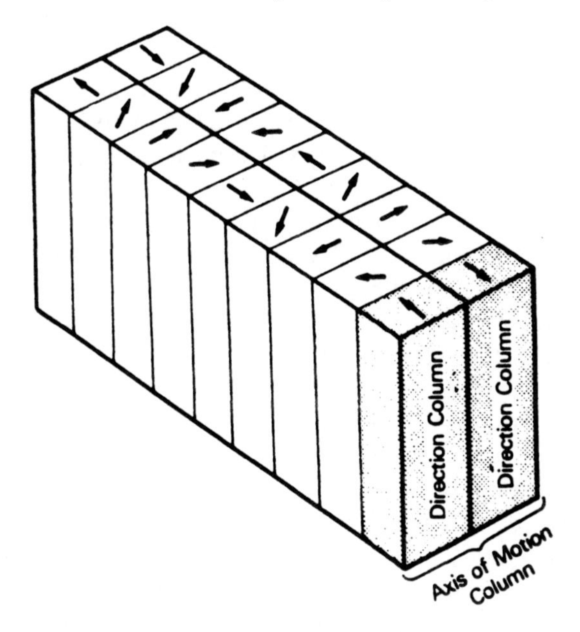

FIGURE 4.19. Model of direction columns in the middle temporal (MT) area of the macaque monkey. Vertical dimension represents cortical depth. Horizontal dimension represents cortical surface location. (From Albright et al., 1984.)

tivity is uncommon in area MT (Zeki, 1978), to which the thick dark stripes project (DeYoe and VanEssen, 1985; Shipp and Zeki, 1985). Thus, the motion and color pathways appear to be quite distinct from one another.

Motion processing seems to be the dominant function of area MT. Direction selectivity is present in 86% of all cells in area MT of the macaque (Maunsell and Van Essen, 1983a). A subgroup of these cells shows strongly enhanced responses during visual tracking (Erickson and Dow, 1989). Lesions of area MT interfere with both motion perception and visual tracking behavior (see Newsome and Wurtz, 1988, for a review).

There are indications that preferred direction is mapped onto area MT in an orderly fashion (Albright et al., 1984). Albright et al. have published a model for the directional map (Fig. 4.19), which nevertheless remains somewhat speculative at the present time. Opposing directions are mapped adjacent to one another along one axis of the map. Direction of motion varies systematically along the other axis.

Depth

Hubel and Wiesel (1962) recognized binocularity as a common property of cells in the cat's visual cortex, and initiated a classification system ranging from 1 to 7, where 1 and 7 indicated monocularity from one or the other eye, 4 indicated complete binocularity, and 2–3 and 5–6 indicated binocularity with dominance of one eye or the other. The existence of a specific neural mechanism for binocular depth perception seems to have originated in P.O. Bishop's laboratory in Australia, as part of Pettigrew's Ph.D. thesis. Several papers published by Pettigrew and collaborators (Barlow et al., 1967; Nikara et al., 1968; Pettigrew et al., 1968) in the late 1960s described the first depth-sensitive neurons. Since the cells were studied in paralyzed, anesthetized animals whose eyes were diverged from a normal fixation position, the cells were not actually involved in binocular vision during the experiment (though receptive fields could be superimposed with a prism). The crucial observation made by Pettigrew and his colleagues was that the monocular receptive field locations on the tangent screen were not always in correspondence (in relation to projections onto the tangent screen of known retinal landmarks like the optic disk or the area centralis). That is, there seemed to be "disparities" between the monocular receptive field locations in some cells. These disparities could not be explained by artifacts like residual eye movements, and were reminiscent of earlier proposals by Wheatstone, Ogle, and others (see Ogle, 1950; Schor and Ciuffreda, 1983), suggesting binocular disparity as the likely mechanism of binocular depth discrimination.

If one imagines a subject fixating on a target at C (Fig. 4.20), that target will be imaged at the center (i.e., the foveal center or foveola) of each retina. Other locations on the arc, AD, which is known as the "horopter" or "Vieth–Muller circle" (see, e.g., Schor and Ciuffreda, 1983), will be imaged at geometrically corresponding points on the two retinas, such as those labeled A and D. Locations in front of or behind the horopter (near, far) will be imaged at noncorresponding points on the retinas (C left and A right for "near," B left and C right for "far"). A cortical cell receiving its input from retinal locations C (left) and A (right) would have monocular receptive fields located at C (left eye) and A (right eye) on the horopter, while a cortical cell receiving its input from retinal locations B (left) and C (right) would have monocular receptive fields located at B (left eye) and C (right eye) on the horopter.

It is apparent from Figure 4.20 that far objects produce "uncrossed" receptive field disparities on the horopter (i.e., with the right eye's receptive field on the right), while near objects produce "crossed" horopter disparities (with the right eye's receptive field on the left).

Later studies by several laboratories (Hubel and Wiesel, 1970; Clarke et al., 1976; Poggio and Fischer, 1977; Ferster, 1981; Maunsell and Van Essen, 1983b; Hubel and Livingstone, 1987) confirmed the existence of binocular depth cells in several visual areas (V1, V2, V3?, and MT) in cats as well as monkeys. In monkey striate cortex Gian Poggio and his colleagues (Poggio and Fischer, 1977; Poggio, 1984; Poggio et al., 1985; see Poggio and Poggio, 1984, for a review) have found four major classes of binocular depth cells, which they call "near," "far," tuned excitatory, and tuned inhibitory. Sensitivity profiles for the four classes of cells in relation to three-dimensional depth in space are illustrated in Figure 4.21. Depth is represented along the horizontal axis in this figure in relation to an

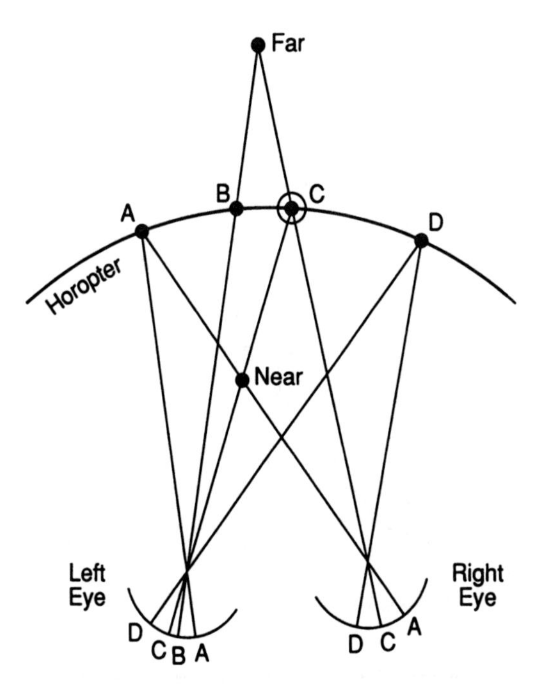

FIGURE 4.20. Geometry of binocular disparity in relation to depth vision. A, B, C, and D indicate objects located on the horopter and their images on the retinas of the two eyes. "Far" and "near" indicate objects located either behind (far) or in front of (near) the horopter. The "far" object is imaged at B (left eye) and C (right eye). The "near" object is imaged at A (right eye) and C (left eye). C (circled) is imaged at the foveal center of the two eyes.

observer located at the left. "Receptive fields" of tuned excitatory cells are located near the horopter (asterisk at the bottom), with near and far cell receptive fields located either closer or further away.

It is clearly of interest to know whether the different types of depth-selective cells are distributed in any organized manner across striate cortex. That is, one would like to know whether depth is "mapped" in striate cortex as other parameters seem to be. At present we have only some preliminary observations by Poggio (1984) that near/far and tuned inhibitory cells are found twice as frequently in the superficial as in the deep layers, while tuned excitatory cells (which tend to be directionally selective) are most commonly found in layers 4B and the layer 5/6 border.

Within area V2 in macaque monkeys the disparity-selective cells are found mostly in the thick CO stripes (Hubel and Livingstone, 1987), with some segregation of zero, near, and far cells. Earlier work in area V2 of the sheep (Clarke et al., 1976), whose wide-set eyes are optimal for studies of disparity, indicated a similar segregation of zero, near, and far groups in different "depth columns." A somewhat theoretical map of orthogonal orientation and disparity columns,

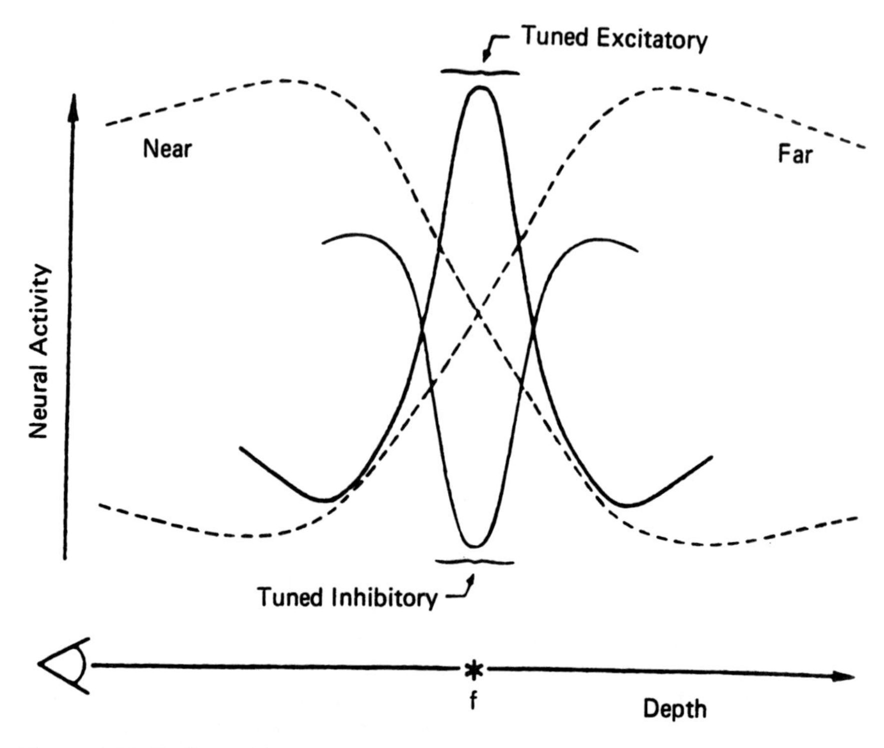

FIGURE 4.21. Outlines of the four basic disparity-sensitive profiles of stereoscopic neurons in visual cortex (areas V1 and V2) of the macaque monkey (from Poggio et al., 1985). The fixation plane (horopter) is indicated by the asterisk at the bottom. The observer is represented by the eye located at the bottom left. The horizontal axis thus represents distance from the observer. (Reprinted with permission from Poggio et al., copyright (1985), Pergamon Press PLC.)

based on the sheep data, is shown in Figure 4.22. Distance along the cortex is represented on the horizontal axis in Figure 4.22A and B, with crossed and uncrossed disparity zones alternating at intervals of about 750 μm (0.75 mm). Lateral movement through the tissue is thus equivalent to back and forth movement in depth along the trajectory ABCDEFGHIJKL indicated in Figure 4.22C.

FIGURE 4.22. Model of the distribution of binocular depth columns and orientation columns in area V2 of the sheep. Orientation columns are represented as parallel stripes along the vertical axis in A (according to the "ice cube" model of Fig. 4.7). Binocular disparity columns are represented as parallel stripes along the horizontal axis in A, with receptive field positions for the two eyes shown in B (upper), horizontal disparities between the two eyes shown in B (lower), and locations in three-dimensional space in C. (From Clarke et al., 1976.)

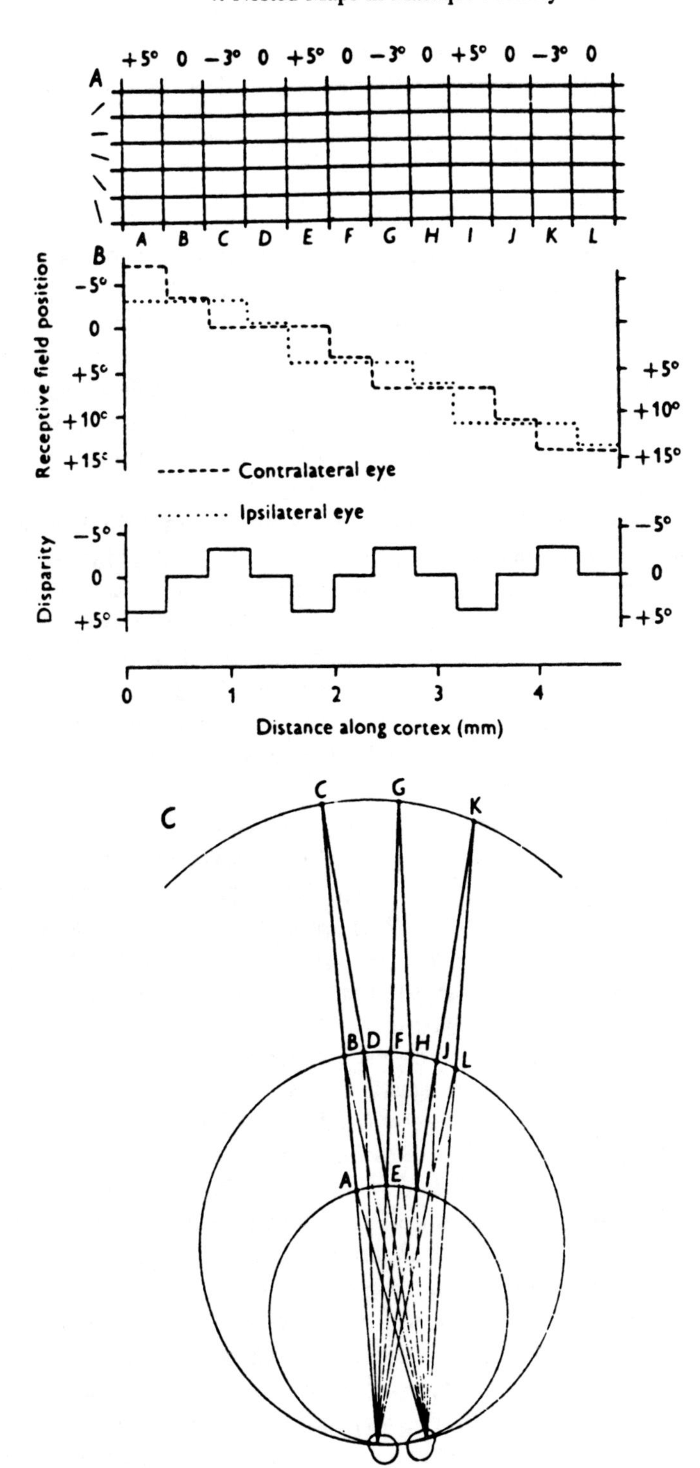
+5° 0 −3° 0 +5° 0 −3° 0 +5° 0 −3° 0
A
A B C D E F G H I J K L
B
Receptive field position
−5°
0
+5°
+10°
+15°
+5°
+10°
+15°
Contralateral eye
Ipsilateral eye
Disparity
−5°
0
+5°
−5°
0
+5°
0
1
2
3
4
Distance along cortex (mm)
C
C G K
B D F H J L
A E I

Disparity-selective cells have been reported in area MT (Maunsell and Van Essen, 1983b), but there is as yet no indication of any "depth map" in area MT.

Nested Maps

This chapter began with a consideration of retinotopy, the oldest and best known example of mapping in the visual system, and then proceeded to describe various models that have been presented for maps of particular visual parameters such as orientation, color, motion, and depth. Although it may be simpler to consider each parameter in isolation, the visual system seems somehow able to map all of them, sometimes even in the same tissue, as is apparently the case for area V1 and area V2. As our knowledge of cortical visual processing continues, we will need to understand how the different maps are nested together.

Livingstone and Hubel (1988) have presented a viewpoint that differs in some respects from the one outlined here. Figure 4.23 illustrates their model, with color localized in the CO blobs of area V1 and in the thin stripes of area V2, with form localized in the CO interblobs of area V1 and in the interstripes of V2, and with movement/stereo localized in layer 4B of area V1 and in the thick stripes of area V2. Such implied segregation of the different modalities contrasts with the present proposal, which suggests that each modality is mapped across the tissue surface. (The segregation model places all colors in each CO blob as a kind of "rainbow" color column).

The apparent restriction of certain colors (i.e., red and blue) to the CO blobs of area V1 and the CO thin stripes of area V2 (Dow and Vautin, 1987; Yoshioka et al., 1988, 1990a,b) provides evidence in favor of the present mapping scheme and against Livingstone and Hubel's (1988) segregation model. Orientation is clearly mapped within the striate cortex (see above). It remains to be determined whether motion and depth are mapped in striate cortex, as they appear to be mapped in extrastriate cortex (see above).

Clearly further data are needed. Horizontal penetrations across V1 and V2, with documentation of preferred color, orientation, depth, and movement direction at each recording site, should help to resolve the dispute. Global metabolic (Tootell et al., 1988a,b) and/or optical studies (Blasdel and Salama, 1986; Grinvald et al., 1988), and computer simulations (Baxter and Dow, 1988, 1989), combined with microelectrode recordings, will be valuable as well. One needs to know not simply where color, form, motion, and depth are processed, but where particular colors, orientations, movement directions, and depths are processed.

If each modality is separately mapped, as suggested here, then information processing would be greatly facilitated, since any particular piece of information (e.g., my blue pen moving to the right across the page as I write these words) would be specifiable as a series of addresses in different maps: "blue" in the color map, "45° clockwise from vertical" in the orientation map, "rightward" in the movement map, "near" in the depth map, and "right horizontal meridian at 0–10°" (depending on how far away I hold my head) in the retinotopic map.

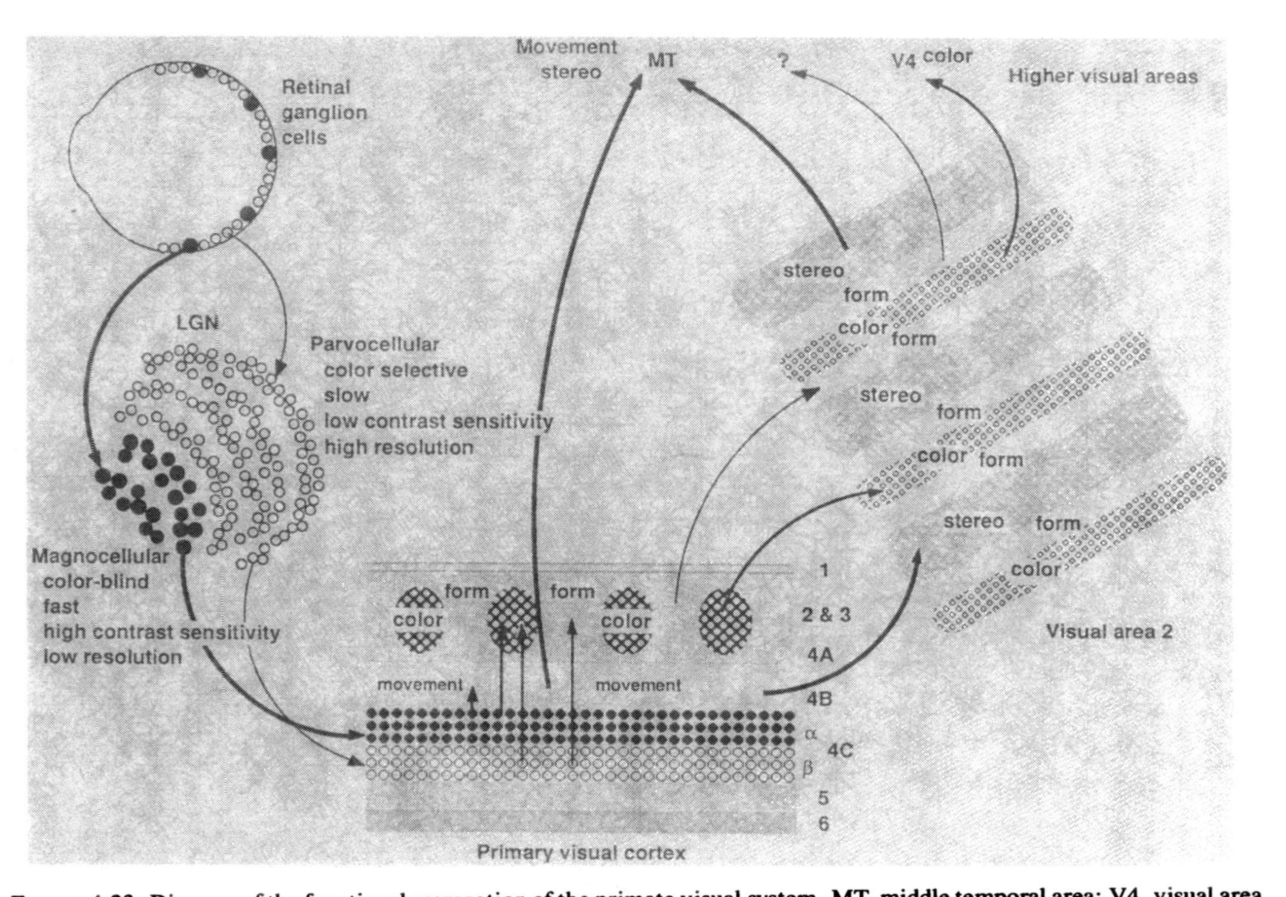

FIGURE 4.23. Diagram of the functional segregation of the primate visual system. MT, middle temporal area; V4, visual area 4; LGN, lateral geniculate nucleus. (From Livingstone and Hubel, 1988. Copyright 1988 by the AAAS.)

This raises the issue as to whether five different locations in the tissue will be specified by these five addresses, or whether somehow a single unique address would suffice. For a single address to suffice, the different maps would ideally need to be orthogonal to one another, to permit blue, for example, to be associated with all possible orientations, directions, depths, and positions. The "ice cube" model in Figure 4.7 has this feature for two parameters, ocular dominance and orientation. The addition of color creates problems for the ice cube model. However, the color model in Figure 4.14 is compatible with the radial orientation models in Figure 4.9, in the sense that colors are represented as concentric rings around cytochrome oxidase nodes in Figure 4.14, while orientations are represented as spokes radiating out from cytochrome oxidase nodes in Figure 4.9. Orientation spokes and concentric color rings would have the desired orthogonality to one another.

Motion and depth would then (ideally) have to be added in such a way as to be orthogonal to color and orientation. In fact, though, only two directions of motion need be specified for each orientation and color, and only three depths (near, far, and zero). These could be provided within layer 4B, or in layers 4B and 6, both of which project to area MT (Lund et al., 1975; Tigges et al., 1981). Layer 4B projects to the thick CO stripes of area V2, as previously mentioned.

Position is handled by simply expanding the cortical representation so that a retinal point maps onto a cortical area, referred to as a "hypercolumn" (Hubel and Wiesel, 1974b), within which all the other parameters are then systematically mapped. The size of hypercolumns varies with eccentricity (Dow et al., 1981; Van Essen et al., 1984; Gattass et al., 1987), being substantially larger in the fovea than in the parafovea, and larger in the parafovea than in the periphery. The concept of "nested" mappings proposed here may explain why V1 hypercolumns are larger in the fovea and smaller in the periphery. Sufficient tissue area is required to allow accurate high resolution addressability for all parameters in foveal striate cortex (and probably in area V2 as well). As resolution requirements for all parameters fall off at higher eccentricities, hypercolumn size diminishes as well.

The concept presented here, then, is that any given stimulus configuration consisting of a particular color, shape, depth, and motion direction seen at a particular spatial locus will have a unique address in striate cortex. The address is not the site of a "grandmother cell" selective for all of the stimulus features, but is the center of a number of different distribution functions, one for each parameter. In such a system any particular cell can in fact be fairly broadly tuned, and thus active much of the time. No particular cell is essential to the functioning of the system. The combined activity of the population is what carries the information.

Such a system of precise addressability within the context of nested mappings is useful for input purposes, as a means of providing "labeled lines" for visual perception. Addressability is also potentially useful for retrieval mechanisms involving selective attention and memory (Dow et al., 1987a,b).

At visual processing levels beyond area V2 there appears to be further specialization. For example, only location, movement, and perhaps depth are likely to be mapped in area MT, and only location, color, and perhaps orientation are likely to be mapped in area V4. It will clearly be of interest, as data accumulate, to compare the mapping principles in these areas with those in areas V1 and V2.

References

Albright TD, Desimone R, Gross CG (1984). Columnar organization of directionally selective cells in visual area MT of the macaque. J. Neurophysiol. **51**:16–31.

Allman JM, Kaas JH (1971). Representation of the visual field in striate and adjoining cortex of the owl monkey (*Aotus trivirgatus*). Brain Res. **35**:89–106.

Baizer JS, Robinson DL, Dow BM (1977). Visual responses of area 18 neurons in awake, behaving monkey. J. Neurophysiol. **40**:1024–1037.

Banks MS, Aslin RN, Letson RD (1975). Sensitive period for the development of human binocular vision. Science **190**:675–677.

Barlow HB, Levick WR (1965). The mechanism of directionally selective units in rabbit's retina. J. Physiol. **178**:477–504.

Barlow, HB, Hill RM, Levick WR (1964). Retinal ganglion cells responding selectively to direction and speed of image motion in the rabbit. J. Physiol. **173**:377–407.

Barlow HB, Blakemore C, Pettigrew JD (1967). The neural mechanism of binocular depth discrimination. J. Physiol. **193**:327–342.

Bauer R (1982). A high probability of an orientation shift between layers 4 and 5 in central parts of the cat striate cortex. Exp. Brain Res. **48**:245–255.

Bauer R (1983). Differences in orientation and receptive field position between supra- and infragranular cells of cat striate cortex and their possible functional implications. Biol. Cybern. **49**:137–148.

Bauer R, Dow BM (1987). Global radial and concentric maps of orientation coding in upper and lower layers of striate cortex in monkey and cat. Neuroscience **22**(Suppl):430.

Bauer R, Dow B (1989). Complementary global maps for orientation coding in upper and lower layers of the monkey's foveal striate cortex. Exp. Brain Res. **76**:503–509.

Bauer R, Fischer WH (1987). Continuity or incontinuity of orientation columns in visual cortex: A critical evaluation of published and unpublished data. Neuroscience **22**: 841–847.

Bauer R, Dow BM, Vautin RG (1980). Laminar distribution of preferred orientations in foveal striate cortex of the monkey. Exp. Brain Res. **41**:54–60.

Bauer R, Dow BM, Snyder AZ, Vautin R (1983). Orientation shift between upper and lower layers in monkey visual cortex. Exp. Brain Res. **50**:133–145.

Bauer R, Eckhorn R, Jordan W (1989). Iso- and cross-orientation columns in cat striate cortex: A re-examination with simultaneous single- and multi-unit recordings. Neuroscience **30**:733–740.

Baxter W, Dow BM (1988). Centric models of the orientation map in primate visual cortex. In *Neural Information Processing Systems—Natural and Synthetic* (DZ Anderson, ed.), pp 62–72. American Institute of Physics Press, New York.

Baxter WT, Dow BM (1989). Horizontal organization of orientation-sensitive cells in primate visual cortex. Biol. Cybern. **61**:171–182.

Baylor DA (1987). Photoreceptor signals and vision. Invest Opthalmol. Vis. Sci. **28**: 34–49.

Baylor DA, Nunn BJ, Schnapf JL (1987). Spectral sensitivity of cones of the monkey macaca fascicularis. J. Physiol. **390**:145–160.

Berlin B, Kay P (1969). *Basic Color Terms*. University of California Press, Berkeley.

Blasdel GG, Salama G (1986). Voltage-sensitive dyes reveal a modular organization in monkey striate cortex. Nature (London) **321**:579–585.

Bowmaker JK, Dartnall HJA (1980). Visual pigments of rods and cones in a human retina. J. Physiol. **298**:501–511.

Boynton RM, Olson CX (1987). Locating basic colors in the OSA space. Color Res. Appl. **12**:94–105.

Boynton RM, Eskew RT Jr, Olson CX (1985). Blue cones contribute to border distinctness. Vision Res. **25**:1349–1352.

Braitenberg V, Braitenberg C (1979). Geometry of orientation columns in visual cortex. Biol. Cybern. **33**:179–186.

Brazier MAB, Petsche H (1978). *Architectonics of the Cerebral Cortex*. IBRO Monograph Series. Raven, New York.

Brown PK, Wald G (1964). Visual pigments in single rods and cones of the human retina. Science **144**:45–51.

Carroll EW, Wong-Riley MTT (1984). Quantitative light and electron microscopic analysis of cytochrome oxidase-rich zones in the striate cortex of the squirrel monkey. J. Comp. Neurol. **222**:1–17.

Clarke PGH, Donaldson IML, Whitteridge D (1976). Binocular visual mechanisms in cortical areas I and II of the sheep. J. Physiol. **256**:509–526.

Cleland BG, Levick WR (1974). Properties of rarely encountered types of ganglion cells in the cat's retina and an overall classification. J. Physiol. **240**:457–492.

Crawford MLJ, Meharg LS, Johnston DA (1982). Structure of columns in monkey striate cortex induced by luminant-contrast and color-contrast stimulation. Proc. Natl. Acad. Sci. U.S.A. **79**:6722–6726.

Daniel PM, Whitteridge D (1961). The representation of the visual field on the cerebral cortex in monkeys. J. Physiol. **159**:203–221.

DeMonasterio FM (1979). Asymmetry of ON- and OFF-pathways of blue-sensitive cones of the retina of macaques. Brain Res. **166**:39–48.

DeMonasterio FM, Gouras P (1975). Functional properties of ganglion cells of the rhesus monkey retina. J. Physiol. **251**:167–195.

Derrington AM, Krauskopf J, Lennie P (1984). Chromatic mechanisms in lateral geniculate nucleus of macaque. J. Physiol. **357**:241–265.

Desimone R, Schein SJ (1987). Visual properties of neurons in area V4 of the macaque: Sensitivity to stimulus form. J. Neurophysiol. **57**:835–868.

DeValois RL, Abramov I, Jacobs GH (1966). Analysis of response patterns of LGN cells. J. Opt. Soc. Am. **56**:966–977.

DeYoe EA, VanEssen DC (1985). Segregation of efferent connections and receptive field properties in visual area V2 of the macaque. Nature (London) **317**:58–61.

Dow BM (1974). Functional classes of cells and their laminar distribution in monkey visual cortex. J. Neurophysiol. **37**:927–946.

Dow BM, Bauer R (1984). Retinotopy and orientation columns in the monkey: A new model. Biol. Cybern. **49**:189–200.

Dow BM, Vautin RG (1987). Horizontal segregation of color information in the middle layers of foveal striate cortex. J. Neurophysiol. **57**:712–739.

Dow BM, Snyder AZ, Vautin RG, Bauer R (1981). Magnification factor and receptive field size in foveal striate cortex of the monkey. Exp. Brain Res. **44**:213–228.

Dow BM, Bauer R, Snyder AZ, Vautin RG (1984). Receptive fields and orientation shifts in foveal striate cortex of the awake monkey. In *Dynamic Aspects of Neocortical Function* (GM Edelman, WE Gall, WM Cowan, eds.), pp. 41–65. Wiley, New York.

Dow BM, Vautin RG, Bauer R (1985). The mapping of visual space onto foveal striate cortex in the macaque monkey. J. Neurosci. **5**:890–902.

Dow BM, Yoshioka T, Vautin RG (1987a). Change in responsiveness of macaque V1, V2 and V4 cells in relation to fixation point color during a color matching task. Invest. Ophthalmol. Vis. Sci. 28 (Suppl):196.

Dow BM, Yoshioka T, Vautin RG (1987b). Selective suppression of responses of foveal V1, V2 and V4 cells in the macaque during a color matching task. Soc. Neurosci. Abst. **13**:624.

Dreher B, Fukada Y, Rodieck RW (1976). Identification, classification and anatomical segregation of cells with x-like and y-like properties in the lateral geniculate nucleus of old-world primates. J. Physiol. **258**:433–452.

Dubner R, Zeki SM (1971). Response properties and receptive fields of cells in an anatomically defined region of the superior temporal sulcus in the monkey. Brain Res. **35**:528–532.

Eisner A, MacLeod IA (1980). Blue-sensitive cones do not contribute to luminance. J. Opt. Soc. Am. **70**:121–123.

Erickson RG, Dow BM (1989). Foveal tracking cells in the superior temporal sulcus of the macaque monkey. Exp. Brain Res. **78**:113–131.

Estevez O (1979). On the fundamental data-base of normal and dichromatic color vision. Ph.D. Thesis, University of Amsterdam, Krips Repro Meppel Amsterdam.

Ferster D (1981). A comparison of binocular depth mechanisms in areas 17 and 18 of the cat visual cortex. J. Physiol. **311**:623–655.

Gattass R, Gross CG (1981). Visual topography of striate projection zone (MT) in posterior superior temporal sulcus of the macaque. J. Neurophysiol. **46**:621–638.

Gattass R, Gross CG, Sandell JH (1981). Visual topography of V2 in the macaque. J. Comp. Neurol. **201**:519–539.

Gattass R, Sousa APB, Rosa MGP (1987). Visual topography of V1 in the Cebus monkey. J. Comp. Neurol. **259**:529–548.

Gattass R, Sousa APB, and Gross CG (1988). Visuotopic organization and extent of V3 and V4 of the macaque. J. Neurosci. **8**:1831–1845.

Gotz KG (1987). Do "d-blob" and "1-blob" hypercolumns tessellate the monkey visual cortex? Biol. Cybern. **56**:107–109.

Gotz KG (1988). Cortical templates for the self-orientation of orientation-specific d- and 1-hypercolumns in monkeys and cats. Biol. Cybern. **58**:213–223.

Grinvald A, Frostig RD, Lieke E, Hildesheim R (1988). Optical imaging of neuronal activity. Physiol. Rev. **68**:1285–1366.

Guld C, Bertulis A (1976). Representation of fovea in the striate cortex of vervet monkey, cercopithecus aethiops pygerythrus. Vision Res. **16**:629–631.

Hammond P, MacKay DM (1977). Differential responsiveness of simple and complex cells in cat striate cortex to visual texture. Exp. Brain Res. **30**:P275–296.

Hawken MJ, Parker AJ, Lund JS (1988). Laminar organization and contrast sensitivity of direction-selective cells in the striate cortex of the Old World monkey. J. Neurosci. **8**: 3541–3548.

Helmholtz HV (1924). *Physiological Optics*, (JPC Southall, ed.), Optical Society of America, Rochester, NY (reprinted, Dover, New York, 1962).

Hendrickson AE, Hunt SP, Wu JY (1981). Immunocytochemical localization of glutamic acid decarboxylase in monkey striate cortex. Nature (London) **292**:605–607.

Hendry SHC, Jones EG, Hockfield S, McKay RDG (1988). Neuronal populations stained with the monoclonal antibody cat-301 in the mammalian cerebral cortex and thalamus. J. Neurosci. **8**:518–542.

Hering E (1878). *Outlines of a Theory of the Light Sense* (LM Hurvich and D Jameson, trans., 1964). Harvard University Press, Cambridge, MA.

Horton JC (1984). Cytochrome oxidase patches: A new cytoarchitectonic feature of monkey visual cortex. Phil. Trans. R. Soc. London **304**:199–253.

Horton JC, Hedley-Whyte ET (1984). Mapping of cytochrome oxidase patches and ocular dominance columns in human visual cortex. Phil. Trans. R. Soc. London B **304**: 255–272.

Horton JC, Hubel DH (1981). Regular patchy distribution of cytochrome oxidase staining in primary visual cortex of macaque monkey. Nature (London) **292**:762–764.

Hubel DH (1982). Exploration of the primary visual cortex, 1955–78. Nature (London) **299**:515–524.

Hubel DH (1988). *Eye, Brain and Vision*. Freeman, New York (Scientific American Library).

Hubel DH, Livingstone MS (1987). Segregation of form, color and stereopsis in primate area 18. J. Neurosci. **7**:3378–3415.

Hubel DH, Wiesel TN (1959). Receptive fields of single neurons in the cat's striate cortex. J. Physiol. (London) **148**:574–591.

Hubel DH, Wiesel TN (1962). Receptive fields, binocular interaction and functional architecture in the cat's visual cortex. J. Physiol. **160**:106–154.

Hubel DH, Wiesel TN (1963). Shape and arrangement of columns in cat's striate cortex. J. Physiol. **165**:559–568.

Hubel DH, Wiesel TN (1965). Receptive fields and functional architecture in two non-striate visual areas (18 and 190) of the cat. J. Neurophysiol. **28**:229–289.

Hubel DH, Wiesel TN (1968). Receptive fields and functional architecture of monkey striate cortex. J. Physiol. (London) **195**:215–243.

Hubel DH, Wiesel TN (1970). Stereoscopic vision in macaque monkey. Nature (London) **225**:41–42.

Hubel DH, Wiesel TN (1972). Laminar and columnar distribution of geniculo-cortical fibers in the macaque monkey. J. Comp. Neurol. **146**:421–450.

Hubel DH, Wiesel TN (1974a). Sequence regularity and geometry of orientation columns in the monkey striate cortex. J. Comp. Neurol. **158**:267–294.

Hubel DH, Wiesel TN (1974b). Uniformity of monkey striate cortex: A parallel relationship between field size, scatter, and magnification factor. J. Comp. Neurol. **158**:295–306.

Hubel DH, Wiesel TN (1977). Functional architecture of macaque monkey visual cortex. Proc. R. Soc. London (Biol.) **198**:1–59.

Hubel DH, Wiesel TN, LeVay S (1977). Plasticity of ocular dominance columns in monkey striate cortex. Phil. Trans. R. Soc. London B **278**:377–409.

Hubel DH, Wiesel TN, Stryker MP (1978). Anatomical demonstration of orientation columns in macaque monkey. J. Comp. Neurol. **177**:361–379.

Kaiser PK, Boynton RM (1985). Role of the blue mechanism in wavelength discrimination. Vision Res. **25**:523–529.

Kaneko A (1970). Physiological and morphological identification of horizontal, bipolar and amacrine cells in goldfish retina. J. Physiol. **207**:623–633.

Kaneko A, Tachibana M (1981). Retinal bipolar cells with double colour-opponent receptive fields. Nature (London) **293**:220–222.

Knudsen EI, duLac S, Esterly SD (1987). Computational maps in the brain. Annu. Rev. Neurosci. **10**:41–65.

Kolb H (1970). Organization of the outer plexiform layer of the primate retina: Electron microscopy of Golgi-impregnated cells. Phil. Trans. R. Soc. London B **258**:261–283.

Kolb H, DeKorver L (1988). Synaptic input to midget ganglion cells of the human retina. Invest. Ophthalmol. Vis. Sci. **29** (Suppl):326.

Konishi M (1986). Centrally synthesized maps of sensory space. Trends Neurosci. **9**: 163–168.

Kruger J, Bach M (1982). Independent systems of orientation columns in upper and lower layers of monkey visual cortex. Neurosci. Lett. **31**:255–230.

Lazareva NA, Novikova RV, Tikhomirov AS, Shevelev IA (1986). Different properties of two groups of orientation discriminators in the cat visual cortex. Neurophysiologia **18**:68–74.

Lee BB, Valberg A, Tigwell DA, Tryti J (1987). An account of responses of spectrally opponent neurons in macaque lateral geniculate nucleus to successive contrast. Proc. R. Soc. London B **230**:293–314.

Lennox-Buchthal MS (1962). Single units in monkey, *Cercocebus torquatus atys*, cortex with narrow spectral responsiveness. Vision Res. **2**:1–15.

LeVay S, Hubel DH, Wiesel TN (1975). The pattern of ocular dominance columns in macaque visual cortex revealed by a reduced silver stain. J. Comp. Neurol. **159**:559–575.

LeVay S, Wiesel TN, Hubel DH (1980). The development of ocular dominance columns in normal and visually deprived monkeys. J. Comp. Neurol. **191**:1–51.

LeVay S, McConnell SK (1982). ON and OFF layers in the lateral geniculate nucleus of the mink. Nature (London) **300**:350–351.

LeVay S, McConnell SK, Luskin MB (1987). Functional organization of primary visual cortex in the mink (*Mustela vison*), and a comparison with the cat. J. Comp. Neurol. **257**:422–441.

Leventhal AG (1983). Relationship between preferred orientation and receptive field position of neurons in cat striate cortex. J. Comp. Neurol. **220**:476–483.

Leventhal AG, Schall JD (1983). Structural basis of orientation sensitivity of cat retinal ganglion cells. J. Comp. Neurol. **220**:465–475.

Leventhal AG, Schall JD, Wallace W (1984). Relationship between preferred orientation and receptive field position of neurons in extrastriate cortex (area 19) in the cat. J. Comp. Neurol. **222**:445–451.

Livingstone MS, Hubel DH (1984). Anatomy and physiology of a color system in the primate visual cortex. J. Neurosci. **4**:309–356.

Livingstone MS, Hubel DH (1987a). Connections between layer 4B of area 17 and thick cytochrome oxidase stripes of area 18 in the squirrel monkey. J. Neurosci. **7**:3371–3377.

Livingstone MS, Hubel DH (1987b). Psychophysical evidence for separate channels for the perception of form, color, movement, and depth. J. Neurosci. **7**:3416–3468.

Livingstone MS, Hubel DH (1988). Segregation of form, color, movement, and depth: Anatomy, physiology, and perception. Science **240**:740–749.

Lund JS (1988). Anatomical organization of macaque monkey striate visual cortex. Annu. Rev. Neurosci. **11**:253–288.

Lund JS, Lund RD, Hendrickson AE, Bunt AH, Fuchs AF (1975). The origin of efferent pathways from the primary visual cortex, area 17, of the macaque monkey as shown by retrograde transport of horseradish peroxidase. J. Comp. Neurol. **164**:287–304.

MacNichol EF Jr, Svaetichin G (1958). Electric responses from the isolated retinas of fishes. Am. J. Ophthalmol. **46**:26–46.

Maguire WM, Baizer JS (1984). Visuotopic organization of the prelunate gyrus in rhesus monkey. J. Neurosci. **4**:1690–1704.

Malpeli JG, Baker FH (1975). The representation of the visual field in the lateral geniculate nucleus of *Macaca mulatta*. J. Comp. Neurol. **161**:569–594.
Malpeli JG, Schiller PH (1978). Lack of blue OFF-center cells in the visual system of the monkey. Brain Res. **141**:385–389.
Marks WB, Dobelle WH, MacNichol EF Jr (1964). Visual pigments of single primate cones. Science **143**:1181–1183.
Martin KAC (1988). From enzymes to visual perception: A bridge too far? Trends. Neurosci. **11**:380–388.
Maunsell JHR, Van Essen DC (1983a). Functional properties of neurons in the middle temporal visual area of the macaque monkey. I. Selectivity for stimulus direction, speed, and orientation. J. Neurophysiol. **49**:1127–1147.
Maunsell JHR, Van Essen DC (1983b). Functional properties of neurons in the middle temporal visual area of the macaque monkey. II. Binocular interactions and sensitivity to binocular disparity. J. Neurophysiol. **49**:1148–1167.
McConnell SK, LeVay S (1984). Segregation of on- and off-center afferents in mink visual cortex. Proc. Natl. Acad. Sci. U.S.A. **81**:1590–1593.
McIlwain JT (1986). Point images in the visual system: New interest in an old idea. Trends Neurosci. **9**:354–359.
Michael CR (1981). Columnar organization of color cells in monkey's striate cortex. J. Neurophysiol. **46**:587–604.
Michael CR (1988). Retinal afferent arborization patterns, dendritic field orientations, and the segregation of function in the lateral geniculate nucleus of the monkey. Proc. Natl. Acad. Sci. U.S.A. **85**:4914–4918.
Motokawa K, Taira N, Okuda J (1962). Spectral responses of single units in the primate visual cortex. Tohoku J. Exp. Med. **78**:320–337.
Murphy PC, Sillito AM (1986). Continuity of orientation columns between superficial and deep laminae of the cat primary visual cortex. J. Physiol. **381**:95–110.
Myerson J, Manis PB, Miezen FM, Allman JM (1977). Magnification in striate cortex and retinal ganglion cell layer of owl monkey: A quantitative comparison. Science **198**: 855–857.
Newsome WT, Wurtz RH (1988). Probing visual cortical function with discrete chemical lesions. Trends Neurosci. **11**:394–400.
Nikara T, Bishop PO, Pettigrew JD (1968). Analysis of retinal correspondence by studying receptive fields of binocular single units in cat striate cortex. Exp. Brain Res. **6**: 353–372.
Ogle KN (1950). *Researches in Binocular Vision*. Saunders, Philadelphia.
Pettigrew JD, Nikara T, Bishop PO (1968). Binocular interaction on single units in cat striate cortex: Simultaneous stimulation by single moving slit with receptive fields in correspondence. Exp. Brain Res. **6**:391–410.
Poggio GF (1984). Processing of stereoscopic information in primate visual cortex. In *Dynamic Aspects of Neocortical Function* (GM Edelman, WE Gall, WM Cowan, eds.), pp. 613–635. Wiley, New York.
Poggio GF, Fischer B (1977). Binocular interaction and depth sensitivity in striate and prestriate cortex of behaving rhesus monkey. J. Neurophysiol. **40**:1405.
Poggio GF, Poggio T (1984). The analysis of stereopsis. Annu. Rev. Neurosci. 7:379–412.
Poggio GF, Motter BC, Squatrito S, Trotter Y (1985). Responses of neurons in visual cortex (V1 and V2) of the alert macaque to dynamic random-dot stereograms. Vision Res. **25**:397–406.
Polyak S (1957). *The Vertebrate Visual System*. University of Chicago Press, Chicago.

Schall JD, Perry VH, Leventhal AG (1986a). Retinal ganglion cell dendritic fields in old-world monkeys are oriented radially. Brain Res. **368**:18–23.

Schall JD, Viteck DJ, Leventhal AG (1986b). Retinal constraints on orientation specificity in cat visual cortex. J. Neurosci. **6**:823–836.

Schiller PH, Malpeli JG (1978). Functional specificity of lateral geniculate nucleus laminae of the rhesus monkey. J. Neurophysiol. **41**:788–797.

Schor CM, Ciuffreda KJ (1983). *Vergence Eye Movements: Basic and Clinical Aspects*. Butterworths, Boston, MA.

Shipp S, Zeki S (1985). Segregation of pathways leading from area V2 to areas V4 and V5 of macaque monkey visual cortex. Nature (London) **315**:322–325.

Silverman MS, Grosof DH, De Valois RL, Elfar SD (1989). Spatial-frequency organization in primate striate cortex. Proc. Natl. Acad. Sci. U.S.A. **86**:711–715.

Smith VC, Pokorny J (1975). Spectral sensitivities of the foveal cone photopigments between 400 and 500 nm. Vision Res **15**:161–171.

Stell WK, Lightfoot DO (1975). Color-specific interconnections of cones and horizontal cells in the retina of the goldfish. J. Comp. Neurol. **159**:473–502.

Stone J, Fukuda Y (1974). Properties of cat retinal ganglion cells: A comparison of W-cells with X- and Y-cells. J. Neurophysiol. **37**:722–748.

Stryker MP, Zahs KR (1983). ON and OFF sublaminae in the lateral geniculate nucleus of the ferret. J. Neurosci. **3**:1943–1951.

Svaetichin G (1956). Spectral responses from single cones. Acta Physiol. Scand. 39, Suppl **134**:17–46.

Talbot SA, Marshall WH (1941). Physiological studies on neural mechanisms of visual localization and discrimination. Am. J. Ophthalmol. **24**:1255–1263.

Tanaka M, Weber H, Creutzfeldt OD (1986). Visual properties and spatial distribution of neurons in the visual association area of the prelunate gyrus of the awake monkey. Exp. Brain Res. **65**:11–37.

Tansley BW, Boynton RM (1978). Chromatic border perception; the role of red- and green-sensitive cones. Vision Res. **18**:683–697.

Tigges J, Tigges M, Anschel S, Cross NA, Letbetter WD, McBride RL (1981). Areal and laminar distribution of neurons interconnecting the central visual cortical areas 17, 18, 19 and MT in squirrel monkey (Saimiri). J. Comp. Neurol. **202**:539–560.

Tomita T (1965). Electrophysiological study of the mechanisms subserving color coding in the fish retina. Cold Spring Harbor Symp. Quant. Biol. **30**:559–566.

Tootell RBH, Silverman MS, Switkes E, DeValois RL (1982). Deoxyglucose analysis of retinotopic organization in primate striate cortex. Science **218**:902–904.

Tootell RBH, Silverman MS, DeValois RL, Jacobs GH (1983). Functional organization of the second cortical visual area in primates. Science **220**:737–739.

Tootell RBH, Switkes E, Silverman MS, Hamilton SL (1988a). Functional anatomy of macaque striate cortex. II. Retinotopic organization. J. Neurosci. **8**:1531–1568.

Tootell RBH, Silverman MS, Hamilton SL, DeValois RL, Switkes E (1988b). Functional anatomy of macaque striate cortex III: Color. J. Neurosci. **8**:1569–1593.

Ts'o DY, Gilbert CD (1988). The organization of chromatic and spatial interactions in the primate striate cortex. J. Neurosci. **8**:1712–1728.

Valberg A, Lee BB, Tigwell DA (1986). Neurones with strong inhibitory S-cone inputs in the macaque lateral geniculate nucleus. Vision Res. **26**:1061–1064.

Van Essen D (1979). Visual areas of the mammalian cerebral cortex. Annu. Rev. Neurosci. **2**:227–263.

Van Essen DC (1985). Functional organization of primate visual cortex. In *Cerebral Cortex* (A Peters, EG Jones, eds.), Vol 3, pp 259–329. Plenum, New York.

Van Essen DC, Newsome WT, Maunsell HR (1984). The visual field representation in striate cortex of the macaque monkey: Asymmetries, anisotropies, and individual variability. Vision Res. **24**:429–448.

Vautin RG, Dow BM (1985). Color cell groups in foveal striate cortex of the behaving macaque. J. Neurophysiol. **54**:273–292.

von Bekesy G (1960). *Experiments in Hearing*. McGraw-Hill, New York.

Werner JS, Wooten BR (1979a). Opponent chromatic mechanisms: Relation to photopigments and hue naming. J. Opt. Soc. Am. **69**:422–434.

Werner JS, Wooten BR (1979b). Opponent chromatic response functions for an average observer. Percept. Psychophys. **25**:371–374.

Wiesel TN (1982). Postnatal development of the visual cortex and the influence of environment. Nature (London) **299**:583–591.

Wiesel TN, Hubel DH (1966). Spatial and chromatic interactions in the lateral geniculate body of the rhesus monkey. J. Neurophysiol. **29**:1115–1156.

Wong-Riley MTT (1979). Changes in the visual system of monocularly sutured or enucleated cats demonstrable with cytochrome oxidase histochemistry. Brain Res. **171**: 212–214.

Wurtz RH (1969). Visual receptive fields of striate cortex neurons in awake monkeys. J. Neurophysiol. **32**:727–742.

Yoshioka T, Dow BM, Vautin RG (1988). Close correlation of color, orientation and luminance processing in V1, V2, and V4 of the behaving macaque monkey. Soc. Neurosci. Abst. **14**:457.

Yoshioka T, Dow BM, Vautin RG (1990a). Neuronal responses to video colors in macaque monkey visual cortex. I: Color and luminance. Submitted.

Yoshioka T, Dow BM, Vautin RG (1990b). Neuronal responses to video colors in macaque monkey visual cortex. II: Color, orientation, and cytochrome oxidase. Submitted.

Zahs KR, Stryker MP (1988). Segregation of ON and OFF afferents to ferret visual cortex. J. Neurophysiol. **59**:1410–1429.

Zeki SM (1973). Color coding in rhesus monkey prestriate cortex. Brain Res. **53**:422–427.

Zeki SM (1978). Uniformity and diversity of structure and function in rhesus monkey prestriate visual cortex. J. Physiol. **277**:273–290.

Zeki SM (1983). The distribution of wavelength and orientation selective cells in different areas of monkey visual cortex. Proc. R. Soc. London B **217**:449–470.

Zrenner E, Gouras P (1981). Characteristics of the blue sensitive cone mechanism in primate retinal ganglion cells. Vision Res. **21**:1605–1609.

5
Development of Orderly Connections in the Retinotectal System

SUSAN B. UDIN

How does a growing axon know where to terminate? Growing axons must choose the proper pathway, select the appropriate target nucleus or nuclei, and connect to the proper cells in the proper part of each target nucleus. A classic system for studying this problem is the retinotectal projection. There are several advantages for using this system to study questions of pathway guidance and formation of orderly connections.

Some of the advantages of this system include the following:

1. The cell bodies are easy to find. Only one type of cell, the retinal ganglion cell, projects axons into the optic nerve from the eye.
2. The axons are fairly easy to study with anatomical, biochemical, and electrophysiological methods.
3. The normal projection has very precise, orderly, and predictable topography.
4. In frogs and fishes, the projection can be studied in both developing and adult animals. The animals develop in water rather than *in utero* and so are easily manipulated at all stages. Adult animals, which are larger, hardier and often easier to investigate than young ones, are capable of regenerating cut optic axons, and their regrowing axons "recapitulate" some (but not all) aspects of development.

Normal Retinotectal Topography

What do we mean by "retinotectal topography"? This phrase refers to the orderly relationship between the positions of ganglion cell bodies in the retina and the positions of their terminals in the optic tectum. Under normal circumstances, axons that originate from neighboring retinal ganglion cells will terminate close to one another in the tectum; the ganglion cells form essentially a two-dimensional array across the retina, and their terminals distribute themselves in the same sort of two-dimensional array, or map, across the tectum. Many thousands of axons form such maps, but in the interest of simplicity, this chapter will demonstrate the patterns formed by only a few cells. Figure 5.1

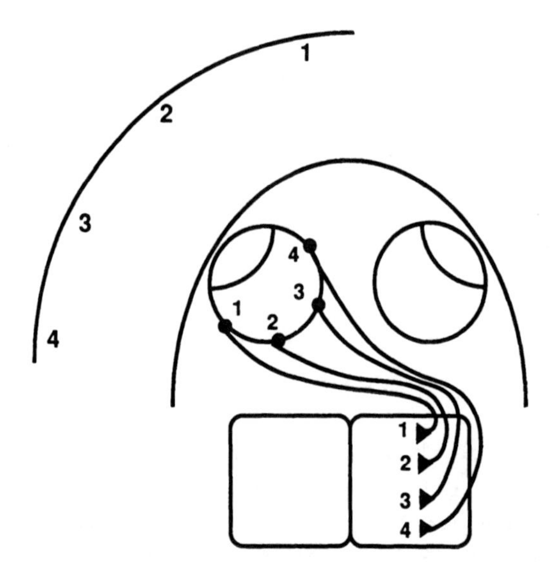

FIGURE 5.1. Schematic illustration of retinotectal topography. Four ganglion cells which see four positions in the visual field send their axons to terminate in an orderly pattern in the opposite tectum.

shows a schematic view of four ganglion cells connecting the eye and the tectum. In this view, we are looking down on the animal so that we are viewing a horizontal cross section of the eyes and are looking down upon the surface of the tectum. The left eye is positioned to view a surface which can be used to display visual stimuli. For example, a dark spot at position **1** on the surface will activate ganglion cell **1**, which connects to a position near the rostral (front) end of the contralateral (opposite) tectum. A stimulus at position **2** will activate ganglion cell **2**, which will project to a slightly more caudal (rearward) position on the tectum, and so forth. Thus, four locations in space are seen by four retinal ganglion cells, which project to four locations in the tectum. The nasal regions of the retina project to the caudal part of the tectum and the temporal regions of the retina project to the rostral parts of the tectum. (These positions represent only the horizontal dimension of the two dimensions of the retinotectal projection; the principles that guide the formation of this horizontal dimension of the map also hold true for the vertical dimension.) A convenient mnemonic for remembering the normal orientation of the retinotectal map is that the front of the visual field is represented on the front of the tectum and the back of the visual field is represented on the back of the tectum.

The Major Hypotheses

The first half of this chapter will present the major hypotheses that were proposed to explain how retinal axons find the proper locations in the tectum. First we will present some of the evidence that led Sperry (1944) to devise his chemoaffinity hypothesis; he proposed that there are chemical markers that direct the growing axons and instruct them to connect to the proper tectal cells. We will then describe some of the experimental results that did not fit that hypothesis and which led Gaze and Keating (1972) to devise the systems matching hypothesis. They suggested that retinal axons obtain most of their information from one another rather than from the tectal cells. These two ideas were intensely debated for about 15 years, until it finally became clear that in fact both groups were partly correct, in that both mechanisms are used by growing axons, and they normally work together to produce orderly retinotectal projections.

Some Methods

One major technique for determining where a given optic axon is terminating in the tectum is electrophysiological mapping. An extracellular electrode is inserted into a given site in the superficial layers of the tectum. The electrode can detect the electrical activity of the axons that branch in that location. When a visual stimulus activates those axons, the electrode will pick up the spike activity. Thus, in a normal frog or fish, one could test four positions as shown in the diagram and, knowing that the image of the visual field is inverted by the lens of the eye, one could then infer which ganglion cells were projecting to those tectal locations.

The relationship between the retina and the tectum can also be assessed by anatomical methods. Many tracers are available to label connections. Some are inserted into parts of the eye or optic nerve and are transported toward the tectum (anterogradely). For example, the enzyme horseradish peroxidase (HRP) can be inserted into a small cut in the eye to label the projections from that region of the eye and thus to show where those axons project. Tracers can also be inserted into the tectum for transport back to the eye (retrogradely). For example, a small injection of a fluorescent dye can be made in part of the tectum; in a few hours or days, that tracer will be visible in the ganglion cells that project to the injection site in the tectum.

The Origin of the Chemospecificity Hypothesis

If one cuts the optic nerve in a frog, the nerve will eventually regenerate, and visual abilities will recover. How do the axons find the correct places to make their connections? Sperry (1944) devised a test to help decide between two quite different mechanisms. One possible mechanism was that each axon finds its proper place by means of some special label or chemical tag on the appropriate target cells. This idea seems fairly plausible now, but it was not as obvious, 40 or 50 years ago, that there could be enough labels such that 20,000 axons could each find their correct sites. [At the time, it was believed that there were 20,000 fibers in the optic nerve; we now know that there are actually about 500,000 in the optic nerve of adult *Rana pipiens* frogs (Maturana, 1960).] The other, very different, mechanism to be tested was one based on learning or experience. According to this idea, each axon wanders around until it finds a location that produces functionally adaptive behavior. To understand this idea, we first must know the function(s) of the tectum.

Ablation (surgical removal) of the tectum demonstrates that it plays an essential role in prey-catching. If a tasty-looking bug were present at position **1** in the visual field, its image would stimulate ganglion cells at retinal position **1**, which in turn would activate cells at the front of the tectum and initiate a snap toward position **1**. Similarly, a bug at position **4** would activate the back of the tectum and initiate a snap toward position **4** (Fig. 5.2A). What would happen if axon **1** regenerated to the back of the tectum, the region that activates turns toward the side? The bug at position **1** would now cause the frog to turn toward position **4**; it would not get the bug; and somehow, axon **1** would be induced to move to a new and perhaps more appropriate position. Experience would reinforce only useful connections.

Sperry devised a very clever test to distinguish which of the two mechanisms was more likely: the mechanism based on chemical tags or the one based on learning. He rotated one eye in an adult frog (Fig. 5.2B). If the optic nerve is not cut, then the animal displays inappropriate behavior because the rotated eye transmits an upside down image. What happens if the nerve is cut and allowed to regenerate? Will the axons return to their original sites and yield maladaptive behavior, or will they grow to new locations that will bring the visual map to tectal sites that will yield appropriate behavior? The actual result is that the regenerated axons return to their original locations, and the poor frog keeps making errors in localizing prey.

This result led Sperry to reject the learning/experience hypothesis in favor of what he named the "chemoaffinity" hypothesis. This idea has been modified many times over the years, but one of the commonest formulations involves the suggestion that there are gradients of some sort of markers in both the retina and the tectum (Fig. 5.3). Consider the matching of a set of ganglion cells along the temporonasal retinotectal axis to the corresponding set of tectal positions along the rostrocaudal tectal axis. Each retinal ganglion cell has an identity conferred by a hypothetical substance, **A**. The axons, which also "know" their identity, reach the tectum and search for the site marked by the appropriate amount of some substance **X** (Fig. 5.3). In a similar manner, another pair of gradients would mark mediolateral positions.

Sperry hypothesized that both the retinal and tectal gradients were not only specific to each locus but also stable. Subsequent work cast doubt on the stability of the markers and led to the so-called systems matching hypothesis.

The Origin of the Systems Matching Hypothesis

One version of this hypothesis states that each retinal position would correspond uniquely to one—and only one—tectal position. How could you test whether there really is unique point-to-point labeling? One way is to prevent some of the

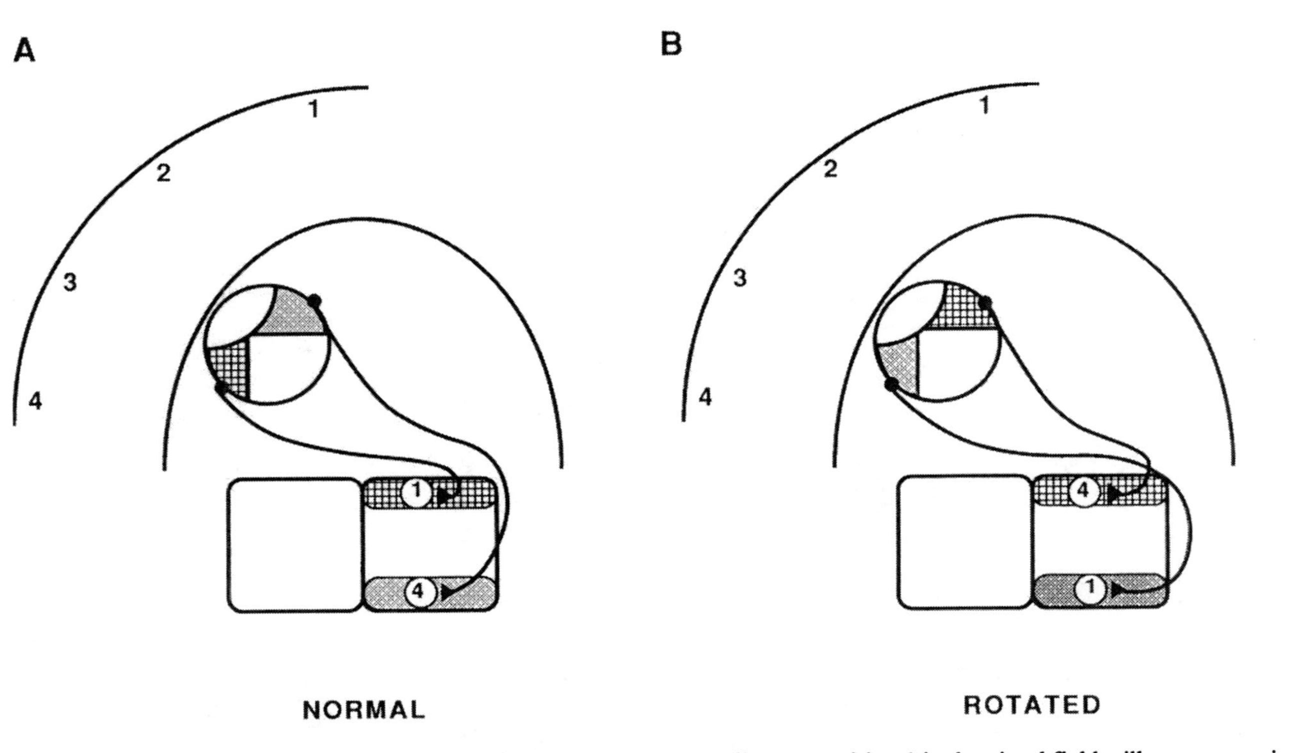

FIGURE 5.2. When the eye is rotated by 180°, the part of the eye that usually sees position 1 in the visual field will now see position 4 and vice versa. If each ganglion cell regenerates to its original location, then the map of the visual field on the tectum will be rotated by 180°, and the frog's prey-catching behavior will be inappropriate. This is in fact what occurs.

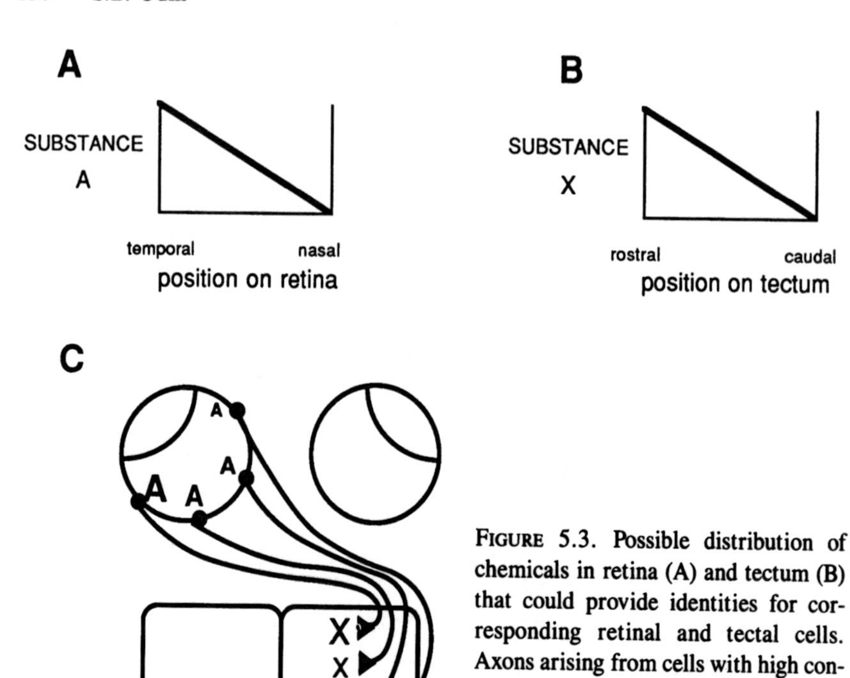

FIGURE 5.3. Possible distribution of chemicals in retina (A) and tectum (B) that could provide identities for corresponding retinal and tectal cells. Axons arising from cells with high concentrations of substance **A** would project to tectal cells with high concentrations of substance **X**.

axons from occupying their normal tectal sites and then to examine whether they do or do not reposition their terminals to other, incorrect sites. One such "mismatch" experiment was performed by Gaze and Sharma (1970). They removed half of the tectum of adult goldfish. After several months, they found that the displaced retinal axons had found new homes. In fact, the whole retina now

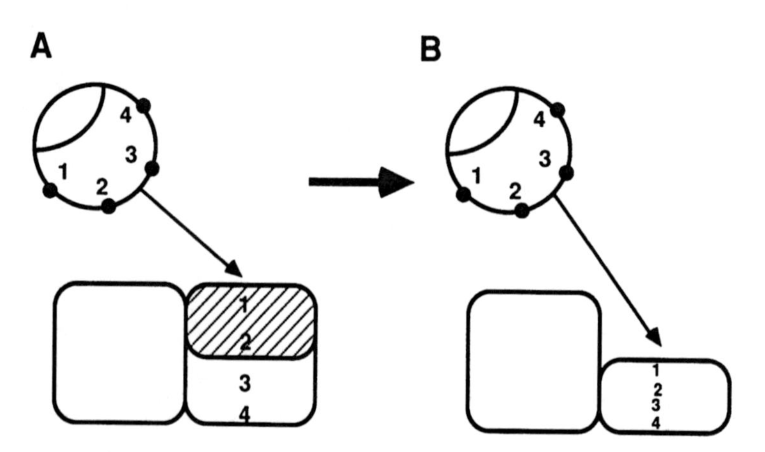

FIGURE 5.4. When the rostral part of the tectum is surgically removed in goldfish (A), the entire extent of the retinotectal projection will eventually compress in an orderly way on the remaining half-tectum (B).

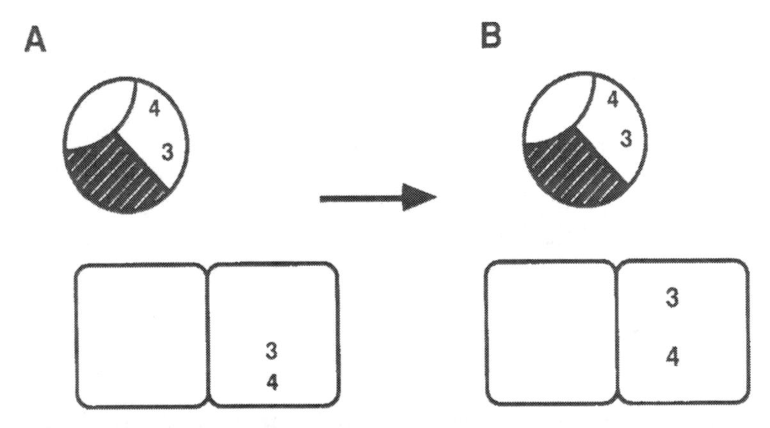

FIGURE 5.5. When half of the retina is surgically removed, the remaining axons expand their zones of termination to fill the tectum in an orderly way.

projected in an orderly way to the remaining tectal fragment. In other words, the whole retinotectal map was compressed onto the half tectum. (see Fig. 5.4). This result demonstrated that there is not a unique, point-to-point matching of retinal positions to tectal positions.

This result, which contradicted Sperry's prediction, was confirmed by other groups (Udin, 1977), and a complementary mismatch experiment by one of the members of Sperry's own lab amplified this type of result. Yoon (1972) removed half the retina in adult goldfish and determined that the remaining retinal axons expanded their projections to cover the whole tectum (Fig. 5.5). Even though the axons still had their normal termination sites available, they wound up at new positions.

Thus, Sperry's idea that there were unique and stable labels was disproved (although the idea of labile labels was still viable). These results prompted a counterhypothesis by Gaze and Keating (1972). Their "systems-matching" hypothesis proposed that retinal axons do not look for unique addresses on tectum. Instead, it was proposed that the major factor was the interaction of axons, not with the tectum, but with one another. Their interactions would ensure that proper neighbor relations would be preserved, or, if disrupted, would be reestablished; each axon would make stable connections only when it was in proximity to other axons from the same region of the retina. The terminals in the tectum would thus reproduce the same neighborhood relations of the cell bodies in the retina. In our schematic representation of the retinotectal projection in Figure 5.1, axon **2** would be stable only when it was near axons **1** and **3** and not if it was overlapping the territory occupied by axon **4**. In addition, the population of axons would adjust to fill all available tectal space. If part of the tectum were missing, the population would compress, and if part of the retina were missing, the remaining axons would expand. The tectum's role would primarily be to provide polarity cues to ensure that axon **1** wound up at the front of the tectum rather than at the back or the side. The retinal axons would then arrange themselves in proper order within the available tectal space.

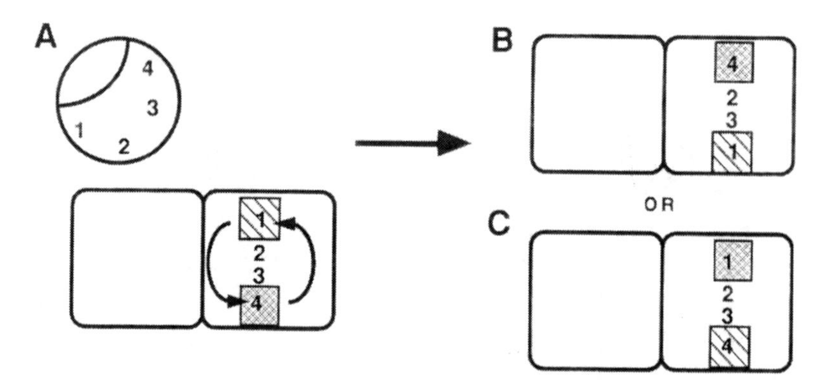

FIGURE 5.6. When two pieces of tectum are surgically removed and are interchanged, two sorts of maps can result. In some cases, the axons terminate in their original tissue and the internal order of the projection is abnormal (B), but in other cases, the axons ignore the identity of the tectal tissue and form a map that is normal in organization (C).

Things were looking bad for Sperry, but then Gaze, among others, did some experiments that supported the idea of tectal markers (Levine and Jacobson, 1974; Gaze and Hope, 1983). They tested for the presence of stable markers in tectal tissue by asking whether axons could recognize a particular piece of tectum even if it were put in an unusual location. Such "graft translocations" could involve switching two chunks of tectum, as illustrated in Figure 5.6. Would the retinal axons recognize their "proper" tectal targets despite the rearrangement, as predicted by the chemospecificity hypothesis (Fig. 5.6B) or would they ignore the origin of the tectal tissue and organize themselves with respect to one another, as predicted by the systems matching hypothesis (Fig. 5.6C)? The result was that the maps from some tecta conformed to the chemospecificity prediction while the maps from other tecta conformed to the systems matching hypothesis!

In other words, sometimes fibers follow tectal labels and sometimes they terminate where they will be near proper neighbors. What can one make of this sort of result? The current thinking is that there must be multiple factors. The tectum evidently does have position labels—weak ones—and there also are fiber-fiber interactions. Normally these two mechanisms work together, but in graft translocations, some subtle factors (probably relating to surgical technique and the details of graft healing) can tip the balance so that one or the other of these factors predominates.

Two Modes of Fiber–Fiber Interactions

What is the nature of the fiber–fiber interactions? How do axons influence one another? There seem to be two sorts of bases for interaction. There is some sort of biochemical distinction between different retinal ganglion cells and their axons, and such distinctions can somehow provide signals that the axons can use to determine if they are touching similar or dissimilar axons. The other major cue comes from correlations of electrical activity. Ganglion cells in the same area of the retina see approximately the same visual stimuli at about the same time and therefore fire in a correlated way; such firing patterns can provide information about whether pairs of axons come from nearby or distant retinal regions.

Evidence for Unique Retinal Axon Identities

First, what is the evidence for markers of retinal axon identity? One can show that ganglion cells have an embryonic identity that they can "remember" no matter where in the retina they are placed. One way to demonstrate this is to alter the physical location of cells from a known part of the eye and to test whether those cells make projections that correspond to their original retinal positions or to their new retinal positions. In frog embryos, "compound eyes" can be surgically constructed by removing a quadrant of one eye (Fig. 5.7, area **4**) and replacing it with a noncorresponding quadrant (area **1**) from another animal's eye (Straznicky et al., 1981; Willshaw and Gaze, 1983). The compound eye now has two embryologically identical sets of cells (area **1**) at opposite positions in the retina. There now is a retina with the normal complement of cells, but those cells do not represent the normal spectrum of cell "identities." What sort of map is made by such a retina? Consider first how the individual components of the compound retina would project if they had not been put together. Positions **1**, **2**, and **3** would spread out to fill the whole extent of the tectum if area **4** had simply been discarded without any replacements (Fig. 5.7A). Similarly, the cells from quadrant 1 would also expand their projection to fill the entire tectum if there were no other cells in the retina (Fig. 5.7B). When those cells are grafted into the host eye, however, the grafted cells' axons do not expand over the whole tectum. Instead, they project to the same region of the tectum—no more and no less—as the host cells from area **1**: they behave as if they remember their original identity and project as if they had never been displaced to an inappropriate region.

Thus, the tendency of axons to terminate with embryologically similar axons can override the tendency of axons to expand to fill the tectum. No one yet has any evidence about the underlying mechanisms that allow axons to recognize the degree of embryological similarity of other axons.

Evidence for Activity Cues for Retinotectal Topography

Let us now consider the other mechanism that assists axons in forming an orderly retinotectal map. There is evidence that axons tend to terminate with other axons that have similar firing patterns, and electrical activity is an essential part of the final refinement of the retinotectal map. One can test this idea by preventing activity during development or regeneration. The axons can form a crude map in the absence of activity but cannot produce completely normal order. Consider

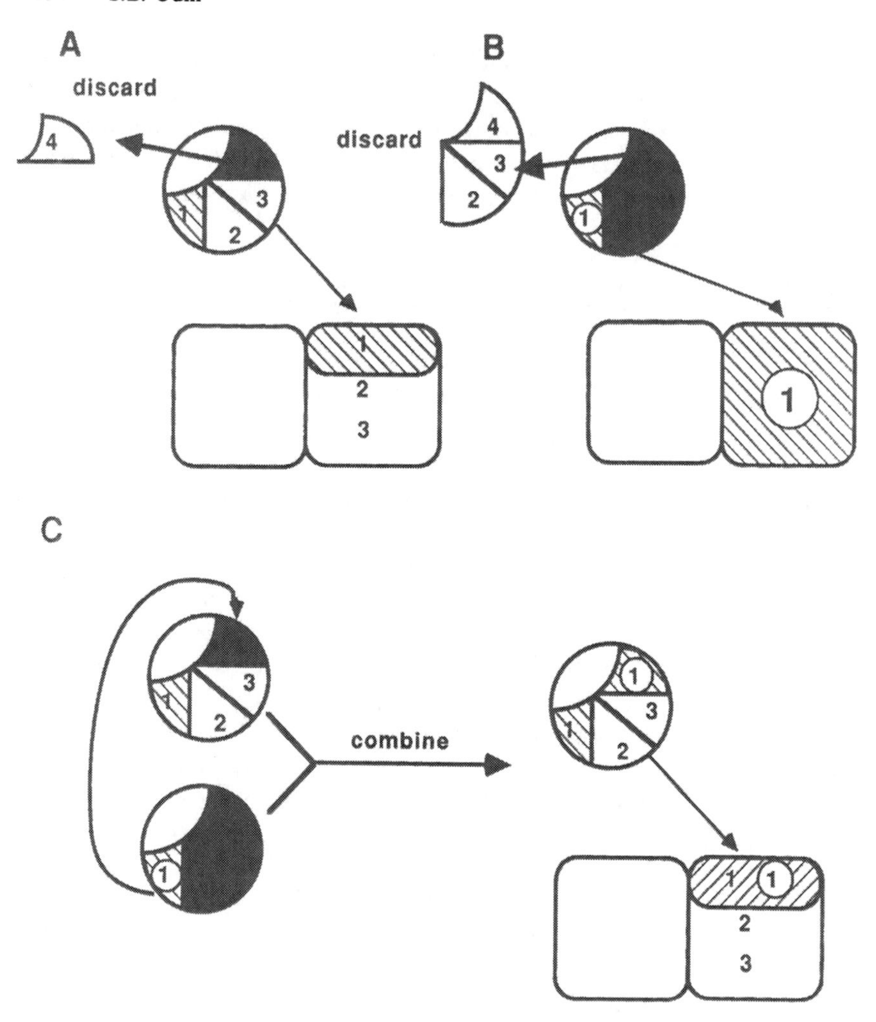

FIGURE 5.7. Surgically constructed compound eyes demonstrate that axons from individual regions of retina have an embryologically derived identity that disposes them to terminate near other endings of axons with the same identity. Thus, removing the nasal quadrant of one eye induces the remaining axons to expand their projection (A). If instead one removes the temporal three-quarters of the retina (B), the remaining nasal quadrant is capable of projecting to the entire tectum. If one inserts a temporal quadrant into an eye lacking the nasal quadrant, then the temporal axons are constrained to map together with the temporal axons from the host (C).

the stages of re-formation of the map after the optic nerve has been crushed. The cut axons can regenerate, and they gradually re-form a map.

Let us consider eight retinal ganglion cells (Fig. 5.8A). Their axons will grow back to make a very messy map at first (Fig. 5.8B), and they will slowly shift their terminals to reach normal locations (Fig. 5.8C). However, they will not refine the map fully without activity. If you block firing of action potentials

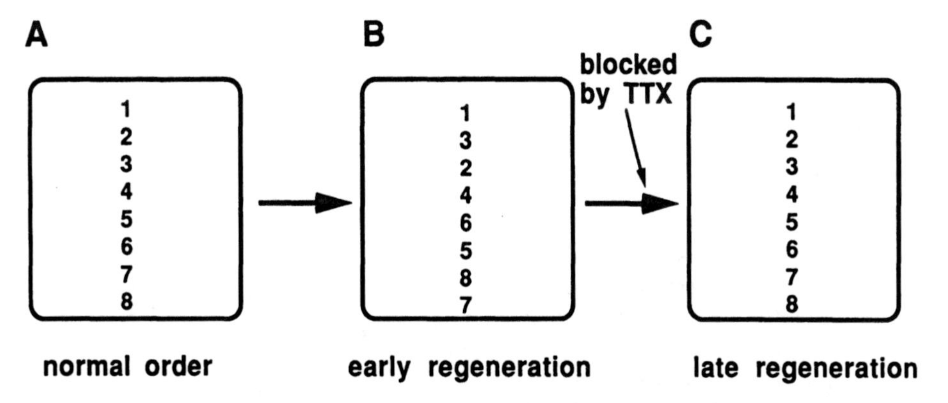

FIGURE 5.8. If the optic nerve is cut, the normal order of the map (A) is restored only partially at first. For example, the positions of axons 2 and 3 are reversed, as are the pairs 5,6 and 7,8 (B). Gradually, full ordered is achieved (C), but that final refinement is prevented by blockers of activity such as tetrodotoxin.

during regeneration, the refinement process is halted (Meyer, 1983; Schmidt and Edwards, 1983) and the map retains the disorder shown in Figure 5.8B. The usual method for abolishing action potentials is to inject tetrodotoxin (TTX) into the eye. (TTX is a drug that blocks voltage-sensitive sodium channels and thus blocks the initiation and propagation of action potentials.)

In summary, these experiments, and many others, have led to the following ideas:

1. Both retinal and tectal cells have markers.
2. Axons find their final locations by interacting with tectal cells and with one another, using both electrical and nonelectrical cues.

Possible Mechanism of Activity-Dependent Axon Interaction

How could these postulated electrical cues operate? How does activity tell fibers whether they are terminating near their proper neighbors? One hypothesis focuses on the role of tectal cells as mediators of interactions among retinotectal axons. According to this notion, if the axons that terminate on a given tectal cell fire at about the same time, they will trigger off cellular changes that stabilize those recently-active terminals.

How would this mechanism help refine the regenerating retinotectal map? Consider a tectal cell that receives many inputs from axons from one part of the eye (**1**) and just a few inputs from axons from other parts of the eye (**2** and **3**) (Fig. 5.9).

When a visual stimulus falls on region **1** of the retina, many of the axons synapsing onto the tectal cell will fire; the tectal cell dendrite will be depolarized sufficiently to trigger some other events, which in turn will stabilize the

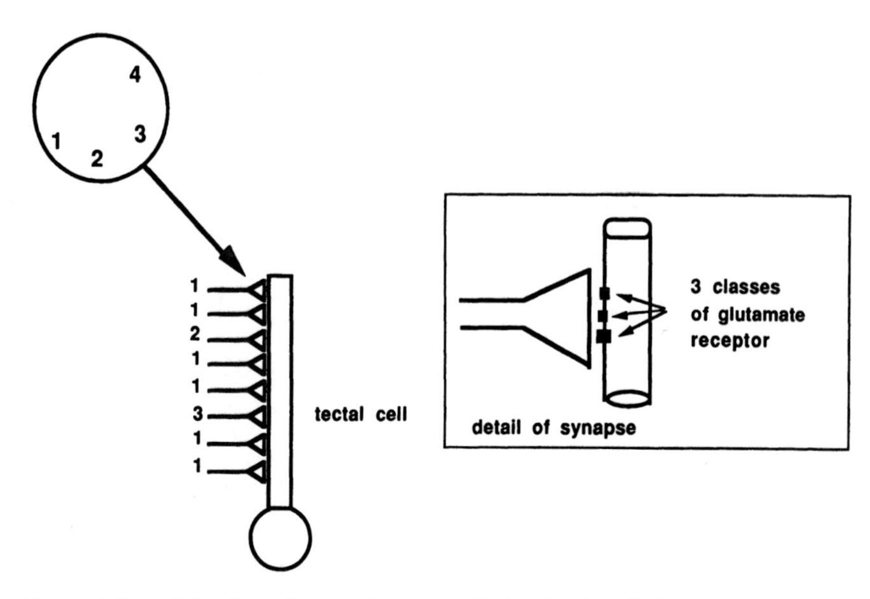

FIGURE 5.9. Activity-dependent mechanisms assist in rejection of mislocated terminals (**2** and **3**). This process involves a special class of glutamate receptor on the tectal cell dendrite.

terminals from retinal region **1** onto that dendrite. If, instead, a stimulus falls on region **2** or region **3** of the retina, only one terminal will be activated. The dendrite will not be depolarized enough and the activated terminal will not be stabilized (or will even be destabilized).

There is now much evidence accumulating to support this idea and to suggest a way that simultaneous activity can produce significant changes via a special class of receptors for glutamate. Glutamate, an amino acid, is probably the transmitter released by optic nerve terminals (Roberts and Yates, 1976; Langdon and Freeman, 1987). There are several classes of glutamate receptor (molecules which respond to glutamate) (Fig. 5.9). When glutamate is released and binds to glutamate receptors, channels in the membrane open up, allow certain cations to pass, and produce a depolarization. One kind of glutamate receptor, however, is usually blocked by Mg^{2+} ions when the dendrite is near resting potential (Nowak et al., 1984) (Fig. 5.10A). When enough inputs are active at the same time, the summated depolarization is enough to kick out the magnesium ions. When that happens, the channels can open up (Fig. 5.10B). These special receptors are known as NMDA (*N*-methyl-D-aspartate) receptors because NMDA is a potent and selective ligand for these receptors (see below).

Why is this important?

1. The extra channels boost the depolarization: the same number of active synapses will induce more depolarization when the NMDA receptor channels open than they would otherwise.

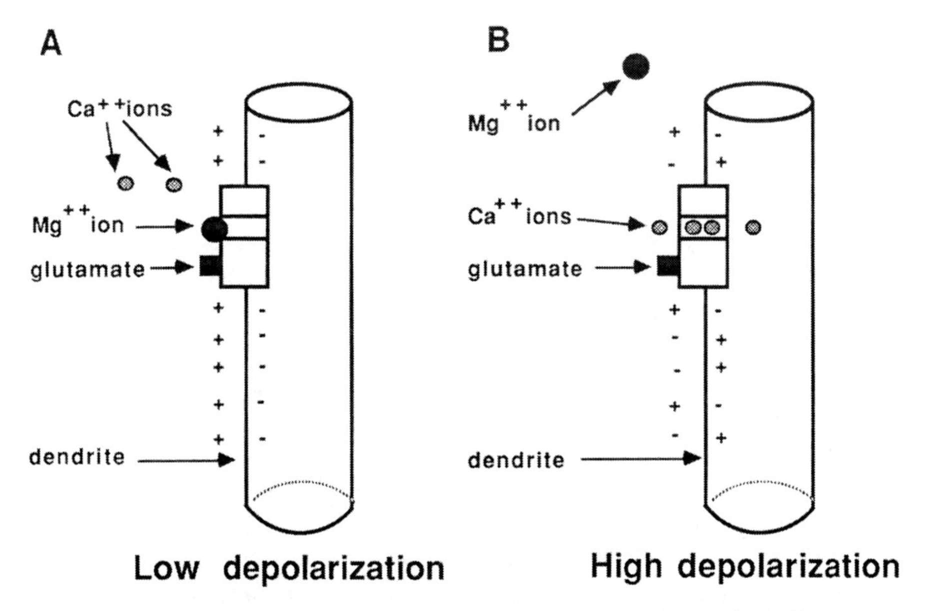

FIGURE 5.10. At resting potential, NMDA-type glutamate channels are blocked by magnesium ions (A). When there is a net depolarization of the dendrite, as by activation of other glutamate channels, then the positively charged magnesium ions are repelled from the membrane, allowing calcium ions to enter the dendrite (B).

2. The NMDA channels allow **calcium** ions to enter the dendrite (MacDermott et al., 1986), and calcium ions are very important for regulating a vast number of structures and processes. Presumably, some event or events then occur to stabilize the synapses that had just fired.

Evidence That NMDA-Type Glutamate Channels Play a Role in Retinotectal Topography

There is evidence that NMDA channels are involved in mediating experience-dependent plasticity in the retinotectal system, where the transmitter does appear to be glutamate. The first such indications came from the work of Cline et al. (1987), who used a somewhat bizarre, but highly revealing preparation, the three-eyed frog.

To understand the experiments involving glutamate, some background information on this preparation is essential. A three-eyed frog is made by implanting an extra eye into a frog embryo. The extra eye usually will project to one of the tecta, where it shares the tectal neuropil with the input from the host contralateral eye. Surprisingly, the two eyes do not project evenly across the surface, but instead they subdivide the tectum into a series of alternating stripes

(Constantine-Paton and Law, 1982) (Fig. 5.11A). The overall topography of each projection is normal relative to the eyes' anatomical markers. Each nasal retinal pole projects caudally, each temporal pole projects rostrally, and so forth, but the two maps interdigitate rather than overlapping. (See Chapter by Dow for a description of stripes in the mammalian visual cortex.)

What causes the axons to produce this pattern? The striping seems to reflect a compromise between two different tendencies:

1. The tendency to terminate in the proper tectal location (chemospecificity);
2. The tendency to terminate only in proximity to other axons with similar activity patterns (systems matching axon interactions).

If the axons were guided solely by the tectal markers, they would have to share territory with axons from the other eye. Why doesn't this occur? The reason lies in the differing visual fields to the two eyes: their placement in two different positions in the head causes them to look at different locations and therefore to have, at any given moment, very different activity patterns. As a result of these two opposing forces, the axons first grow to anatomically "correct" locations but then shift slightly medially or laterally, so that they interact primarily with axons from the same eye. Thus, they increase their proximity to axons with similar activity, albeit at the expense of being, in many cases, slightly mispositioned on the tectum.

How do we know that the segregation phenomenon is really a reflection of an activity mechanism? Evidence for this interpretation comes from experiments in which retinotectal action potentials in three-eyed frogs were blocked by application of tetrodotoxin to the eye or optic nerve. This treatment deprives the axons of any electrical cues by which to distinguish which axons arise from which eye. As a result, the host and graft eyes' axons share the tectum without forming stripes (Constantine-Paton and Reh, 1985) (Fig. 5.11B).

More recent experiments show that stripes can also be abolished by subtler alterations of synaptic function. Instead of blocking all activity, it is possible to block just one component of synaptic transmission by applying NMDA receptor blockers to the tectum (Cline et al., 1987). The axons still fire and can transmit activity to tectum, but stripes do not form. In this situation, there is still normal activity in the optic axons, and they can still transmit information to the tectal cells via the non-NMDA channels, but they cannot produce that critical extra depolarization or calcium influx. Thus, whatever processes are normally triggered by activating the NMDA channels no longer can function, and axons that fire in a correlated pattern are not selectively stabilized.

Cell Death

Another phenomenon that contributes to the formation of orderly retinotectal connections is selective cell death. This process does not seem to play much of a role in frogs or fishes, but is an important factor in the development of mammalian retinotectal projections.

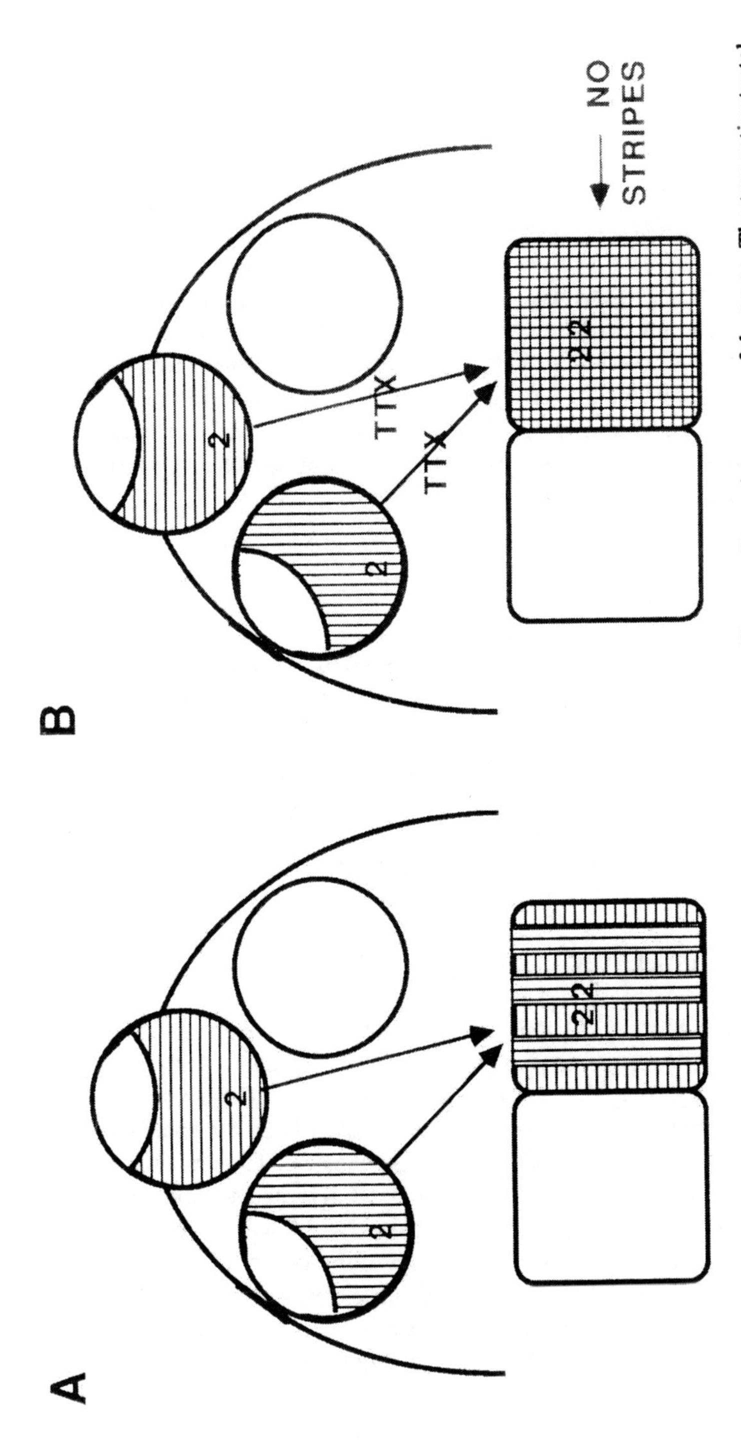

FIGURE 5.11. When a third eye is implanted into a frog embryo, that third eye will generally project to one of the tecta. The two retinotectal projections will each form normally oriented topographic maps that are interleaved in stripes (A). However, if activity is prevented by application of tetrodotoxin, no stripes form (B).

First, let us present a little background on the formation of retinotectal maps in mammals. Are the underlying mechanisms generally the same as those we have seen in cold-blooded vertebrates? One way to test this question is to perform comparable types of experiments, such as retinal or tectal lesions. In most mammals, the retinotectal projection is well ordered by the time of birth and does not show any rearrangements in response to such experimental treatments administered postnatally. However, some mammals, such as rats and hamsters, are born after a very short gestation period and are very immature. Their retinotectal projections are still developing and they show the same sorts of plasticity that have been seen in frogs and fishes. For example, retinal lesions in the first postnatal week cause expansion of the projection from the remaining parts of the eye (Frost and Schneider, 1979). Tectal lesions in the first postnatal week induce compression of the retinotectal map onto the remaining tectal fragment, if it is not too small (Finlay et al., 1979). Thus, many of the same principles of axon–tectum and axon–axon interaction probably guide organization of retinotectal maps in both mammalian and nonmammalian vertebrates.

A Role of Cell Death

Cell death seems to augment the error-correction mechanisms that we have described above. During normal development, many axons grow to the wrong parts of the tectum. Some of these errors are corrected not by rearranging the erroneous connections but rather by eliminating the whole retinal ganglion cell and all of its processes. Cell death is a common part of normal development. It is not unusual to find that 50% of cells in a given structure die. Some (but probably not all) of this process has an error-correction function.

Cell Death and Error Correction in Rats

Let us examine some experiments that illustrate that cell death can assist in correction of erroneous projections in the retinotectal projection of developing rats. O'Leary et al. (1986) used anatomical methods to see which parts of the retina project to a given part of the tectum at different ages. First, they marked cells at early ages to see how accurate the early projections are. Their method involved the use of a dye called fast blue. This substance can be injected into the tectum, where it will be picked up by retinotectal axons, which will transport it back to the retina. One can inject fast blue into part of the tectum on the day of birth (postnatal day 0); after 2 days, the label will have reached the retina. If the rat is killed on day 2, then the pattern of labeled cells reveals information about the topography of the projection at the time of injection. The results demonstrate that the caudal edge of the tectum receives connections not only from cells in the correct (nasal) part of the eye but also from ganglion cells scattered throughout the other parts of the eye (Fig. 5.12A).

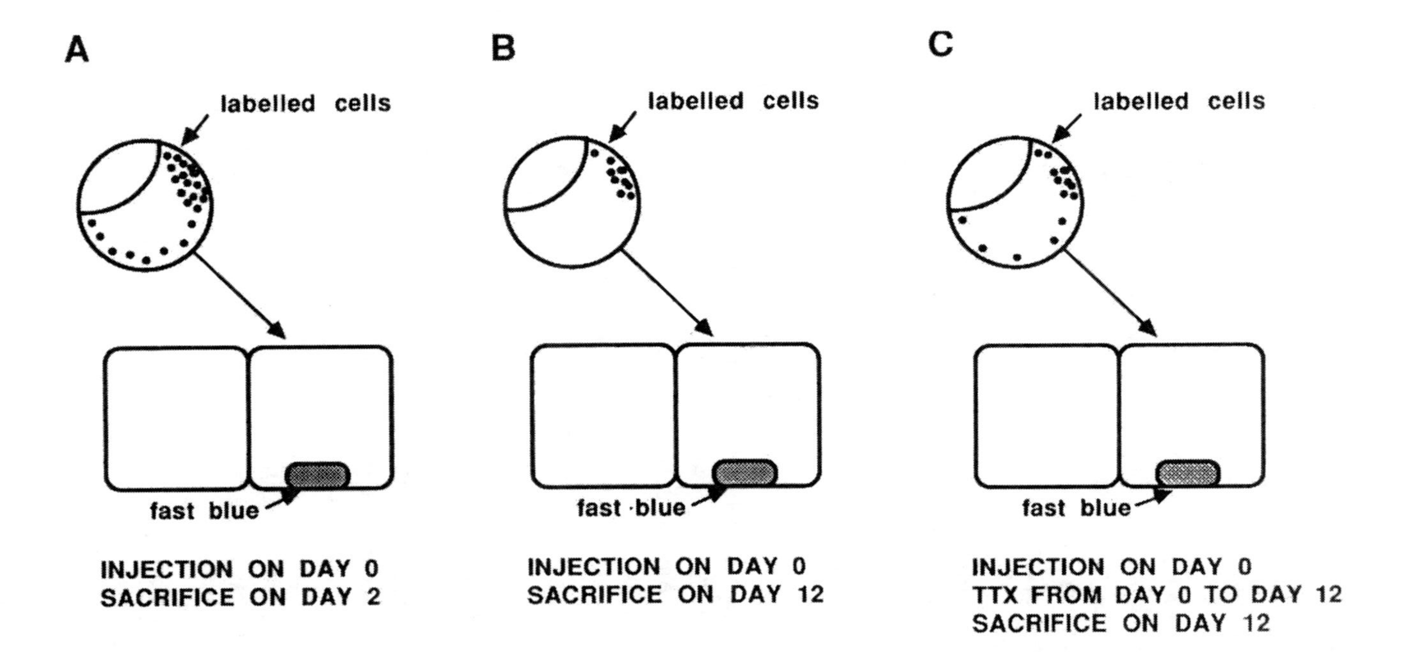

FIGURE 5.12. Retrograde tracing with fast blue indicates that the caudal end of the tectum initially receives input from cells distributed throughout the retina (A). Normally, the incorrectly projecting cells are eliminated from the retina (B). If activity is prevented during development, many cells die, but the cell death is random and does not improve the accuracy of the retinotectal map (C).

The same sort of technique can be used to show that those "wrong" cells are later eliminated. As before, fast blue is injected into the caudal part of the tectum at day 0. This time, however, the animal is not sacrificed until day 12. Examination of the retina after this additional 10 day time span reveals that fast blue is found only in cells in the nasal part of the retina, i.e., only the "correct" location (Fig. 5.12B). The cells in the "wrong" part of the retina have died and been removed.

Why did they die? What mechanism eliminates all of the erroneous cells? Activity may be involved. Sets of axons that originate close together in the retina will have more similar firing patterns than will sets of axons that originate from distant regions of the retina. Most of the cells that originally project to the injection site come from approximately the correct location and will have patterns of activity that correlate well with one another. In contrast, the less numerous aberrant cells have activity patterns that correlate poorly with one another and with the larger population of correct cells. The correct cells will stabilize one another, and the ectopic cells will be out-competed.

One can test this hypothesis by injecting TTX into the eye during the period of cell death (day 0 to day 12) (Fawcett and O'Leary, 1985). The pattern of cell death can be monitored as above, by injecting fast blue into the caudal region of the tectum on day 0 and sacrificing on day 12. The results show that much cell death still occurs, but the distribution of the surviving cells is abnormal. The surviving cells are found not only in the nasal region of the retina but also at other locations (Fig. 5.12C). Cell death in this case seems to be random rather than selective, and the resulting pattern is much more like the normal pattern at day 2 than at day 12. It appears that the "correct" axons have been deprived of their competitive advantage and therefore are no more likely to survive the period of death than are any of the other ganglion cells.

If this is so, then for what are the cells competing? Fawcett and O'Leary (1985) speculate that tectal cells release a trophic factor that retinal cells must receive if they are to survive during development. The factor would be taken up best by retinal axon terminals, and the axons would transport the factor back to the retina. Such growth factors have been shown to exist in many parts of the nervous system. The properly located terminals would be able to pick up more of this postulated substance than would the aberrantly located terminals. If this trophic substance is in short supply, then not all of the axons might be able to acquire enough of the trophic substance to allow them to survive. Axons that were correctly located and that had their synapses stabilized by the activity-dependent processes indicated above might also become more efficient at ingesting the life-giving tonic and thus be favored for survival.

The Role of Activity in the Formation of Binocular Maps in the Tectum

Binocular Maps in Xenopus

Next, we will consider a system where the firing patterns due to visual input are the dominant organizing influence: the binocular system in *Xenopus* frogs. Thus

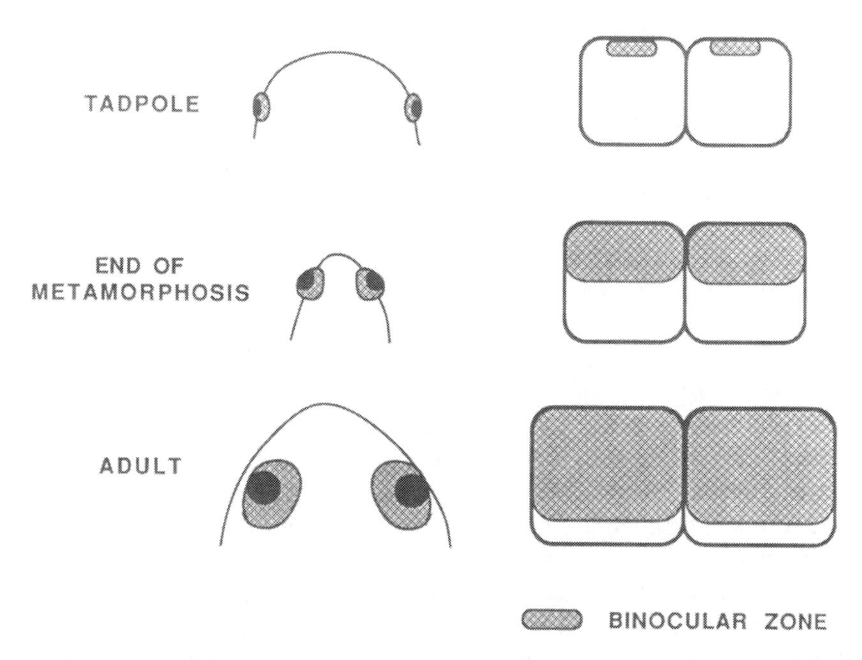

FIGURE 5.13. The relative position of the eyes changes during development in *Xenopus* frogs. The amount of binocular overlap increases with age, and the amount of the tectum (right column) that receives retinotectal input corresponding to the binocular portion of the visual field also increases.

far, we have addressed only the matter of how each eye's map comes to be represented in an orderly way on the opposite tectum, but this contralateral retinotectal map is not the only visual input to the tectum. Each tectum also gets a representation of the ipsilateral eye, and these two separate maps of the visual world normally are brought into register in the tectum, so that each visual field position that is seen by both eyes comes to be represented at a single position in each tectum.

Developmental Eye Movement

In *Xenopus*, the axons that bring the image of the ipsilateral eye to the tectum are very strongly influenced by visual input during development. They make and break connections during development, and new connections are stabilized when they help to bring the images from the two eyes into register (Udin, 1989). This ability is especially important for *Xenopus* because of the unusual and dramatic changes in eye position that occur during the development of this species. The eyes of *Xenopus* tadpoles initially have almost no binocular overlap with one another. With the onset of metamorphosis, changes in body morphology cause the eyes to move dorsally, eventually settling into a position with 170° of binocular field (Fig. 5.13).

The Isthmotectal Pathway

Despite these changes, the maps from the two eyes are in register throughout development (Grant and Keating, 1986). To understand how the wiring patterns change to keep the eyes' maps together, we have to first describe the way that binocular input reaches the tectum. The route is a bit roundabout. Consider the maps on the left lobe of the tectum, as illustrated in Fig. 5.14A. The right eye projects directly to the left tectum in the manner we have seen earlier in this chapter. Here we represent just two visual field positions and two retinal ganglion cells projecting from the right eye. The projections that bring visual information from the left eye to the left tectum are indirect. First, the left eye projects to the right tectum (arrows not shown). The right tectum then projects to a structure called the nucleus isthmi (the homologue of the mammalian parabigeminal nucleus) (arrows not shown). Thus, the right nucleus isthmi receives a map relayed from the left eye. Cells in the right nucleus isthmi then project to the left tectum, where their terminals form an orderly map. Thus, a particular point in the binocular part of the visual field, say, point A, will activate ganglion cells in both eyes, which will relay that information to a single point in each tectum via the illustrated pathways (Gruberg and Udin, 1978). (The right tectum also receives binocular input via connections that are the mirror image of those illustrated for the left tectum.)

The Role of Activity in the Ipsilateral Map

What is the evidence that activity plays a role in establishing this orderly arrangement? One of the strongest indications that visual activity is part of the process whereby the ipsilateral (isthmotectal) projections come into alignment with the contralateral projections is the effect of dark rearing. If *Xenopus* are reared in the dark, the ipsilateral map does not form in an orderly way, and the ipsilateral units are not in register with the contralateral units (Keating and Feldman, 1975).

This result shows that the map cannot form properly without visual input, but it also is possible to show that visual input does not just permit the map to form but actually can instruct it how to form. One way to demonstrate this effect is to purposely misalign the eyes during development. One can rotate one eye in the tadpole, so that one eye is right side up and the other one is upside down (Fig. 5.14B). This mismatch induces the isthmotectal axons to terminate in different positions than normal; the new positions bring the maps from the two eyes into register. For example, an axon that normally would wind up in the caudolateral part of the tectum might now end up at the opposite pole (Udin and Keating, 1981).

To understand this phenomenon, consider first what happens to the retinotectal projection of the rotated eye. As discussed earlier in this chapter, this manipulation does not cause any changes in the connectivity of the retinotectal axons: each axon projects to the same tectal locus as normal. However, the rotated position of the eye causes the ganglion cells to relay a rotated map. Thus, the cell indicated by the asterisk (Fig. 5.14B) in the right retina now relays

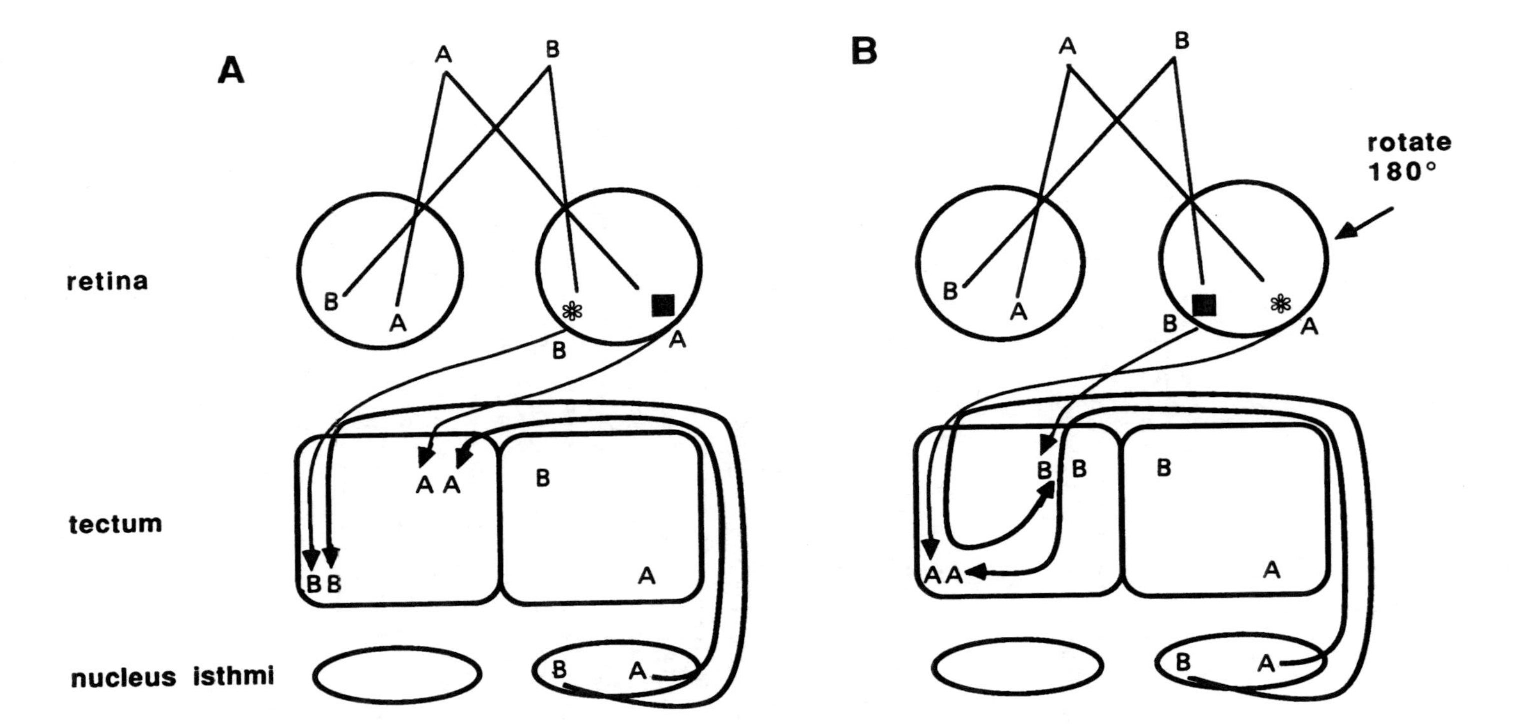

FIGURE 5.14. The nucleus isthmi is the relay nucleus for the ipsilateral eye's representation on the tectum. The isthmotectal map is in register with the contralateral eye's retinotectal map (A). The trajectories of isthmotectal axons will change if one eye is rotated during development, and the new pattern of projections will bring the ipsilateral map into register with the contralateral map.

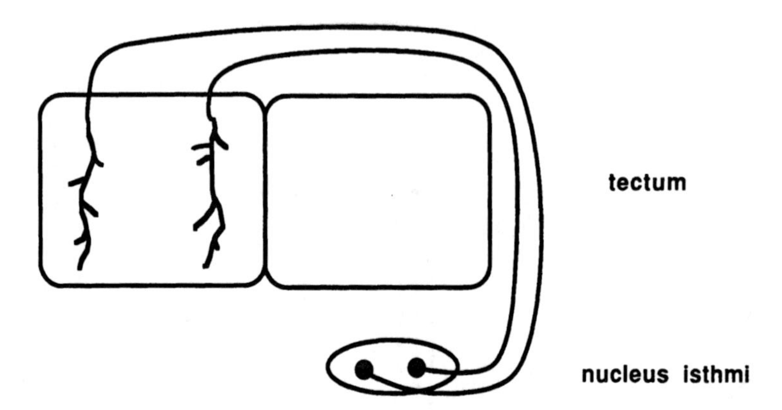

FIGURE 5.15. When isthmotectal axons first reach the tectum, they grow the full rostrocaudal length of the tectum, but they span only about 20% of the mediolateral extent of the tectum.

information about position **A** in the visual field rather about than position **B**. As a result, the retinotectal map from the rotated right eye to the left tectum is correspondingly rotated. What happens to the input from the normally oriented left eye? We see that the retinotectal map from the left eye to the right tectum is normal, as is the tectoisthmic map. However, the axons from the right nucleus isthmi do not have a normal arrangement: they are rerouted so that an isthmic axon that has a receptive field at point **A** in the visual field now projects to caudolateral tectum rather than rostromedial tectum. This new topography brings the isthmotectal map into register with the retinotectal map.

What is the evidence for such rearrangements of connections? First, electrophysiological recordings show altered ipsilateral topography. No matter what the degree of misalignment of the two eyes, each ipsilateral map conforms to the contralateral map on each tectum (Gaze et al., 1970). Second, anatomical labeling with horseradish peroxidase also demonstrates rewiring (Udin and Keating, 1981). If a small injection of HRP is placed in the tectum of an animal with a rotated eye, and if a comparable injection is made in a normal frog's tectum, the isthmic cells that are back-labeled in the two animals will be in different locations. In addition, if HRP is injected into the nucleus isthmi, then the isthmotectal axons will be filled and the abnormalities of the axons in eye-rotated frogs will be very apparent (Udin, 1983). The axons in normal frogs follow a rather orderly rostrocaudal trajectory in the tectum, whereas the axons in experimental frogs follow quite abnormal pathways, with much criss-crossing and turning en route to their final sites of termination.

Isthmotectal Axons' Trajectories during Development

What developmental processes produce these trajectories? In normal *Xenopus*, the axons grow in very early, before the onset of eye migration, and they branch

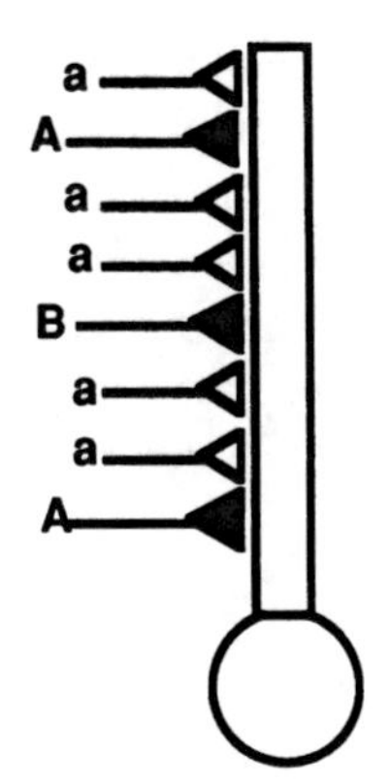

FIGURE 5.16. The terminal of axon B, which has a different receptive field from axons a and A, will be eliminated because its activity does not match that of the retinotectal axons.

along rostrocaudal strips of tectum (Udin, 1989). Each axon follows a rostrocaudal trajectory, with little tendency to wander very far laterally or medially from the position at which it first entered the tectum (Fig. 5.15). When the eyes begin to shift and there is a region of binocular overlap, then these axons begin to make more localized branches in "appropriate" regions. The other branches gradually are resorbed.

How does a branch "know" if it is in the right place? As we have seen before in the formation of the retinotectal map, activity cues from visual input play a role, but it is relatively minor, being important only for the final refinement stages. In contrast, the role of visual activity in the development of the isthmotectal map is crucial: proper binocular activity patterns are essential for the proper formation of the *Xenopus* isthmotectal map. Our hypothesis is that an isthmotectal terminal is stabilized when its activity matches the activity of the nearby retinotectal terminals. If an isthmotectal axon forms a branch at an improper location, then that branch will eventually fail to survive because of its tendency to fire out of synchrony with the nearby retinotectal axons (Fig. 5.16).

Cellular Mechanisms

The isthmotectal system may depend on the same sort of activity related mechanisms that play a role in retinotectal map refinement. As we have seen above, NMDA receptors may be an essential element in determining whether particular sets of synapses are stabilized or not. Our data (Scherer and Udin, 1989a) indicate that chronic treatment with NMDA receptor blockers does not prevent the formation of an ipsilateral map but does prevent the isthmotectal map from coming into register with the retinotectal map.

The Critical Period

The ability of isthmotectal axon to change their locations in response to altered visual input normally is limited to a critical period of development. Rotating the eye after about two months past metamorphosis normally does not lead to a compensatory rotation of the ipsilateral projections. However, plasticity can be restored by chronic application of NMDA. If one eye is rotated by 90° at eight months postmetamorphosis, and chronic NMDA is begun at the same time and is continued for the next three months, the ipsilateral maps *do* rotate. These results imply that the normal loss of plasticity is related to a change in the number or functioning of NMDA receptors (Scherer and Udin, 1989b).

Summary

Retinotectal and isthmotectal maps are formed by the combined action of several mechanisms, including chemoaffinity labels, selective cell death, and matching of activity patterns. Great progress is currently being made in discovering the mechanisms by which activity influences developing synapses by way of NMDA receptors. Progress on identification of the physical identity of chemoaffinity cues has still been elusive, but the many advances in biochemical tools for isolating and characterizing molecules in the brain promise to yield some clues in the next decade. Similarly, we do not yet know what molecules may be involved in the control of selective cell survival in the retinotectal projection, but the daily accumulating tally of survival factors for neurons in the central nervous system also may soon bring us some convincing evidence of the identity of the molecules and processes controlling neuronal survival.

References

Cline HT, Debski EA, Constantine-Paton M (1987). N-Methyl-D-aspartate receptor antagonist desegregates eye-specific stripes. Proc. Natl. Acad. Sci. U.S.A. **84**:4342–4345.

Constantine-Paton M, Law MI (1982). The development of maps and stripes in the brain. Sci. Am. **247**:62–70.

Constantine-Paton M, Reh TA (1985). Dynamic synaptic interactions during the formation of a retinotopic map. In *Neurobiology: Molecular Biological Approaches to Understanding Neuronal Function and Development* (P. O'Lague, ed.), Liss, pp. 151–168. Liss, New York.

Fawcett JW, O'Leary DDM (1985). The role of electrical activity in the formation of topographic maps in the nervous system. Trends Neurosci. **8**:201–206.

Finlay BL, Schneps SE, Schneider GE (1979). Orderly compression of the retinotectal projection following partial tectal ablation in the newborn hamster. Nature (London) **280**:153–155.

Frost DO, Schneider GE (1979). Plasticity of retinofugal projections after partial lesions of the retina in newborn Syrian hamsters. J. Comp. Neurol. **185**:517–567.

Gaze RM, Hope RA (1983). The visuotectal projection following translocation of grafts within an optic tectum in the goldfish. J. Physiol. **344**:257–275.

Gaze RM, Keating MJ (1972). The visual system and "neuronal specificity." Nature (London) **237**:375–378.

Gaze RM, Sharma SC (1970). Axial differences in the reinnervation of the goldfish optic tectum by regenerating optic nerve fibres. J. Physiol. **10**:171–181.

Gaze RM, Keating MJ, Szekely G, Beazley L (1970). Binocular interaction in the formation of specific intertectal neuronal connexions. Proc. R. Soc. London B **175**:107–147.

Grant S, Keating MJ (1986). Normal maturation involves systematic changes in binocular visual connections in *Xenopus laevis*. Nature (London) **322**:258–261.

Gruberg ER, Udin SB (1978). Topographic projections between the nucleus isthmi and the tectum of the frog *Rana pipiens*. J. Comp. Neurol. **179**:487–500.

Keating MJ, Feldman J (1975). Visual deprivation and intertectal neuronal connections in *Xenopus laevis*. Proc. R. Soc. London B. **191**:467–474.

Langdon RB, Freeman JA (1987). Pharmacology of retinotectal transmission in the goldfish: Effects of nicotinic ligands, strychnine, and kynurenic acid. J. Neurosci. **7**:760–773.

Levine R, Jacobson M (1974). Deployment of optic nerve fibers is determined by positional markers in the frog's tectum. Exp. Neurol. **43**:527–538.

MacDermott AB, Mayer ML, Westbrook GL, Smith SJ, Barker JL (1986). NMDA-receptor activation increases cytoplasmic calcium concentration in cultured spinal cord neurons. Nature (London) **321**:519–522.

Maturana HR (1960). The fine anatomy of the optic nerve of anurans. J. Biophys. Biochem. Cytol. **7**:107–121.

Meyer RL (1983). Tetrodotoxin inhibits the formation of refined retinotopography in goldfish. Dev. Brain Res. **6**:293–298.

Nowak L, Bregestovski P, Ascher P, Hebert A, Prochiantz A (1984). Magnesium gates glutamate-activated channels in mouse central neurones. Nature (London) **302**:462–465.

O'Leary DDM, Fawcett JW, Cowan WM (1986). Topographic targeting errors in the retinocollicular projection and their elimination by selective ganglion cell death. J. Neurosci., **6**:3692–3705.

Roberts PJ, Yates RA (1976). Tectal deafferentation in the frog: Selective loss of L-glutamate and γ-aminobutyrate. Neuroscience **1**:371–374.

Scherer WJ, Udin SB (1988). The role of NMDA receptors in the development of binocular maps in *Xenopus* tectum. Soc. Neurosci. Abst. **14**:675.

Scherer WJ, Udin SB (1989a). N-methyl-D-aspartate antagonists prevent interaction of binocular maps in *Xenopus* tectum. J. Neurosci. **9**:3837–3843.

Scherer WJ, Udin SB (1989b). NMDA restores plasticity of binocular tectal maps in post-critical period *Xenopus*. Soc. Neurosci. Abst. **15**:1212.

Schmidt JT, Edwards DL (1983). Activity sharpens the map during the regeneration of the retinotectal projection in goldfish. Brain Res. **269**:29–39.

Sperry RW (1944). Optic nerve regeneration with return of vision in anurans. J. Neurophysiol. **7**:57–69.

Straznicky C, Gaze RM, Keating MJ (1981). The development of the retinotectal projections from compound eyes in *Xenopus*. J. Embryol. Exp. Morphol. **62**:13–35.

Udin SB (1977). Rearrangements of the retinotectal projection in *Rana pipiens* after unilateral caudal half-tectum ablation. J. Comp. Neurol. **173**:561–583.

Udin SB (1983). Abnormal visual input leads to development of abnormal axon trajectories in frogs. Nature (London) **301**:336–338.

Udin SB (1989). The development of the nucleus isthmi in *Xenopus*. II. Branching patterns of contralaterally projecting isthmotectal axons during maturation of binocular maps. Vis. Neurosci. **2**:153–163.

Udin SB, Keating MJ (1981). Plasticity in a central nervous pathway of *Xenopus*: Anatomical changes in the isthmotectal projection after Larval eye rotation. J. Comp. Neurol. **203**:575–594.

Willshaw DJ, Gaze RM (1983). The visuotectal projections made by *Xenopus* 'pie slice' compound eyes. J. Embryol. Exp. Morphol. **74**:121–137.

Yoon M (1972). Synaptic plasticities of the retina and of the optic tectum in goldfish. Am. Zool. **12**:106.

Part 2
Psychophysics

6
Perceptual Aspects of Spatial and Temporal Relationships

K.N. LEIBOVIC

Introduction

Our immediate experience of the world around us is that space and time are continuous. As we move we do not encounter gaps in space, and our time flows without interruption. Based on what we see and hear we believe we can localize events precisely in space and time. But from a comparison with scientific observation and inference we conclude that our sensory performance is subject to some limitations. In the following pages we discuss some of these limitations. Unlike the chapters which take a more specialized approach we now look at the broader picture and search for possible connections between seemingly disparate observations.

Eye Movements, Photoreceptor Mosaic, and Visual Persistence

There are many examples illustrating the differences between belief based on perception and inferred "reality": one of these occurs in reading. When we read a printed page we may believe that our eyes move smoothly over each line and then return to the start of the next. In fact, it is well known that our eyes jump from point to point dwelling on each point of a line for a while, and then jump back to the beginning of the next line (Yarbus, 1967). The number of jumps and the dwell time at each point depend on the difficulty of the text. Figure 6.1 shows two examples of eye movements during reading. The first was conceptually more difficult for the reader than the second. It contains more fixation points and each line takes longer to read. Another example is found when we view a photograph or a scene. Figure 6.2 shows eye movements while looking at a portrait. Again we may think that we scan the scene in a smooth and continuous manner. But, in fact, our eyes go briefly over numerous fixation points separated by jumps and concentrated on those features conveying salient information.

How is this behavior of our visual system related to what is known about structure and function?

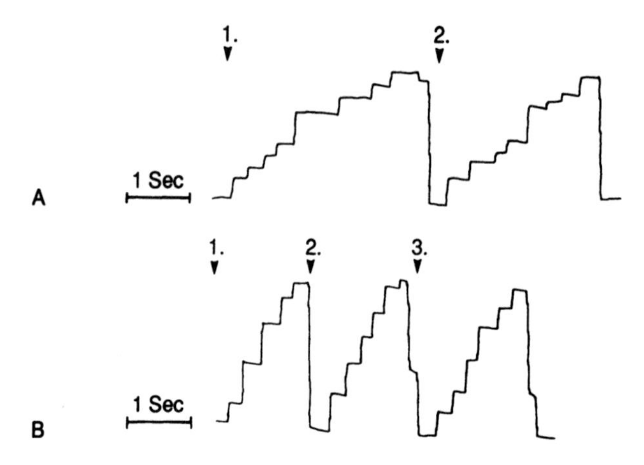

FIGURE 6.1. Two examples of eye movements during reading. The horizontal segments represent durations during which the eyes dwell on a "fixation point." The vertical segments represent the movements of the eyes along the line, in short steps while reading the line and in a long, quick step back to the beginning of the next line. The text in A was more difficult to comprehend than in B. (Modified from Pöppel 1985 Figs. 1 and 2, © Deutsche Verlags-Anstalt, with permission.)

Beginning with the eye there are physical limitations due to the optics and the spacing of photoreceptors. This is discussed in more detail in Chapter 10. If two separate points of light were projected on two adjacent cones the effect would be the same as of a larger single patch of light restricted to these same cells. Thus, to distinguish two point sources as separate there must be at least one cell which is stimulated noticeably less than its two neighbors. But due to diffraction two point sources must also be a least distance apart in order to be resolved by any optical system. It turns out that under optimum conditions cone separation matches the diffraction limit of the eye. This reinforces our belief in the optimal design of biological systems. At the same time it points up a limit to spatial resolution.

Our retina containing rods and cones is anything but uniform (see, e.g., Pirenne, 1962a). In the central retina there is the fovea, about 0.5 to 1.0 mm in diameter, with a cone density of 2×10^5 per mm^2. Within the fovea a 0.3- to 0.5-mm-diameter rod free area contains 30,000 to 40,000 cones. These figures correspond to a circle of some 5 μm^2 or 2.5 μm diameter or about 30 sec. of visual angle allocated per cone. From the fovea out the rod density increases from zero to a peak of some 150,000 per mm^2 at 18° from the fovea and then declines again to less than 40,000 per mm^2 toward the edge of the retina or ora serrata. The rods mediate our vision in dim illumination. It needs only one photon to evoke a rod response and only about a dozen photons in a local area to have the sensation of light in total darkness. The cones mediate daylight vision when light intensity is some 6 or 7 orders of magnitude above the absolute rod threshold and up to another 7 or 8 orders of magnitude above that in bright sunlight (see, e.g., Riggs,

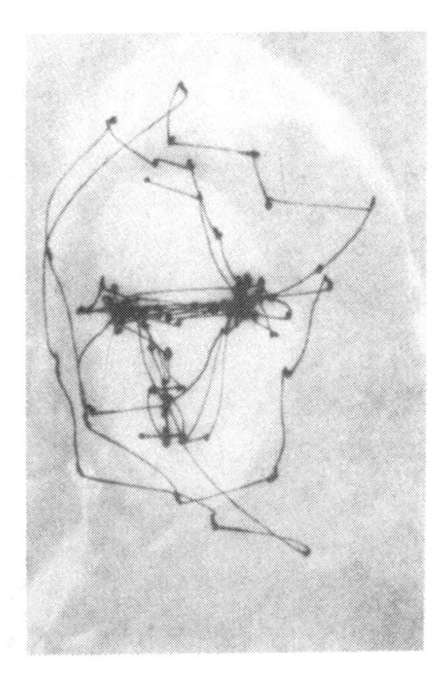

FIGURE 6.2. An example of eye movements while examining a portrait. (Yarbus 1967, Fig. 114 © Plenum Press, with permission.)

1965). Thus, rods are more sensitive, but color vision and high spatial and temporal acuity are achieved in cone vision and involve the fovea. Of course, color vision is not confined to the fovea. After all, there are more cones outside than within it. But extrafoveal cones are more sparsely distributed. Peripheral vision including both rods and cones plays an important role in visual search and influences directed eye movements (see, e.g., Ch. 9 in Haber and Hershenson, 1973). Thus central vision enables us to see detail, while peripheral vision is sensitively attuned to change. But foveal vision is restricted: 1 mm on the retina only corresponds to some 200 min of visual angle. This means that the image of a 20-mm-diameter circle at a reading distance of 40 cm will more or less cover the whole fovea. Hence the need to scan a scene which extends beyond that field of view, as we saw in Figures 6.1 and 6.2.

Let us now take another look at eye movements, this time at involuntary eye movements. For, as we may expect from the discrepancies between what we think and what we actually do, when reading or viewing a scene our eye movements are not under strict voluntary control. In fact, it can be demonstrated that our eyes move involuntarily even when we fix our gaze on a marker in the visual field as shown in Figure 6.3a. These involuntary movements contain several components. There is a high frequency tremor, a slow drift, and a rapid saccade. The durations and amplitudes of these movements vary considerably. The mean amplitudes and durations are approximately as follows: 20″ amplitude and 40–80

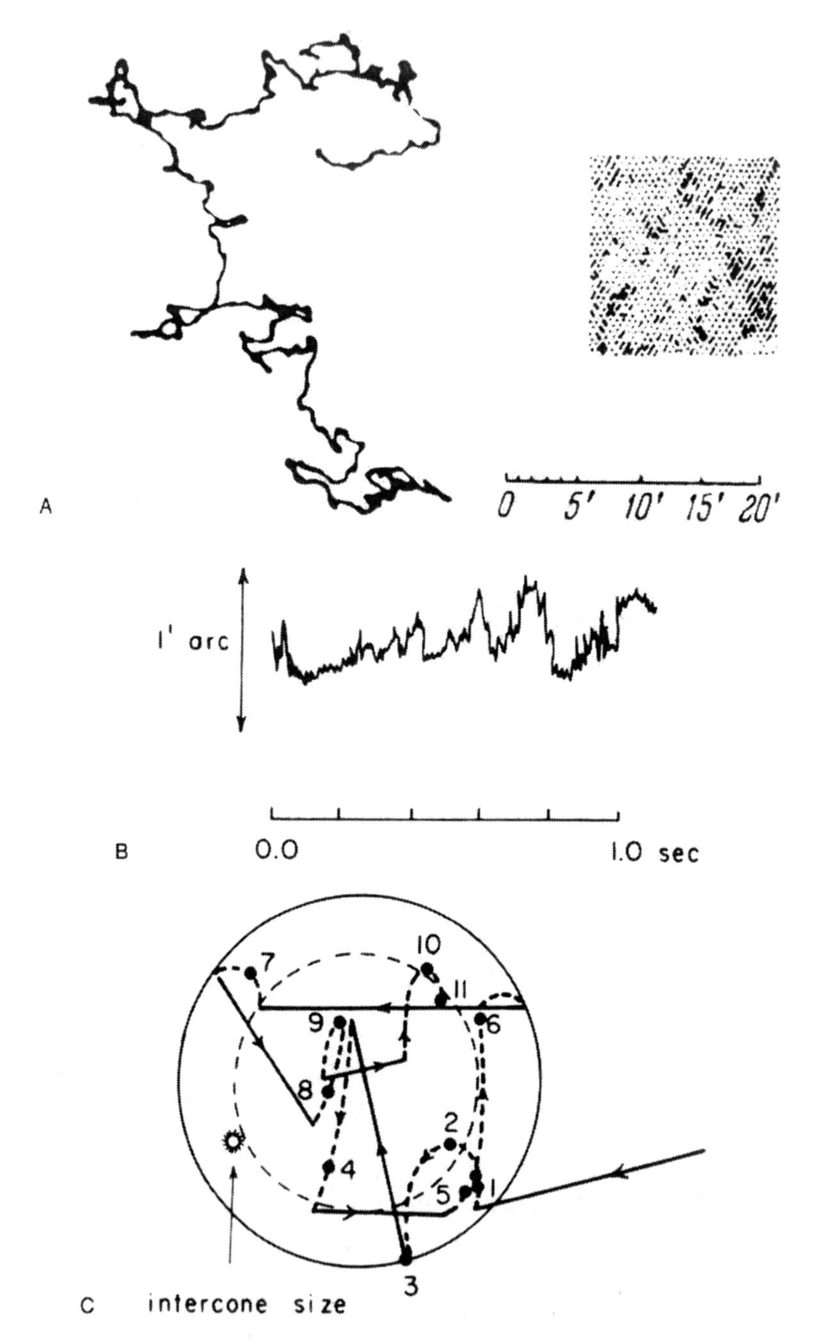

FIGURE 6.3. (a) A record of eye movements while a subject "fixated" on a marker for 1 min. The distribution of retinal cones is illustrated on the right and below it is a scale in minutes of visual angle. (From Yarbus, 1967, Fig. 54b, © Plenum Press, with permission.) (b) Enlarged record of high-frequency tremor. (c) Record of slow drift (dotted) and saccade (full lines). The numbered dots are separated by 200 msec. The large circle corresponds to 10′ of visual angle (b and c from Ditchburn, 1961, © Chemical Publishing, with permission.)

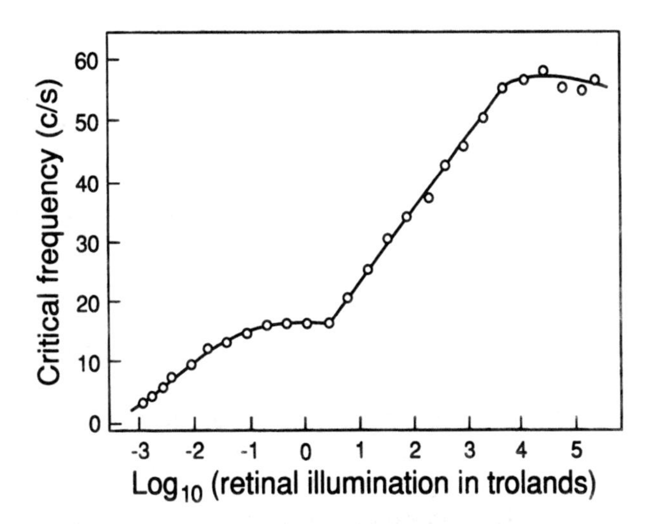

FIGURE 6.4. Variation of flicker fusion frequency with illumination. A centrally fixated 19° field was flashed on and off and the frequency was determined at which the light appeared to be steady. The lower limb of the curve applies to rod vision, the upper limb to cone vision. (After Hecht and Smith, 1936, J. Gen. Physiol. **19**:979–989, by copyright permission of the Rockefeller University Press.)

Hz frequency for the high-frequency component, 2′ to 5′ amplitude and 200–300 msec duration for the slow drift, and 5′ to 6′ amplitude and 30 msec duration for the saccade (Ditchburn, 1961, Yarbus, 1967; but see also Steinman et al., 1973). These components are illustrated in Figure 6.3b and c. They form the fine grain of the pattern of involuntary eye movements. They are not revealed in the coarser measurements of Figures 6.1 and 6.2 where they only appear as "fixation points." When we compare the amplitudes of these movements with foveal cone dimensions we find that the tremor corresponds to somewhat less than a cone diameter and the drift and the saccade to some 4 to 10 cone diameters. As for the durations, it is instructive to consider some temporal parameters in vision.

We know that the apparent motion we see on a cinema screen is due to the rapid succession of a series of still pictures. Similarly, when a flickering light is presented to a subject it will appear as a continuous light above a certain frequency, the so called critical fusion frequency (CFF) (see, e.g., Pirenne, 1962b). This frequency depends on the state of adaptation as illustrated in Figure 6.4. There a centrally fixated field was alternately flashed on and off. The retinal illumination is the average over a light–dark cycle, and as it increases, so the critical frequency increases. A representative critical frequency in photopic vision may be taken as about 30 Hz. We would therefore want to be at this limit if we were to view a scene, acquiring a series of samples while preserving a sense of continuity.

There is another phenomenon which should be mentioned. It is visual persistence, a kind of short-term visual memory. We experience this in the afterimages of bright presentations, but these are complicated by interactions with slow adaptive mechanisms (cf. review by Ganz, 1975). A better measure of persistence in a steady-state environment may be the fading of a stabilized image. It is possible experimentally to produce such stabilized images which remain stationary on the retina even as the eye moves (Ditchburn and Ginsborg, 1952; Riggs et al., 1953; Barlow, 1963; Yarbus, 1967). When this is done one finds that the image disappears within about 1–3 sec and does not reappear. Thus, 1–3 sec can be taken as a measure of this kind of short-term visual memory. Beyond that, eye movements are essential to acquire and retain visual input.

A Model for the Acquisition of Visual Information

In surveying the above phenomena there emerges a model for the acquisition of visual information (Leibovic, 1965, 1968, 1972).

First, when two stimuli occur during an interval corresponding to the critical flicker fusion frequency they appear as contemporaneous. This interval can thus be considered as one "frame" or single presentation in perception. Second, for a "good look," to perceive and evaluate an elementary pattern we need some 200–300 msec. (Riggs, 1964), the same as the time spent, on the average, on a "fixation point" during reading or in examining a visual scene. It is also the mean duration of the slow involuntary drift of the eyes. Therefore, if we take, say, 30 msec for the duration of one frame, we acquire between 6 and 10 frames during a good look. Moreover, during this time the image of any object point has moved over some 4–10 cone diameters, remembering that the latter occupy about 30″ of visual angle and that the mean excursion of the slow drift is between 2′ to 5′. It is as if a package of some one half to one dozen images were impressed onto shifted, overlapping sets of photoreceptors. Of the two other components of the involuntary eye movements, the tremor can serve to remove high frequency spatial noise, as is sometimes done when blurring a photograph to smooth its appearance. As regards the saccades, some authors dispute their obligatory presence (see, e.g., Steinman et al., 1973; Kowler and Steinman, 1980) and do not ascribe any useful visual functions to them. Others have suggested that their function is to keep the eyes on target after the more or less random sampling motion of the drifts (see Ditchburn, 1980). There is evidence that they probably do not contribute to information acquisition: visual acuity during these microsaccades seems to be depressed (Beeler, 1967) as it is in the larger, voluntary saccades (Volkman, 1962) as well as for experimental image displacements with stimuli as brief as a few microseconds (MacKay, 1970). In any event their duration is so short that they correspond to at most one frame smeared over a displacement of some 4–10 cones. Finally, the package of impressed images sampled during the drift is held in a short-term visual memory for 1–3 sec and then, as in a stabilized image, fades away.

Although the model I have proposed is somewhat speculative, it has the merit of pulling together a number of spatiotempral observations, and it presents a coherent picture of a well-coordinated system for information acquisition. The numbers we have quoted should be seen as orders of magnitude and no more. They will vary with subject, task, and experimental conditions.

Space Perception

Continuity is the absence of gaps. Perceived temporal gaps cannot be smaller than the critical frequency for flicker fusion and spatial gaps between point sources cannot be smaller than two receptor diameters. However, other geometric forms can yield a higher resolution known as hyperacuity, which must involve network mechanisms (Westheimer, 1979, 1981). For example, there is vernier acuity: when two vertical lines are separated by a gap of between 2′ to 5′ of visual angle, a sideways displacement of 6″ of one line relative to the other can be detected. This is an order of magnitude better than the discrimination of two point sources. Stereoscopic acuity is of the same order of magnitude: disparities of less than 10″ in the two eyes give rise to perceived depth. This level of performance is clearly not limited by the spacing of receptors on the retina. The neural basis of hyperacuity is not fully understood. Presumably it results from some cooperative interactions in neural assemblies. It is not clear whether the micromovements such as the image blurring produced by the high-frequency eye tremor or the sampling during a drifting motion contribute to this capability. It appears from some data that vernier acuity is hardly different in stabilized and normal viewing (see Westheimer, 1981). There are two interesting observations which are consistent with our model for information acquisition (see Westheimer, 1981). The first is that two targets in a hyperacuity test have to be presented "simultaneously," i.e., within some 30 msec to be detected and when the interval is extended, performance is appreciably degraded. The second is that a dim vernier target has a high spatial threshold, i.e., the separation between the two vernier segments must be relatively large to be detected. But when the same target is flashed repeatedly for 200 msec, the threshold is normal. Moreover, the successive target presentations need not be in the same retinal location, but can sweep over 2′ to 4′ of visual angle and still give the same threshold. These results agree with the idea of a 30 msec frame of simultaneity and of an integrated 200 msec information packet for a "good look" during a 2′ to 4′ slow drift of the eye, as presented above.

Due to the different positions of the two eyes the image of an external point on the left retina will be displaced relative to that on the right retina. A spot of light to the left of the fixation point, for example, is displaced temporally from the fovea in the right eye and nasally in the left eye. It is this disparity which is the basis of binocular space perception (see also Dow, Chapter 4). By comparison, the angular positions of the eyes and accommodation to a near or distant focus are insignificant cues (see, e.g., Graham, 1965).

Figure 6.5 shows how the external world, with reference to a fixation point, is represented in the visual cortex. We are all aware that an object will appear double unless it is within a certain domain around the fixation point. This is Panum's area. Thus, we normally have single vision in only a restricted, sausage shaped portion of our surroundings, namely Panum's area, using a small subset of our retinal mosaic, namely the central blob we call the fovea, to view our environment.

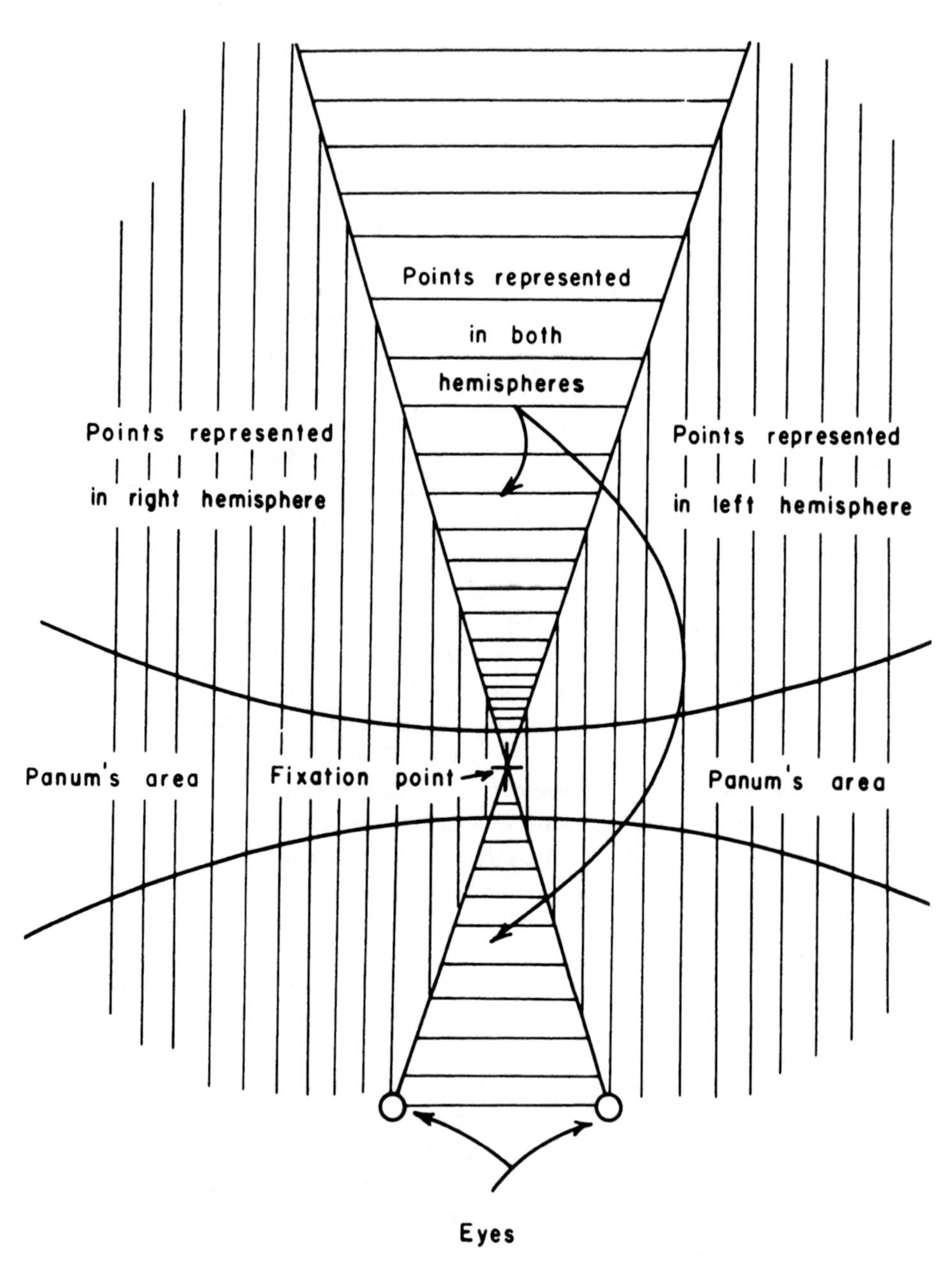

FIGURE 6.5. The projection of visual space onto the eyes and its mapping onto visual cortex V1. (From Leibovic, 1970, © Academic Press, with permission.)

Although there are monocular cues which inform us about spatial relationships, binocular vision is fundamental to our perception of space.

In everyday discourse we speak of absolute distances in our surroundings and parallel lines which never meet. This is Euclidean geometry. It may therefore come as a surprise when we find that our perceived space is non-Euclidean. For Euclidean geometry seems the "natural" geometry and it stood supreme for more than 2000 years until it was challenged by mathematicians like Bolyai, Gauss, and Lobachevsky who developed hyperbolic geometry. A more general description of physical space is provided by Riemannian geometry. Fundamental to the latter is a distance metric which is a function of the coordinates. The problem with perceived space is that the distance between any two points depends on the objects in the environment. The metric would have to include this feature and in a changing environment it could become very complicated. In any event the data to describe such a metric are not yet available. Even in the simplest situations, the parameters necessary to quantitate our perceptions remain unknown. This is illustrated in the following discussion, which highlights the need of further research in this field.

In reduced cue situations an experimental subject is exposed to minimal stimulation in an attempt to isolate a particular set of variables. Such is the case when a set of point sources of light in an otherwise blank environment is available. For example, a subject sitting in the dark with his eyes slightly above the level of a horizontal table may be asked to align a set of lights on the table, perpendicular to his line of regard. When the lights are set up at different distances from the subject they form a set of perceived frontoparallel lines. It is found that instead of setting the lights along straight lines, subjects arrange them on curves which are concave toward them at near distances and convex at far distances. Now, as stated earlier, the only significant cue in such a situation is presumed to be disparity, in other words the relative positions of the images in the two eyes. Thus, if one knows one frontoparallel line one should be able to predict all the others (Leibovic et al., 1971). For one such line determines the positions of the retinal images of the lights which have been set up. To set up another frontoparallel line the subject moves his fixation point to a new distance. This involves a rotation of the eyes and if we now backproject the retinal image points out into space we can predict where the lights need to be to produce these same retinal images. When one performs the necessary calculations one finds that the theoretical frontoparallel line differs significantly from the actual one. This is shown in Figure 6.6. While the subject fixated a point 18″ in front of him, he set up the "frontoparallel" points along the curve E–E, passing through $x=18''$. When these points are projected to produce the same retinal images for the fixation point at $x=54''$, they yield the curve T-T through $x=54''$. As can be seen these points are quite different from the experimental ones on the curve E–E through $x=54''$. In turn, the latter, projected to the fixation point through $x=18''$, are also different from the experimental points at that fixation. We must conclude that disparity is not the only cue in these conditions. Whatever other cue there is, it must enter into the neural computations.

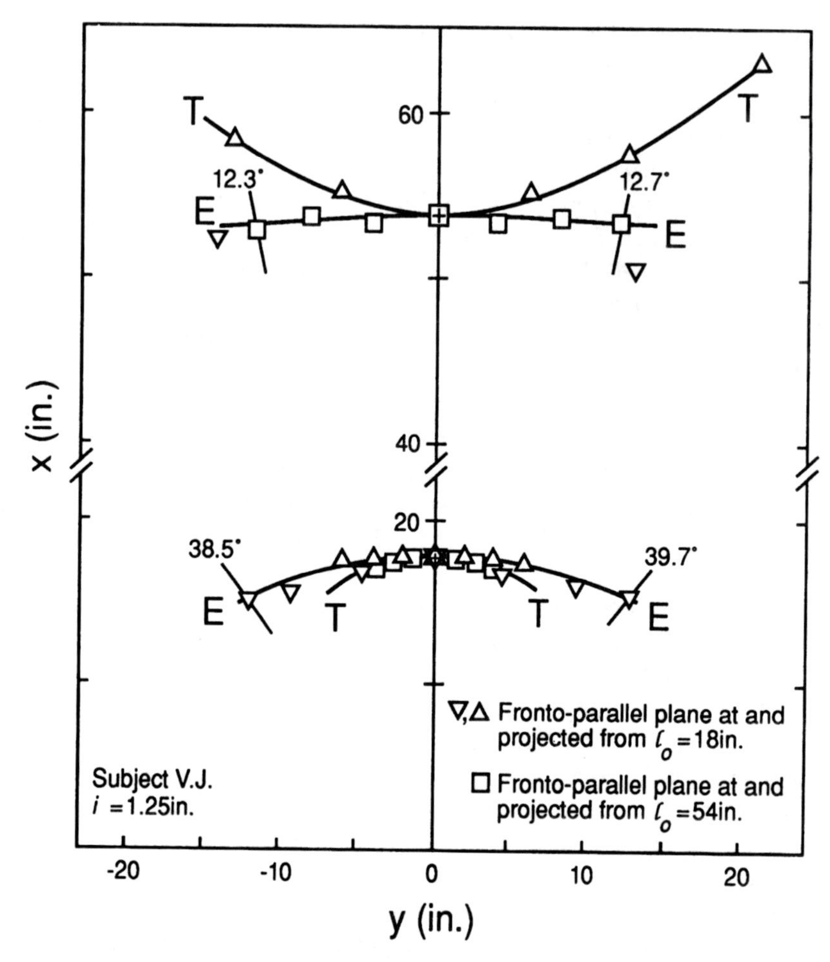

FIGURE 6.6. Experimental (E) and theoretical (T) frontoparallel lines at 18 in. and 54 in. from the subject. The theoretical points at 18 in. (54 in.) are the projections from the experimental points at 54 in. (18 in.). The curves are drawn by eye through the experimental and theoretical points. (Modified from Leibovic et al., 1970, © Academic Press, by permission.)

It is natural to search for a coordinate system for perceived space. Having one we could describe quantitatively our judgments of orientation, distance, shape, and movement. A major attempt in this regard was made by Luneburg (1947) who deduced from available data in reduced cue experiments such as we have discussed that perceived space could be described by a hyperbolic geometry (see also Hardy et al., 1953). For the reasons indicated above, hyperbolic geometry being a special case of Riemannian geometry, this turns out to be an inadequate description (Leibovic et al., 1971). But although a perceptual coordinate system remains elusive, there is progress. A basic fact to remember is that we refer space to ourselves, as the observer, at the origin of coordinates; and as we move through space it is perceived distance and direction, or orientation, which are of special relevance. It would not be surprising, then, to find that our perceived space relied on a kind of polar coordinate system (Leibovic et al., 1971). Indeed, there is support for this proposition on theoretical as well as psychophysical and neurophysiological grounds (see Richards, 1975 for further discussion). First, it is possible theoretically to derive a set of curves in space whose retinal images remain invariant as the fixation point moves along such a curve (Leibovic et al., 1971). These curves include a subset which is perceived to be equidistant from the observer and another which is perceived to have a constant radial orientation, essentially forming a subjective polar coordinate system. Second, eye movements from one fixation point to another take place in two stages: a change of orientation and a change of distance (Westheimer and Mitchell, 1956; see also Richards, 1975). This suggests separate mechanisms for processing orientation and distance. Third, single cell recordings have shown that there are cortical cells which are tuned to disparity or position (Barlow et al., 1967) and, apparently, to direction or orientation (Blakemore, 1970; Hänny et al., 1980) with respect to the observer. This provides a neural basis for our polar coordinates. The three lines of evidence suggest the outlines for a possible egocentric frame of reference.

Neural Representations of Percepts

The above example of the frontoparallel lines, interesting in itself, has a bearing on a long-standing question: Are percepts encoded in single cells, or in fixed cell assemblies or in certain activity patterns analogous to computer programs?

We have already seen in previous chapters how cells are combined into receptive fields and converging onto a target cell, make the latter respond selectively to a restricted class of stimuli (see Chapters 3 and 4). Some cells in visual cortex respond best to edges or bars in a particular orientation; others have varying degrees of ocular dominance; some binocular cells are tuned to different positions in space in relation to the point of fixation and others appear to be tuned to visual direction suggesting a kind of mapping of space onto the cortex as in a polar coordinate system. There have even been some reports of cells in temporal cortex which respond best when hands and faces are viewed (Gross et al., 1972, Bruce et al., 1971; Kendrick and Baldwin, 1987). On examination, however, we find that the receptive field tuning is so broad that single cells could never yield the clear and precise judgments which we can make in visual perception. In addition, the tuning is multivariable, for the cells respond to more than one stimulus parameter. This is discussed in more detail in Chapter 10.

We know that the retina is mapped topologically, point for point, onto V1 and from there further pathways lead to other cortical and subcortical areas (see Chapter 4). This means, in particular, that the retinal images of a frontoparallel line project onto fixed cell assemblies in cortex. But as we change the point of fixation that same set of cell assemblies no longer gives us the percept of a frontoparallel line. This argues against the representation of percepts in cell assemblies at least at the level of the spatial map. A more likely basis for percepts is something like a computational program of neural activity which is generated or regenerated from or activated by the appropriate input or cue. A hard wired encoding of percepts in fixed cell assemblies would not only require a close correspondence with the retinal image, but it would also constrain processes such as generalization, conditioning, and adjustment to change. A computational program, on the other hand, is logic based and, using a set of operations, can generate a variety of output depending on the input. The percept then becomes synonymous with a defined logical program of operations. The deviations of the experimental from the theoretical frontoparallel lines argue that there are other than disparity cues which contribute to the percept. These may be weak cues, like eye convergence, which nevertheless enter into the input to the computational program generating the neural activity which corresponds to the perceived frontoparallel line.

Convergence and Divergence and Mapping of Visual Space

The presence of receptive fields implies a convergence of cells from one level in the nervous system onto another. This is a necessity if one is to detect features such as edges or movement or any extensive variable. Figure 6.7 shows an example of very simple contrast or movement detectors illustrating how such features could be computed in converging networks. A receptive field can be pictured as containing a collection of such elementary networks side by side. At the same time as convergence there is a divergence in the nervous system, so that one cell sends signals to a collection of cells which we may call its responsive field. At first sight such a convergence and divergence would seemingly have to degrade the message and one might then ask how we can achieve the high resolution of which the visual system is capable?

It turns out that there need be no image degradation in a system with converging and diverging transmission as shown in Chapter 10. For example, light itself is propagated in this way. From any point on a wavefront, light radiates into all directions and any point on a screen receives radiation from all the points on the wavefront. But the image of an object can be faithfully reproduced by interposing the appropriate optics. What our analysis in Chapter 10 shows is that at the same time as computing relevant image parameters, the light (and color) intensities of each image point can be preserved in transmission. The computations of contrast, movement and so on are not made in a vacuum, but with reference to the image on the retina from which adjacent neighborhoods project onto adjacent neighborhoods in the cortex. This seems to be the significance of the mapping from retina through LGN onto cortex: the computations performed in the neural networks are embedded in a map of the external world as projected onto our retinas and from there to later processing stages.

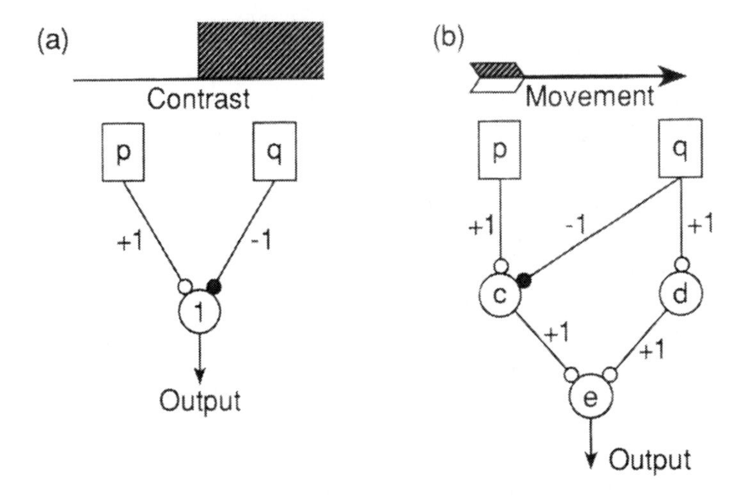

FIGURE 6.7. (a) One sided contrast detector. The output unit has a threshold of 1 and responds only if there is an input from p but not from q. When p and q are stimulated p produces an input of $+1$ and q of -1 to the output unit. All units produce a single output pulse. (b) Directional movement detector. d, c respectively produce short and long lasting excitation in e. Inhibition of c by q is somewhat longer lasting than excitation by p. c and d have a threshold of 1 and e a threshold of 2. The output strengths of the units are as shown.

A Sense of the Present Time

We have examined our sense of space from the minimum detectable to its global representation. We can see detail within a larger scene and this also enables us to move back and forth in space. By contrast, in time we are bound to the present moment. We can remember the past and project our thoughts into the future but what we experience is the present. This experience is not of an arbitrarily short physical "point" in time. Rather it is a fuzzy temporal "moment" which is conditioned by our sensory faculties, including vision. What is the basis for our visual experience of the present moment?

There are both retinal and cortical factors. One can say, as a general principle, that peripheral sense organs are matched to the central nervous system and vice versa. Thus, the characteristics of photoreceptors are mirrored in higher centers and a whole range of psychophysical phenomena can be traced back to them.

We have already seen that an increase in its flicker frequency makes a light appear continuous. This fact, which has long been known to psychophysics, has its counterpart in photoreceptor responses as discussed in Chapter 2. The data presented there are from *Bufo* rods. But the statement is also true for primates and cones as well as rods (Baron and Boynton, 1975). Psychophysical results mirror those obtained in photoreceptor recordings. Photoreceptors cannot follow changes in frequency beyond the critical fusion frequency (see Fig. 2.13,

Chapter 2). There is an integration time during which the light energy is summed to produce a response. When two light pulses follow each other during this integration time they are not seen as separate.

Another demonstration of such an integration time is the following. When either a long or short flash, both containing the same number of photons, is delivered within a certain interval, the photoreceptor response waveforms are the same. This is illustrated in Figure 6.8. The corresponding psychophysical phenomenon is known as Bloch's law. In one form it states that at the threshold of vision, up to a critical duration *T*,

$$\text{Flash Intensity} \times \text{Duration} = \text{Constant}$$

For example, suppose a 100 msec flash of intensity 100 photons/cm²/sec is just detectable at threshold under specified conditions, then a flash intensity of 1000 photons/cm²/sec will be required in a 10 msec flash to produce the same effect. The number of photons delivered in the two flashes is the same and they will be perceptually equivalent up to a flash duration *T*. In man, *T* may be as long as 100 msec. It is interesting that a form of Bloch's Law holds in the spatial as well as the temporal domain. If a given spatial contrast can just be perceived when it is presented for, say, 25 msec, then the contrast can be halved if the presentation time is doubled to 50 msec. Again, this holds up to an upper limit of stimulus duration, which is about 100 msec in man. As the eye adapts to higher luminances, *T* shortens. This parallels the increase in flicker fusion frequency under the same conditions (Roufs, 1972). In fact, *T* can be deduced from CFF data (Cornsweet, 1970) implying that they are generated by the same mechanisms.

FIGURE 6.8. Stimulus summation in *Bufo* rods (K.N. Leibovic and Y.Y. Kim unpublished data). (a) Intracellularly recorded responses in the presence of a background of nominal ND (neutral density) 6. In each row are superimposed responses to flashes differing in duration by a factor of 10. The flash durations are shown on the right. In each column the number of photons in a flash is constant. It is given in terms of the ND of the longer flash at the top and bottom of the figure. To the left of the slanting line the responses superimpose, but to the right they deviate from each other significantly. (b) Stimulus summation as a function of stimulus intensity and the state of adaptation. The ordinate is the flash intensity above threshold. The abscissa is the flash duration. The lines are regression lines based on experimental data recorded either intracellularly or with a suction electrode. DA, dark adapted; B10 and B30 are 10% and 30% bleached rhodopsin, respectively; I6 and I5 are backgrounds of ND6 and 5, respectively. To the left of the regression lines there is stimulus summation. For example, with a background of ND5 and a flash containing about 10 times the number of photons as a threshold flash (i.e., $I_F - I_T = 1$), the waveforms superimpose up to a flash duration of 150 msec.

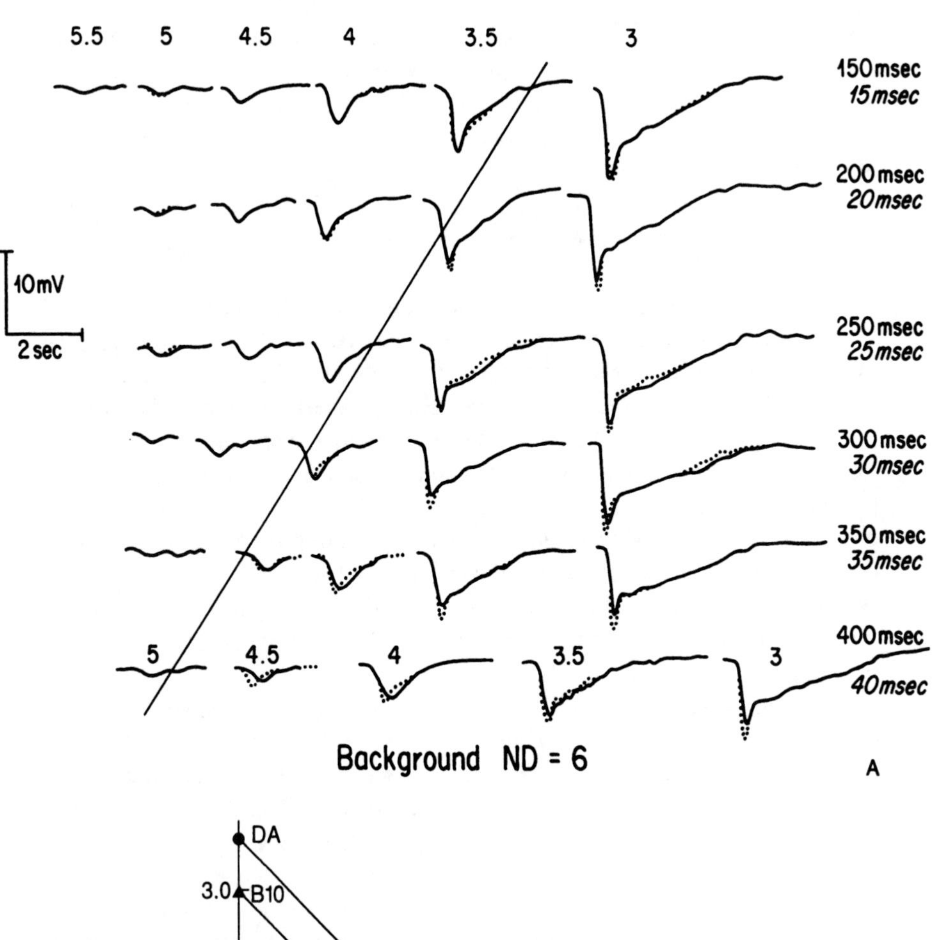
5.5
5
4.5
4
3.5
3
150 msec
15 msec
200 msec
20 msec
250 msec
25 msec
300 msec
30 msec
350 msec
35 msec
400 msec
40 msec
10 mV
2 sec
Background ND = 6
A

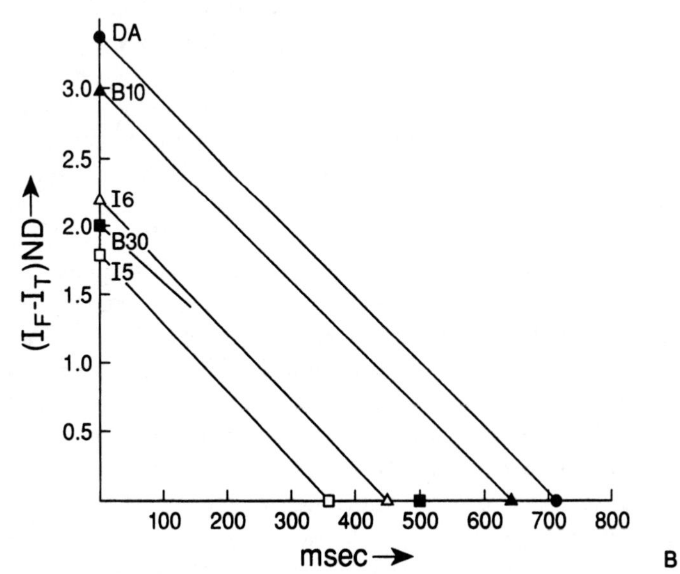
DA
B10
I6
B30
I5
$(I_F - I_T)$ND→
msec→
100
200
300
400
500
600
700
800
0.5
1.0
1.5
2.0
2.5
3.0
B

Figure 6.8 shows that stimulus summation can be demonstrated at the cellular level in photoreceptors. In these experiments we are not restricted to threshold conditions as in the psychophysical case. We can see how the critical duration T depends on flash intensity and the state of adaptation and whether we have bleaching or background adaptation (Fig. 6.8b; compare also the discussion of equivalence, in Chapter 2). By comparison with human psychophysics it should be remembered that in the cold-blooded *Bufo* the responses are much slower.

The above data underline the caveat made earlier that the figures we have used in our theoretical arguments should be considered as ballpark values and no more. Our purpose is to establish possible connections in the larger design. In this spirit we have proposed that elementary 30 msec frames are combined into a 200 msec information packet. When reading or examining a photograph or viewing a collection of objects we are processing and combining elementary features in a perceptual synthesis. As our eyes are held for 200 msec or so in a drifting mode on a small neighborhood of a printed page or a point of a visual scene we acquire more information than we could in the instant of a brief flash. This information synthesis provides us with our "good look." But the fact that we are not aware of the time spent on a "good look" is consistent with the proposition that perceived time has not yet receded into the past.

The above observations can help explain the meaning of simultaneous and successive events and our sense of what is "now." When two successive flashes produce the same photoreceptor response as a single flash of the same energy they can be perceived only as simultaneous or as one flash. This happens when the flash frequency is above the critical flicker fusion frequency. Just below that frequency we may observe a difference, as we observe a flow of events in a cinematograph. But if two such successive events are presented close together we are unable to say which came first, although we may know that two occurred. There is a gray area between the perception of simultaneity and succession. This gives us the notion that time flows continuously, which Newton believed was the nature of physical time. Within this flow of time there is the irreducible peripheral moment corresponding to the critical fusion frequency or the flash integration time. But a sense of "now" must involve more complex processes as well. To recognize a simple pattern it may be enough to flash it on a screen for less than the 30 msec or so of the critical fusion interval. But for anything with more content we need some 200 msec for a "good look" as we scan a photograph or read a book. It takes no conscious time as the "frames" of a "good look" are integrated in what seems we are "now" engaged in, at this present moment.

There is a yet higher order of integration which seems to be involved in our experience separating the present from past and future. Such a separation cannot have precise, perceptual boundaries. For, if we could say that at this exact physical (in the Newtonian sense) instant we entered from the past into the present, then this instant would have no temporal extent. At the same time we would experience this instant as the present "now" and, moreover, be able to distinguish it from any subsequent instant, however close, after having entered the present and having left the past behind. A similar argument would apply to a boundary between present and future. This is clearly not the case. But our experience of the present encompasses more than our sense of what is "now." It shares with the latter an integration of events into a perceptual whole or temporal "Gestalt." It has been suggested that we can get a measure of this from the perceived flipping of ambiguous figures such as a Necker cube [see the delightful book by Ernst Pöppel (1985) with which many ideas in this chapter have a kinship]. As one views this figure, which is a line drawing of a three-dimensional cube, one sees after a while what appeared to be the front face suddenly flipping over to become the rear face and vice versa. This process is largely involuntary and repeats itself every 1–3 sec on average. It occurs both monocularly as well as binocularly. If one tries to either prevent the cube from flipping over or to speed up the rate at which it does so, there are upper and lower limits to the duration when the cube appears stable. Another example which may be relevant to this problem is the fading of a stabilized image which, as mentioned earlier, also occurs within 1–3 sec. This fading may not be a simple peripheral phenomenon. There are reports that the image fades piecewise and that the nonfaded parts which disappear later are perceptual rather than sensory units (Pritchard et al., 1960). Yet another observation concerns the reproducibility of visual stimulus durations (Vierordt, 1868 quoted by Pöppel, 1985). A subject is presented with a stimulus of given duration and he is asked to reproduce this duration. It is found that up to about 3 sec, the durations are slightly overestimated, but beyond that they are significantly underestimated. It is as if the visual scene of the stimulus loses its temporal coherence after about 3 sec.

These phenomena suggest that there is an elementary unit of attention shaped by our sensory apparatus and registered centrally, as if a screen were to scroll by holding about 1–3 sec of input and renewing it after that. For as long as a stabilized image persists or an ambiguous configuration is stable or a visual scene is temporally coherent we are processing visually in the present time.

Our experience of time thus turns out to be a hierarchy of complex, interrelated processes subserved by sensory and integrative mechanisms in the brain. It includes the notions of simultaneous and successive events, a sense of what is "now," and the experience of the present. As we enquire into these processes by biophysical, physiological, and psychological techniques we gain a better understanding of the characteristics and limitations of perceptual experience. Our first and persisting view of the world is given to us by this experience. By knowing our limitations we may be able to appreciate our place in this world in a better perspective.

Acknowledgments. I am particularly grateful to Drs. Robert Boynton, Naomi Weisstein, Henk Spekreijse, and Bill Maguire for their valuable and constructive comments on this chapter.

References

Barlow HB (1963). Slippage of contact lenses and other artifacts in relation to fading and regeneration of supposedly stable retinal images. Q. J. Exp. Psychol. **15**:36–51.

Barlow HB, Blakemore C, Pettigrew JD (1967). The neural mechanism of binocular depth discrimination. J. Physiol. **193**:327–342.

Baron WS, Boynton RM (1975). Response of primate cones to sinusoidally flickering homochromatic stimuli. J. Physiol. **246**:311–331.

Beeler GN (1967). Visual threshold changes resulting from spontaneous saccadic eye movements. Vision Res. **7**:769–775.

Blakemore C (1970). The representation of three dimensional visual space in the cat's striate cortex. J. Physiol. **209**:155–178.

Cornsweet TN (1970). *Visual Perception*. Academic Press, New York.

Ditchburn RW (1961). Eye movements in relation to perception of color. In *National Physical Laboratory Symposium on Visual Problems of Color*, Vol. 2. Chemical Publishing Co., New York.

Ditchburn RW (1980). The function of small saccades. Vision Res. **20**:271–272.

Ditchburn RW, Ginsborg BL (1952). Vision with stabilized retinal image. Nature (London) **170**:36–37.

Ganz L (1975). Temporal factors in visual perception. In *Handbook of Perception* (EC Carterett, MP Friedman, eds), Vol. 5. Academic Press, New York.

Graham CH (1965). Visual space perception. In *Vision and Visual Perception* (CH Graham, ed.), Chap. 18. Wiley, New York.

Haber RN, Hershenson M (1973). The *Psychology of Visual Perception*. Holt, Rinehart & Winston, New York.

Hänny P, von der Heydt R, Poggio GF (1980). Binocular neuron responses to tilt in depth in the monkey visual cortex. Evidence for orientation disparity processing. Exp. Brain Res. **41**:A26.

Hardy LH, Rand G, Rittler MC, Blank AA, Boeder P (1953). The Geometry of Binocular Space Perception Report to USONR Project NR 143-638, Contract N6onr27119. Columbia University, New York.

Hecht S, Smith EL (1936). Intermittent stimulation by light. J. Gen. Physiol. **19**:979–989.

Kowler E, Steinman RM (1980). Small saccades serve no useful purpose. Reply to a letter by RW Ditchburn. Vision Res. **20**:273–276.

Leibovic KN (1965). A note on visual integrative mechanisms. Bull. Math. Biophys. **27**: 305–310.

Leibovic KN (1968). Observations on some experiments on visual masking. J. Gen. Psychol. **78**:19–26.

Leibovic KN (1972). *Nervous System Theory*. Academic Press, New York.

Leibovic KN, Balslev E, Mathieson TA (1970). Binocular space and its representation by cellular assemblies in the brain. J. Theoret. Biol. **28**:513–529.

Leibovic KN, Balslev E, Mathieson TA (1971). Binocular vision and pattern recognition. Kybernetik **8**(1):14–23.

Luneburg RK (1947). *Mathematical Analysis of Binocular Vision*. Princeton University Press, Princeton, NJ.

MacKay DM (1970). Elevation of visual threshold by displacement of the image. Nature (London) **225**:90–92.

Pirenne MH (1962a). Rods and cones. In *The Eye* (H Davson, ed.), Vol. 2, Chap. 2. Academic Press, New York.

Pirenne MH (1962b). Flicker and afterimages. In *The Eye* (H Davson, ed.), Vol. 2, Chap. 11. Academic Press, New York.

Pöppel E (1985). *Grenzen des Bewustseins*. Deutsche Verlags-Anstalt, Stuttgart.

Pritchard RM, Heron W, Hebb DO (1960). Visual perception approached by the method of stabilized images. Can. J. Psychol. **14**(2):67–77.

Richards W (1975). Visual space perception. In *Handbook of Perception* (EC Carterette, MP Friedman, eds.), Vol. 5, Chap. 10, Academic Press, New York.

Riggs LA (1964). Human vision in relation to eye movements. In *The Physiological Basis of Form Discrimination*. NIH Symposium, Brown University, Providence, LI.

Riggs LA (1965). Light as a stimulus for vision. In *Vision and Visual Perception* (CH Graham, ed.), Chap. 1. Wiley, New York.

Riggs LA, Ratliff F, Cornsweet JC, Cornsweet TN (1953). The disappearance of steadily fixated test objects. J. Opt. Soc. Am. **43**:495–501.

Roufs JAJ (1972). Dynamic properties of vision I: Experimental relationships between flicker and flash thresholds. Vision Res. **12**:261–278.

Steinman RM, Haddad GM, Skavenski AA, Wyman D (1973). Miniature eye movement. Science **181**:810–819.

Volkman FC (1962). Vision during voluntary saccadic eye movements. J. Opt. Soc. Am. **52**(5):571–578.

Westheimer G (1979). The spatial sense of the eye. Invest. Ophthalmol. Vis. Sci. **18**(9):893–912.

Westheimer G (1981). Visual hyperacuity. *Progress in Sensory Physiology*, Vol. 1, pp. 1–30, Springer-Verlag, Berlin, Heidelberg, New York.

Westheimer G, Mitchell AM (1956). Eye movement responses to convergence and stimuli. Arch. Ophthalmol. **55**:848–856.

Yarbus AL (1967). *Eye Movements and Vision*. Plenum, New York.

7
Classical and Modern Psychophysical Studies of Dark and Light Adaptation and Their Relationship to Underlying Retinal Function

THOMAS E. FRUMKES

Classical Psychophysical Approach to Adaptation

Visual adaptation is the change in visual processing resulting from exposure to darkness (dark adaptation) or light (light adaptation). Although some understanding developed in the eighteenth and nineteenth centuries, the outlook still reported in most textbooks today evolved in the first half of the twentieth century and largely reflects the influence of S. Hecht, B.H. Crawford, W.A.H. Rushton, and W.S. Stiles. These and other visual psychophysicists usually restricted their interest in adaptation to changes in detection threshold, i.e., the least intense light that can be detected. To be sure, adaptation induces changes in visual perception such as color, brightness, and the ability to see temporal or spatial details that are subjectively more obvious to the typical observer and are probably of greater importance. But absolute sensitivity was most carefully studied as it is easiest to study objectively. As stressed in the latter half of this chapter, this choice has had a considerable effect, some of it deleterious, upon current views of adaptation.

In typical adaptation experiments, a small, brief test flash of light is presented repetitively to the particular region of the retina of interest. The observer adjusts the intensity (or more correctly, the retinal illuminance) of this flash until it is just detected. At least six assumptions were made by classical psychophysicists, usually implicitly, in interpreting such threshold data. These are listed numerically at the outset as they are referred to again later in this chapter.

1. The test flash itself has no long-term effect on the visual system and its threshold serves as a probe indicating the overall sensitivity of the visual system.
2. Vision is most likely mediated by several different parallel processes referred to by a variety of terms, most commonly systems, channels, or mechanisms (see McGuire et al., Chapter 9). These parallel processes may be anatomically defined structures (rods vs. cones), physiologically defined processes with no known anatomical property as was originally the case for on- and off-pathways

(see Slaughter, Chapter 3), or processes that were inferred on the basis of psychophysical data (e.g., spatial frequency channels).

3. At any given level of neural organization, the most sensitive of the parallel processes alone determines test threshold.
4. Parallel processes adapt independently of each other.
5. All processes are most sensitive in their dark-adapted state: light adaptation serves to desensitize a process.
6. All processes responding to a test flash can produce only one type of neural signal. This is true even of mechanisms with opponent process characteristics such as the color mechanisms discussed by Boynton in his chapter on human color perception.

Figure 7.1 shows typical dark and light adaptation data. To collect the dark adaptation data on the left, the observer's eye was first presented with an extremely intense "white light" source for 30 sec to light adapt the eye. Then a test probe of light was presented once every 4 sec to the observer. This stimulus was presented to a particular area of the retina (7° from the fovea) where there are both rods and cones; this had a particular wavelength (512 nm), size and shape (a disc 40′ of arc in diameter), and duration (500 msec). The observer continually adjusted the illuminance of the test flash until it was just visible. Accordingly, threshold is plotted as a function of time in the dark. As is well known, threshold for the test flash decreases monotonically during the ensuing period of darkness.

For the light adaptation data on the right, the observer was first adapted to darkness by being placed in a totally dark room for about a half hour. Then, a large, continuously exposed background field was presented at many different illuminance levels. Test threshold (plotted on the ordinate) was obtained as a function of background illuminance (along the abscissa): test threshold increases with background illuminance. Notice the obvious kink in both light and dark adaptation functions. Light adaptation functions such as shown in Figure 7.1B are often referred to by a variety of other names in the literature, e.g., increment threshold functions, or threshold vs. illuminance (tvi) curves.

Photochemical Theory: A Partial Explanation for Dark Adaptation But Not for Light Adaptation

Although data such as Figure 7.1 were known for many years, Selig Hecht (1937) was probably the first investigator to have a direct impact on scientific thought today. Like several of his predecessors (e.g., Aubert, 1865; Kohlrausch, 1922), Hecht asserted that the upper limb of the dark and light adaptation curves of Figure 7.1 pertained to the functioning of cones and the lower limbs to the functioning of rods. This follows directly from the "duplicity theory" of Schultze, who in 1866 proposed that vertebrate vision was mediated by two classes of photoreceptors that functioned under either dimmer (rods) or brighter (cones) levels of illumination. Since the time of Hecht's 1937 review, both electrophysiological

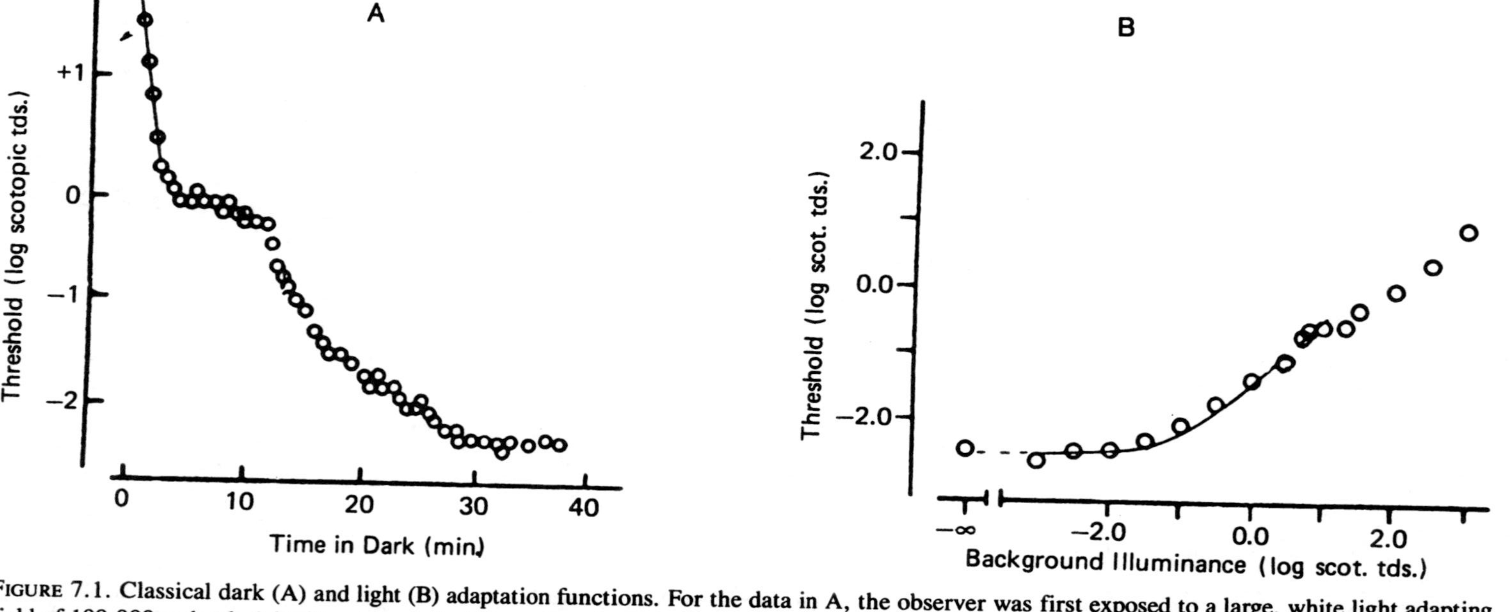

FIGURE 7.1. Classical dark (A) and light (B) adaptation functions. For the data in A, the observer was first exposed to a large, white light adapting field of 100,000 trolands (tds) illuminance for 30 sec. Starting at time zero (on the abscissa), a 40′ of arc diameter, 500 msec duration test flash of light of 512 nm wavelength was presented once every 4 sec to a position on the retina that was 7° nasal to the fovea. The observer continually adjusted the illuminance of the flash until it could just be observed. The abscissa values indicate the adjusted threshold illuminance value obtained every 30 sec. Part B shows thresholds of a test flash with identical spatial, temporal, and wavelength characteristics. This test flash was presented to the totally dark adapted eye (the minus infinity abscissa value) or in the presence of a continually exposed, concentric, 14° diameter, 490 nm adapting field with the illuminance value indicated on the abscissa. The ordinate indicates the minimal intensity necessary to just detect the test flash. Both sets of data were obtained from the same observer. (From Bauer et al., 1983a.)

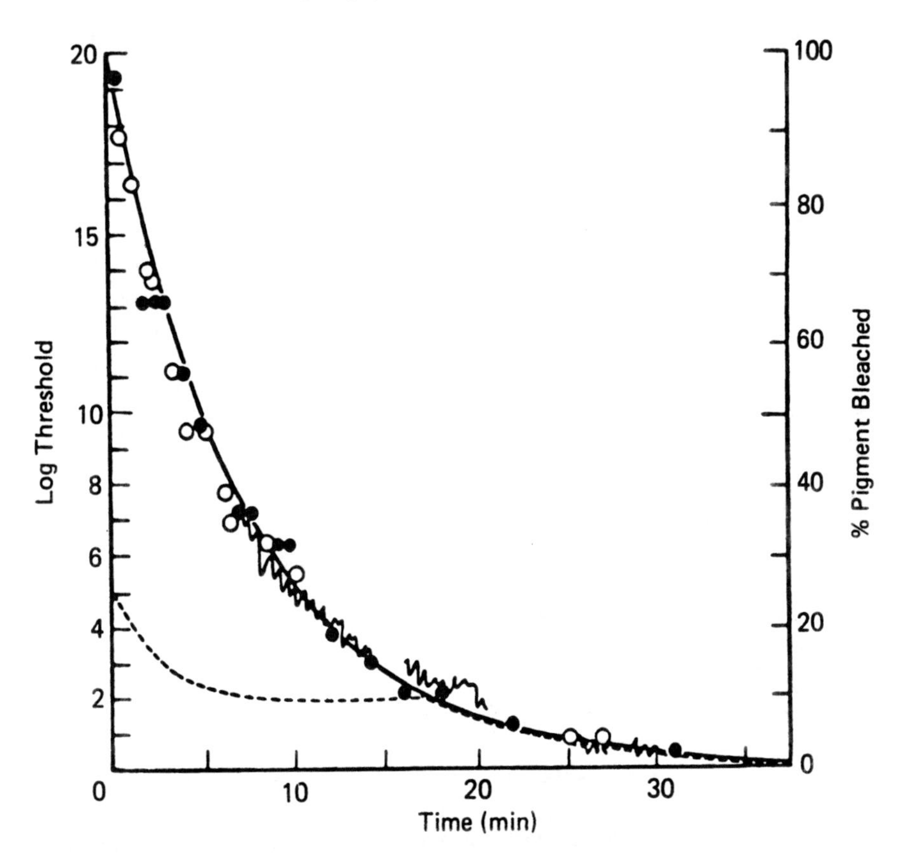

FIGURE 7.2. Dark adaptation functions obtained under conditions where rods are responding the simuli. The continuous functions show thresholds of a test flash of light and relate to the left ordinate: this is expressed in arbitrary log units such that 0 is the threshold in the totally dark adapted eye. The lowest indicated function (dashed lines) are data from a normal observer and shows clear cut rod and cone limbs; the squiggly line is data from a rod monochromat. The right ordinate and data points indicate the percentage of rhodopsin that is bleached as a function of time in the dark. The open circles are data obtained from the normal eye while the closed circles are data from the rod monochromat. The fitted continuous smooth curve is an exponential with a time constant of 7.5 min. (From Rushton, 1965.)

and psychophysical data collected from animals or anomalous human observers with only rods or only cones have made the distinction between rod and cone functioning obvious. For example, one dark adaptation curve shown in Figure 7.2 was obtained from a "rod-monochromat." This is an individual with a very rare type of hereditary condition in which only rod photoreceptors are functioning. Notice that the rod threshold data from the normal subject (the lower limb of the dashed function and left ordinate) agree with the threshold data from this subject (squiggly line and left ordinate). However, for the rod monochromat there is no rod–cone break and one smooth curve describes threshold data throughout the entire period of time in the dark.

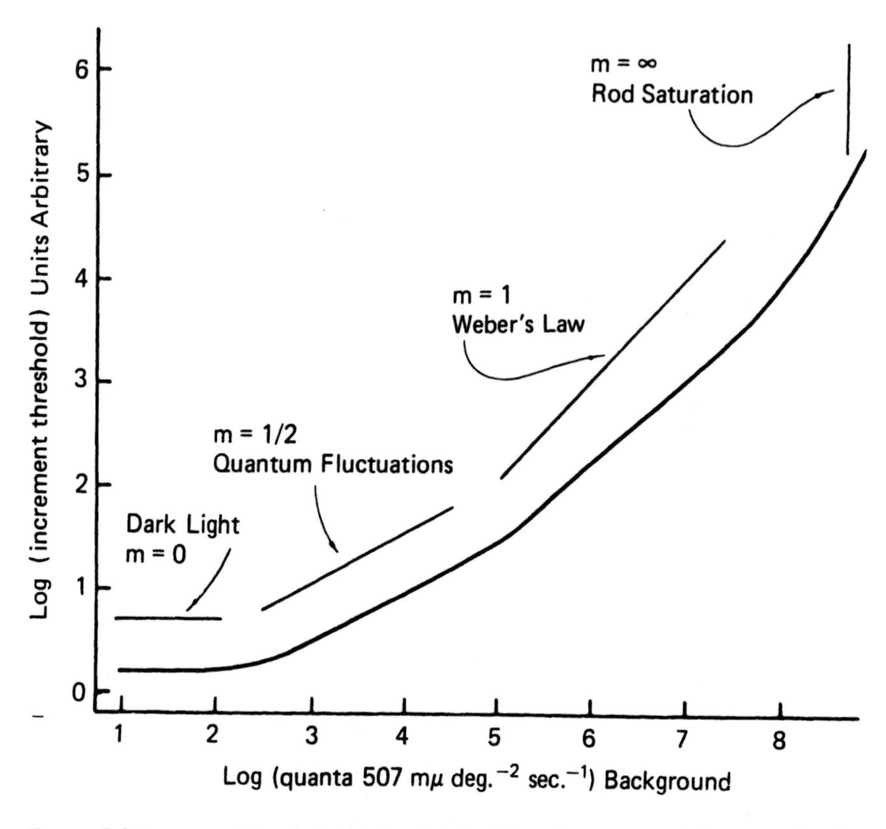

FIGURE 7.3. Increment threshold (light adaptation) data from a normal observer using the "rod-isolation" procedure developed by Aguilar and Stiles (1954). Thresholds of a test flash of light (45′ diameter, 8 msec duration, 507 nm wavelength) presented 10° from the fovea were determined as a function of the illuminance of a 10° diameter, continually exposed adapting field of approximately 600 nm wavelength. The smooth curve can roughly be approximated by four straight lines with a slope (m) of 0, ½, 1, and infinity. The significance of the first three slopes is explained in the text. An infinite slope ("rod saturation") indicates conditions where rod vision can no longer detect the test flash. (From Barlow, 1965.)

Using special procedures, light adaptation data can be collected in normal observers reflecting just rod or cone function. The continuous curve in Figure 7.3 shows such data obtained in a normal subject under special "rod isolation conditions" first described by Aguilar and Stiles (1954). Once again, notice the lack of a rod–cone break.

I will refer to the data in Figures 7.2 and 7.3 throughout this chapter. Here, I should stress that under more usual circumstances in normal observers and as shown in Figure 7.1, the difference in the range of illuminance over which rods and cones apparently operate adheres to the duplicity theory. These data provide some support for assumptions 2, 4, and 5 listed at the outset of this chapter. According to this classical interpretation and to these assumptions, there is a scotopic range of illuminance over which rod functioning is evident, and a photopic range where cone functioning is evident. Some authors also refer to a "mesopic" range over which the activity of either type of photoreceptor may become apparent. This term is very poorly defined in the literature, and if it refers to the overlap in the overall operating ranges of rod and cone photoreceptors, it exceeds 3 $\log_{10}$ units (see Bauer et al., 1983b). Even under mesopic conditions, threshold is assumed to be mediated by the more sensitive type of photoreceptor and is uninfluenced by the activity of other photoreceptors. In the case of the dark adaptation functions in Figures 7.1A and 7.2 in normal observers, rods are much less sensitive than cones during the first 5 min or so of dark adaptation. Hence, threshold is determined by the activity of the more sensitive cones. Only some time after cone adaptation is complete will rods of normal observers become more sensitive than cones and mediate threshold.

Hecht (1937) went much further than his predecessors, and attempted to explain these data in terms of hypothetical photochemical mechanisms that, during his lifetime, were not amenable to verification by procedures other than psychophysics. Specifically, Hecht posited that visual sensations were related to the bleaching of receptor photopigments, with the amount of unbleached pigment being the determinant of visual sensitivity. If there were no process of pigment regeneration, continually exposed light would decrease the amount of unbleached photopigment and, hence, visual sensitivity would be lost. In the dark and in the light as well, however, photopigment would gradually regenerate and visual sensitivity would be regained. With the particular chemical kinetics he postulated, Hecht predicted that visual threshold would be directly proportional to the amount of photopigment in the bleached state. The differences between rods and cones reflected the sensitivity differences in the visual pigments that responded to light, and to differences in the time constants that governed their regeneration. In other respects, though, he believed that rod and cone functioning were quite similar.

Neural Theory: Nonphotochemical Factors That Contribute to Adaptation

After Hecht's death, a number of investigators developed the technique of retinal densitometry, which makes it possible to measure the amount of bleached vs. unbleached photopigment in the living eye and compare such data to sensitivity measures. W.A.H. Rushton (1961) was the first to actually relate the amount of rhodopsin (rod photopigment) bleached to psychophysical threshold in humans; at approximately the same time, John Dowling (1960) related the sensitivity of the electroretinogram b-wave in rats to the amount of bleached rhodopsin. Both sets of data showed that the state of long-term visual sensitivity generally related quite well to the amount of bleached photopigment and adhered to the expression,

$$\log(I_t/I_0) = kB \tag{1}$$

where I_t is the test flash threshold intensity in a given state of adaptation, I_0 is test threshold intensity in the completely dark adapted eye, B is the percentage of photopigment that is bleached, and k is a proportionality constant. For humans, the constant k is about 3 for cones and about 19 for rods. I_0 depends on the specifics of the experimental situation (see Rushton, 1963, 1965; Dowling, 1987). Equation (1) is usually referred to as the Dowling–Rushton relationship. In Figure 7.2, the data points and right abscissa refer to the amount of rod photopigment in a bleached state in both a rod-monochromat (closed circles) and normal (open circles) observer. Notice the close correspondence between the logarithm of the psychophysical threshold of the rod monochromat and these photopigment data.

The data of Dowling and Rushton indicate one glaring error in Hecht's original photochemical theory. Specifically, Eq. (1) and Figure 7.2 indicate that the relationship between visual threshold and bleached photopigment is logarithmic, not linear as stated by Hecht. Moreover, the constants I_0 and k in Eq. (1) are such that very little bleaching is needed to result in a large change in visual sensitivity. For example, it follows from Figure 7.2 that a 5% bleach of photopigment will produce a 10-fold (one log unit) change in visual threshold, and elsewhere Rushton (1963) claimed even greater changes in threshold resulting from small percentage bleaches. Such findings would be very difficult to account for by any simple modification of Hecht's scheme for chemical kinetics. In fact, the underlying reason for such drastic nonlinear changes is just beginning to be understood in terms of transduction biochemistry in photoreceptors (see Leibovic, Chapter 2).

Selig Hecht died in 1947. The death knell to his photochemical theory is attributable to psychophysical data published the same year by B.H. Crawford. To put Crawford's data in the clearest context, I will first refer briefly to some simple, relevant photochemistry. In a monomolecular, irreversible photochemical reaction, the amount of a photoproduct (P) resulting from exposure to light of intensity I is predicted by the relationship,

$$I \times t \propto P \tag{2}$$

where t is the duration of the constant intensity light. Equation (2) dates from the nineteenth century and is referred to as the Bunsen–Roscoe photochemical law. Using a variety of procedures including retinal densitometry, Rushton (1965) showed that the Bunsen–Roscoe law accurately described the amount of photopigment bleached by flashes of different durations up to at least several seconds.

In the early twentieth century, a large number of investigations related the threshold intensity of a light flash, I_t to its duration. Up to some critical duration, threshold is determined by the product of I_t and its duration t (for brief review, see Frumkes, 1979). This relationship is referred to as Bloch's law. However, the critical duration that limits Bloch's law is approximately 0.1 sec. For predicting the amount of bleached rhodopsin (B) as measured by retinal densitometry, the relevant critical duration is usually several orders of magnitude longer (Rushton, 1965; but for some paradoxical results with very short duration flashes, see Pugh, 1975). This discrepancy underscores the importance of some nonphotochemical factor(s) that must influence visual sensitivity.

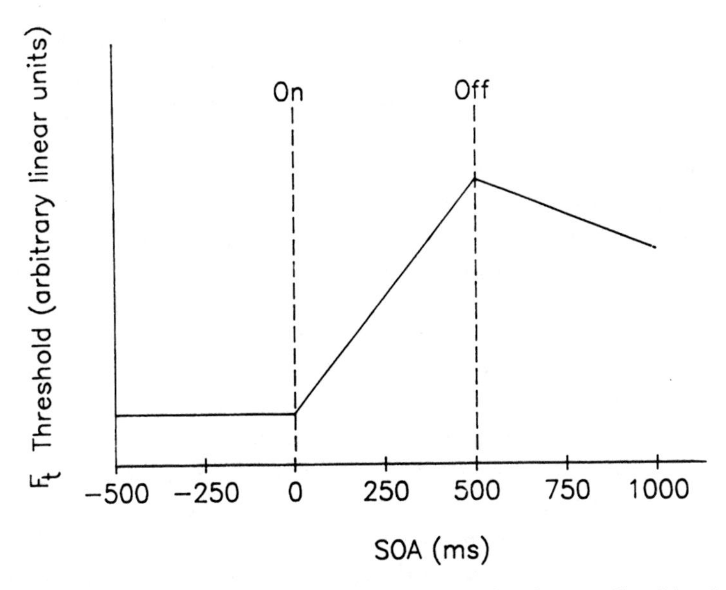

FIGURE 7.4. Hypothetical "rapid light and dark" adaptation data predicted by the photochemical theory of Hecht. Thresholds for a small, brief test flash of light are plotted on the ordinate in arbitrary linear units as a function of the stimulus onset asynchrony for a concentric, large adapting flash of 500 msec duration. The time for adapting flash onset and offset are indicated by the vertical dashed lines labeled "On" and "Off." Notice that test threshold is constant prior to the onset of the adapting flash. See text for further explanation.

In Crawford's (1947) study, an adapting flash of light of approximately 0.5 sec duration was presented every few seconds to his observers. A smaller, briefer, test flash of light was presented at various delay intervals with respect to the time of onset of the adapting flash. Hence, the threshold was determined as a function of the interval between the onset of the adapting flash and the onset of the test flash, an interval often referred to in the modern psychophysical literature as the stimulus-onset asynchrony interval or SOA. A negative SOA value means the test flash onset precedes the adapting flash onset, and a positive value means the test flash onset follows the adapting flash onset. The major methodological distinction between Crawford's rapid adaptation technique and that used to generate more classical data such as illustrated in Figure 7.1A is that he measured adaptation in intervals measured in milliseconds, not in minutes.

Before proceeding to actual data, I will first consider results predicted by a photochemical theory such as Hecht's, which is indicated schematically in Figure 7.4. Test threshold would not rise until the adapting stimulus was presented, since only then would photopigment be bleached. Test threshold would rise throughout the time period that the adapting stimulus was presented. According to Hecht's specific chemical scheme, threshold should rise throughout the entire time of presentation of the adapting stimulus (unless all photopigment was bleached away): this rise would be entirely linear if some photopigment regeneration did not occur in the light. Threshold of the test flash should begin to fall just at the time of adapting flash offset. Recovery of threshold would depend on the rate of regeneration of photopigment. Recovery should probably be quite slow judging by the time course of classical dark-adaptation curves (e.g., Figs. 7.1A and 7.2). In fact, more modern data show that the time course for photopigment regeneration is an exponential function that can be characterized by the time for regeneration of 50% of bleached photopigment. This halftime is about 400 sec for rod and 100 sec for cone photopigment (Rushton, 1965; Dowling, 1987).

Data similar to that collected by Crawford (1947) are shown in Figure 7.5. The illustrated data were actually collected by F. Naarendorp as controls for experiments discussed later in this chapter, under conditions where cone (upper coordinates) and rod (lower coordinates) vision were studied separately. The data that can be attributed to the functioning of cones, like Crawford's original data, differ from the hypothetical results of Figure 7.4 in five major respects. First, test flash threshold begins to rise before the adapting flash is presented. This "backward masking" effect cannot be explained by any modification of a strictly photochemical adaptation theory. From the present perspective, this failure of photochemical theory is the most important conclusion to be drawn from the "backward masking" portion of data such as Figure 7.5.

A number of alternative "neural hypotheses" have been proposed to account for backward masking. It was known since the nineteenth century that the latency of a photic response decreases as the intensity of a light flash increases. Since the adapting flash is usually more intense than the test flash, its elicited response will arrive at the brain before the response elicited by the test flash. According to this explanation, then, "backward masking" is still really "forward masking" within the nervous system. But backward masking has been observed under situations in which the adapting flash is actually dimmer than the test flash (Boynton, 1961), ruling this out as the exclusive explanation. Second, it has been suggested that "psychological" or "mental" time is divided into indivisible "temporal quanta" of approximately 100–200 msec duration (e.g., Stroud, 1956; see also Leibovic, Chapter 6). Although a test flash may precede an adapting flash, it may be presented in the same quantal time period. Obviously, the closer the onset of the test and adapting flashes are together in time (i.e., the less negative the SOA value), the more likely they will arrive in the brain during the same quantal time period and on a statistical basis, the higher test threshold. This type of argument was pursued in the late 1960s and early 1970s by individuals who would now call themselves cognitive psychologists. Third, evoked potential studies have shown that in cortex, discrete light flashes elicit a primary response and for several hundred milliseconds thereafter, a secondary discharge: although this secondary discharge has a higher threshold, its presence may be necessary for detection of a flash of light. Sturr and Battersby (1966) showed in the cortex of the cat that at negative SOA values, adapting flash presentation decreases or eliminated the secondary discharge to the test flash and they related this to backward masking.

Second, threshold of the test stimulus is maximal at the time of onset of the adapting flash. Third, while the adapting field is still present, the test threshold decreases and, hence, the change in test threshold is in the opposite direction

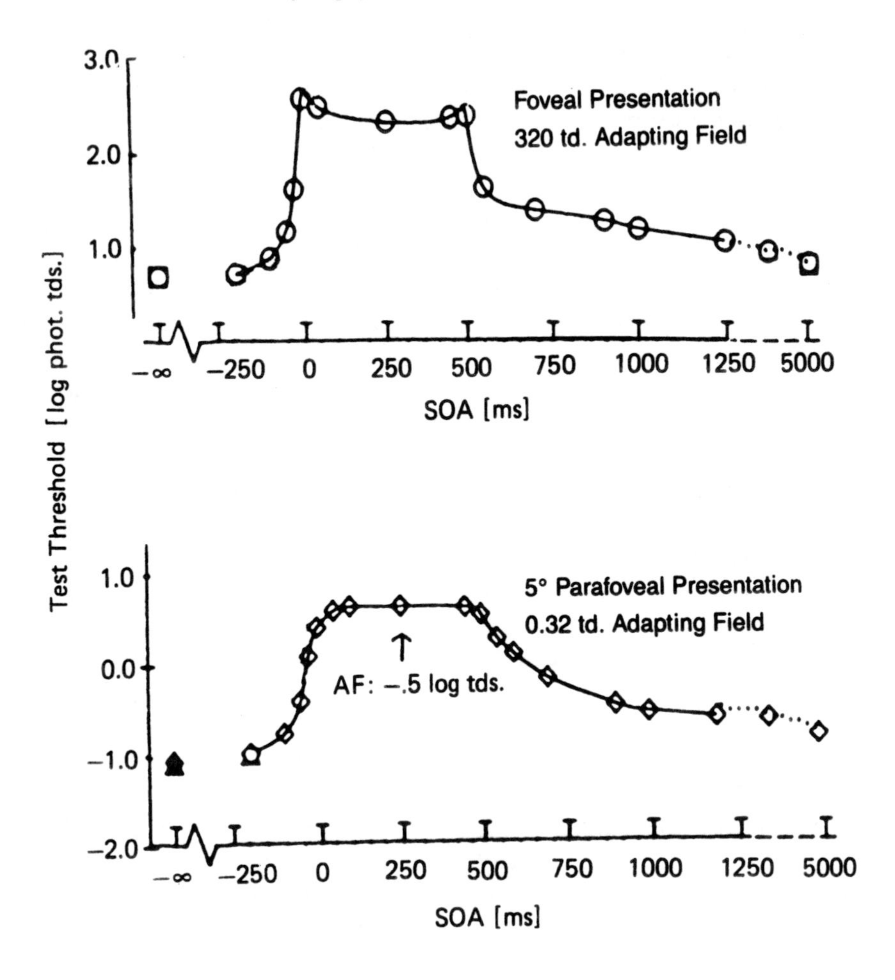

FIGURE 7.5. Rapid light and dark adaptation functions obtained for either cone or rod vision using the procedure of Crawford (1947). Thresholds of a 54′ diameter, 10 msec duration test disc of "white" light were determined in respect to the time of onset of a concentric 500 msec duration, 20° diameter, white light adapting flash. Notice that threshold is plotted on the ordinate in logarithmic units. For the data represented in the top figure, the test flash was presented foveally and the adapting flash was 320 trolands in illuminance: unillustrated control experiments determined that these data reflect the functioning of the cone-related visual system. For the data represented in the lower figure, the test flash was presented 5° from the fovea and the adapting flash was 0.32 trolands in illuminance: unillustrated control experiments determined that these data reflect the functioning of the rod-related visual system. (From Naarendorp, 1988.)

predicted by Hecht's theory! Under the conditions shown in Figure 7.5, this decrease is slight. With longer duration and more intense adapting flashes than those used for the illustrated data, this decrease can continue for about 1–2 sec and exceed 1 log unit; thereafter, test threshold remains quite stable (see Baker,

1963). Fourth, test threshold often rises around the time of the offset of the adapting flash, a factor that is also at odds with photochemical theory. Fifth, unless the adapting flash is extremely intense, threshold of the test stimulus returns to control level within 200–2000 msec following stimulus offset. Since modern photopigment measures show that the halftimes for rod and cone photopigment regeneration are well over 1 min, threshold recovery is much, much faster than the time course of photopigment regeneration. The bottom of Figure 7.5 shows analogous data for conditions under which rod vision is examined. Although the time course is somewhat slower, results are qualitatively similar to those obtained for cone vision.

Crawford's psychophysical data and Rushton's photochemical data showed that something other than photopigment kinetics plays a very large role in visual adaptation. It is probably still accurate to state that photochemical considerations predict many aspects of data obtained during the "slower" phases of dark adaptation such as those indicated in Figures 7.1A and 7.2. To a first approximation, photopigment concentration as described by the Dowling–Rushton relationship predicts visual detection threshold within a few seconds after an observer is placed in total darkness. This first approximation has been questioned on theoretical grounds and better mathematical fits have been obtained between photopigment concentration and threshold (e.g., Lamb, 1981). Moreover, neural factors must play a considerable role in long-term dark adaptation as indicated by three types of examples. First, using neurophysiological procedures under conditions where bleached photopigment does not regenerate, rod sensitivity nevertheless recovers for many minutes following the offset of an adapting field (e.g., Pepperberg et al., 1978; Leibovic et al., 1987). Although sensitivity never approaches the levels observed under conditions where photopigment does regenerate, this indicates that "slow" and "photochemical" are not necessarily synonymous terms. Second, in some types of retinal pathology, there is a considerable dissociation between the time course of photopigment regeneration during dark adaptation and the time course for recovery of psychophysical threshold sensitivity (for example, see Krill and Archer, 1977). Third, in normal observers, light adapting one eye can alter the time course of dark adaptation in the other eye (Lansford and Baker, 1969). Presenting the contralateral adapting stimulus shifts a long-term dark adaptation curve (such as shown in Fig. 7.1A) to the left by several minutes without altering its shape! The magnitude of this effect is heavily dependent on a variety of spatial parameters including the shape and size of both test and adapting stimuli.

Under some circumstances, many visual scientists refer to an "equivalence of light adaptation and dark adaptation." Accordingly, using test flash threshold as an equating variable, it is possible to equate the "time in the dark" in Figure 7.1A to "background illuminance" in Figure 7.1B. Such an equation can have considerable predictive value (for discussion, see Crawford, 1947; Hart, 1987; Leibovic et al., 1987; Pugh, 1988). It was used in the original publication from which Figure 7.1 was derived (Bauer et al., 1983a), and is discussed further in Chapter 2 by Leibovic in this volume. From the standpoint of photochemistry, however, light and dark adaptation are not at all equivalent. For example, although a 5% bleach of rod photopigment raises absolute threshold 10-fold, a 10-fold increase in rod threshold can be produced by a continually exposed background that bleaches an immeasurably small amount of rod photopigment. It would take well over 5 min for human rods to fully recover from a 5% bleach: in contrast, absolute threshold sensitivity is achieved within a few seconds after removing the continually exposed "equivalent background."

The Functions of Visual–Neural Adaptation

There are some aspects of visual adaptation that are neither photochemical nor neural in origin. In lower vertebrates, there are several different types of retinomotor movements associated with adaptation that change synaptic efficacy. Furthermore, exposure to light will influence the activity of various types of neurons, which will greatly change the extracellular concentration of several different ions, particularly K^+ and Ca^{2+}. This, in turn, can affect the excitability of other neurons (for a discussion as it affects retinal adaptation, see Bauer et al., 1983b; Dowling, 1987). However, most aspects of adaptation unrelated to photopigment concentration are "neural" in origin. I will first discuss the overall functions of visual adaptation to put these in a better context. Unlike some authors, I maintain that there is more than one function.

Change in Operating Range

The overall range of operation of the human visual system has been estimated to be anywhere between 8 and 12 $\log_{10}$ units. However, the range of stimulus intensities over which any neuron, including photoreceptors, can provide differential responses is considerably smaller, i.e., about 3 $\log_{10}$ units (for discussion, see Barlow, 1981; Pugh, 1988; Leibovic, Chapter 2). The most fundamental function for visual adaptation is to adjust the restricted operating range of individual neurons so that a graded signal is provided over the entire range of intensities the organism experiences.

Two obvious means partially restrict the range over which the photoreceptors themselves must operate. First, the size of the pupil decreases as the ambient level of illumination increases and, consequently, less light is admitted into the eye. However, human pupillary diameter changes from about 8–9 mm in total darkness to about 1–2 mm in the presence of continuous bright illumination. This corresponds to a change in pupillary area by a factor of much less than 100, and therefore alters retinal illumination by less than 2 $\log_{10}$ units, a small fraction of the total range of visual experience (Thompson, 1987). Second, humans and most other vertebrates have two different types of photoreceptors that are best equipped to operate over either dimmer (rods) or brighter (cones) ranges of illumination. But even under experimental conditions over which pupil diameter does not restrict retinal illumination, the overall range of operation of rods

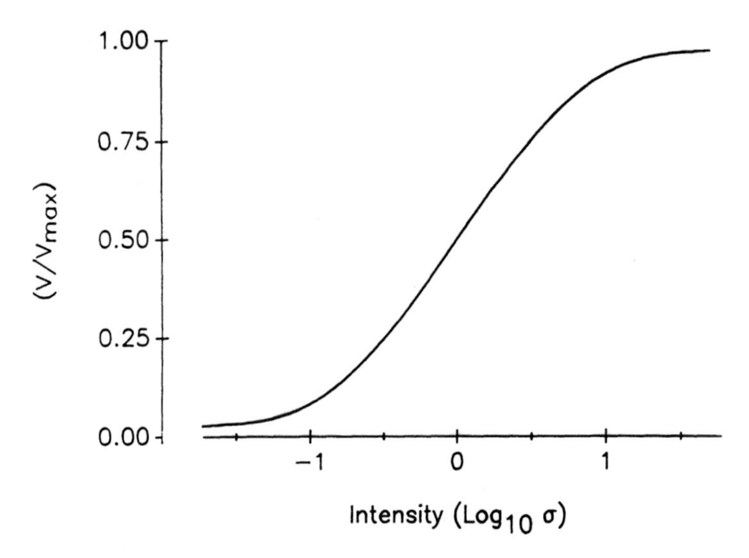

FIGURE 7.6. Amplitude of the response of a hypothetical distal retinal neuron as a function of the intensity of the light flash used to evoke the response. Light flash intensity is expressed along the abscissa as a function of the log of the semisaturation constant, σ. The indicated functional relationship is that adhering to Eq. (3) under the condition where $n = 1.0$.

exceeds 5 log units, and the range for cones is even greater. In the remainder of this section, I largely restrict discussion to one type of photoreceptor, the rods, although a large body of modern psychophysical literature also describes adaptation within cones (see MacLeod, 1978).

To put the problem in clearer context, first consider the operation of a typical neuron in the distal retina, i.e., a photoreceptor, horizontal, or bipolar cell. When dark adapted, the response (V) to a test flash of intensity I_t can be expressed in the equation,

$$\frac{V}{V_{max}} = \frac{I_t^n}{(I_t^n + \sigma^n)} \tag{3}$$

where V_{max} is the response to an extremely bright flash that saturates the receptor, σ is a semisaturation constant (that value of I_t leading to a response that is 50% of V_{max}), and n is a constant with a value that for photoreceptors is usually unity but is sometimes as low as 0.7

Figure 7.6 is a plot of Eq. (3) for the condition where n is equal to 1.0 and σ is a constant. If this neuron (or its input) did not in any way adapt, the practical range of operation would be 3 log units: test flashes 1.5 log units less intense than σ or dimmer would produce negligible responses; those 1.5 log units more intense than σ or brighter would produce essentially the same, maximal amplitude (saturating) response.

If there is "adaptation," the range of operation over which this neuron is able to respond differentially ($\pm 1.5 \log \sigma$) must change with ambient illumination. In recent years, the manner in which this is accomplished has been considered in two different fashions. Many psychophysicists and researchers who record from retinal ganglion cells regard the visual system as a multistage information processing system and speak about neural adaptation in terms of a unitary process, i.e., a change in "gain control" (see Adelson, 1982a,b; Shapley and Enroth-Cugell, 1984). On the other hand, the neurophysiological literature describing the properties of photoreceptors considers neural adaptation to reflect two separate processes (e.g., Dowling, 1987). To provide continuity with Chapter 2 of the present volume, I adopt the latter viewpoint.

One adaptation process involves a change in the value of σ. This is tantamount to saying that the photoreceptor puts on different darkness "sunglasses," which are most appropriate for the ambient level of illumination. This is commonly referred to as a "change in set point," as adaptation would shift the entire plotted function in Figure 7.6 to the left or right without changing its shape. According to a second process, V_{max} in Eq. (3) changes. Decreasing V_{max} decreases V and would cause the function plotted in Figure 7.6 to appear shifted to the right and have a lesser slope. Decreases of V_{max} with adaptation are commonly referred to as "response compression." Data such as Figure 7.3 show that increment thresholds for human rod vision change gradually over at least a 5 log unit range of adapting field intensity levels. Since the rod-related visual system responds differentially over a 5 log unit range, it must somehow adapt.

Adelson (1982a) studied some of the dynamic properties of "neural" adaptation. In his experiments, he combined features of usual increment threshold and Crawford rapid adaptation procedures using rod isolation conditions in normal observers. Accordingly, he obtained many increment threshold functions of the type shown in Figure 7.3, but used adapting flashes of variable duration; the test flash onset was always presented 200 msec before the offset of the adapting flash. In Figure 7.7, first consider the increment threshold function obtained when test and adapting flash onset were presented synchronously (the closed circles). Under this condition, the entire range of adapting field intensities over which test flash threshold changed gradually and monotonically was about 3 log units. This presumably represents the range of operation over which the rod system can respond differentially without adaptation. With longer duration adapting flashes (and, hence, progressively longer SOA values), some adaptation occurs prior to the presentation of the test flash. Therefore, the visual system was less influenced by presenting dimmer adapting flashes, but the range of adapting flash illuminance over which the test flash can be detected was extended to include higher illuminance levels. The amount of adaptation increases systematically with the duration to the adapting flash. For the limiting condition when the adapting flash is continually presented (the open squares), the increment threshold range is about 5 log units. Except for some minor differences caused by specific stimulus features, this limiting curve in Figure 7.7 replicates the smooth function in Figure 7.3. The reader should also understand the relationship between the data

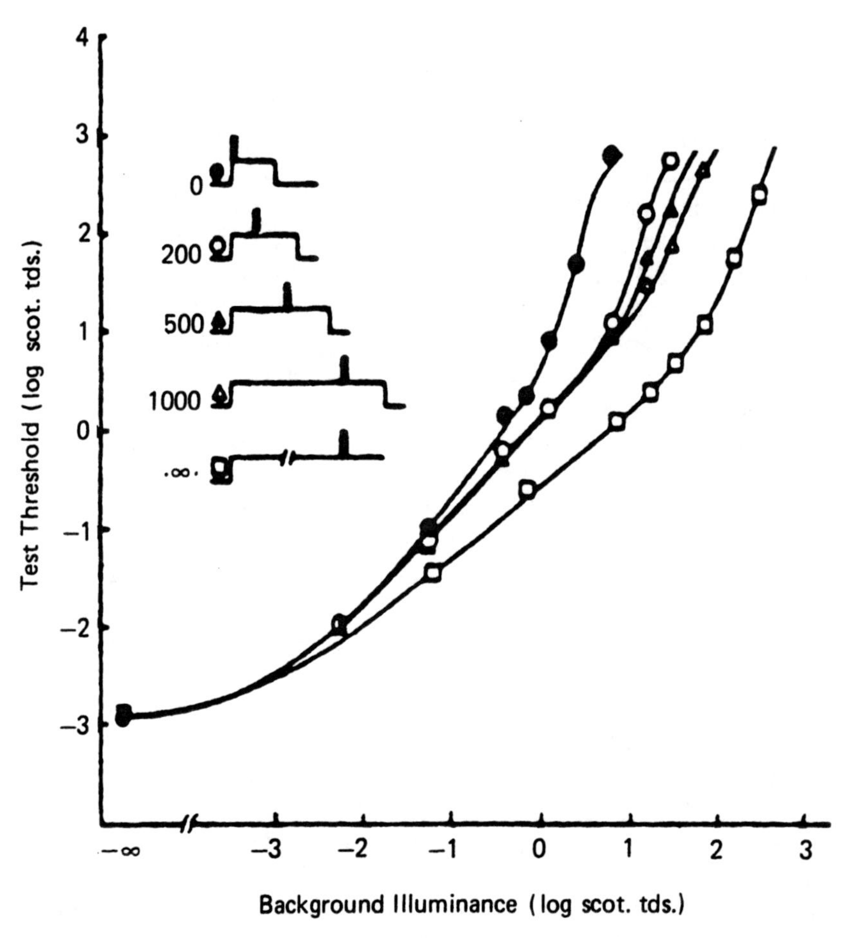

FIGURE 7.7. Increment threshold (light adaptation) functions obtained under rod-isolation conditions. Thresholds of a 30 msec duration, 4.5° square test flash of 480 nm wavelength as a function of the illuminance of a 11° diameter, red background flash. The adapting field duration is as indicated by the symbols presented in the upper left. Except for the data plotted by open squares, the test flash was always presented 200 msec before the offset of any duration adapting flash. The data plotted with open squares were obtained using a continually exposed adapting field and, hence, involved stimulus conditions similar to those used for collecting the data indicated in Figure 7.3. (From Adelson, 1982a, copyright [1982], Pergamon Press plc.)

in Figure 7.7 and the rod-related function presented in the lower portion of Figure 7.5. If stimulus parameters had been the same, the slight decline in threshold obtained after adapting flash onset in Figure 7.5 (i.e., SOA values of between 0 and 500 msec) would have been much greater. Thus, Figure 7.5 would be the equivalent of single points along a vertical slice cut through the functions comprising Figure 7.7.

Data such as Figure 7.7 can readily be explained in terms of a change in "set point" of the visual system (or in Adelson's terminology, a change in "gain control"). Although no comparable experiment has been done as carefully, the same kind of explanations can account for the decrease in threshold after the onset of the adapting flash for human cone vision. Undoubtedly, changes in rod or cone vision occurring within a few seconds after the offset of the adapting flash in part reflect the time course over which neurons readapt to the dark. But as discussed later in this chapter, Adelson (1982b) as well as Geisler (1979) stress the role of an additional factor that determines threshold during the earlier stages of dark adaptation.

The actual "site" of visual adaptation, which is reflected in data such as Figures 7.3, 7.5, and 7.7, is unknown. A number of different techniques have been used to measure response of a variety of different types of photoreceptor cells. Intracellular recordings from cold-blooded vertebrates indicate that all photoreceptors show response compression, while some photoreceptors additionally show a change in set point (for references, see Enroth-Cugell and Shapley, 1984; Pugh, 1988). Measurements of photocurrent generated by primate rod outer segments indicate that primate rods show only response compression (Baylor et al., 1984), a situation quite different than that for toad rods as described by Leibovic in Chapter 2 of this volume. If the results of Baylor et al. in primates are confirmed by other techniques, this would suggest that the site of neural adaptation is not as much within the rods themselves than at a subsequent neural locus or loci within the visual system. On the other hand, adaptation data from cat retinal ganglion cells seem very similar to those observed in humans by psychophysical means (for review, see Shapley and Enroth-Cugell, 1984), suggesting that the neural locus or loci are within the retina. Although adaptation has been studied by means of recordings from neurons lying between the photoreceptors and ganglion cells in submammalian vertebrates (e.g., Thibos and Werblin, 1978; Dowling, 1987), the relationship of these results to mammalian data is unclear (for discussion, see Shapley and Enroth-Cugell, 1984).

Parenthetically, it should be noted that a variety of psychophysical data has been offered over the years as evidence for a role of postretinal processing in light adaptation. These usually show that adapting lights in one eye influence visual sensitivity to test flashes in the other eye. Such interaction is highly unlikely to be purely retinal in origin and the best known candidate in primates for mediating such postchiasmal interaction is the visual cortex. One example of such results is the work of Landsford and Baker (1969) referred to earlier in this chapter. Another example is the work of Battersby and Wagman (1962) using Crawford's (1947) short-term adaptation paradigm. In general, large interocular effects are found only with stimuli with very specific spatial features. Since these data in many ways relate more to spatial processing than ambient level of illumination, they probably do not fit the definition of adaptation presented at the outset of this chapter.

Summation or Discrimination

Under extreme conditions of dark adaptation, the rods of humans and other primates respond to every absorbed photon of light (i.e., the smallest quantity of light that is physically possible). However, a threshold sensation requires summation of the input from several different rods (Hecht et al., 1942; Barlow, 1981; Baylor et al., 1984). For these reasons, it is advantageous to be able to summate the signals stemming from photons absorbed by a large number of rods over a long period of time.

There is some evidence that under totally dark adapted conditions, brightness sensations relate to the total number of photons absorbed by rods (Sakitt, 1972); in more classical terms, threshold is determined by the total amount of energy distributed over a considerable area of the retina and for a relatively long time period (Graham and Margaria, 1935). The evolutionary advantage of this mode of operation is clear. Since the variation in energy at any point in time or space is very great with respect to the average amount of energy at that point, it is impossible to provide accurate information regarding specific temporal/spatial features of stimuli. Rather than attempting the impossible, the visual system is designed to "count every quantum" and provide only a measure related to the overall energy content of a stimulus of a rather extended duration and area.

Under these extreme dark adapted conditions, most authors assume that the detection of dim lights is made on the basis of signal-to-noise discriminations. Accordingly, there is activity within the nervous system called noise. Presentation of a test flash of light produces a signal in the form of additional activity within the nervous system. The only method for discrimination is on the basis of the overall level of activity. If this exceeds a certain criterion level, the observer deduces that the signal was present in addition to noise (Barlow, 1956; deVries, 1943; Rose, 1942).

Under extreme dark adapted conditions with no background present, the only source of noise is indigenous biological activity. For rod vision, Barlow (1956) was the first to suggest that this indigenous noise reflected thermal decomposition of rhodopsin. His supposition is supported by behavioral experiments in which the effect of body temperature on visual sensitivity was investigated in cold-blooded vertebrates (Aho et al., 1988, but also see Leibovic Chapter 10). The additional noise introduced by presenting a very dim background (adapting) field is inconsequential in respect to this indigenous activity. Hence, with very dim adapting fields, test threshold is observed to be independent of adapting intensity, a functional relationship with a slope (*m*) of zero in Figure 7.3. But as the background increases in intensity, it produces a noise level that eventually becomes considerably larger than the indigenous noise. Under these conditions, which still involve dim lights, the variability in stimulus intensity can be described by a Poisson distribution. It is usually assumed that the activity level within the nervous system introduced by both the adapting stimulus and test flashes is proportional to the intensity of these stimuli (e.g., for Eq. (3), if n is equal to 1 and light flashes are much dimmer than σ, $V = (V_{max} I_t)/\sigma$, or $V \propto I_t$).

It follows that test flash intensity at threshold should be proportional to the measure of variability in the background intensity, its standard deviation, which for a Poisson distribution is its square root. On logarithmic coordinates, this would be indicated by a functional relationship with a slope of ½ (see Fig. 7.3).

At higher background intensities, the slope of increment threshold functions such as Figure 7.3 approximates 1.0, indicating that the visual system is no longer operating in the same fashion. The higher value slope indicates that performance has to some degree deteriorated since an ideal detector mechanism would yield lower thresholds (i.e., greater sensitivity). But as indicated below, it also indicates that the visual system has become geared to a mode of functioning providing more information about the form and shape of stimuli by responding to contrast independent of illumination level (see next section).

To summarize, the visual system responds under very dim light conditions to the total energy content in a stimulus. At somewhat higher levels of illumination, it responds in a manner that provides information regarding the distribution of energy in that stimulus with respect to space and time. Clearly, the change from one mode to the other is a second function of adaptation.

Brightness Constancy and Weber's Law

Under background illumination conditions corresponding to the increment threshold function with slope of ½ in Figure 7.3, the apparent brightness of a stimulus is largely determined by the number of photons absorbed by photoreceptors (Sakitt, 1972; MacLeod, 1974). As the ambient level of illumination increases to even a modest level, brightness sensations can be quite independent of either the number of photons absorbed or even the total flux of photons per unit area (Heinemann, 1955; Land and McCann, 1971; Whittle and Challands, 1969). Instead, apparent brightness is determined by the ratio of the luminance coming from a stimulus with respect to the luminance from surrounding areas, or by "contrast." In Figure 7.3, this range of illumination corresponds to the situation in which the logarithmic increment threshold plot has a slope of 1. Although not indicated graphically here, logarithmic increment threshold functions with slopes of 1 and contrast-dependent apparent brightness estimates are even more characteristic of cone vision.

This new mode of functioning with higher levels of illuminations is discussed usually in introductory psychology textbooks in two separate contexts. A logarithmic increment threshold function with unity slope is referred to usually as an example of "Weber's law," [i.e., there is a constant ratio (Weber's constant) between ambient intensity (the "I" of introductory texts) and the "just noticeable difference" in intensity ("jnd," or "ΔI")]. This means that the visual system is no longer limited by quantal fluctuations and signal-to-noise discriminations; it is now comparing the amplitude of two different signals stemming from different regions of the retina. Such a mode of operation is impossible near absolute threshold, since the distribution of available quanta is not sufficiently stable for even the most sophisticated detecting device to specify what is "figure" and what is "ground."

In a separate section, the same introductory psychology text is likely to discuss brightness constancy. Most objects viewed under normal viewing conditions are not sources of illumination but merely reflect light incident on them. The amount of light they provide to the eye of the observer depends on both the amount of light incident on them and their reflectance. Reflectance for most objects does not change with incident illumination as does reflected light. Therefore, there generally is a constant ratio between the amount of light reflected from a "test object" of interest (reflecting a luminance of $I + \Delta I$) and its "background" (reflecting a luminance of I). A measure that considers the ratio of the luminance provided by a test object in respect to background luminance provides brightness constancy, the modus operandi of the visual system under all but the dimmest and brightest levels of illumination. According to this explanation, Weber's law and brightness constancy are manifestations of the same underlying process.

To summarize, a third function of adaptation is to provide brightness constancy at all but the dimmest levels of illumination. Brightness perception and related chromatic factors are discusses in the chapter by Boynton in this volume.

Improved Perceptual Discriminations

A wide variety of discriminative processing is known to improve as the ambient level of adaptation increases. For example, as noted by Leibovic in Chapter 2, critical flicker frequency (CFF), the highest rate of flicker that can be discriminated from a steady light source, increases with the level of illumination as shown here in the top of Figure 7.8. Notice that with the largest stimulus employed (19° diameter), there is an approximately linear relationship between CFF and the logarithm of stimulus intensity under conditions in which either rods or cones function. This logarithmic relationship is often referred to as the Ferry–Porter law. Similarly, spatial acuity, the ability to make border distinctions, increases with ambient level of illumination as exemplified by the data on the bottom of Figure 7.8. A large body of literature exists that relates visual performance measures such as reading rate to ambient level of illumination. Although perhaps not characterized as well by any specific psychophysical function, the reader can probably think

---

►

FIGURE 7.8. Changes in critical flicker frequency (on top, data from Hecht and Smith, 1936) or visual acuity (at bottom, data from Shlaer, 1938) as a function of stimulus illuminance (plotted along the abscissae in log trolands. For the data on top, on–off square-wave flickering stimuli of the indicated diameter were presented foveally. Notice the rod–cone transition in the functions obtained with the two largest diameter stimuli. For the data at the bottom, visual acuity was obtained using a "Landolt C." Accordingly, the ordinate represents the smallest gap that could be detected in a circle with a variable width gap. The units on the ordinate are such that the ability to discern a gap of 1 min of arc corresponds to 0.0 log units, the ability to discern 10 min of arc corresponds to a visual acuity of −1.0, and so forth. In collecting these data, no fixation was used and the observer was instructed to alter his fixation position until acuity was maximal. Notice the rod–cone break in the function.

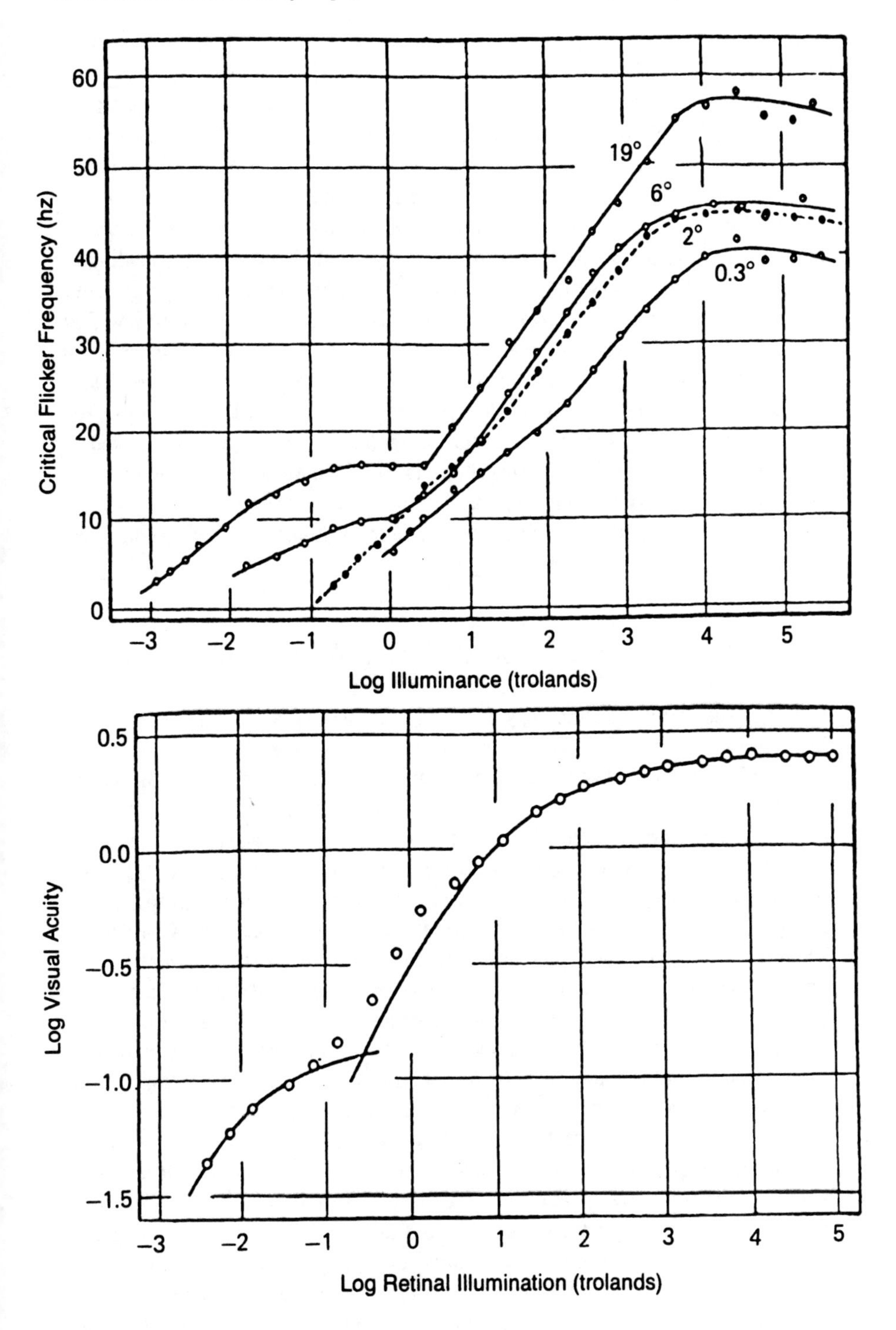

Critical Flicker Frequency (hz)
60
50
40
30
20
10
0
19°
6°
2°
0.3°
−3
−2
−1
0
1
2
3
4
5
Log Illuminance (trolands)
Log Visual Acuity
0.5
0.0
−0.5
−1.0
−1.5
Log Retinal Illumination (trolands)

of a number of perceptual abilities that change with level of illumination. Such changes in visual discriminations are a fourth function for visual adaptation.

In addition to data such as Figure 7.8 that reflect discriminative capacity, a large body of psychophysical literature shows that the ability of the visual system to summate energy over time and space decreases with ambient illumination (e.g., Graham and Margaria, 1935). In general, discrimination improves and summation decreases as the ambient level of illumination increases. Undoubtedly, the underlying factors producing these different changes are responding to the same set of physical limitations determined by ambient level of illumination. But unless one channel conveys all visual information, it does not follow that these two sets of psychophysical observations are determined by a common underlying physiological mechanism. In fact, virtually all modern theories of perception posit the existence of many parallel information channels (see MaGuire et al., Chapter 9). On a more pragmatic note, I would have liked to include in this chapter data relating illumination-dependent changes in temporal and spatial summation to changes in discriminative capacity of the types indicated by Figure 7.8. But I am unaware of empirical functions relating these two sets of illumination-dependent changes.

Newer Psychophysical Approaches to Adaptation

In the 1970s, John Dowling suggested that neural adaptation can be separated into photoreceptor vs. "network" properties (for review, see Dowling, 1987, Chapter 7). In most of his studies, he and his co-workers examined the influence of adapting fields on responses to spatially diffuse test flashes in the retina of the skate and several other cold-blooded vertebrates. They found that the influence of light adaptation on the response of horizontal cells was very similar to that on the photoreceptors' response. (For a description of the response properties of horizontal cells and other types of retinal neurons see Slaughter, Chapter 3). Therefore, he concluded that the adaption-induced alterations observed in horizontal cells reflect changes in the photoreceptors and not intrinsic horizontal cell properties. In contrast, the influence of adapting fields on the response of amacrine and ganglion cells was much greater than on the photoreceptors. Dowling suggested that the neural "network" within the inner plexiform layer of the retina provides a distinct, adaptation-induced change in vision.

Although Dowling's contextual distinction is original, "network" induced adaptational changes in vision were known much earlier. For example, it has been known since the work of Ricco in 1877 that up to some limiting area, psychophysical threshold depends on the total luminous flux (the product of retinal illuminance and area) provided by a stimulus, not just its retinal illuminance (flux/unit area): the size of the limiting area over which spatial summation occurs increases markedly with dark adaptation (Graham and Margaria, 1935). More recently, Barlow et al. (1957) noted that the receptive field properties of individual ganglion cells in the cat changed with dark adaptation. They stressed that the antagonism produced by the receptive field surround mechanism becomes much less obvious (and was originally reported to disappear) with dark adaptation. Presumably, psychophysical changes in spatial summation reflect the change in the neural network also observed in ganglion cells (Barlow, 1965).

Retinal ganglion cells are the only conduit of information by which the perceptual apparatus receives visual information. If their spatial coding property changes so drastically with visual adaptation, it would seem that there should be large changes in the qualitative properties of visual sensations with adaptation. Anyone who has even emerged from a dark movie theater during daytime cannot be surprised by this conclusion. Changes in the ability to perceive color, form, and spatial and temporal details are but a few of the phenomenological changes that have been noted in the literature. Therefore, it would seem that to some degree, the psychophysicist should abandon a view of adaptation that is restricted largely to a consideration of changes in visual detection threshold, i.e., those exclusively based on the type of data presented in Figures 7.1 through 7.7.

I suggest that the classical approach to visual adaptation described has two principal limitations. First, in spite of distinctions between photochemical vs. neural properties, or photoreceptor vs. network properties, adaptation is almost always still considered to produce one unitary effect. As indicated, I believe that adaptation has a number of independent functions, and I suggest these are accomplished by different mechanisms that may be partially or totally independent of each other. Changes in detection threshold do not necessarily reflect changes in other aspects of visual perception.

Second, the six assumptions made at the outset of this chapter must be reexamined. The first of these (that the test flash has in itself no effect on the visual system but is only a probe) is still made by most psychophysicists and many physiologists today. To be sure, it is not always correct. For example, under special circumstances using a Crawford adaptation procedure, a dim, small adapting flash less intense that the test flash can be shown to influence test threshold for several hundred milliseconds (Boynton, 1961). It logically follows that the test flash must perturb the visual system for at least as long a time period as the adapting flash. But such exceptions are quite rare: more generally, one could argue that presentation of the test flash at threshold disturbs the visual system no more than a voltmeter or oscilloscope used as a monitor disturbs a low current, complicated electronic circuit. The second assumption (that vision is mediated by several parallel processes) has probably never been questioned seriously. But as indicated below, the other four assumptions are often untenable.

In the following sections, I will consider psychophysical studies that examine the influence of visual adaptation on visual attributes other than detection threshold. I will first consider an issue that has been of particular interest to me recently, a phenomenon called suppressive rod–cone interaction. Then I will present a variety of other phenomena that all suggest limitations to conclusions obtained by more classical approaches to adaptation.

Suppressive Rod–Cone Interaction

A growing amount of evidence has shown that dark adapted, photically unstimulated rods tonically inhibit cone-mediated vision: by removing this inhibition, rod-light adaptation enhances cone-mediated sensations. This phenomenon is referred to as suppressive rod–cone interaction (SRCI). Although related phenomena were reported earlier (e.g., by Lythgoe and Tansley, 1929; Lie, 1963), current interest pertains to results published since 1983 (e.g., Alexander and Fishman, 1985; Coletta and Adams, 1984; Goldberg, Frumkes, and Nygaard, 1983). Here, I largely restrict discussion to results from my own laboratory.

In our laboratory, we initially discovered SRCI while studying the influence of dark adaptation on flicker sensitivity using procedures similar to those described at the outset of this chapter. The observer's eye was first exposed to an intense adapting field for 1 min. Then throughout the time period of adaptation, the observer adjusted the illuminance of a small flickering test probe until flicker could just be perceived. The left portion of Figure 7.9 shows the minimal intensity necessary to see flicker as a function of time in the dark. When the flickering test probe was of 5 Hz and of medium or short wavelengths (in Fig. 7.9, a stimulus stemming from a green light emitting diode or LED), threshold decreased throughout the entire period of dark adaptation. This function looks very similar to a typical dark adaptation curve such as shown in the left hand portion of Figure 7.1 and has obvious rod and cone limbs. With the same flicker test probe, we also determined the intensity necessary to discern flicker as a function of the illuminance of a large background field, and these data (not illustrated here) were similar to the classical increment threshold function shown in the right of Figure 7.1. Such data are entirely in agreement with classical concepts, i.e., the six assumptions listed at the outset of the chapter, and the view that adaptation serves one unitary function.

Rod adaptation, however, has a rather striking influence on cone-mediated flicker sensations. To show this, we used test probes that produced flicker that could be detected only by cones. Such conditions were obtained by two different means. First, in some unillustrated experiments, we used shorter wavelength flicker test probes with flicker frequencies ≥ 15 Hz: according to classical data

---

►

FIGURE 7.9. Influence of rod adaptation on cone-mediated flicker sensitivity. For both sets of coordinates, a 2°20′ diameter test probe was presented 7° from the fovea and was sinusoidally flickered at the indicated frequencies with a constant modulation depth of about 80%. The observer adjusted test illuminance (the values indicated on the abscissa) until flicker was just detected. For the data presented on the left coordinates, flicker illuminance was adjusted throughout the time period of dark adaptation following 1 min exposure to a 40,000 td white light bleaching stimulus. Flicker was generated by a green LED of 5 Hz, or a red LED of the indicated frequency. For the data presented on the right coordinates, flicker illuminance thresholds (using a red LED source flickering of the indicated frequency) were determined as a function of the illuminance of a continually exposed, 512 nm adapting field of 28° diameter. (From Goldberg et al., 1972. Copyright 1972 by the AAAS.)

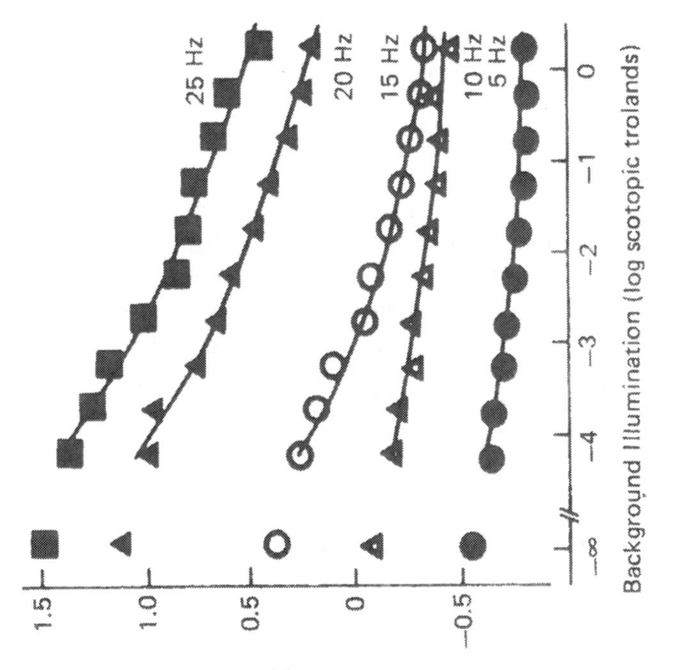
25 Hz
20 Hz
15 Hz
10 Hz
5 Hz
Test Illumination (log photopic trolands)
1.5
1.0
0.5
0
−0.5
Background Illumination (log scotopic trolands)
−∞
−4
−3
−2
−1
0

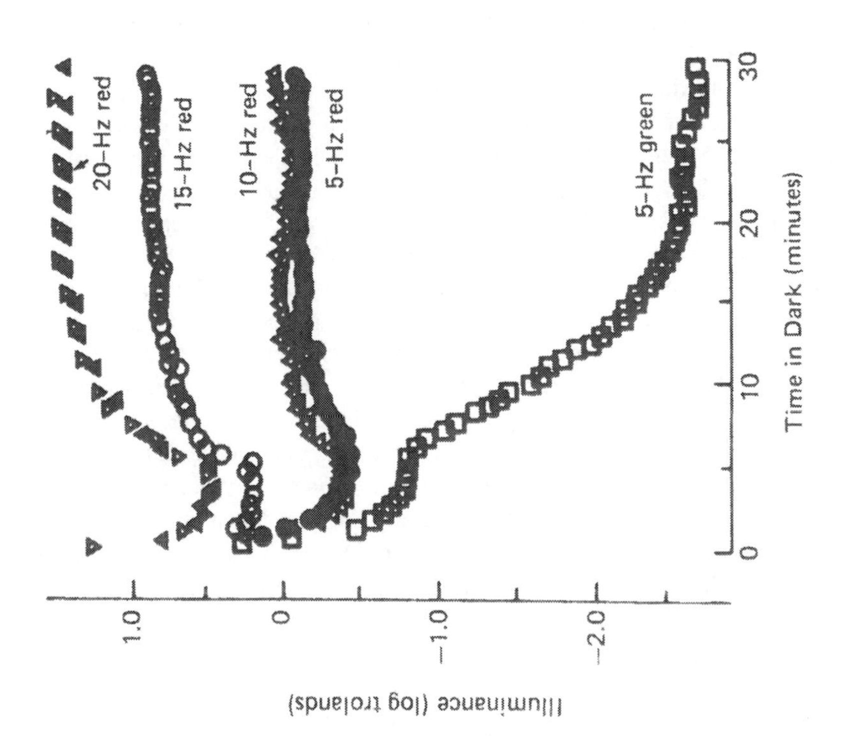
20-Hz red
15-Hz red
10-Hz red
5-Hz red
5-Hz green
Illuminance (log trolands)
1.0
0
−1.0
−2.0
Time in Dark (minutes)
0
10
20
30

such as at the top of Figure 7.8, such stimuli are beyond the frequency following capacity of rods (data not illustrated in Fig. 7.9). Second, we used test probes of longer wavelengths (for the data illustrated in Fig. 7.9, flicker stemming from a red LED). For wavelengths longer than 630 nm, cones are at least as sensitive as rods at all adaptation levels (Wald, 1945) so that flickering red LEDs would be likely to stimulate cones. When the sensitivity to such cone-mediated flicker was assessed, the intensity threshold necessary to see flicker decreased throughout the cone recovery stage of adaptation. But threshold increased throughout the rod recovery stage of adaptation as shown in Figure 7.9 on the left. This effect was quite small with 5 Hz flicker, but increased with flicker frequency to an almost 1 log unit effect with the highest illustrated frequency, 20 Hz.

We similarly studied the influence of light adaptation on cone-mediated flicker sensitivity. The right portion of Figure 7.9 shows the intensity of the same small stimulus just perceived as flickering as a function of the intensity of a large, steadily exposed background. Since this background was too dim to directly influence cones, it selectively lights adapted rods. The intensity necessary to see flicker decreased with increasing intensity of the rod-stimulating adapting field. Once again, this effect is barely evident with low-frequency flicker, but increases to an effect which exceeds one log units with 25 Hz flicker. A number of control experiments additionally ensured that for the experiments summarized in Figure 7.9, cones alone detected flicker when the frequency approached 15 Hz (Goldberg, 1983; Coletta and Adams, 1984). In addition, the dark adaptation changes obtained after 5 min in the dark and all the light adaptation data shown in Figure 7.9 relate to the adapted state of rods (Goldberg, 1983; Goldberg et al., 1983).

Now consider several of the assumptions made at the outset of this chapter. Since cone-mediated flicker sensitivity is greater when rods are somewhat light adapted, assumption 4 (independent adaptation of parallel mechanisms) and assumption 5 (that a mechanism is most sensitive in the most dark adapted state) are invalid. These conclusions become more evident by comparing results from normal observers with those of one type of "colorblind" observer in which these assumptions are apparently valid. As discussed in the chapter by Boynton, human color vision is mediated by three different types of cones with maximal spectral sensitivity at the short, middle, and long wavelength end of the visual spectrum. Protanopia is a genetic type of "colorblindness" inherited in an X-linked recessive fashion and shared by approximately 1% of the male population, in which the normal long wavelength cone photopigment is missing. Figure 7.10 compares cone-mediated flicker sensitivity changes during dark adaptation in a normal observer and in a protanope. Results from the protanope and normal subject are the same throughout the initial cone recovery stage of adaptation. But throughout the rod-recovery period of adaptation, flicker sensitivity for the protanope remains high (threshold is low) while that for the normal observer decreases. A similar lack of SRCI has been observed in three other protanopes (Coletta and Adams, 1985; Frumkes et al., 1988), but this result may not extend to all protanopes (K. Alexander, personal communication). These findings in protanopes, as well as other sorts of data in normal observers (Coletta and Adams, 1984),

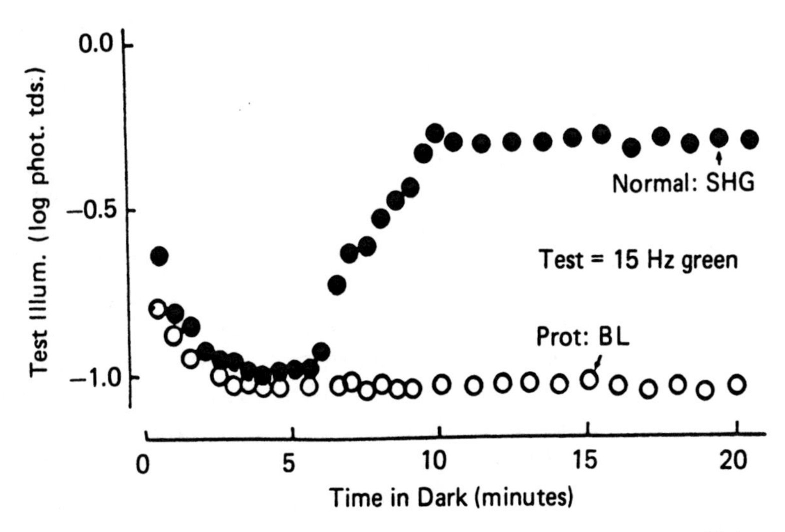

FIGURE 7.10. Influence of rod dark adaptation on the sensitivity to 15 Hz flicker in the same normal observer whose data are shown in Figure 7.9 (closed symbols), or a protanope (open symbols). All conditions are as indicated in the legend to Figure 7.9 except that 15 Hz flicker stemmed from a green LED source. (From Goldberg, 1983.)

suggest that SRCI is specifically related to the red-cone mechanism. It is of interest that deuteranopes, hereditary "color-blind" individuals lacking the normal middle wavelength cone photopigment, exhibit SRCI in the same manner as normal observers (Goldberg, 1983; Coletta and Adams, 1985). The exact reason for the failure to observe SRCI in protanopes is not entirely clear since in addition to their photopigment abnormality, they probably have several neuroretinal anomalies that are not yet understood.

More recently, we have also studied SRCI using physiological procedures and have examined the influence of selective rod light adaptation upon the amplitude of cone-mediated responses to rapid flicker. These experiments have involved ERG procedures in humans (Arden and Frumkes, 1986), or intracellular recordings from the retina of various amphibian species (Frumkes and Eysteinsson, 1987, 1988; Eysteinsson and Frumkes, 1989). Pflug and Nelson (1986) have obtained similar findings by means of intracellular recordings from cat retinal horizontal cells. All these data are remarkably similar to our human psychophysical findings suggesting that they reflect the same or a very similar phenomenon. Probably the most valuable of these physiological data are results in amphibians that show that SRCI can be documented by recordings from most types of cone-driven retinal neurons. In addition, pharmacological agents that prevent retinal horizontal cells from responding to photic stimulation of rods block SRCI in all types of neurons. Of related interest, specific types of distal retinal abnormalities in humans either enhance or abolish SRCI (e.g., Alexander and Fishman, 1985; Arden and Hogg, 1985; for review see Frumkes, 1990). Collectively, these findings show that SRCI reflects an inhibitory influence of retinal horizontal cells on cone pathways which is modulated by rods.

Rod adaptation influences other aspects of cone-mediated vision in humans. For example, Lie (1963) studied the "specific threshold" for many different wavelength stimuli, i.e., the intensity that is necessary to correctly identify the color of a stimulus. Specific and detection thresholds were nearly identical throughout the cone-recovery period of dark adaptation. But during the rod-recovery period of dark adaptation, specific threshold intensity increased while threshold for detection decreased. Two types of considerations indicate that this "Lie effect" reflects a somewhat different process than SRCI using flicker. First, the Lie effect undoubtedly relates to the activity of several different types of cones, while SRCI using flicker particularly relates to the activity of the long wavelength cone mechanism (Coletta and Adams, 1984; Frumkes et al., 1988). Second, in certain nightblind individuals with abnormal rod vision, the Lie effect is absent but SRCI as measured with flicker remains quite evident (Alexander et al., 1988).

Recently, Naarendrop, Denny, and Frumkes (1988) reported an analogous influence of rod adaptation on cone-mediated spatial vision. In these experiments, the intensity just necessary to detect the orientation of a short duration test grating was determined as a function of the state of rod light and dark adaptation. As shown on the left of Figure 7.11, if the spatial frequency was less than 3 cycles per degree, the intensity necessary to detect the grating decreased throughout dark adaptation, or increased as the intensity of a background field increased: these data have clear rod and cone limbs. However, with spatial frequencies greater than 3 cycles per degree, which recent evidence indicates to be beyond the capability of rod-mediated vision (e.g., see Hess and Nordsby, 1986), sensitivity to gratings appears to decrease as rods dark adapt or increase as rods are light adapted. This effect increases dramatically to about a 1 log unit effect with the highest illustrated spatial frequency, 21 cycles per degree. The right of Figure 7.11 shows corresponding light adaptation data. This rod adaptation influence on cone spatial vision is absent in the one protanope who has been studied (Frumkes et al., 1988), suggesting this effect might reflect the same underlying process as SRCI using flicker. In addition,

---

▶

FIGURE 7.11. Influence of rod adaptation on cone-mediated grating sensitivity. For both sets of coordinates, a 6° diameter squarewave (100% modulation) grating of 500 msec duration was centered 3° from the fovea. The observer adjusted grating luminance until the orientation of the bars making up the grating could just be detected (values indicated on the abscissae). For the data on the left, grating intensity was adjusted throughout the time period of dark adaptation following 1 min exposure to a 40,000 td white light bleaching stimulus. Except for the open square function that was generated using a 512 nm grating, the grating appeared red in color (wratten 29 filter) and was of the indicated spatial frequency in cycles per degrees. For the data on the right, such thresholds were obtained using either red appearing (wratten 29 filter) or 512 nm wavelength gratings as a function of the luminance of a 14° diameter continually exposed background field of 480 nm wavelength. (From Naarendorp et al., 1988. Copyright [1988], Pergamon Press plc.)

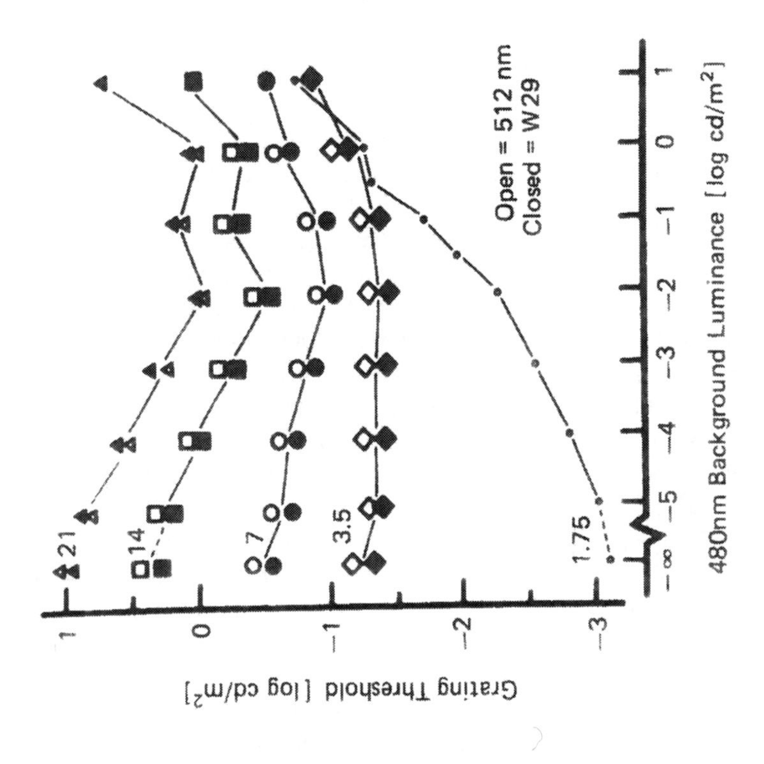

Open = 512 nm
Closed = W29
21
14
7
3.5
1.75
1
0
−1
−2
−3
Grating Threshold [log cd/m²]
−∞
−5
−4
−3
−2
−1
0
1
480nm Background Luminance [log cd/m²]

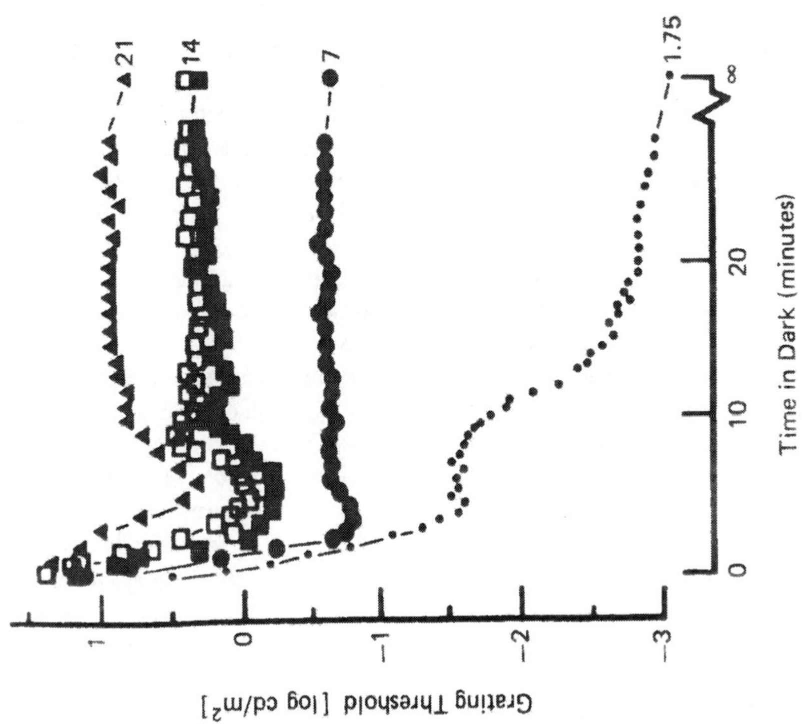

21
14
7
1.75
1
0
−1
−2
−3
Grating Threshold [log cd/m²]
0
10
20
∞
Time in Dark (minutes)

a large number of control experiments show that the rods that influence both flicker (Fig. 7.9) and grating acuity (Fig. 7.11) are those in a retinal area adjacent to (rather than coincident with) the retinal area on which the test stimulus is focused (Goldberg, 1983; Naarendorp, 1988).

Naarendrop (1988) also compared the time course for developing this grating acuity effect with the time course of changes in detection threshold during early light and dark adaptation. He used a modification of Crawford's procedure, and had the observer adjust the intensity just necessary to detect a 20 cycle per degree grating as a function of the time of onset of an adapting stimulus. The open symbols in Figure 7.12 show the influence of a dim, rod-stimulating background field (one too dim to directly influence cones) of 500 msec duration on grating detectability in the center of the fovea. (The format in Fig. 7.12 is such as to foster comparison with Fig. 7.5.) Notice that the threshold of the test grating decreases below control levels, an effect that begins only after the adapting stimulus if presented. The recovery from this supersensitive stage occurs fairly rapidly after the dim adapting field is removed. The closed symbols in Figure 7.12 show results of a similar experiment in which the adapting stimulus is 100 times more intense and directly influences cones. In this case, the intensity necessary to see the test grating increases during presentation of the adapting stimulus. But about 400 msec following adapting stimulus offset, sensitivity to this grating is considerably greater than control sensitivity. This period of supersensitivity gradually wanes to a negligible effect about a second and a half later. The lower coordinates shows similar data obtained with parafoveal stimulus presentation.

The meaning of these findings is emphasized by replotting some of the data in Figures 7.5 and 7.12 on the same coordinates. On the bottom of Figure 7.13, changes in rod detection threshold and changes in grating visibility produced by the same rod-stimulating background are compared directly. The time course of these two effects are quite different at SOA intervals near the onset of the background flash. This difference reinforces the assertion that light adaptation-induced changes in detection threshold and in spatial vision involve different events with independent time courses. However, the time courses for the recovery of detection threshold and spatial sensitivity are similar. The top of Figure 7.13 similarly shows changes in cone detection threshold and grating visibility produced by the same bright adapting stimulus. At light onset, both types of changes are quite similar but at light offset, the enhancement of spatial acuity bears no obvious relationship to the adaptation induced changes in detection threshold. At present, it is impossible to specify in detail the underlying reason for these differences in time course. However, it would seem that detection thresholds most probably reflect a "straightthrough" pathway from cones to bipolar cells. In contrast, changes in spatial vision must additionally involve the activity of horizontal cells, the distal retinal neuron involved in lateral spatial interactions. If this type of explanation is correct, it would seem as if alterations in horizontal cell response properties produced by adaptation are considerably more sluggish than those occurring within the photoreceptors themselves. At the time of writing, an almost complete lack of published, relevant neurophysiological data limits the usefulness of further speculation.

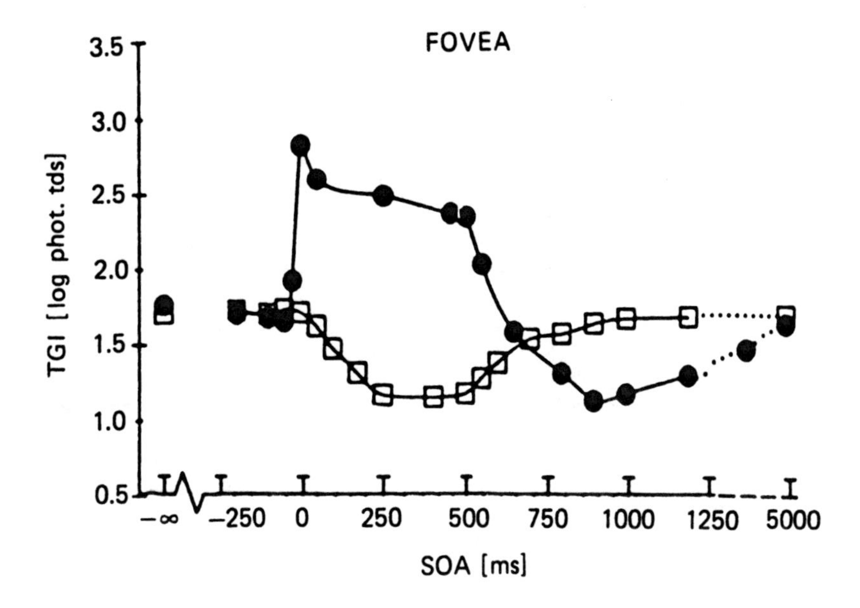

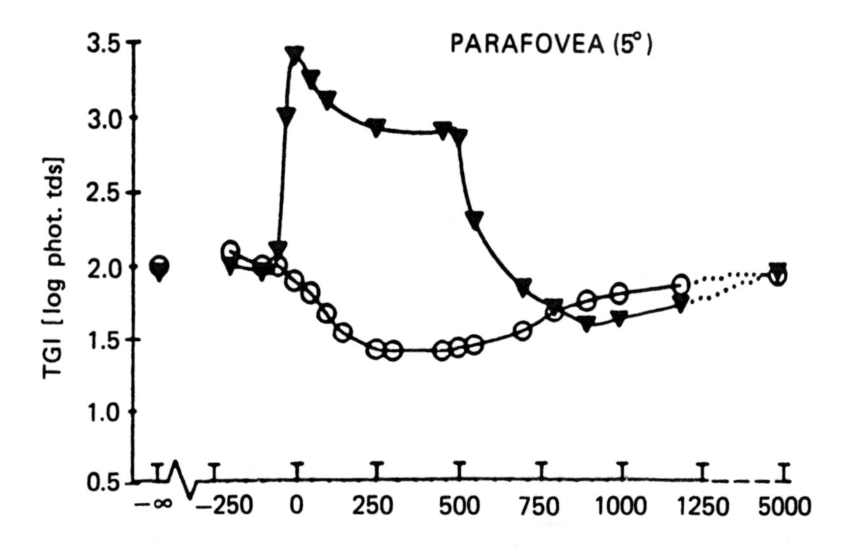

FIGURE 7.12. Influence of 20° diameter, 500 msec adapting flashes of either 0.32 td (open symbols) or 320 td (closed symbols) on the minimal illuminance necessary to test the orientation of a test grating (i.e., the threshold grating illuminance or TGI plotted on the ordinate). The adapting flash has the same parameters as indicated in the legend to Figure 7.5. The test grating was an on–off squarewave of 20 cycles per degree and 54′ overall extent presented either foveally (upper figure) or 5° parafoveally (lower figure). The minus infinity condition along the abscissa is TGI obtained with no adapting flash present. Notice that the dim adapting flashes decrease TGI at SOA values of between 0 and 1000 msec: the brighter adapting flash either increases TGI (SOA values of −50 to +500 msec) or decreases TGI (SOA values between −600 and +2000 msec). (From Naarendorp, 1988.)

FOVEA

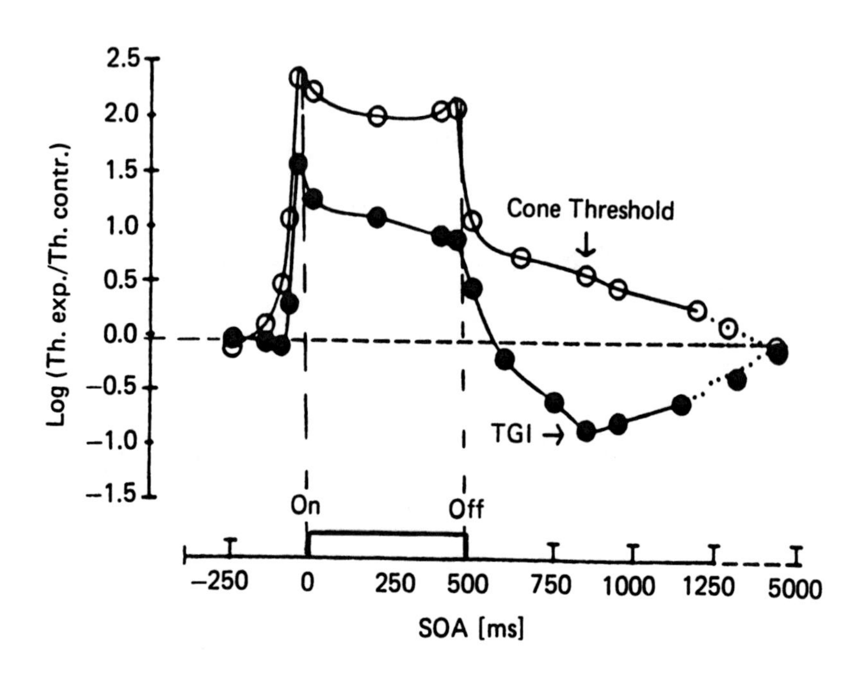
Log (Th. exp./Th. contr.)
2.5
2.0
1.5
1.0
0.5
0.0
−0.5
−1.0
−1.5
Cone Threshold
TGI
On
Off
−250
0
250
500
750
1000
1250
5000
SOA [ms]

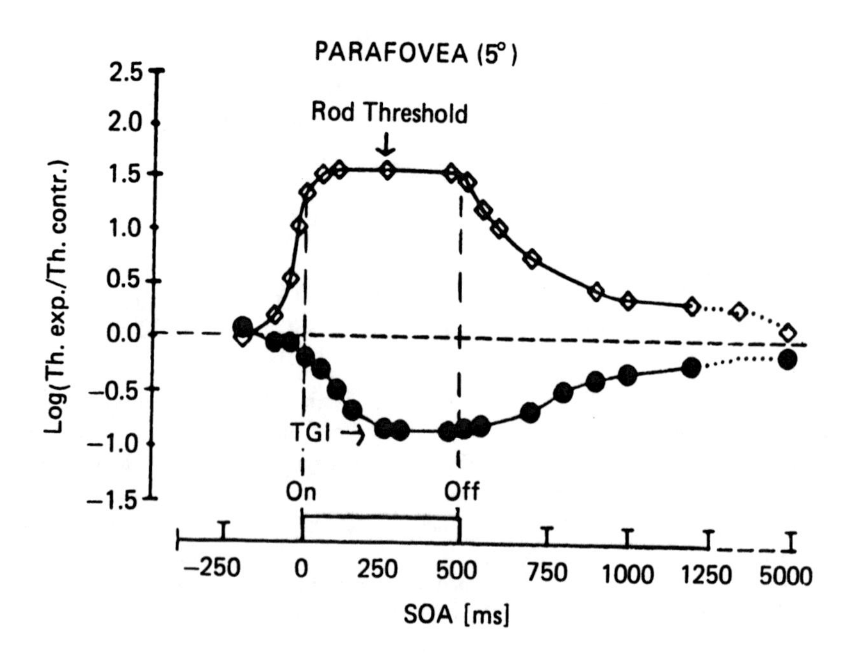
PARAFOVEA (5°)
Rod Threshold
Log(Th. exp./Th. contr.)
2.5
2.0
1.5
1.0
0.5
0.0
−0.5
−1.0
−1.5
TGI
On
Off
−250
0
250
500
750
1000
1250
5000
SOA [ms]

Other Phenomena

Other types of research also point to limitations in the classical psychophysical approach to adaptation. Boynton et al., (1964) first showed that different cone mechanisms can interact in a complex manner to determine threshold. Frumkes et al. (1973) used a similar approach to that of Boynton et al. to show that rod- and cone-related signals could summate together to determine absolute threshold. Both of these observations indicate that under at least some circumstances, test threshold detection is mediated by the cooperative action of more than one mechanism. Thus, the third assumption listed at the outset (that the most sensitive mechanism alone determines test threshold) is incorrect. Since these psychophysical data were collected, modern neurophysiological data show that signals from different photoreceptor types (different classes of cones, or rods and cones) summate together at many different loci within the retina including within the photoreceptors themselves (for review, see Sterling, 1983).

Color vision researchers have known since the work of Hering and Helmholtz in the nineteenth century that adaptation of the separate color mechanisms interrelates in a complex fashion. A relatively newly discovered phenomenon, transient tritanopia, provides an example. As indicated above and discussed in much greater detail in the chapter by Boynton in this volume, color vision in normal humans is mediated by three different types of cones. Tritanopia is an extremely rare hereditary condition corresponding to a loss in function of the short wavelength sensitive cones. To produce transient tritanopia, the eye of the normal observer is first strongly adapted to a yellow light that maximally adapts the pigment of the long and middle wavelength cones, but has relatively less influence on the short wavelength sensitive cones. For a short time period following the termination of the yellow adapting light, sensitivity to the short wavelength end of the spectrum is severely depressed as in the case of hereditary tritanopia. Transient tritanopia has been explained in several different ways (see Mollen and Polden, 1977, 1979).

◄

FIGURE 7.13. Comparison of the influence of short-term visual adaptation on detectability and grating visibility. For both sets of coordinates, the open symbols represent the detection threshold for a solid disc stimulus (as in Fig. 7.5) while the closed symbols represent TGI (as in Fig. 7.12). The ordinate indicates the deviation in either type threshold from control value, which is also represented by the horizontal dashed line. The upper figure represents data obtained in the fovea using an adapting flash of 320 td that directly influences cones. Although changes in both types of threshold are initially similar, notice that threshold grating illuminance is greatly decreased below control levels shortly after adapting flash offset. The lower figure represents data obtained 5° from the fovea using an adapting flash of 0.32 log td, which can only directly influence rods. Rod adaptation has a considerably different influence on both the time course and the type of change in sensitivity for these differing tests of visual processing. See legends to Figures 7.5 and 7.12 and text for further explanation. (From Naarendorp, 1988.)

Extracellular recordings from monkey ganglion cells suggest that transient tritanopia reflects in part the difference in latency of the center- and surround-mechanism of color-opponent neurons coding blue/yellow (Zrenner, 1983; see also Chapter 4 in this volume). Although a detailed explanation is a matter of some dispute, transient tritanopia questions the validity for assumption 4. Hence, the adapted state of the short wavelength-sensitive cone mechanism must depend on the adapted state of at least one other cone mechanism.

Finally, some psychophysical data indicate a lack of validity for the last assumption listed at the outset of this chapter (i.e., that all processes produce only one type of signal). Geisler (1979) and Adelson (1982B) combined the increment threshold approach to light adaptation (that shown in Fig. 7.1b) with Crawford's approach to light adaptation, emphasizing situations in which the test flash is presented after the offset of an adapting stimulus. After presenting stronger adapting flashes, they found that observers could not detect the primary image produced by any intensity test flash but could detect its afterimage. In fact, the increment threshold functions related to the primary image and afterimage were quite distinct as shown on the top of Figure 7.14. Their explanation for this observation rests on the time course of photoreceptor responses. For example, the bottom of Figure 7.14 shows data obtained from a turtle cone (recorded by Baylor et al., 1974) to 10 msec duration light flashes. Fourteen different intensity flashes were used differing from each other by 0.32 log unit steps. The eight dimmest flashes produced responses that are distinctly different from each other throughout their time course. The brighter

---

▶

FIGURE 7.14. The importance of cone response waveform for visual processing. The upper psychophysical data show thresholds of a test flash as a function of the illuminance of an adapting flash (background) of the indicated duration. Conditions were designed such that cones alone detect all stimuli. When adapting stimuli are 2 log td or dimmer, observers report that at threshold they detect the primary image produced by the test flash: in the presence of more intense adapting flashes, they detect the positive afterimage of the test flash. For adapting flashes brighter than 2 log td, the reaction time necessary for responding to the test flash increases systematically (data not shown). The lower data show intracellular responses of a turtle cone to 10 msec flashes of 14 different intensities increasing in 0.32 log unit steps. Notice that the dimmest 9 flashes all produce responses that are distinguishable throughout their time course. For the 5 most intense flashes, however, the saturating peak response is nearly identical and responses differ only during the "after potential" of their responses. The arrows indicate the time period at which responses to adjacent intensity flashes differs most. Notice that for the 6 most intense flashes, this time period of greatest difference increases systematically with stimulus intensity. Geisler (1979) suggests that psychophysical reaction time and threshold data specifically relate to the functional relationship between photoreceptor response waveform and stimulus intensity. (From Geisler, 1979, 1980. Copyright [1980], Pergamon Press plc; Baylor et al., 1974.)

flashes all produced the same "saturating" amplitude response. However, after the flash is extinguished, the time course for recovery of the cell was different for each of the different intensity flashes. The time period for this "after potential" is reminiscent of the positive after image evoked by a similar flash in human observers. Thus, although the peak amplitudes produced by the brighter flashes are indiscriminable, different afterpotential signals can be presumably discriminated. Geisler and Adelson suggest that the photoreceptors can produce a primary and afterimage signal with discriminable perceptual properties and with considerably different adaptive properties.

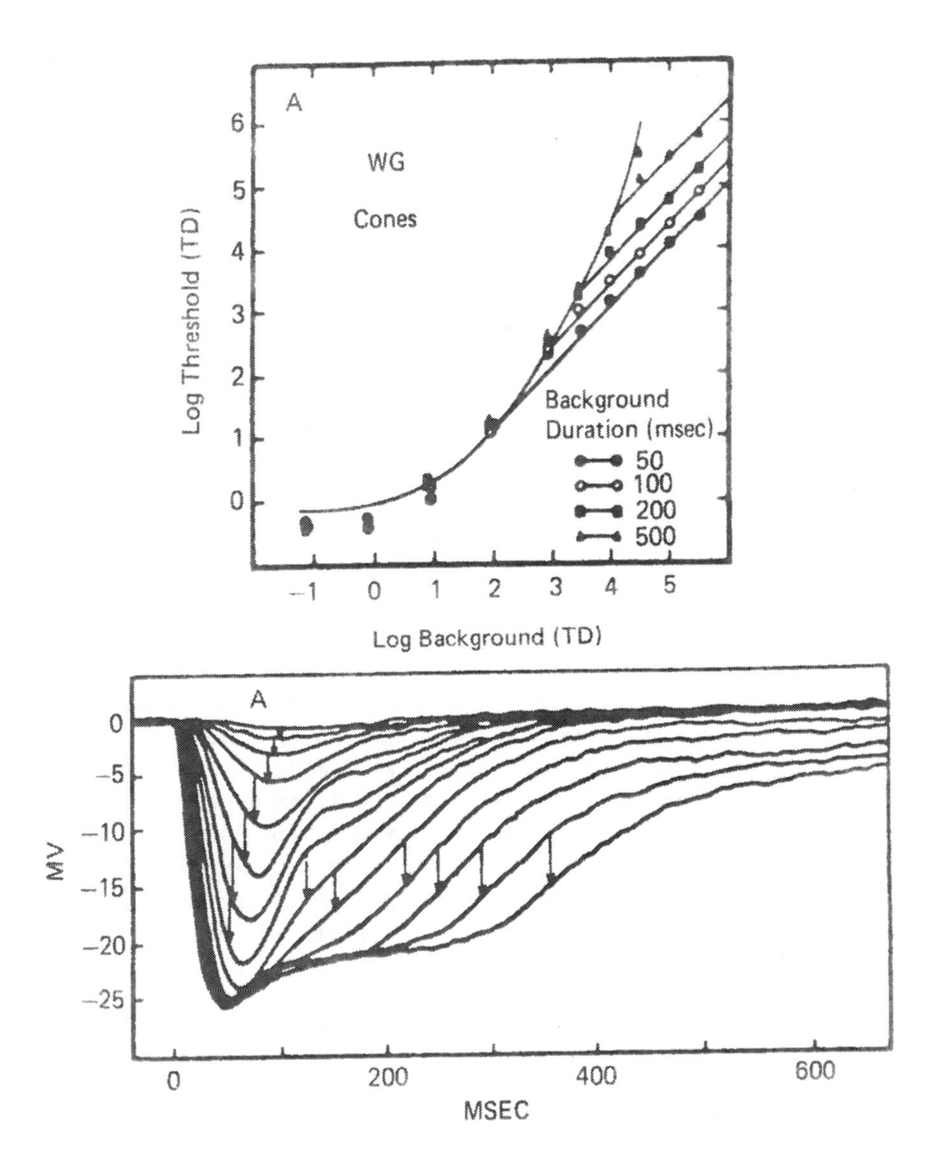

Summary and Closing Perspective

At the outset of this chapter, visual adaptation was defined very generally as the change in perception resulting from exposure to dark or light. The overview of classical data made it clear that adaptation involves photochemical factors and neural changes that occur both within the photoreceptors themselves as well as within postreceptor retinal neurons. A variety of data, particularly more modern results, indicate that dark and light adaptation influence different aspects of visual perception in addition to the threshold for detection, namely certain aspects of color, time, and space perception. At least in some cases, these different types of changes occur independently of one another.

Many more striking examples of perceptual alteration could be cited depending on the definition of the term adaptation. In the first half of this century, the Gestalt psychologists documented a large number of visual aftereffects. Such phenomena included demonstrations that exposure to one direction of motion alters sensitivity to motion perception in another direction, or that exposure to one orientation tends to "tilt" the percept evoked by stimuli presented in another orientation. Although interest in such demonstrations waned in the immediate post war era, cognitive psychologists have recently become concerned again with a growing number of such phenomena. Probably the most widely studied example is the "McCullough effect." McCullough (1965) first showed that exposure to colored gratings in one spatial orientation alters sensitivity to the color of gratings in other orientations. Thus, after fixating on a red vertical grating, achromatic horizontal gratings often appear to be slightly green.

Although this type of research is quite interesting, it can be distinguished from that covered in the present overview in three major respects. First, these after effects can be extremely long lasting. The McCullough effect often lasts weeks, sometimes months, and there are some claims that is never extinguishes. Second, such phenomena do not result from a change in overall illumination and can be demonstrated when overall illumination remains constant. Thus, these are not changes in light or dark adaptation. Third, these phenomena often remain unexplained in spite of extensive research. The McCullough effect has been described in literally hundreds of published articles but has never been related specifically to the activity of any neural locus. For example, Livingston and Hubel have recently attempted to relate different types of perceptual phenomena to specific visual pathways within the striate area of visual cortex (primary visual cortex, Brodmann area 17, or V1 cortex). Their earlier work distinguished three pathways on the basis of morphological and physiological criteria. In a recent article (Livingston and Hubel, 1987) that has evoked considerable controversy within the scientific community, they attribute over 20 different optical illusions specifically to the activity of one of their three functional subdivisions of striate cortex or of nearby Brodmann cortical area 18. Although their speculation goes considerably beyond available data, they specifically cite the McCullough effect as a case that totally defies their classification scheme and that most probably involves processing in unknown brain areas.

In closing, the present review emphasized instances in which the behavioral and neural scientist described successfully the same phenomena, and presented their results in comparable terms. This ability to interrelate these two different sets of observations makes adaptation a topic of continual interest to the general scientific community.

Acknowledgments. Some unpublished research referred to in this review were supported in part by Grants EY01802 and EY05984 from the National Institutes of Health. I gratefully acknowledge comments on earlier versions of this manuscript by Iwona Beczkowska, R.M. Boynton, Gudrun Lange, R. Lanson, K.N. Leibovic, E. Pugh, and S.M. Wu.

References

Adelson EH (1982a). Saturation and adaptation in the rod system. Vision Res. **22**: 1299–1982.

Adelson EH (1982b). The delayed rod afterimage. Vision Res. **22**:1313–1328.

Aguilar M, Stiles WS (1954). Saturation of the rod mechanism of the retina at high levels of illumination. Opt. Acta **1**:59–65.

Aho A-C, Donner K, Hyden C, Larsen LO, Reuter T (1988). Low retinal noise in animals with low body temperature allows high visual sensitivity. Nature (London) **334**:348–350.

Alexander KR, Fishman GA (1985). Rod-cone interaction in flicker perimetry: Evidence for a distal retinal locus. Documenta Ophthalmol. **60**:3–36.

Alexander KR, Fishman GA, Derlacki DJ (1988). Mechanisms of rod-cone interaction: evidence from congenital stationary night blindness. Vision Res. **28**:575–583.

Arden GB, Frumkes TE (1986). Stimulation of rods can increase cone flicker ERGs in man. Vision Res. **26**:711–721.

Arden GB, Hogg CR (1985). Rod-cone interactions and analysis of retinal disease. Br. J. Ophthalmol. **69**:405–415.

Aubert (1865). *Physiologie der Netzhaut*. Morgenstern, Breslau.

Baker HD (1963). Initial stages of light and dark adaptation. J. Opt. Soc. Am. **53**:98–103.

Barlow HB (1956). Retinal noise and absolute threshold. J. Opt. Soc. Am. **46**:634–639.

Barlow HB (1965). Optic nerve impulses and Weber's law. Cold Spring Harbor Symp. Quant. Biol. **30**:539–546.

Barlow HB (1981). Critical limiting factors in the design of the eye and visual cortex. Proc. R. Soc. London B **212**:1–34.

Barlow HB, Fitzhugh R, Kuffler SW (1957). Changes of organization of the receptive fields of the cat's retina during dark adaptation. J. Physiol. **137**:338–354.

Baylor DA, Hodgkin AL, Lamb TD (1974). The electrical response of turtle cones to flashes and steps of light. J. Physiol. **242**:685–727.

Baylor DA, Nunn BJ, Schnapf JL (1984). The photocurrent, noise and spectral sensitivity of rods of the monkey, *Macaca fasciularis*. J. Physiol. **357**:575–607.

Battersby WS, Wagmen IH (1962). Neural limitations of visual excitability: IV. Spatial determinants of retrochiasmal interaction. Am. J. Physiol. **203**:359–365.

Bauer GM, Frumkes TE, Holstein GR (1983a). The influence of rod light and dark adaptation upon rod-cone interaction. J. Physiol. **337**:121–135.

Bauer GM, Frumkes TE, Nygaard RW (1983b). The signal-to-noise characteristics of rod-cone interaction. J. Physiol. **337**:101–119.

Boynton RM (1961). Some temporal factors in vision. In *Sensory Communication* (RA Rosenblith, ed.), pp. 739–756. MIT Press, Cambridge.

Boynton RM, Ikeda M, Stiles WS (1964). Interactions among chromatic mechanisms as inferred from positive and negative increment thresholds. Vision Res. **4**:87–117.

Coletta NJ, Adams AJ (1984). Rod-cone interactions in flicker detection. Vision Res. **24**:1333–1340.

Coletta NJ, Adams AJ (1985). Loss of flicker sensitivity on dim backgrounds in normal and dichromatic observers. Invest. Ophthalmol. Visual Sci. Suppl. **26**:187.

Crawford BH (1947). Visual adaptation in relation to brief conditioning stimuli. Proc. R. Soc. London B **134**:283–302.

deVries H (1943). The quantum character of light and its bearing upon threshold of vision, the differential sensitivity and visual acuity of the eye. Physica **10**:553–564.

Dowling JE (1960). The chemistry of visual adaptation in the rat. Nature (London) **188**:114–118.

Dowling JE (1987). *The Retina: An Approachable Part of the Brain*. Harvard University Press, Cambridge, MA.

Eysteinsson T, Frumkes TE (1989). Physiology and pharmacological analysis of suppressive rod-cone interaction in *Necturus* retina. J. Neurophysiol. In Press.

Frumkes TE (1979). Temporal summation in frogs and men. Behav. Brain Sci. **2**:261–263.

Frumkes TE (1990). Suppressive rod-cone interaction. In *Handbook of Clinical Electrophysiology of Vision Testing* (GB Arden, Heckenlively, eds.). Mosby, St. Louis. In press.

Frumkes TE, Eysteinsson T (1987). Suppressive rod-cone interaction in distal vertebrate retina: Intracellular records from *Xenopus* and *Necturus*. J. Neurophysiol. **57**:1361–1382.

Frumkes TE, Eysteinsson T (1988). The cellular basis for suppressive rod-cone interaction. Visual Neurosci. **1**:263–273.

Frumkes TE, Sekuler MD, Barris MC, Reiss EH, Chalupa LM (1973). Rod-cone interaction in human scotopic vision. I: Temporal analysis. Vision Res. **13**:1269–1282.

Frumkes TE, Naarendorp F, Goldberg SH (1986). The influence of cone stimulation upon flicker sensitivity mediated by adjacent rods. Vision Res. **26**:1167–1176.

Frumkes TE, Naarendorp F, Goldberg SH (1988). Abnormalities in retinal neurocircuitry in protanopes: Evidence provided by psychophysical investigation of temporal-spatial interactions. Invest. Opthalmol. Visual Sci. Suppl. **29**:163 (abstr.).

Geisler WH (1979). Initial image and afterimage discrimination in the human rod and cone systems. J. Physiol. **294**:165–179.

Geisler WS (1980). Increment threshold and detection latency in the rod and cone systems. Vision Res. **20**:981–994.

Goldberg SH (1983). The influence of dark adapted rods on cone mediated flicker sensations. Unpublished portions of a Doctoral Dissertation, City University of New York.

Goldberg SH, Frumkes TE, Nygaard RW (1983). Inhibitory influence of unstimulated rods in human retina: Evidence provided by examining cone flicker. Science **221**: 180–182.

Graham CH, Margaria R (1935). Area and the intensity-time in the peripheral retina. Am. J. Physiol. **113**:299–305.

Hart WM Jr (1987). Visual adaptation. In *Adler's Physiology of the Eye, 8th ed.* (RA Moses, WM Hart, eds.), pp. 389–414. Mosby, St. Louis.

Hecht S (1937). Rods, cones and the chemical basis of vision. Physiol. Rev. **17**:239–290.

Hecht S, Smith EL (1936). Intermittent stimulation by light. J. Gen. Physiol. **19**:979–989.

Hecht S, Shlaer S, Pirenne MH (1942). Energy, quanta, and vision. J. Gen. Physiol. **25**: 819–840.

Heinemann EG (1955). Simultaneous brightness induction as a function of inducing a test field luminances. J. Exp. Psychol. **50**:89–96.

Hess RF, Nordsby K (1986). Spatial and temporal limits of vision in the achromat. J. Physiol. **371**:365–385.

Kohlrausch A (1922). Untersuchungen mit farbigen Schwellen prueflichtern ueber den Dunkeladaptationsuerlauf des normales Auges. Pflueg. Arch. Ges. Physiol. **196**:113–117.

Krill AE, Archer DB (1977). *Hereditary Retinal and Choroidal Diseases, Vol. II: Clinical Characteristics*. Harper & Row, Hagerstown.

Lamb TD (1981). The involvement of rod photoreceptors in dark adaptation. Vision Res. **21**:1773–1782.

Land EH, McCann JJ (1971). Lightness and retinex theory. J. Opt. Soc. Am. **61**:1–11.

Lansford TG, Baker HD (1969). Dark adaptation: An interocular light-adaptation effect. Science **164**:1307–1309.

Leibovic KN, Dowling JE, Kim YY (1987). Background and bleaching equivalence in steady-state adaptation of vertebrate rods. J. Neurosci. **7**:1056–1063.

Lie I (1963). Dark adaptation and the photochromatic interval. Documenta Ophthalmol. **17**:411–510.

Livingston MS, Hubel DH (1987). Psychophysical evidence for separate channels for the perception of form, color, movement, and depth. J. Neurosci. **7**:3416–3468.

Lythgoe RJ, Tansley K (1929). The relation of the critical frequency of flicker to the adaptation of the eye. Proc. R. Soc. London B **105**:60–92.

MacLeod DIA (1974). Psychophysical studies of signals from rods and cones. Unpublished Doctoral Dissertation, University of Cambridge.

MacLeod DIA (1978). Visual sensitivity. Annu. Rev. Psychol. **29**:613–645.

McCollough C (1965). Color adaptation of edge-detectors in the human visual system. Science **149**:1115–1116.

Mollon JD, Polden PG (1977). An anomaly in the response of the eye to light of short wavelengths. Proc. R. Soc. London B **278**:207–240.

Mollon JD, Polden PG (1979). Post-receptoral adaptation. Vision Res. **19**:435–440.

Naarendorp F (1988). The influence of rod and cone adaptation upon cone-mediated spatial acuity. Unpublished doctoral dissertation, City University of New York.

Naarendorp F, Denny N, Frumkes TE (1988). Rod light and dark adaptation influence cone-mediated spatial acuity. Vision Res. **28**:67–74.

Pepperberg DR, Brown PK, Lurie M, Dowling JE (1978). Visual pigment and photoreceptor sensitivity in the isolated skate retina. J. Gen. Physiol. **71**:369–396.

Pflug R, Nelson R (1986). Enhancement of red cone flicker by rod selective background in cat horizontal cells. Neurosci. Abst. **16**:406.

Pugh EN Jr (1975). Rushton's paradox: Rod dark adaptation after flash photolysis. J. Physiol. **248**:413–431.

Pugh EN Jr (1988). Vision: Physics and retinal physiology. In *Stevens Handbook of*

Experimental Psychology, 2nd ed., Vol. 1: Perception and Motivation (RC Atkinson, RJ Herrnstein, G Lindzey, RD Luce, eds.), pp. 75–163. Wiley, New York.

Rose A (1942). The sensitivity performance of the human eye on an absolute scale. J. Opt. Soc. Am. **38**:196–208.

Rushton WAH (1961). Rhodopsin measurement and dark-adaptation in a subject deficient in cone vision. J. Physiol. **156**:193–205.

Rushton WAH (1963). Increment threshold and dark adaptation. J. Opt. Soc. Am. **53**:104–109.

Rushton WAH (1965). Visual adaptation. Proc. R. Soc. London B **162**:20–46.

Sakitt B (1972). Counting every quantum. J. Physiol. **223**:131–150.

Shapley R, Enroth-Cugell C (1984). Visual adaptation and retinal gain controls. In *Progress in Retinal Research*, Vol. 3 (NN Osbourne, GJ Chader Jr, eds.), pp. 263–346. Pergamon, New York.

Schlaer S (1938). The relation between visual acuity and illumination. J. Gen. Physiol. **21**:165–188.

Sterling P (1983). Microcircuitry of the cat retina. Annu. Rev. Neurosci. **6**:149–185.

Stroud JM (1956). The fine structure of psychological time. In *Information Theory in Psychology* (H Quastler, ed.), pp. 174–207. Free Press, Glencoe IL.

Sturr JF, Battersby WS (1966). Neural limitations of visual excitability. VIII: Binocular convergence in cat geniculate and cortex. Vision Res. **6**:401–418.

Thibos LN, Werblin FS (1978). The response properties of the steady antagonistic surround in the mudpuppy retina. J. Physiol. **278**:79–99.

Thompson HS (1987). The pupil. In *Adler's Physiology of the Eye, 8th ed.* (RA Moses, WM Hart Jr, eds.), pp. 311–338. Mosby, St. Louis.

Triesman M (1965). Signal detection theory and Crozier's law: Derivation of a new sensory scaling procedure. J. Math. Psychol. **2**:205–218.

Wald G (1945). Human vision and the spectrum. Science **101**:287–291.

Whittle P, Challands PDC (1969). The effect of background luminance on the brightness of flashes. Vision Res. **9**:1095–1110.

Zrenner E (1983). *Neurophysiological Aspects of Color Vision in Primates*. Springer-Verlag, New York.

8
Human Color Perception

ROBERT M. BOYNTON

Definition of Color

Common sense would suggest that the word color refers to the special quality that color photography, television, or printing adds to black-and-white, or colorless, versions of the same scene. However, in a technical sense the word "color" is also used to refer to variations in lightness, implying that color exists also in black-and-white reproductions. Wyszecki and Stiles's (1982) *Color Science* (a massive, authoritative handbook concerned mainly with the physical basis of color vision) offers the following definition (p. 487):

> *Color* is the aspect of visual perception by which an observer may distinguish differences between two structure-free fields of view of the same size and shape, such as *may* be caused by differences in the spectral composition of the radiant energy concerned in the observation. [italics mine]

It is convenient to distinguish between two aspects of color. The first, chromatic color, is what color reproduction adds to black-and-white. The second aspect bears the seemingly oxymoronic label achromatic color. Let us examine the difference between these aspects of color as they relate to the "structure-free fields of view" of the Wyszecki–Stiles definition, assuming that the two fields are precisely joined. Unless they are of equal intensity, or very nearly so, the two fields will be distinguished as different, and a contour will be formed between them. If the two fields differ in radiance (physical intensity per unit area) but have the same relative distributions of radiance as a function of wavelength, there will be an achromatic color difference between them. If the radiances of two such fields are equated, or differ by a physical amount too small to be discriminated, the fields will match exactly and the contour separating them will disappear.

Chromatic color differences occur when the relative spectral distributions of the two fields are dissimilar. In this case, there is ordinarily no relative radiance of the two fields that will cause them to appear identical or that, in general, will eliminate the contour formed between them. There are important exceptions. Some fields that differ spectrally nevertheless match (see the section on

Metamerism) and some fields that do not match at any relative radiance nevertheless fail to exhibit an obvious contour between them (Tansley and Boynton, 1978; Boynton et al., 1985).

Most of the time, in the real world, achromatic and chromatic color differences occur together and most contours are formed by a combination of the two (Frome et al., 1981). As evidenced by the typically minor loss of perceptible contour in black-and-white reproductions, achromatic contours vastly predominate.

Given the technical definition of color, which will be adopted here, all vision then becomes color vision from the viewpoint that the visual field is made up of regions of more-or-less homogeneous colors separated by boundaries that are formed by color differences. When the regions are very small, as in color television, complex and realistic visual scenes can be represented by an array comprised of a very large number of tiny and homogeneous colored dots, called pixels.

This chapter will emphasize chromatic color perception. When the word "color" is used without a modifier, chromatic color is the intended concept.

Functions of Color

Broadly defined, visual perception is concerned with the extraction of useful information from the external physical world. The adequate stimulus for vision, of course, is light. Each quantum of light, or photon, has associated with it a specific and unalterable energy, one which is inversely related to its wavelength. Chromatic color vision occurs under so-called photopic conditions, where illumination is strong, and many millions of photons are available. Typically these conditions are provided by sources exhibiting a broad and continuous distribution of wavelengths across the visible spectrum, which, for example, is the case for sunlight or incandescent illumination. Our perceptual apparatus has evolved to make good use of this spectral information if sufficient light is available. Vision mediated by rods, at scotopic radiances too low for cones to function, is achromatic. An impoverished form of color vision occurs at intermediate, or mesopic levels of illumination.

That color is important aesthetically seems undeniable: almost everyone agrees that color adds beauty and sparkle to the world, and that it affects mood. However, attempts to deal with such aspects of color scientifically have been of very limited value (Kaiser, 1984). Because these subjective aspects of color belong more properly to the worlds of art and fashion than to science, we will not attempt to deal with them in this chapter.

Of those aspects of color that can be discerned and studied quantitatively, two stand out. One of these, related to the formation of contour, has already been mentioned in the previous section. Although achromatic contours do predominate, there are occasions when they disappear and the remaining chromatic contours provide the sole basis for contour perception. This is called an isoluminant condition (Gregory, 1977), referring to the fact that two or more regions in the visual field are of equal luminance (see section on Luminance, Brightness, and Lightness). Contours formed in this way, which are rare in everyday life, exhibit some peculiar properties. More commonly, chromatic differences add to the normal perception of contours already present because of luminance differences. For some recent studies of isoluminant phenomena, see Cavanagh and Favreau (1985), Mullen and Baker (1985), and Grinberg and Williams (1985).

A second function of color vision, related to categorical perception, is probably more important. This aspect of color perception ordinarily applies to objects that are not spatially contiguous. Despite our perception that color is an inherent property of the surface of an object, it actually depends not only on a property of the object, but also on an appropriate source of illumination and an observer with color vision. When perceived colors differ, many tasks are facilitated: for example, we are better able to discriminate members of opposing football teams, to judge the ripeness of a banana, or to search for a red automobile in a parking lot. This second function of chromatic color vision will be emphasized in this chapter.

Requirements for Human Color Perception

For human color perception to be possible, certain conditions must be met that fall naturally into three categories. Briefly stated, normal color perception can occur only if (1) the observer has normal color vision, (2) the objects and surfaces in the environment have variable spectral reflectance characteristics, and (3) a source of light is available that includes wavelengths throughout the visible spectrum. In this section, these requirements will be elaborated in a preliminary fashion; some will be described in more detail in later sections of this chapter.

The Physiological Requirement

For nearly 200 years, it has been generally understood that the initial basis for color vision lies in the differential excitation of three different classes of cone photoreceptors in the retina. Almost all sources of visible radiation in the real world are effective on all three kinds of cones, so the discrimination of differences in color depends on an evaluation, by later stages of visual processing, of the ratios of excitations among the three cone types (called L, M, and S cones according to whether they are maximally sensitive in the long, middle, or short wavelengths of the physical light spectrum).

Within the visible part of the electromagnetic spectrum, the spectral tuning of the three classes of cones is very broad (see Fig. 8.1). In fact, two of them (L and M) are responsive to all wavelengths of the visible spectrum, but S cones are virtually insensitive to the long wavelength region. The action spectra of the cones are such that each wavelength yields a unique ratio of excitation among the three cone types; this ratio, as well as the perceived color associated with it, both change continuously and without repetition as the wavelength of light is gradually varied from one end of the spectrum to the other. For a detailed discussion of the relation between hue and wavelength, see Boynton (1975).

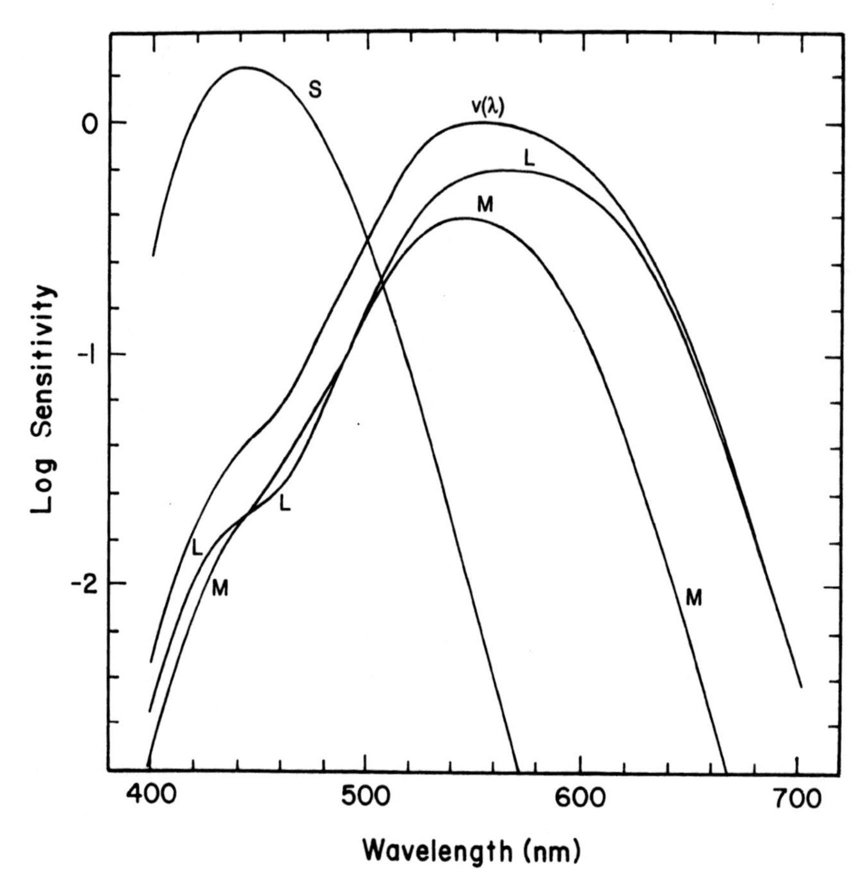

FIGURE 8.1. The action spectra of human cones as estimated by Smith and Pokorny (1975). The curves are based on equations derived by Boynton and Wisowaty (1980). The L and M functions have been positioned vertically, relative to each other, so that their sum is proportional to the $V(\lambda)$ function. The vertical positioning of the S function is arbitrary.

Because the signals that a receptor generates are not "tagged" according to the wavelength of the light absorbed (called the "principle of univariance" by Naka and Rushton, 1966), any two lights, if suitably adjusted for intensity, can produce exactly the same kind and amount of response from a given class of photoreceptor. This is exactly what happens under dim illumination, where perception is mediated by rods, and this is why such scotopic vision is achromatic.

Most so-called color-blind individuals still enjoy a reduced form of color vision mediated predominantly by signals generated by S cones and one of the other two kinds (Pokorny et al., 1979). There have been a few cone monochromats reported, who seem to possess only one kind of cone in their retinas, and as a result they are totally color-blind. Much more common, though still rare, are the so-called rod monochromats who have only rod vision.

Surfaces Exhibiting Differential Spectral Reflection

If every surface in the environment were covered with gray paint from the same batch, there would be no physical basis for color vision in that environment. Painting the world gray, which would be an expensive way to achieve achromatic reproduction of a scene despite using color film in the camera, would not result in the disappearance of objects. Contours would remain, dividing areas of varying lightness that would continue to result from shadows and variations related to the distance of surfaces from, and their angular relation to, sources of light in the environment.

Of course most real environments are not like this at all. Instead, surfaces vary widely in their reflectance characteristics. Of the greatest importance for human color perception is the diffuse spectral reflectance of a surface, which is the percentage of diffusely reflected light measured as a function of wavelength across the visible spectrum. Most surfaces reflect light at all visible wavelengths, but differentially so, and their "spectral reflectance curves" are continuous, without rapid change of slope anywhere in the spectrum. Some typical examples are given in Figure 8.2.

In addition to their diffuse reflectance characteristics, most surfaces also exhibit a specular component as well. The limiting example of specularity is a perfectly reflecting, dust-free plane mirror. If suitably framed, this close approximation to an ideal specular surface cannot be perceived as a surface; instead, as we all know from common experience, the reflected images of other surfaces are perceived as if they were behind the surface of the mirror. An intermediate example is the highly polished surface of a new automobile. Specular reflection occurs from its surface, which acts like a mirror. However, much of the incident light penetrates the polished surface and is diffusely reflected from the substrate (see the section on the Physical Basis of Surface Colors). We learn to discern the chromatic color of the surface by positioning our eyes, relative to the source of illumination, to avoid the specular component of reflection, which, unlike the diffuse reflection from underneath, is very directional. When handling small objects, we can accomplish this by movement of the object instead. Such adjustments are commonplace and are made almost automatically.

Appropriate Light Sources

It is safe to assume that human vision evolved in sunlight. The spectrum of sunlight is not a simple matter, and it varies as a function of many parameters, including latitude, cloud conditions, time of year, time of day, and clarity of the atmosphere. A few of these complications are represented in Figure 8.3. Sunlight exhibits two properties that are very important for color vision. First, energy is available at all wavelengths across the visible spectrum, and the balance of this energy is not outrageously skewed. Second, there are no gaps in the spectrum.

From the foregoing considerations related to human observers and reflecting surfaces, the importance of the balanced and continuous nature of sunlight should

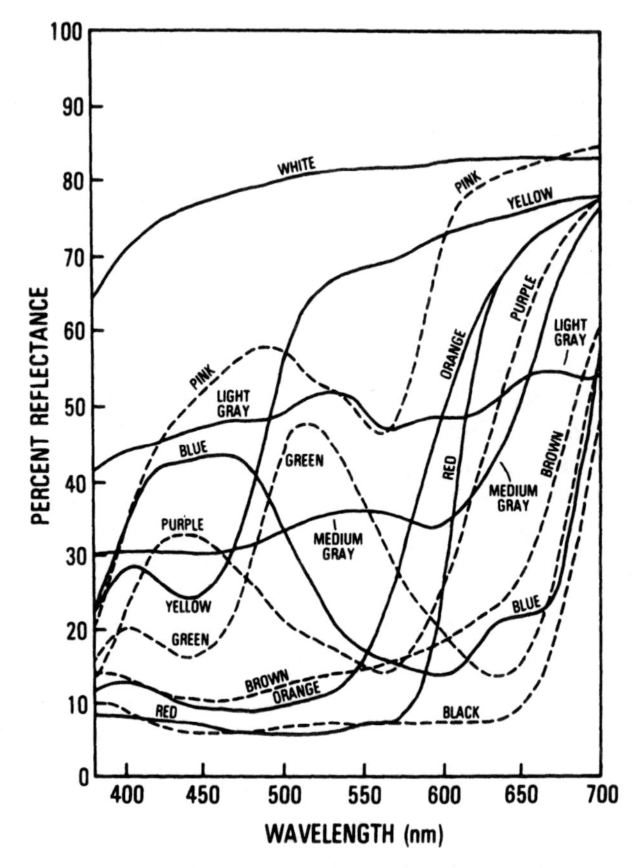

FIGURE 8.2. Spectral reflectance curves of ordinary construction papers used in an experiment by Stefurak and Boynton (1986). Percent diffuse reflectance is plotted as a function of wavelength (measured by the Hemmendinger Color Laboratory, Belvadere, NJ). The names of the colors, as reported by the subjects, are indicated for each curve.

be evident. The point can be emphasized by considering what happens when the source of illumination is monochromatic. In this case, no matter what the spectral reflectances of environmental surfaces, and regardless of the capacity of the human visual apparatus to mediate color perception under sunlight, there can be no color perception in the sense of distinguishing different chromatic colors. Surfaces can reflect only more or less light of whatever wavelength is generated by the monochromatic source. (Fluorescence, which is an important exception to this principle, will be discussed later.)

This monochromatic condition has been closely approximated by low-pressure sodium sources. Although not strictly monochromatic, the two spectral bands in the yellow part of the spectrum are nearly coincident, and there is a negligible amount of radiation in other spectral regions. In a parking lot bathed

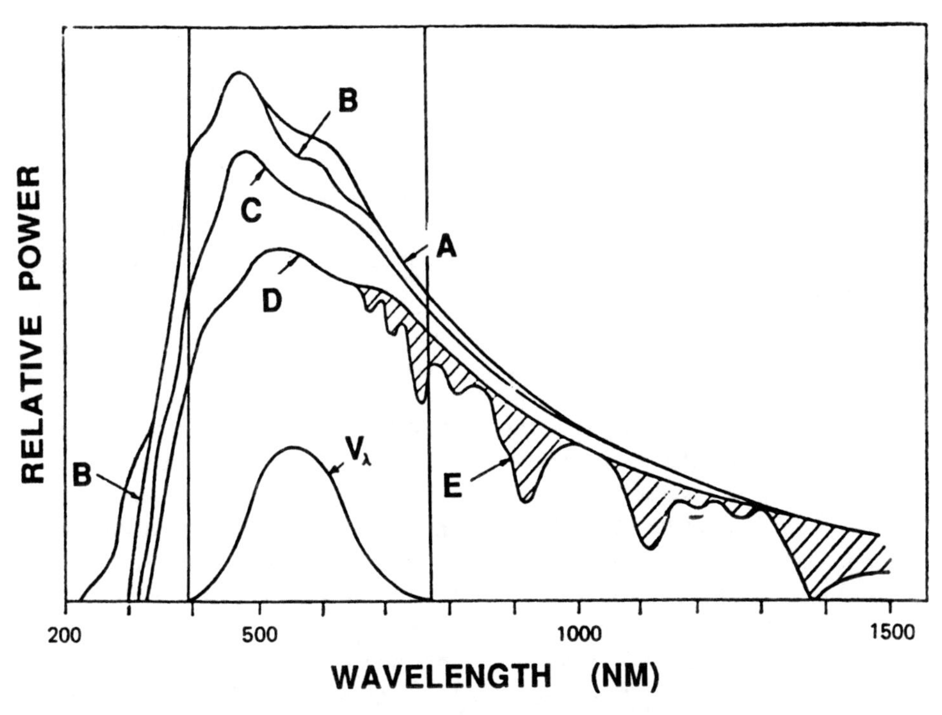

FIGURE 8.3. The topmost curve (A) shows the spectral distribution of sunlight as it enters the Earth's atmosphere. The other curves show how the amount of light is reduced, and its spectral distribution altered, by additional factors: B=after ozone absorption; C=after molecular scattering; D=after aerosol scattering; E=after water and oxygen absorptions: terrestrial sunlight. The relative spectral sensitivity of the eye [$V(\lambda)$] is shown at the bottom. (From Boynton, 1979, after Henderson, 1977.)

by this light, automobiles cannot be discriminated by their chromatic color. All other sources of artificial light are better for color rendering; the best is tungsten-incandescent illumination, which is smoothly continuous across the visible spectrum.

In concluding this introduction to what is required for human color perception, it should be stressed that all three of the requirements discussed—appropriate illumination, differential surface reflectances, and an observer with color vision—must be met for chromatic color perception to occur. In none of these cases, however, is the situation all-or-none. There are illuminants that are poor for color rendering without completely destroying it. There are natural environments in which, although nothing may be painted gray, black-and-white predominates. And there are individuals—a good many of them in fact—who lack normal color vision but who are not literally color-blind (Pokorny et al., 1979).

Sources of Illumination

Illumination is prerequisite for all visual perception. In this section we will discuss the nature of some of the illuminants in common use today. A discussion of the implications of these sources for color vision will conclude the section. Because the spectral distribution of a light source is especially critical for color vision, attention will be focused on this characteristic, although it must be kept in mind that a sufficient intensity of illumination is always required for good color perception.

Sunlight

Sunlight begins with gamma radiations whose wavelengths are well outside the visible range on the short-wavelength side; this radiation is spread to longer wavelengths by processes of absorption and reemission that take place throughout the bulk of the sun (Henderson, 1977, p. 15). Further modification of the spectrum of sunlight occurs, as shown in Figure 8.3, as a result of absorption and scatter in the Earth's atmosphere. The exact form of the resulting spectrum depends on many factors, but curve E in Figure 8.3 is a reasonable approximation of what reaches the Earth. Note that the maximum available energy from sunlight peaks in the same region of the spectrum where the eye is most sensitive. This coincidence is probably not accidental, but is more likely the product of biological evolution. Although most of the energy of sunlight in the infrared region of the spectrum is wasted for vision, it is a vital source of heat. Photosensitive receptor pigments with peak sensitivities at wavelengths much beyond 700 nm have not evolved in any species; if they existed, the high level of spontaneous, thermally induced activity, especially that associated with their long-wavelength limbs, would probably make them too noisy to be of much use for vision (see Brindley, 1960, p. 251).

On a cloudy day, sunlight is scattered by the large particles that make up atmospheric overcast, yielding illumination that is geometrically diffuse, coming from all parts of the sky. On a clear day, small-particle scatter, which is wavelength dependent, favors short wavelengths and provides the physical basis for the blue sky. On such a day, the light falling on unshaded objects consists of direct sunlight, which of course is quite directional (the sun subtends only about 0.5° at the Earth's surface), mixed with scattered skylight, which is important for filling in shadows.

Because a surface does not have a color that is independent of the illuminant, the "correct" rendering of its color is arbitrary. Nevertheless, sunlight is commonly regarded as optimal for the rendering of the color of surfaces, and it therefore provides a standard against which artificial sources are typically gauged.

However, sunlight is a highly variable standard. Judd et al. (1964) undertook an analysis of the spectral distributions of 622 samples of daylight, which included sunlight and sunlight plus skylight. These measurements had been made by previous investigators at several locations, and at various times of the day and year. Judd and his colleagues concluded that natural light could be well represented by a suitable combination of the two basis functions shown at the top in Figure 8.4. When these are assigned appropriate weights, the lower set of

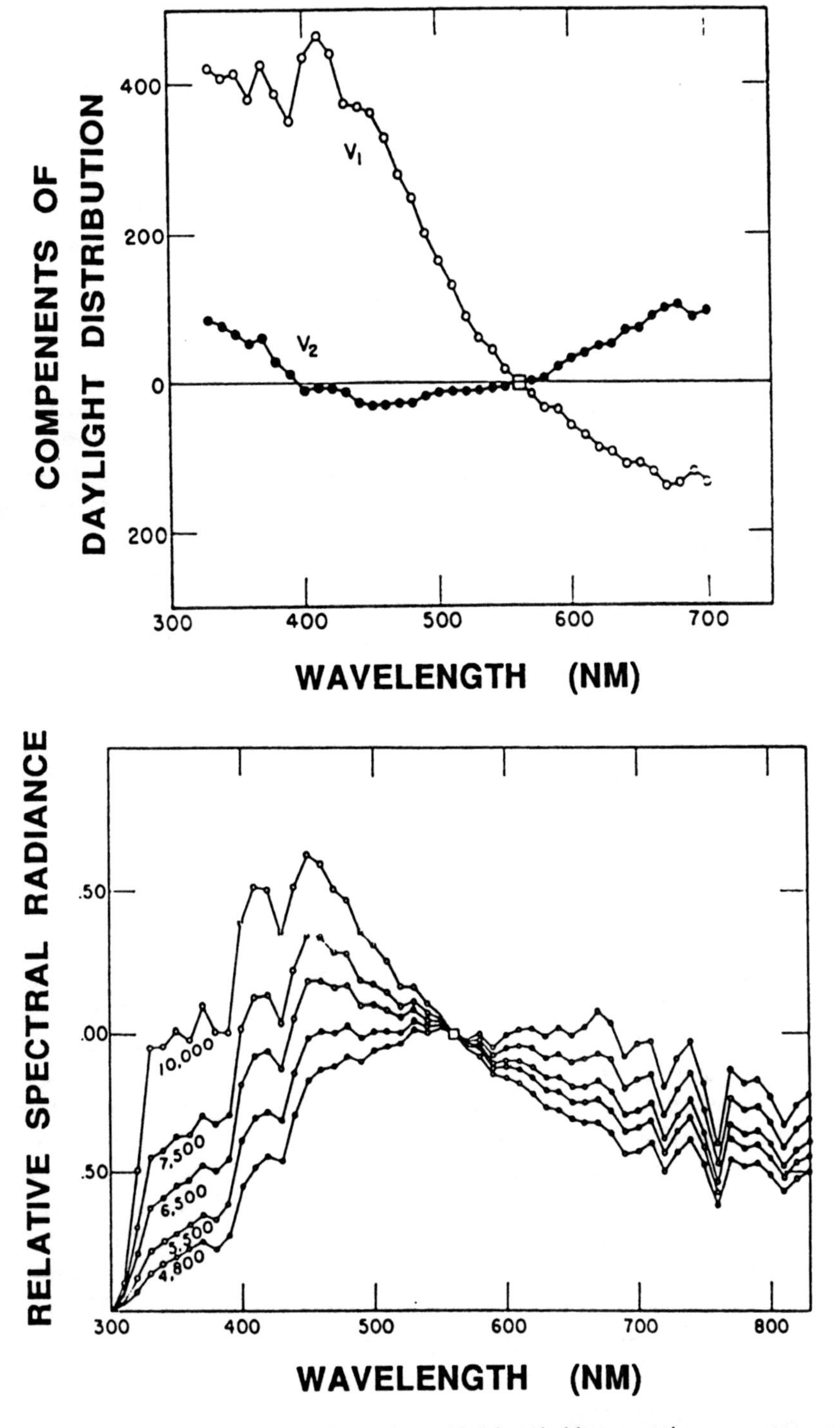

FIGURE 8.4. Top: Basis functions that, when added in suitable proportions, can account for the spectral distribution of typical daylight. Bottom: Calculated spectral distributions of daylight for various correlated color temperatures (where the chromaticity of daylight is approximately equal to that of a blackbody radiator). (From Judd et al., 1964.)

curves of Figure 8.4 results, each of which closely corresponds to a subset of the measured spectral distributions of sunlight, which can vary depending on the time of day and atmospheric conditions. Each of these provides a close visual match to light emitted from a blackbody radiator (to be defined in the next section) of the color temperature shown; the similarity between curve D of Figure 8.3 and the 7500 K curve of Figure 8.4 should be noted.

Incandescence

An ideal surface in thermodynamic equilibrium, which absorbs all of the radiant energy falling on it and reemits it, is called a "blackbody." The spectral distribution of the ideal blackbody radiator can be approximated by constructing an object having a cavity with a relatively small opening, heating it, and observing the radiant energy emitted from the cavity through the opening. There is no visible light emitted at low temperatures. As temperature is increased, a dim red light is first observed, which, as it brightens, also becomes yellow, then white, and finally bluish. This is based on a gradually changing spectral distribution that can be calculated from a formula developed by Planck (see Perrin, 1953, p. 192).

Incandescent light, whose spectral distribution approximates that of a blackbody at the same temperature, is both spatially and temporally incoherent, meaning that there is a random distribution of photons of different wavelengths, where even those of the same wavelength are not necessarily in phase at any given point in space relative to the source.

Flame sources, beginning with firelight and improved with the invention of candles, gas burners, and gas-fired mantles, were widely used before the development of the tungsten-incandescent lamp, which is in such common use today. The melting point of tungsten, about 3650 K, does not allow it to simulate daylight, whose color temperature ranges from about 5000 K upward. There is a tradeoff between the temperature at which a tungsten source is operated and the life of the lamp. Photoflood lamps, which may have a life expectancy of only 10 hr, can provide a color temperature as high as 3400 K. Tungsten-halogen bulbs, commonly used in slide projectors, induce evaporated tungsten to combine with the halogen gas that fills a quartz envelope. The compound gas circulates inside the lamp; when its molecules contact the filament, its heat dissociates the compound. The halogen is freed and the tungsten is deposited on the filament, thus restoring it to some extent. These lamps operate in the 2700 to 3050 K temperature range. Ordinary household lamps operate from about 2400 to 2600 K and the emitted light is distinctly yellowish in comparison with daylight.

Other Artificial Sources

Many nonincandescent artificial light sources are gas discharge lamps. When present as a gas or vapor and electrically activated at low pressure, chemical elements radiate characteristic "lines" of nearly monochromatic light that relate to specific transitions of energy levels within the atoms of that element. As the pressure inside the gas discharge tube is increased, the lines "smear out" and appreciable energy is emitted in the side bands. At very high pressure, all visible wavelengths are represented, although very unevenly.

Low-pressure sodium lighting is confined almost entirely to the sodium doublet at 589.00 and 589.59 nm, which (as noted earlier) prohibits chromatic color vision. Although much better, high-pressure sodium lighting is still a poor source for color rendering. A variety of other kinds of gas discharge lamps are also used for commercial and street lighting.

In fluorescent light sources, which are in very widespread use, a phosphor coats the inside of a tube filled with mercury vapor. When energized, this vapor emits mostly invisible ultraviolet radiation. The fluorescent coating absorbs this radiation and emits light in the visible spectrum. Depending on the phosphors that coat the inside of the lamp, a variety of spectral distributions can be produced, some of which are acceptable for color rendering.

The simulation of daylight has been achieved most closely by the high-pressure xenon arc lamp, in common use in vision laboratories and commercial motion picture projectors. In the latter case it has replaced the carbon arc, which was used experimentally as a source of artificial outdoor lighting before Edison and Swan independently developed practical incandescent lamps more than a century ago.

Figure 8.5 brings together the spectral distributions of several light sources in common use. Each of these has a "color rendering index" (CRI), given in the figure legend, which attempts to capture, in a single number, the quality of the source for this purpose, taking 100 as the optimal index provided by tungsten light (or sunlight). Color rendering is actually much too complicated a matter to be specified along a single dimension and, in fact, the CRI is based on calculated shifts of color appearance for a limited set of reflecting samples.

Physical Basis of Surface Colors

Assuming a suitable light source and an observer with normal color vision, the color appearance of an object depends on the interaction of the incident illumination with the substance of the illuminated object. This is a very complex subject, to which Nassau (1983) has devoted an entire book. Even so, he states in his introduction (p. x) that "There is no one single approach that explains all the causes of color, and at least five theories or formalisms are invoked. A rigorous treatment of any one of these would require at least a full volume of its own, but significant insight can nevertheless be gained from the descriptive account presented here."

While implicitly assuming a human observer with normal color vision, Nassau lists 15 "causes of color" that relate to the nature of light and its interaction with matter (Table 8.1). Where color perception is concerned, the interaction of organic compounds with light (No. 6 on Nassau's list) is of paramount importance and will be discussed here first.

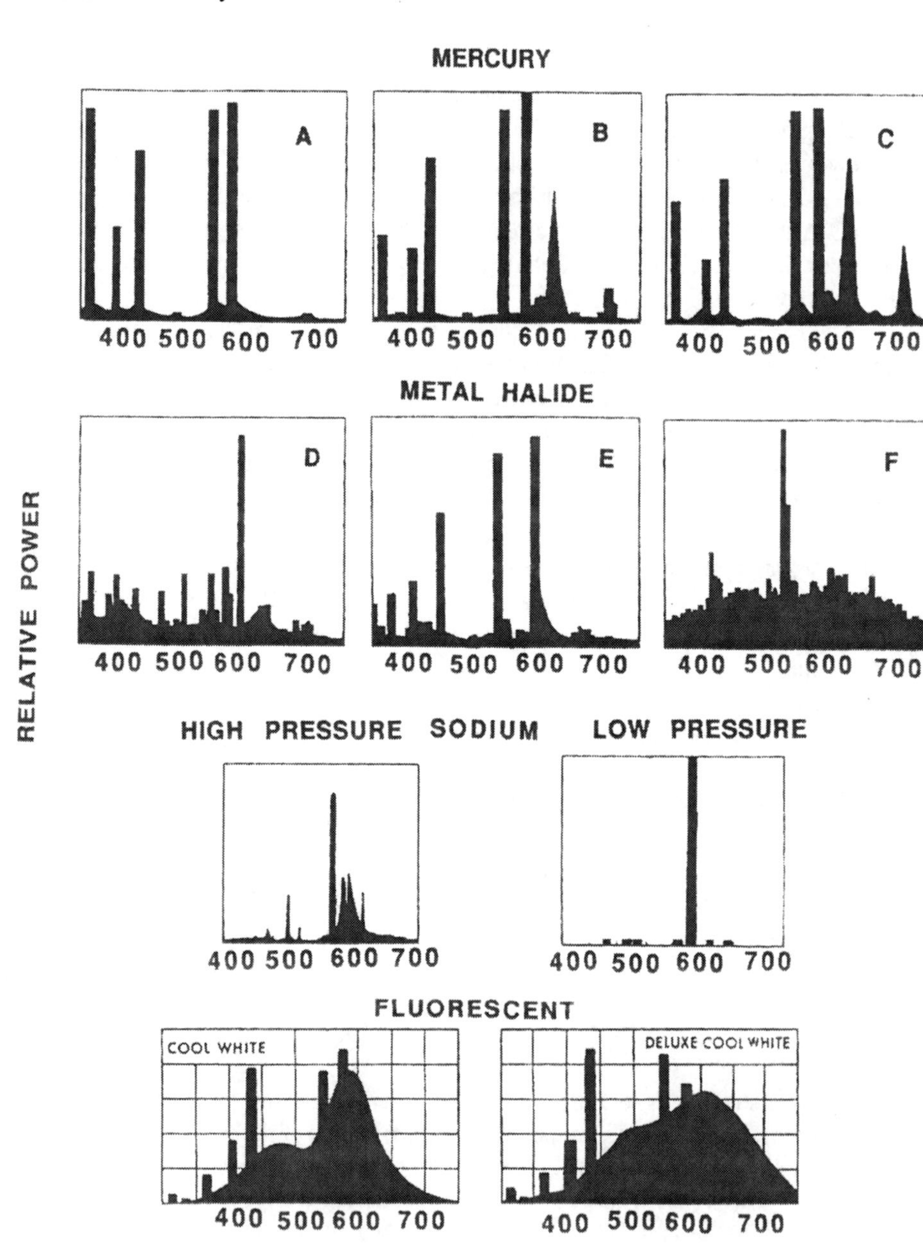

FIGURE 8.5. Spectral power curves (radiance as a function of wavelength) for a variety of light sources. The spectral lines are actually much narrow than those depicted. The color-rendering indices for typical sources of these various types are clear mercury (A), 15; improved mercury (B,C), 32; metal halide (D,E,F), 60; high-pressure sodium, 21; low-pressure sodium, −44; cool white fluorescent, 62; deluxe cool white fluorescent, 89. (From Kaufman, 1981.)

TABLE 8.1. Fifteen "causes of color."[a]

Vibrations and Simple Excitations

1. Incandescence: Flames, lamps, carbon arc, limelight
2. Gas excitations: Vapor lamps, lightning, auroras, some lasers
3. Vibrations and rotations: Water, ice, iodine, blue gas flame

Transitions Involving Ligand Field Effects

4. Transition metal compounds: Turquoise, many pigments, some fluorescence, lasers, and phosphors
5. Transition metal impurities: Ruby, emerald, red iron ore, some fluorescence and lasers

Transitions between Molecular Orbitals

6. Organic compounds: Most dyes, most biological colorations, some fluorescence and lasers
7. Charge transfer: Blue sapphire, magnetite, lapis lazuli, many pigments

Transitions Involving Energy Bands

8. Metals: Copper, silver, gold, iron, brass, "ruby" glass
9. Pure semiconductors: Silicon, galena, cinnabar, diamond
10. Doped or activated semiconductors: Blue and yellow diamond, light-emitting diodes, some lasers and phosphors
11. Color centers: Amethyst, smoky quartz, desert "amethyst" glass, some fluorescence and lasers

Geometrical and Physical Optics

12. Dispersive refraction, polarization, etc.: Rainbow, halos, sun dogs, green flash of sun, "fire" in gemstones
13. Scattering: Blue sky, red sunset, blue moon, moonstone, Raman scattering, blue eyes and some other biological colors
14. Interference: Oil slick on water, soap bubbles, coating on camera lenses, some biological colors
15. Diffraction: Aureole, glory, diffraction gratings, opal, some biological colors, most liquid crystals

[a] From Nassau (1983). Copyright © 1983 John Wiley & Sons, Inc. Reprinted by permission of John Wiley & Sons, Inc.

Organic Compounds

Consider the application of a glossy red paint to a neutral surface. If one coat of the paint does not suffice to "hide" the surface, this fact implies that some of the incident light must have passed twice through the layer of paint. Assuming that a second coat of paint solves this problem, then the paint layer is opaque and we may assume that additional coats of the red paint would not alter the appearance of the surface.

The perceived color of the painted surface now depends, though only in part, on the nature of pigment particles, suspended in a binder that, when dry, adheres to the surface of the painted object. The particles, which constitute one kind of colorant, may be either inorganic or organic in nature, depending on the paint, and literally centuries of experimentation have gone into their development.

Nassau's category called "Transitions between Molecular Orbitals" refers to an aspect of molecular and atomic theory capable of accounting, at least in principle, for the spectral selectivity of the absorbing particles. Viewed at the molecular level, the reflection of light is not just a matter of photons bouncing off a surface. Instead, all of the photons that are not transmitted through a material are absorbed and some of these are selectively reradiated with a probability that varies depending on their wavelength (Weisskopf, 1968).

The relative indices of refraction of the particles of colorant and the binder help to determine the degree to which the incident light will be transmitted, absorbed, or diffusely scattered by the paint layer. In the example of the red paint, measurement would show a selective reflection of the longer visible wavelengths, similar to the curve labeled "RED" in Figure 8.2, with the light coming off the surface in all directions.

If one applies a glossy paint, its hardened surface will exhibit specular reflection, so that a certain percentage of the light, instead of entering into the paint layer to interact with the particles within, will be directionally reflected from the surface. This reflection is not wavelength dependent, and although directional, the reflection is more diffuse than that from a plane mirror. A "flat" paint lacks the smooth surface, and exhibits little specular reflectance. Intermediate degrees of directionality of surface reflectance are possible, which give a sheen, rather than a shine, to the surface (see Hunter and Harold, 1987).

These varieties of reflection provide important cues that allow one to predict the felt temperature and hardness of surfaces through vision alone and, with experience, this permits educated guesses about the materials that constitute the objects that are seen.

Pigments constitute one of two classes of colorants, the other being dyes. Dyes differ from pigments in a variety of ways: (1) most dyes are water soluble, whereas pigments are not, (2) most dyes are organic, whereas most pigments are inorganic, (3) most dyes are much more transparent than pigments, and (4) dyes have a natural affinity for the material to which they become attached, whereas pigments require a special binder.

Plastics raise some interesting problems. Billmeyer and Saltzman write (1981, p. 115): "In the case of a plastic that is integrally colored by dispersing an insoluble colorant into the resin, we hedge a little and call the plastic the binder. If a soluble dye is used, we call the same plastic the substrate."

Metals

Metals differ in a number of ways from other substances. From the standpoint of molecular orbital theory, the periodic and regular array of the atoms is what makes metals special. From the standpoint of color vision, the crucial characteristic of a metal is that, unlike specular reflection of the sort that occurs from the surface of glass or water, the reflected light is somewhat wavelength selective, giving the metal its distinctive hue, and the reflected light also exhibits a directionality that is more or less specular depending on the metal and how the surface is treated. Nassau (1983, p. 164) puts it this way:

When light falls onto a metal it is so intensely absorbed that it can penetrate to a depth of only a few hundred atoms, typically less than a single wavelength. Since the metal is a conductor of electricity, this absorbed light, which is, after all, an electromagnetic wave, will induce alternating electrical currents on the metal surface. These currents immediately reemit the light out of the metal, thus providing the strong reflection.

Geometrical and Physical Optics

Nassau lists other causes of color under this heading. Although important in their own right, most of these phenomena are not of direct importance for object color perception. Dispersive refraction is what Newton observed in his famous experiments on the physical basis of color. The prisms he used were in fact manufactured for use in chandeliers. However, the prismatic color observed in such decorations do not seem so much a property of the glass elements as of the light refracted by them. The same perhaps can be said for the chromatic sparkle that refraction imparts to a diamond ring, the interference colors of soap bubbles and oil slicks, and the colors that one observes from a diffraction grating, or, more commonly today, from the surface of a compact disk.

Something of an exception occurs in the case of certain spectral effects due to scattering and interference. Scatter is itself a very complicated topic; generically, it refers to light that is neither transmitted, reflected, nor absorbed as it passes through a medium, but which instead interacts with the particles of the medium in such a way as to change the direction in which light travels, and in some cases its wavelength. Interference occurs because of the wave nature of light. Sometimes scatter and interference occur so as to cause the resulting color to seem attached to surfaces. For example, both scatter and interference occur on the surface of many animals as a result of very highly specialized and precisely arranged layers. Also, the color of the iris of a blue-eyed animal or person results from scatter in a medium that is essentially free of colorant.

Fluorescence

The light reflected from most surfaces that contain dyes or pigments is of the same wavelength as the incident radiation. The perceived color of the surface, which depends on its diffuse spectral reflectance, results because the probability of a photon being reflected depends on its wavelength. Incident white light, which typically contains photons of all visible wavelengths, is thereby selectively returned, as if it were nothing more than light bouncing off a surface. But as we have seen, reflection is more complicated than this, involving absorption and reradiation of photons. For most reflected photons, no energy is lost, as in an elastic collision, so the wavelength of the light is unchanged. By contrast, in fluorescence, some of the photon energy is lost in absorption, causing the reemitted photon to be of longer wavelength than the incident one. As a result, incident ultraviolet light, which is itself invisible, can produce reflected light that is visually effective, and this is what gives surfaces coated with fluorescent materials their unusual brightness (see Grum, 1980).

Measurement of Surface Reflectance

It is important not to leave the subject of surface reflectance without discussing the physical measurement of surface characteristics that are important for color perception. Where surface color perception is concerned, it is relatively unimportant to know exactly why a particular surface selectively reflects light in the way it does as a function of direction and wavelength. Whatever its cause, diffuse spectral reflectance, which gives a surface its color over a wide range of viewing angles, is relatively easy to measure. Nevertheless, the resulting function depends on the exact conditions of measurement. To some extent these are arbitrary, so that for valid comparisons of measurements to be made among different laboratories and manufacturers, standard procedures are necessary. The CIE has recommended four procedures, the most sophisticated of which is represented in Figure 8.6. The incident beam of white light is nearly normal to the surface and the specularly reflected component is picked off by the light trap at the top. A baffle keeps the diffusely reflected light from the surface from directly reaching the right-hand side of the integrating sphere, which becomes diffusely illuminated by multiple internal reflections from a neutral white surface. The detector at the left receives light from the opposite side of the sphere, which results from the initial diffuse reflection in all directions. Many variations on this theme are possible, some of which may deliberately include the specular component or measurements of reflectance as a function of angle (goniophotometry).

Current Trends in Chromatic Psychophysiology

On the first of 68 pages of an extensive review article concerned with the relations between the psychophysics and physiology of color vision, Lennie and D'Zmura (1988) state that "physiological observations have, on the whole, contributed little to our present understanding of color vision, although recent advances have begun to yield clear indications of the kind of analysis undertaken at successive stages in the visual pathway." On the last page of their paper they conclude that

> The low-level mechanisms of color vision postulated on the bases of psychophysical observations have, in the main, been found in subsequent physiological investigations of neurons in the retina and LGN [but] the properties of higher-level chromatic mechanisms, especially those involved in the analysis of color signals from complex stimuli such as natural scenes, have not yet been thoroughly studied in perceptual experiments. . . It is at this complex level, particularly, that physiologists need strong guidance as to the kinds of mechanisms they should expect to find.

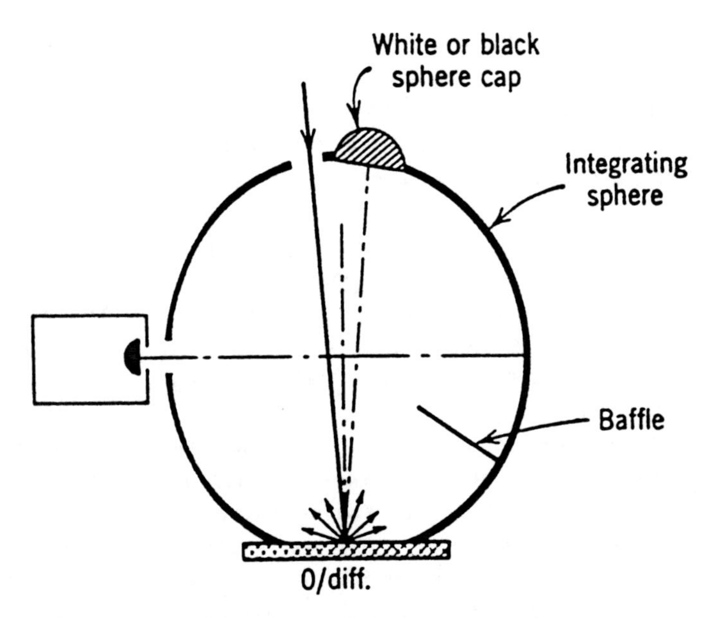

FIGURE 8.6. Schematic diagram showing one of the CIE standard illuminating and viewing geometries for reflectance measurements. (From Wyszecki and Stiles, 1982, p. 156, copyright © 1982 John Wiley & Sons, Inc.)

It is important that these conservative viewpoints be seen in an appropriate historical context. Forty years ago, although visual psychophysics and physiological optics were already well developed, virtually nothing was known about the physiology of color vision. With improved techniques, there have by now been hundreds of studies in this area; these, along with those yet to come, will provide the factual base against which future theories of chromatic processing, more coherent and better supported than those currently available, will likely develop.

The purpose of this section is merely to sketch a few highlights of current progress. More detailed information about receptor function and cortical mechanisms will be found in chapters by Liebovic and Dow in this volume, and extensive references to other relevant literature (263 references) will be found in the Lennie–D'Zmura review.

The most fundamental facts about human color vision are that (1) stimuli that differ physically may nevertheless match for color, and (2) exactly three primaries are needed to match all perceived colors. These observations provided the initial basis for the idea of trichromatic color vision based on overlapping action spectra for the three classes of cones, which are now known with considerable precision. Although the general form of these functions was known almost 100 years ago (see Helmholtz, 1924, Vol. 2, p. 143), the specification of their exact characteristics has proved crucial to the understanding of the subsequent stages of visual processing. Recently, the findings (or inferences, some would say) of psychophysics have been verified by the more direct methods of microspectrophotometry (Dartnall et al., 1983) and electrophysiology (Schnapf et al., 1987). In one sense, consistent with the first statement by Lennie and D'Zmura, these recent studies have added little to what was already known. On the other hand, physiological inferences from psychophysics alone can never be perfectly secure, and direct confirmation, which had to await the development of some very sophisticated experimental techniques, has been both welcome and important.

Just as trichromacy was discovered initially through the application of psychophysical methods, it is likely (as emphasized in the last statement of Lennie and D'Zmura) that the data and theory of psychophysics will pave the way for the future understanding of the higher stages of chromatic visual processing, about which little is fully understood despite the substantial body of empirical evidence that has already accumulated.

The most relevant electrophysiology has been done with macaque monkeys, which by various criteria probably see colors very much as we do. One might suppose that signals from the three types of cones would be kept separate from retina to brain. Instead, conversion to an opponent-color code, the advantages of which have been shown theoretically by Buchsbaum and Gottschalk (1983), is already reflected by activity recorded from retinal ganglion cells (DeValois and DeValois, 1975; Wiesel and Hubel, 1966), and these signals are transmitted with little modification to the lateral geniculate nucleus (LGN), which should be regarded as a very early stage of visual processing.

The facile identification of signals recorded from cells in the LGN with various aspects of color appearance, which is now deeply rooted in many introductory textbooks, is both premature and misleading. There is definitely a red-green opponent color code represented in the activity of many ganglion cells, but there is also an antagonistic center-surround arrangement that is at least as important for the processing of achromatic as chromatic information. As originally shown by Wiesel and Hubel (1966), the receptive fields of these cells are excitatory to long wavelengths in their centers and inhibitory to middle wavelengths in their surrounds (or vice versa). These cells vastly predominate in the LGN, especially for those areas receiving input from the central part of the visual field. As shown most clearly by Ingling and Martinez-Uriegas (1985), these cells will respond as well to achromatic contours. Moreover they respond well to the highest visible spatial frequencies and clearly play the dominant role in spatial vision where small details are concerned.

There are far fewer cells in the retina and LGN that could be designated as yellow-blue (Derrington et al., 1984); such cells also appear somehow to be involved, in some unknown way, in the signaling of redness as well as blueness.

Thus, contrary to earlier thinking, achromatic signals are multiplexed with chromatic ones at the LGN stage and must be sorted out later in ways that are still speculative (Ingling and Martinez-Uriegas, 1985). Finally, we have no idea, really, of what kind of activity in the brain constitutes the immediate underlying precursors of color sensations, and the discovery of ever more areas of the brain that are concerned with vision (Van Essen et al., 1984) leaves entirely open the question of how this scattered activity, which to some degree seems to be specialized in separate areas for color, form, and motion (Zeki, 1985), gives rise to our cohesive and integrated sensory experiences.

Metamerism

The concept of metamerism has two related, but nevertheless different meanings in color science. The first meaning relates to color matches, and the second to color appearances. In the first case, two lights or surfaces are said to be metameric if, despite physical differences between them, they nevertheless match. From this point of view, the fact that such matches can occur is the manifestation of the phenomenon of metamerism.

All colors change in appearance with a change in illuminant, although this is often not very noticeable because of the limitations of memory for color and the mechanisms of color constancy (to be discussed later). In general, metameric matches experienced under one illuminant fail under another, which is proof positive that color constancy is not perfect. Given a shift from one illuminant to another (for example, from daylight to incandescent light) some spectral reflectance distributions yield colors that are chromatically more stable than others. The second meaning of metamerism refers, in a rather qualitative way, to the degree to which a material appears to change color when viewed successively under different illuminants. Those surfaces showing the greatest changes are said to be the most metameric. Two surfaces that are physically identical will of course change color equally as the illumination is varied; such surfaces are sometimes called isomeric.

The fundamental basis of metameric matches has already been discussed: Any two stimuli that produce exactly the same rate of photon absorption for each of the corresponding types of the three cone photopigments will be indistinguishable, starting at this first stage of visual processing and continuing through all subsequent stages as well. However, metameric matches will in general fail except under very restricted conditions. Consider, for example, a bipartite (split) field, subtending 1° of visual angle overall, where the two halves match when seen by a particular observer despite substantial physical differences. The match will probably fail for any of the following reasons:

- The field is enlarged.
- A different observer makes the match.
- The fields are too bright or too dim.

Nevertheless, the match will probably hold even if, within limits

- The intensity of the fields is varied.
- Surrounds or backgrounds are added that change the appearance of the fields.

Matches remain despite variations in intensity provided that the cone action spectra do not change and rod intrusion does not occur. With a 1° field, rods make an insignificant contribution, and color matches are then based only on the relative probabilities of photon absorption by the three types of cones. These probabilities do not change, no matter how many photons are being absorbed, until a significant amount of photopigment bleaching occurs. Bleaching alters the action spectra of the cones, which upsets the match (Burns and Elsner, 1985).

At very low intensity levels, matches will remain valid for small, rod-free fields until threshold is reached, below which matches are obviously impossible. Larger fields bring rods into play. Matches are also affected by inhomogeneities in the field due to the macular pigment, which covers only the central few degrees of the retina. A screening pigment of this sort, along with selective spectral absorption in the eye media more generally, alters the action spectra of the cone photopigments as measured at the cornea relative to what it would be at the level of the receptors. In general, stimuli that are metameric in the fovea, where the macular pigment is dense, cannot also be metameric in the parafovea beyond the limits of the macular pigment. In practice, perceptual mechanisms that are not well understood tend to average color appearance even over large areas, but careful observation, particularly when two fields that are highly metameric are seen side by side, will reveal that the fields do not in fact match over their entire extent.

Another problem of large fields is rod intrusion, which has been most systematically studied by Trezona (1976). In general, two fields that match exactly for the three kinds of cones will not do so for the rods. Rods do make a contribution to color appearance at intermediate and low levels, by feeding signals through the same pathways shared by cones. This is discussed in much more detail in the chapter by Frumkes. Matches remain trichromatic under these conditions but, because of differential rod input, matches will differ somewhat from those obtained under rod-free conditions.

The outer segments of the cone photoreceptors have antenna-like properties that produce waveguide effects which in turn are spectrally selective (Enoch, 1963). These tend to average out for metamers over the extent of a field, even if it is a rather small one. Nevertheless, waveguide effects, combined with variations in macular pigment, photopigment density, or other differences between the two parts of the retina on which fields to be matched are imaged, can leave a lingering residue of uncertainty that these fields, matched as well as possible, are truly identical. In critical color matching and discrimination experiments, it is commonplace for subjects sometimes to report that they perceive differences between stimuli that are physically identical. Indeed, when fields are enlarged, and the homogenizing effects of steady fixation are minimized, metameric matches seldom seem exact, although the best possible match can usually be repeated with good precision.

Finally, any two observers are unlikely to agree exactly concerning color matches. Webster and MacLeod (1988) have analyzed the interobserver sources of variability inherent in a large set of color matching data. Although a match made by one observer will not generally be fully acceptable to another, the differences between them generally will not be large. This so-called observer metamerism places a definite limit on any trivariate system of color specification. The CIE system of colorimetry, to be described later, is therefore forced to define a standard observer that does not exactly resemble any real human being, but that nevertheless gives an adequate account of metamerism for most practical purposes.

Luminance, Brightness, and Lightness

The three concepts to be discussed in this section all relate to aspects of visual intensity. However, each has a distinctly different meaning. Luminance refers to a physically measurable property of the visual stimulus. Although brightness and lightness both refer to aspects of subjective intensity, these aspects are very different.

Luminance

The radiance of a surface is proportional to the number of ergs per second per unit projected area of the surface that are emitted in a particular direction. This quantity can be measured with a radiometer (a device that records total energy received regardless of its spectral distribution) containing an optical system that allows the operator to view an image of the precise section of surface area being measured. The measured value will be an average over the demarcated area. As the instrument is aimed progressively more obliquely at the surface, a larger physical area of the surface will be included within the subtense of the viewing angle that defines the projected area.

The basic reason why radiance is defined in terms of projected area, rather than surface area, relates to the nature of diffuse surfaces. A "perfectly diffuse" surface will give the same radiance reading regardless of the angle at which it is viewed; it also has a radiance that is uniform over its entire surface. Such surfaces can be approximated with laboratory materials; a uniformly illuminated piece of blotting paper would come close. Radiance remains constant with viewing angle because the intensity of each elemental point of reflected light from a perfectly diffuse surface falls off as a cosine function of the angle of view. At the same time, the area of the surface contained within the viewing angle increases as a cosine function of the same angle. Thus, as the viewing angle becomes more oblique, less energy is received from each element of the surface, but a compensatory larger surface area is included in the measurement. Most real surfaces, which are only partially diffuse, exhibit a reduction in radiance as a function of viewing angle.

Radiance readings also are independent of the distance from the surface at which the measurement is made. This happens because the energy received from each elemental area of the surface decreases according to the inverse square law, while the surface area subtended by the viewing angle increases in proportion to the square of the distance.

Luminance is a measure of visually effective radiance, and it has exactly the same geometric properties. The human eye, rather than a man-made optical system, does the viewing. Energy measurements alone will not suffice to predict the efficacy of light as a visual stimulus, because the eye is insensitive to most of the electromagnetic spectrum and not equally sensitive to all of the visible portion. Luminance is defined in such a way that, for any fixed spectral distribution, a measure of luminance is directly proportional to a measure of radiance. The proportionality factor is a function of wavelength, one which weights radiance at each wavelength according to its visual effectiveness. This factor, called $V(\lambda)$, is defined by the right-hand function shown in Figure 8.7 [also shown is the function $V'(\lambda)$, which applies to rod vision]. This so-called photopic luminous efficiency function, which was standardized by the CIE in 1924, is based principally on the experimental operation of flicker photometry, which was deliberately chosen for the purpose because the resulting data exhibit the necessary properties of additivity that allow luminance to be defined by the following equation:

$$L = K \int E(\lambda) R(\lambda) V(\lambda)\, d\lambda$$

where $E(\lambda)$ is the spectral emittance of the light source, $R(\lambda)$ is the diffuse spectral reflectance of the surface, and K is a constant that depends on the units of measurement.

The $V(\lambda)$ function is an international standard that applies to no real observer, but approximates reality for most. The CIE has, in fact, in defining luminance, replaced the human observer by an ideal detector having the spectral sensitivity $V(\lambda)$. Using a radiometer with a spectrally nonselective radiation detector and a linear response as a function of radiance, where the incoming energy is attenuated with a filter whose transmittance is proportional to $V(\lambda)$, valid luminance readings will be obtained for any possible light input. Many radiometer/photometers accomplish this.

Brightness

As mentioned, brightness is a concept related to a subjective aspect of vision. Strictly speaking, it applies only to stimuli seen in the so-called aperture mode of viewing, where an isolated patch of light is seen in an otherwise completely dark visual field. Colors seen in this way are also called unrelated colors, because they do not relate to, and are unaffected by, other stimuli in the field of view. Isolated signal lights seen on a dark night would meet this criterion.

Luminance and brightness are positively correlated and brightness is a monotonic function of luminance. That is, as the luminance of an isolated field is gradually increased, the field will appear progressively brighter. The quantitative relation between luminance and brightness, however, is more difficult to specify. There has been more than a century of dispute about whether the measurement of the quantity of a sensation, such as brightness, is even possible (Boynton, 1986). Nevertheless, a variety of methods for attempting to do so have been developed, and although the results based on different methods disagree in detail, all agree that the relation between luminance and brightness is compressively nonlinear. To a first approximation, one must increase luminance by a factor of 10 to double the sensation of brightness.

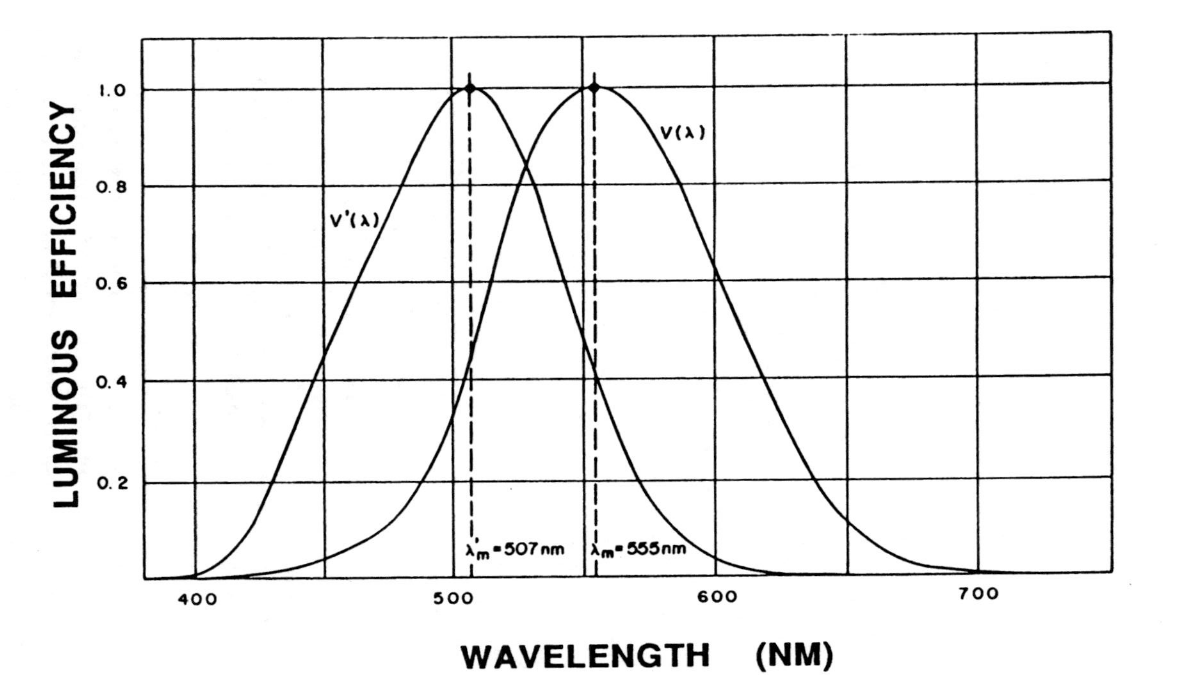

FIGURE 8.7. Spectral luminous efficiency functions for photopic vision [$V(\lambda)$] and scotopic vision [$V'(\lambda)$] as standardized by the CIE. Each function is independently normalized at the indicated wavelength of maximum sensitivity. (From Wyszecki and Stiles, 1982, p. 257, copyright © 1982 John Wiley & Sons, Inc.)

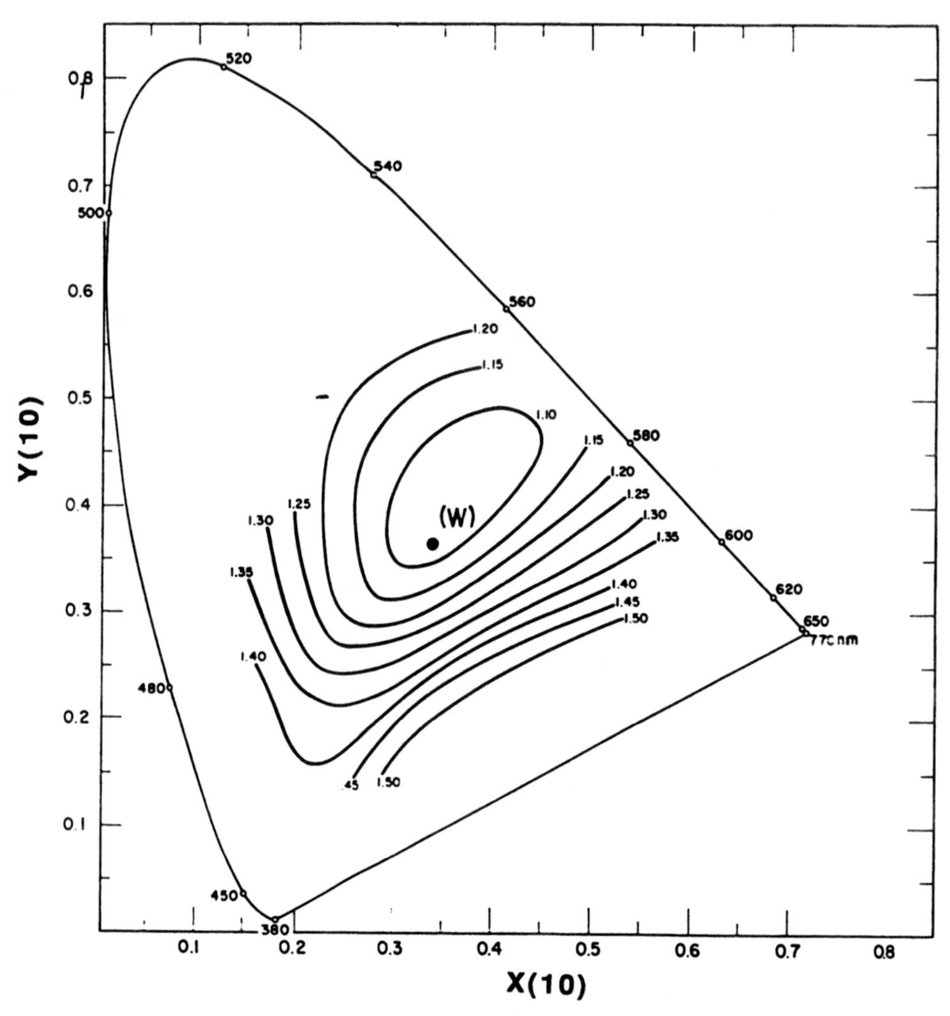

FIGURE 8.8. The CIE chromaticity diagram, showing the loci of constant values of the luminance of white light (W) relative to stimuli on a given locus, required for a brightness match. The coordinates of the diagram are explained in the text. The locus of spectrum colors is also shown, with the positions of certain wavelengths indicated. (From Wyszecki and Stiles, 1982, p. 414, copyright © 1982 John Wiley & Sons, Inc.)

Brightnesses are generally nonadditive. This can be demonstrated using an experimental paradigm that, because it demands only that relative brightnesses be assessed, circumvents at least part of the subjective measurement problem. The experiment does assume, however, that one is able to abstract and compare brightnesses between two fields that differ in chromatic color. The classic experimental example of nonadditivity of brightnesses is this: In a small, isolated, but divided field, fill one-half with white light and the other with red. Ask the subject to adjust the luminance of the white field until that field matches the red one for brightness. Now replace the red field with a green one, and determine the matching luminance of the white field for this case. Next, add the matching luminances of the two white fields together by optical superposition. At the same time, add the adjacent red and green fields together to produce a yellow mixture field. When these operations are performed, it becomes immediately obvious that the yellow field is substantially less bright than the adjacent white one, although both fields are necessarily equal in luminance. To restore a brightness match, it is necessary either to reduce the luminance of the white field or increase that of the yellow one; the amount of such luminance adjustment provides a quantitative measure of the additivity failure.

Figure 8.8, from Wyszecki and Stiles (1982, p. 414), shows the factors by which stimuli, which plot in the CIE chromaticity diagram along the contours shown, must be reduced in luminance to match a white stimulus (W) for brightness. The meaning of this diagram, whose use is widespread in color science, will be explained below. The contours in this figure are based on only one experiment. Ware and Cowen (1983) have examined the data from many such experiments and have developed predictive equations.

Lightness

Lightness refers to a property of surfaces as they appear in complex scenes. It is correlated with the reflectance of a surface, which is the relative amount of light returned from it. For example, a black object retains its blackness whether seen outdoors under sunlight, or indoors under dim illumination. In general, a chromatically neutral surface that reflects more than about 70% of the incident light appears white, one that reflects less than 3 or 4% appears black, with intermediate reflectances appearing as various shades of gray. Colors seen in this mode, which includes most of those of everyday life, are called related colors. Lightness is correlated with radiance only for a particular surface in a particular scene, assuming an eye in a fixed state of adaptation. Thus, although the radiance of a piece of paper that reflects 80% of the incident light will be much less indoors under dim illumination than outdoors under sunlight, it will nevertheless appear white under both circumstances.

Chromaticity

The standard method for representing color is the CIE chromaticity diagram (see Fig. 8.8). The letters stand for International Commission on Illumination (Commission Internationale de l'Éclairage). Developed by the CIE in 1933, long before the cone action spectra were known, a different set of functions of the wavelength λ, called $\bar{x}$, $\bar{y}$, $\bar{z}$, is used, which predicts, at least approximately, the same color matches. So-called tristimulus values, called X, Y, and Z, result from integrating the CIE functions, $\bar{x}$, $\bar{y}$, and $\bar{z}$, with the spectral energy distribution, $E(\lambda)$, of the stimulus:

$$X = \int E(\lambda)\,\bar{x}(\lambda)\,d\lambda$$
$$Y = \int E(\lambda)\,\bar{y}(\lambda)\,d\lambda$$
$$Z = \int E(\lambda)\,\bar{z}(\lambda)\,d\lambda$$

A chromaticity diagram plots the relative values of X and Y:

$$x = X/(X+Y+Z);\;\; y = Y/(X+Y+Z);\;\; z = Z/(X+Y+Z)$$

Conventionally, y is plotted as a function of x. Although intensity information is lost, all colors otherwise are represented on this diagram. The value of z, though not directly represented, is implicit as $1 - (x+y)$. The locus of spectrum colors, which is roughly U-shaped, forms the outer boundary of all real colors, which plot inside the U and a line connecting the extreme ends of the spectrum. A given point in the diagram represents all possible metamers having the specified values of x and y, no matter what their physical characteristics. The diagram is strictly appropriate only for an observer whose color matches are exactly predicted by the $\bar{x}$, $\bar{y}$, and $\bar{z}$ functions. For most purposes, normal observers do not differ enough from this to matter. An advantage of the CIE system is that $\bar{y}$ is proportional to $V(\lambda)$, the photopic spectral luminosity function, which is a measure of the relative luminous efficiencies of different wavelengths in vision.

It is possible to build an instrument to make measurements of chromaticity. Imagine that instead of having a photometer with a single detector of sensitivity $V(\lambda)$, one has a triple photometer with three detectors each equally sensitive to all wavelengths of the visible spectrum. In front of each detector are filters whose relative transmittances are proportional to $\bar{x}$, $\bar{y}$, and $\bar{z}$. Any two stimuli that produce equivalent readings for all three detectors are metameric for the triple photometer, and they will match for the standard observer of the CIE system. The outputs from the three detectors can be digitized and processed by computer chips to give the chromaticity coordinates directly.

Systems of colorimetry based instead on cone excitations have been proposed (e.g., Boynton, 1984) but are not standardized or in common use. Thus it is possible to replace the filters of the triple photometer with filters that mimic cone sensitivities and that predict exactly the same color matches as $\bar{x}$, $\bar{y}$, and $\bar{z}$.

The chromaticity diagram provides a remarkable representation of the relations among colors, based exclusively on the matching operation. As chromaticity is gradually changed, color appearance also gradually changes, and there are no two points in the chromaticity plane (provided they are separated by a discriminable difference) whose represented colors will appear the same. Because a given point in the chromaticity diagram represents the location of all possible metamers for that color (except for some monochromatic stimuli whose chromaticities can be achieved in only one way) chromaticity does not reveal the spectral character of the represented stimulus. A major advantage of chromaticity diagrams is that additive mixtures of any two colors plot along a straight line connecting their chromaticities.

Specifying colors by their chromaticities factors out brightness and lightness. A surface with a particular chromaticity can have many appearances, depending on surround conditions and the state of adaptation of the eye. For this reason, chromaticity is inadequate for the specification of the appearance of surface colors. For example, a chocolate bar and an orange could have the same chromaticity, differing only in reflectance. Normally, of course, these two objects will appear quite different, but they can be made to match visually by selectively increasing the illumination on the chocolate bar and viewing the two objects through small holes in a large screen, so that sources of illumination are hidden. This is the aperture mode of viewing of unrelated colors, so called because their normal surrounds have been eliminated.

Color Discrimination versus Color Categorization

Discrimination

How many different colors are there? Surprisingly, the answer to this question can vary from more than a million to less than a dozen, depending on the operations invoked to assess color differences. The much larger figure is based on the ability of human observers to determine whether two colors match, assuming normal color vision, large patches of color, and optimal lighting conditions. Actually, to demonstrate that there are a million colors, no two of which would be judged to match, would require about 500 billion separate comparisons, so it goes without saying that the experiment has not been done. The estimate of a million or more colors that differ is instead an extrapolation from limited sets of measurements.

The classic experiment in color discrimination is that of MacAdam (1942). Around each of 25 chromaticities, well scattered and of equal luminance, he arranged several pairs of stimuli that, when mixed in suitable proportions, would match the test chromaticity. For a given pair of stimuli, his apparatus allowed an equiluminant and graded substitution of one stimulus for another. With the test chromaticity positioned on one side of a divided field of view, the subject was asked to set the ratio of the comparison stimulus components, which caused its chromaticity to vary continuously along the connecting line, until the two fields appeared to match. This was done many times, using a single control knob, and a scatter of matches was found. The standard deviation of the matches was taken as an index of discriminability along each line. The chromaticity difference corresponding to one standard deviation of such settings was found to vary systematically as a function of the tested angle through the match point, and it also varied greatly depending on which of the 25 chromaticity locations was under test. MacAdam fitted his experimental data with ellipses, one of which is shown in Figure 8.9, and developed procedures for interpolating and extrapolating his results to arbitrary regions of the chromaticity plane. The work was later extended, using a modified method, to include lightness variations as

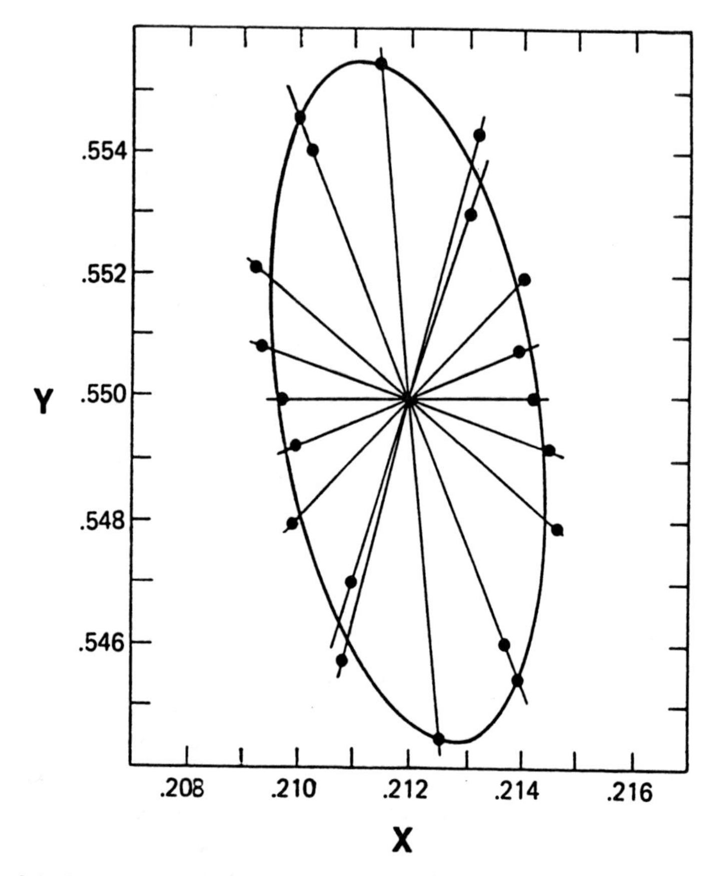

FIGURE 8.9. One of the 25 chromatic discrimination ellipses of MacAdam. The chromaticity of the reference field, which subjects attempted to match, is in the center. Points are experimental measurements (standard deviations of repeated matches) made along the axes shown, all of which pass through the reference chromaticity. The point representing each measurement is plotted twice, on both ends of an axis, to give a symmetrical array through which an ellipse is fitted. (From Boynton, 1979, p. 280, after MacAdam, 1942.)

well; variations in the lightness dimension must be included to reach the estimate of a million or more discriminably different colors.

If two stimuli that differ by only a standard deviation along a line in the chromaticity diagram (sometimes called a MacAdam unit) are observed under ideal conditions, they will not actually look different. MacAdam estimated that about three of his units would be needed to produce a difference that could be discriminated reliably.

This kind of study has been replicated and extended, using other measures of chromatic difference, by others (see Nagy et al., 1987). There is considerable evidence (Krauskopf et al., 1982) that there are three critical dimensions of chromatic variation, related to (1) exchange of excitation of L and M cones at equal luminance, (2) change of excitation of S cones (which does not affect luminance), and (3) change of luminance at constant ratio of L, M, and S cone excitation. By considering the discrimination problem in cone excitation space, certain relations become clear that tend to be obscured in CIE color space (see Nagy et al., 1987, where additional references can be found). By examining discrimination behavior for each of these components individually, and then testing interactions when they vary together, the discrimination problem can be better understood.

Categorization

If there are a million discriminably different colors, how then can it be said that, by some other criterion, there are less than a dozen colors? The answer relates to the natural tendency for categorizing perceived objects and events and the need for remembering and communicating about them. Consider for example the categories "dog" and "cat." No two dogs, if placed side by side, would look exactly alike. Within each category of animal, there are substantial variations in size, color, furriness, and other characteristics. There is overlap between categories for most characteristics, but not all. For example, the two species are genetically distinct and cannot interbreed, and cats have sharp claws that enable them to climb trees, whereas dogs do not. There is not a graded series of animals ranging from cat to dog, with intermediate specimens that would be difficult to classify.

Categorization occurs for colors also. For example, the word "red" refers not to just a single color that differs from all others, but rather to a category of colors, probably hundreds of thousands of discriminably different ones, all of which tend to be called red under appropriate viewing conditions. The word "green" refers to another group of colors, no one of which could possibly be mistaken for any of those in the "red" category. But unlike cats and dogs, any two of these colors, one drawn from each category, are continuously linked by a long series of just-discriminable colors, which can be visualized as falling along a connecting line in the chromaticity diagram, with red gradually turning to reddish orange, to orange, to yellowish orange, to orangy yellow, to yellow, to greenish yellow, to yellowish green, and finally to green. Thus, when tested locally, there are no discontinuities in the domain of color.

And yet color categorization obviously occurs, and each color category is given a name. We then ask how many color names are there? A dictionary of color names prepared by Kelly and Judd (1976) lists about 7500 names in 3 columns on more than 70 pages. An impression of the kinds of names listed can be gleaned by arbitrarily selecting the tenth name listed in the middle column of pages whose numbers are divisible by 5. These names are alamo, crayon blue, bonnie blue, celadon tint, desert sand, golden rapture, lumiere green, infanta, mars orange, light olive-gray, light pinkish lilac, reddish gray, shadow lime, turquoise green, and amber yellow. Note that most of these names are compounds, some of them (e.g., infanta) are probably unfamiliar to most people, and that many of the others refer to objects or commonly experienced places or events. (There was no particular reason for selecting this particular group of names; it is likely that any random selection of 15 names from this dictionary would produce terms having similar qualities.)

By contrast, Berlin and Kay (1969) studied the use of color terms in about 100 languages and concluded that there are only 11 basic color terms in fully developed languages. In English, these names are red, yellow, green, blue, orange, purple, brown, pink, white, gray, and black. Note that these are single words, and that they do not (except for orange) refer to specific objects, places, or events. Other research shows that these names are listed first, used more reliably, with greater consensus, and shorter response times than any other terms (Boynton and Olson, 1987). These 11 basic color terms also translate easily between languages and are commonly learned by the age of five, at which time very few nonbasic terms are used. None of these special attributes applies to the color terms that were sampled from the color dictionary.

Color Order

A variety of color-order systems have been developed. For some of these, sets of color samples have been commercially produced. Three of these systems will be described here.

The "Munsell" colors are depicted in a cylindrical coordinate space whose central vertical axis, representing variations in lightness, runs from black at the bottom to white at the top (Fig. 8.10). Colors of equal lightness are represented in each of nine horizontal planes. Within each plane, the color circle is divided into five segments representing red, yellow, green, blue, and purple. Colors become more saturated as one moves along any radius from the central gray toward the outside of the color space. Saturated colors are strongly chromatic, whereas unsaturated ones have a pastel appearance, being predominantly white with just a suggestion of hue. At any distance from the center, the saturations of all colors are intended to be equal, as are the color differences as one moves around the color circle at any fixed radius. Saturation increases outward from the center. (The Munsell system uses the terms hue, value, and chroma to refer to what others call hue, lightness, and saturation.)

The "ISCC-NBS" method for designating colors (Kelly and Judd, 1976) is based on the Munsell system, but in addition attempts to divide the whole of color space into 267 blocks, each of which has a name made up from a limited set of color terms and modifiers. Twenty-one of these are depicted in Figure 8.11, which is a slice through the purple region that also includes the central achromatic series. In addition to the basic hue terms, the system employs the nonbasic term "olive" to designate one color category. Each of the color names in the system can take one or two modifiers, one of which can be a color term, with the other being chosen from brilliant, light, moderate, pale, strong, very, or vivid (examples of these can be seen in Fig. 8.11).

Translations into the NBS-ISCC system of the terms previously selected from Kelly and Judd's color dictionary will help to further illustrate how the system attempts to reduce 7500 words into a much smaller and more systematic set.

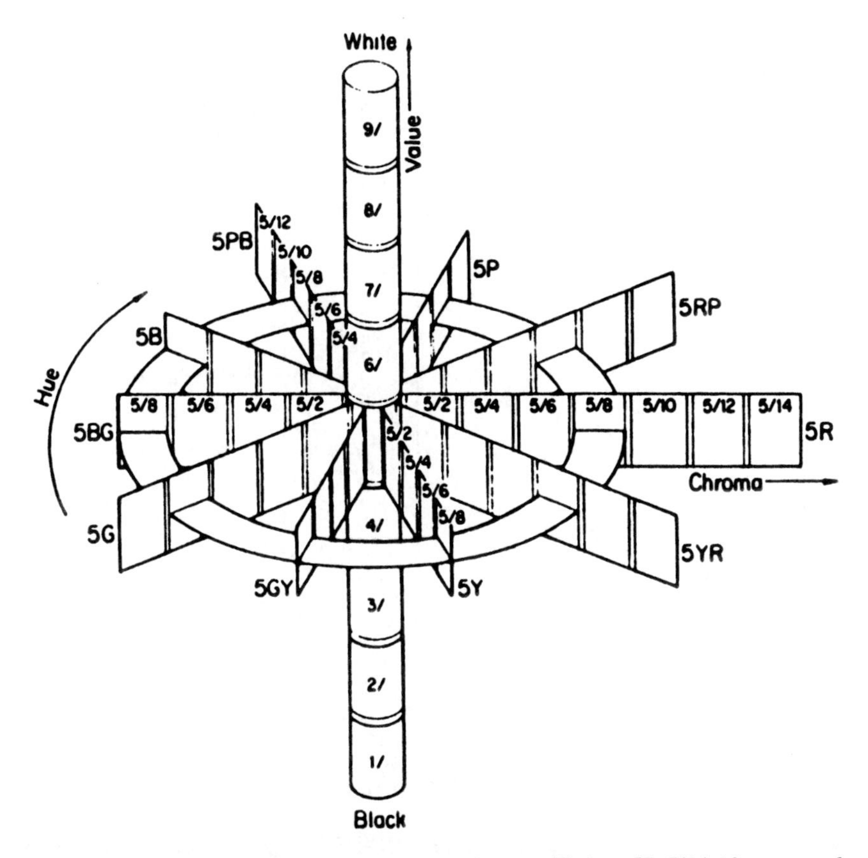

FIGURE 8.10. The Munsell color system. A color specified as 5R 5/14 (the one at the extreme right) is located in the center of the red region at a value (lightness) of 5 (middle of the range) and a strong saturation (chroma) of 14. Each of the five chromatic regions that form the hue circle includes 10 steps numbered 1–10, where 5 is intended to represent the best example of that color. (From Chamberlin and Chamberlin, 1980, p. 23. Reproduced with permission of John Wiley & Sons, Ltd.)

Alamo:	strong brown
Crayon blue:	grayish blue
Bonnie blue:	vivid blue, or vivid purplish blue
Celadon tint:	very pale green
Desert sand:	light yellowish brown
Golden rapture:	brilliant yellow
Lumiere green:	light yellowish green, or moderate yellowish green
Infanta:	moderate blue
Mars orange:	moderate reddish orange
Light olive-gray:	light grayish olive
Light pinkish lilac:	light purplish pink

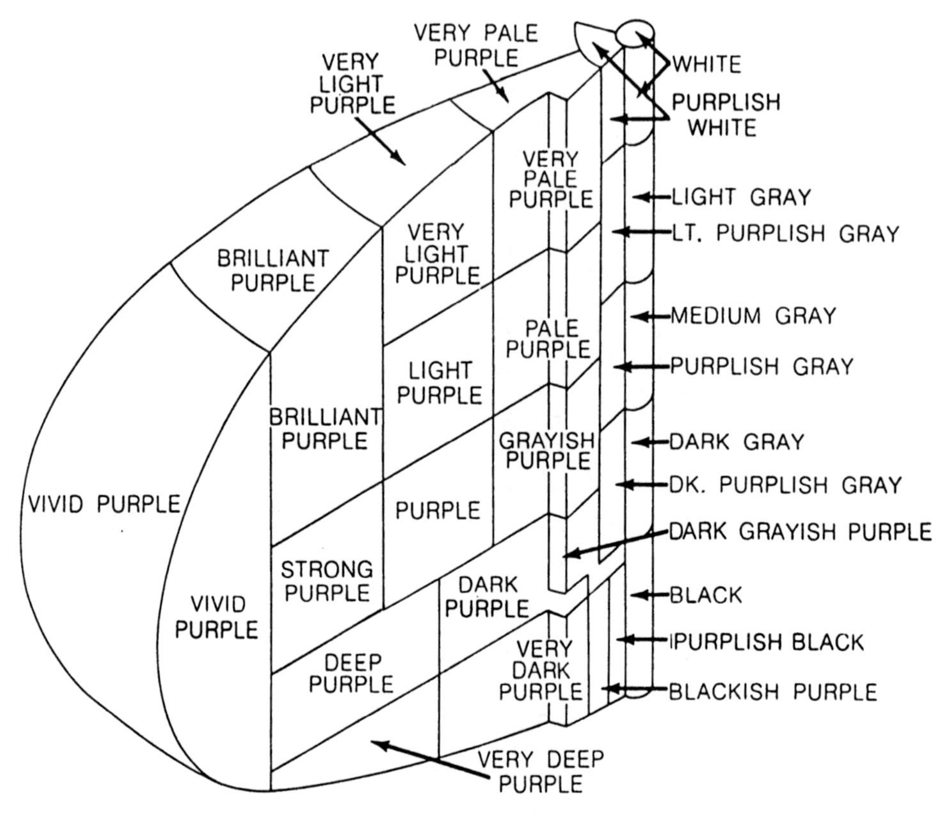

FIGURE 8.11. Illustration of a section of the ISCC-NBS color-name chart in the purple region. (From Kelly and Judd, 1976, p. A-5.)

Reddish gray:	pinkish gray, light grayish-reddish brown, light grayish brown, or light brownish gray
Shadow lime:	light yellow-green
Turquoise green:	strong bluish green, light bluish green, moderate bluish green, or light greenish blue
Amber yellow:	brilliant yellow, light yellow, moderate yellow, or strong yellow.

OSA Uniform Color Scales

A fundamentally different approach to dividing up color space resulted from the work undertaken by a committee of the Optical Society of America (OSA). The aim was to produce a set of colors such that each member of the set would differ from each of its nearest neighbors by the same number of discriminable steps (about 20 MacAdam units). A cartesian, rather than cylindrical, coordinate system was chosen. In this system, illustrated in Figure 8.12, there are 13 levels of lightness along the vertical axis L. Within each lightness level the colors are specified

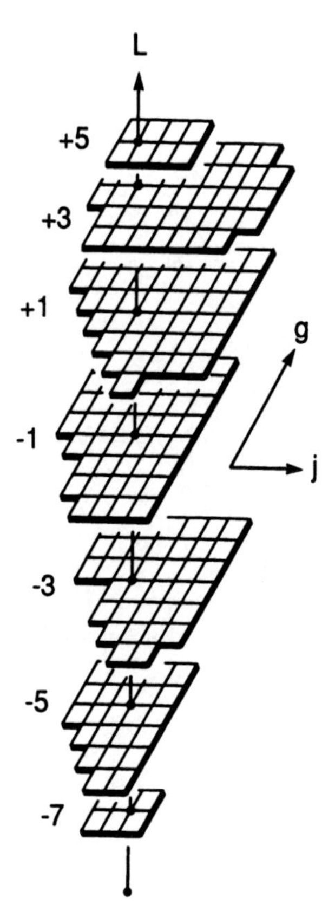

FIGURE 8.12. Arrangement of colors in the system developed by the Optical Society of America. Colors are arranged in an *L,J,g* coordinate system as shown. Only odd-numbered lightness levels are depicted here; six even-numbered levels are also included in the system.

along two chromatic dimensions, one of which (*j*) is roughly yellow-blue, the other (*g*) is red-green. Colors are designated by their positions along the three coordinate axes, with a central gray located at 0,0,0 near the middle of the color solid. The use of color names was deliberately avoided in setting up the system. In my laboratory, the 424 color samples have been named by 27 observers under carefully controlled conditions. Figure 8.13 shows how basic color terms were used by 10 or more subjects at level $L=0$. The chromaticities of the 49 OSA specimens that exist at this level are shown on the CIE chromaticity diagram, with connecting lines that reveal the slightly curvilinear relation between the CIE and the OSA spaces. In this chart, the shaded regions show where the indicated color name was predominantly used. The numbers are labels for the OSA coordinates. The *j* axes are oriented from lower left to upper right and the *g* axes from upper left to lower right. The central gray at $L=j=g=0$ is labeled *N*; along either OSA axis, values are positive upward or negative downward from this central point.)

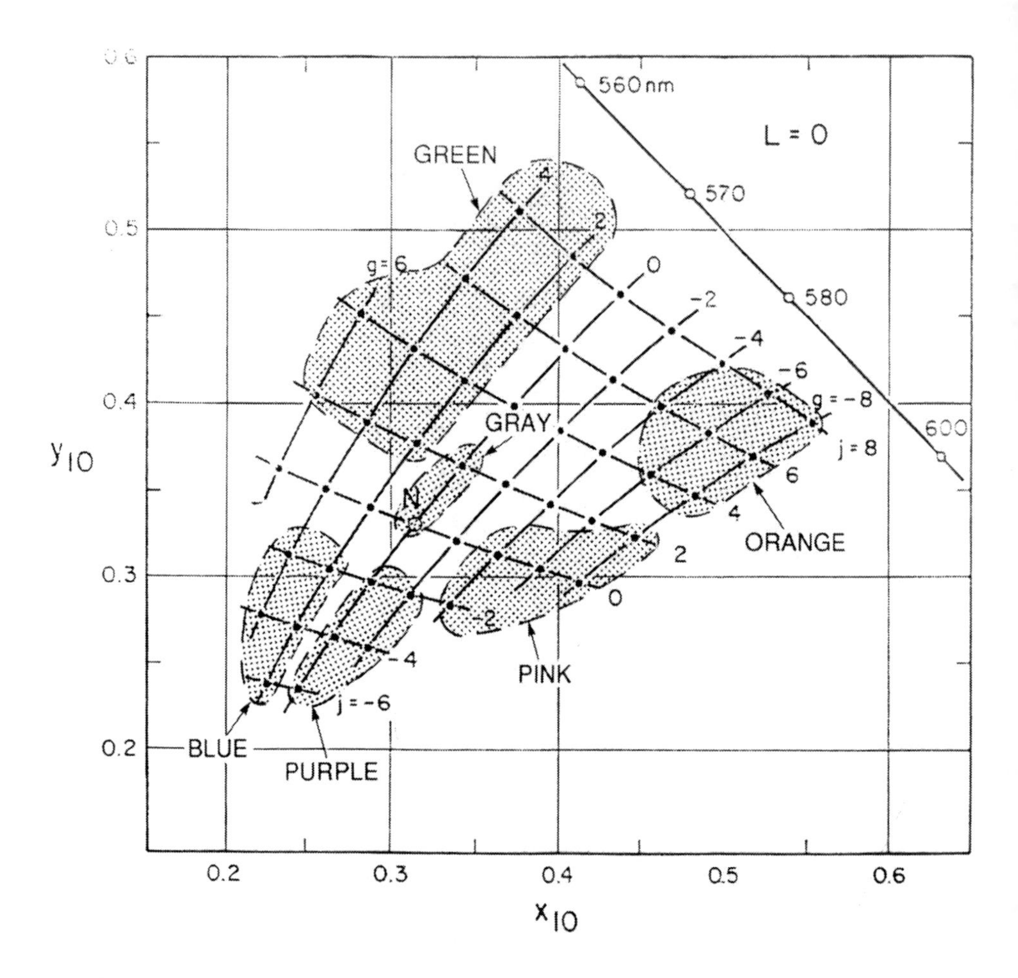

FIGURE 8.13. This diagram shows, at lightness level L=0, the locations of the OSA color samples on the CIE chromaticity diagram (after Wyzecki and Stiles, 1982, copyright © John Wiley & Sons, Inc.), at the intersection of the coordinate axes of that system, which are labeled with the single digits. The shaded areas include color samples called by the indicated names on both presentations by at least 10 of 27 subjects. Of the 11 basic color names, only green, orange, gray, pink, purple, and blue meet that criterion as shown at this level, and the regions within which the criterion is met are shaded. Of the remaining basic colors, red, brown, and black meet the criterion only at lower levels and yellow and white only at higher ones.

In summary, color space is continuous. Starting at any point within it, color appearance changes gradually and continuously as one moves in an arbitrarily selected direction. Yet at the same time, basic color terms are applied in a very natural way to designate regions of colors within which colors seem more alike than colors located in other similarly sized regions of color space that bridge between two of the categorically distinct regions. Global color space is therefore arranged categorically despite the fact that all local color changes occur continuously.

Color Context

For many years, color science was heavily influenced by physicists for whom color appearance was a matter of secondary interest. This influence led to a great deal of important work based on techniques such as color matching and thresholds, while ignoring crucial aspects of color perception related to the relations among colors in complex, real-world environments. That very substantial progress can be made using the physical approach is well documented in *Color Science* (Wyszecki and Stiles, 1982) and in a pair of books by MacAdam (1970, 1981). These researchers have emphasized what Brindley (1960) called "Class A" observations, of which color matching is the classic example. Estimates of brightness exemplify "Class B."

Along the way, there have been voices in the wilderness, crying out that color appearance is what color vision is really all about, and that it simply cannot be ignored. Long ago, Hering (translated by Hurvich and Jameson, 1964) made his fundamental observations on opponent colors. When mixed additively, as on a color wheel, with superimposed projected lights, beam splitters, or by the use of many tiny pixels, colors opposite in the color circle do not blend, but instead they cancel and produce white. Adjacent colors do blend. Thus, for example, bluish greens are readily seen, but reddish greens are impossible sensations. These observations are inexplicable in terms of trichromatic theory, which cannot, for example, explain where yellow comes from or why saturated colors can combine to produce white. Color induction, which is the influence on color perception of surrounding colors, was also investigated by Hering, as well as by Helmholtz and many others. Starting in the 1950s, Hurvich and Jameson (1955) revived interest in Hering's theory, and they performed the first systematic experiments designed to evaluate the contributions of the two sets of opponent mechanisms (red-green and yellow-blue) to the appearance of spectrum colors. Others—especially Helson, Judd, and Evans—also made important contributions; among the physicists it is fair to say that Wright has revealed the broadest interest in color in all its aspects.

But the major impetus to renewed interest in color appearance is due to Edwin Land, the brilliant inventor of the Polaroid process, also founder and first President of the corporation that bears the name of his invention. Land, who had already developed a process of instant black-and-white photography, became interested in color during the period of development of an instant color process. It had long been known that color reproduction could be achieved by taking black-and-white pictures of the same scene separately through red, green, and blue color filters, and then projecting the resulting positive images through the same filters. If only red and green filters are used, a wider gamut of colors can be achieved than one might expect; two-color motion pictures of this kind were fairly common at one time.

What Land (1959a,b) discovered was that if the pictures taken through red and green filters were projected using a red filter in front of one projector and *no* filter in front of the other, a surprisingly wide variety of colors could be created. This worked especially well with natural scenes in which short-wavelength visible radiation was in limited supply, but his demonstrations were later extended to the investigation of Mondrians (arrays of rectangles, named after the painter) in which the rectangles of color were arbitrarily located (see Land, 1986, for a recent summary of this work, which began much earlier).

From the beginning, Land has been adept at publicizing his work, and absolutely brilliant at demonstrating it. Optimal effects depend on perfect registration of the two images; this in turn requires projectors that are very precisely positioned, and the use of carefully selected matching lenses. In his reports, Land exaggerated the nature of the colors in his images by claiming that "a full gamut of color" had been produced; while this was true in the sense that all of the basic color terms were used to identify various regions of the two-color reproduction, the vividness and saturation of many of the colors did not even approach that achievable in a trichromatic reproduction (including, ironically, the colors reproduced by Polaroid color film, which operates according to a trichromatic process).

Not a conventional student of color science, Land instead is an extraordinarily clever inventor who rediscovered certain effects, especially those related to color induction and assimilation, that had actually long been known and previously studied. Induction occurs when a surrounding color "induces" the opponent hue into an area of interest. Assimilation occurs when the hue of the surround color is "assimilated" into an area of interest. These opposite effects depend on the spatial frequency of the pattern. Low spatial frequencies (large areas) favor induction; high spatial frequencies (small, detailed patterns) favor assimilation.

After Land's initial demonstrations, two major articles quickly appeared (Judd, 1960; Walls, 1960) calling attention to many aspects of Land's demonstration that were not new, while noting his conspicuous lack of references to the pertinent literature.

To explain his effects, Land developed his so-called retinex theory of color vision, and for the past 30 years he has been adjusting and modifying it in an effort to improve the predictions that it makes (see Land, 1986, for the latest version). The theory specifically takes account of the influence of colors throughout the spatial extent of a scene as these affect the appearance of any particular region of color within it. However, the theory is not physiologically plausible, because most of the required computations take place independently in three channels, whereas it is well established, as noted earlier, that in the real visual system there is an almost immediate transfer to an opponent-color code (see Brainard and Wandell, 1986, for additional discussion of these issues).

To Land's credit, his work spearheaded a renewed interest concerning the contextual effects of color, which are very real and powerful. Interest in such effects had been flagging in the 1950s despite scholarly publications of the kind documented in the critiques of Judd and Walls. Today, the problems of color context are mostly being treated within the new field of computer vision, within which Land's work definitely fits.

Color Constancy

If one problem stands out above all others, where contextual effects of color are concerned, it is that of color constancy, which is the tendency of the perceived color of a surface to remain constant despite changes in the spectral distribution of the source of illumination. As pointed out early in this chapter, the spectral distribution of light reaching the eye, called the "color signal" by some, is the product of two spectral distributions (those related to source emission and surface reflectance) whereas the perceived color of a surface tends to be related, at least approximately, only to the latter. For example, a book whose surface selectively reflects long wavelengths, and which appears red on the bookshelf when illuminated by incandescent light, remains red and does not seem to change very much when taken outside under sunlight, although the color signal obviously is different.

The full specification of the physical character of sources and surfaces requires that continuous functions of wavelength be measured and plotted. Except for sources that approximate blackbody radiators, simple analytic representations of such functions do not exist. Therefore, if one wished, for example, to store such information in a computer, a table of values containing a very large number of entries would be required. This requirement can be simplified in many cases following an examination of the properties of real sources and surfaces. A reasonable approximation of the spectral reflectances of most surfaces can be achieved by summing only three basis functions of wavelength in various proportions; this allows the specification of a surface reflectance by only three numbers (Cohen, 1964; Maloney and Wandell, 1986). Moreover, as we have seen, a reasonable approximation of the spectral distribution of many light sources can be achieved by summing only two basis functions. Although there are important exceptions, the restricted nature of most emission and reflectance functions simplifies the task of the visual system, which can derive much more information about the probable physical nature of the visual environment than would be possible if no ecological constraints existed regarding the possible shapes of emission and reflectance functions.

It is important to point out that color constancy is by no means perfect, although it is usually good enough that most objects do not shift from one category to another under changes in illumination, provided that the latter are not too extreme. The conventional (and reasonable) approach to the problem of color constancy is to suppose that the problem is solved, to the extent that it is, by a process that somehow abstracts, from the color signal, information about the spectral distribution of the illuminant, and uses this to recover the spectral reflectance distribution of the surface. Clearly, if the spectral distribution of the illuminant is fully known, it is mathematically possible to do this. The question then becomes, how could the visual system obtain such information? The visual system cannot do this directly, because even if one were able to observe the source of illumination, only its color (as related to chromaticity) can be observed. To take an extreme case, it is possible to produce a white light using an additive mixture of monochromatic lights that appear red, green, and blue, respectively, which will match exactly the more conventional full-spectrum white of a tungsten source. Although they match if directly viewed, or projected on a surface that reflects all visible wavelengths equally, two such sources will render surface colors in very different ways, meaning that if the perceived color of the illuminant could be somehow used by the visual system to deduce spectral reflectances under one illuminant, it surely could not be used for the other one. A partial solution to this problem lies in the fact that the kinds of sources for which color constancy works best are those that can be approximated by the weighted sums of only two basis functions. For sources of this kind, reasonable inferences about the spectral distribution of the source can in fact be made on the basis of chromaticity alone. A mathematical treatment of this subject is given by Maloney and Wandell (1986).

But the problem is much deeper than this. Suppose, for example, that the color sample to be judged is illuminated by hidden projectors in an otherwise dark void, so that it is seen as an unrelated color. Logically, there is no possible way, under this circumstance, to disentangle the spectral distributions related to source and surface. In more complicated visual scenes, there must be something else that allows one to gauge the appearance of the illuminant. What could this be?

One possibility is that the source of illumination is directly viewed as a part of the overall scene. This sometimes happens, as when ceiling lights or the sun are seen in peripheral vision, and it is likely that specular highlights from objects in the field are also useful for this purpose (D'Zmura and Lennie, 1986). Most of the time, however, light sources are not in the field of view, and color constancy does not cease to occur in the absence of specular reflections. The major mechanism that remains is that of selective chromatic adaptation, called von Kries adaptation after the investigator who first suggested it (see Worthey and Brill, 1986). When the eye is adapted, for example, to long-wavelength visible light, the sensitivity of the L cones is reduced more than that of the M or S cones. Therefore, if an object is illuminated with reddish light, which would make it appear redder than normal if viewed in isolation, the reddish light reflected from surrounding surfaces selectively reduces the sensitivity of L cones, causing objects to appear less red than they otherwise would.

If there were no selective chromatic adaptation, it is nevertheless possible to show, as Maloney and Wandell (1986) have done, that information based on light reflected from other surfaces in the visual field can be used to partially gauge the character of the illuminant and recover information about the chromaticity of test color. As noted earlier, this is possible only if the gamut of possible illuminants and surface reflectances is considerably restricted.

Summary

The content of this chapter is too broad to permit a detailed summary in a reasonable space, so this will not be attempted. Suffice it to say that an effort has been made here to deal with all three of the factors that are required for color perception: the light, the surface, and the perceiving organism. The remainder of these summary remarks are oriented toward the future.

Human color perception is an extremely complicated subject. A great deal is now understood from a variety of perspectives, associated with the approaches of psychophysics, anatomy, physiology, and computational modeling. Of course there is much left to learn, and doubtless there always will be. For example, psychophysical experiments have only begun to test the limits of color constancy, even though the nature of the problem has been generally understood for more than a century. As a result, the computational modelers concerned with color constancy cannot know precisely what it is they are trying to explain. At a physiological level, we know a great deal about color coding in the retina, lateral geniculate, and cortex, but we do not yet understood what kind of activity in the brain immediately underlies sensations of color such as red or green, or where this activity is located.

Psychophysicists and electrophysiologists can now deal with much more complex conditions than previously, due to the use of computer-controlled color monitors that allow the display of stimuli having almost any combination of spatial, temporal, and chromatic variation. But no matter how clever the stimuli, the fundamental mind–body problems cannot be solved by psychophysics alone, or by research with anesthetized animals who obviously are seeing no color. Fortunately, some researchers now record single unit activity from the brains of awake, behaving animals. This kind of work, especially when undertaken with old world monkeys, would seem to constitute the most relevant physiological approach to problems of human color perception. Significant advances from this kind of research have already begun to be achieved, for example by Dow (Chapter 4, this volume), and we can look forward expectantly to the future fruits of such labors.

Even where peripheral color processing is concerned, where matters are much simpler than for complex perception, it has not proved easy to get clear answers from electrophysiology, even when such efforts have been combined with the best anatomical techniques and experimental designs. No two neurons respond exactly alike, so hundreds, or even thousands, must be examined. The problems of classification, sampling, and choice of experimental parameters have led to results from different laboratories that often have been confusing, and sometimes flatly contradictory. Yet there is no cause for despair. Viewed in the context of the experience of this writer, who began graduate work when virtually nothing was directly known about the physiology of color vision, the advances of the last 40 years seem very encouraging. Technical accomplishments have often been spectacular, and it is heartening that electrophysiologists have shown a progressive tendency to adopt the kinds of controls and experimental designs that were pioneered by psychophysicists and that were once their special province.

In the past, the psychophysical approach with human subjects has led the way toward the understanding of things such as color matches, color appearance, and parallel processing. It is a safe prediction that psychophysics will continue to play an important role in the future of color science, including the effort to better understand the more complex aspects of human color perception. In part this is because psychophysical data, properly treated, will continue to provide the basic quantitative facts that physiology tries to explain.

At this writing, the computational approach is very hot. There has been a tendency toward naive optimism in this area in the past, where, for example, the problems of machine translation and recognition of human handwriting seemed easier to solve 20 or 30 years ago than they do now. Nevertheless, the approach seems necessary for the production of the kinds of very explicit theories that can be tested in the laboratory. Although many such models will be found to be too limited, physiologically unrealistic, or just plain wrong, science advances best by the rejection of plausible theories that are explicit enough to suffer this otherwise unkind fate.

Acknowledgments. Preparation of this chapter was supported by a research grant from the National Eye Institute, which has also supported by research for many years. I thank Lee Fargo, Ethan Montag, Harvey Smallman, and the editor for helpful comments on the initial draft.

References

Berlin B, Kay P (1969). *Basic Color Terms: Their Universality and Evolution*. University of California Press, Berkeley, CA.

Billmeyer FW, Saltzman M (1981). *Principles of Color Technology*. Wiley, New York.

Boynton RM (1975). Color, hue, and wavelength. In *Handbook of Perception* EC Carterette, MP Friedman, eds. Academic Press, New York.

Boynton RM (1979). *Human Color Vision*. Holt, Rinehart, & Winston, New York.

Boynton RM (1984). Psychophysics. In CJ Bartleson, F Grum, eds. *Visual Measurement, Optical Radiation Measurements*. Vol. 5:, pp. 335–366. Academic Press, New York.

Boynton, RM (1986). A system of photometry and colorimetry based on cone excitations. Color Res. Appl. **11**:144–252.

Boynton RM, Olson CX (1987). Locating basic colors in the OSA space. Color Res. Appl. **12**:94–105.

Boynton RM, Wisowaty JJ (1980). Equations for chromatic discrimination models. J. Opt. Soc.Am. **70**:1471–1476.

Boynton RM, Eskew RT Jr, Olson CX (1985). Blue cones contribute to border distinctness. Vision Res. **25**:1349–1352.

Brainard DH, Wandell BA (1986). Analysis of the retinex theory of color vision. J. Opt. Soc. Am. **A3**:1651–1661.

Brindley GS (1960). *Physiology of the Retina and the Visual Pathway*, 1st ed. Edward Arnold, London.

Buchsbaum G, Gottschalk A (1983). Trichromacy, opponent colours coding and optimum colour information transmission in the retina. Proc. R. Soc. London B **220**:89–113.

Burns SA, Elsner AE (1985). Color matching at high illuminances: The color match-area effect and photopigment bleaching. J. Opt. Soc. Am. **A2**:698–704.

Cavanagh P, Favreau OE (1985). Color and luminance share a common motion pathway. Vision Res. **25**:1595–1601.

Chamberlin GJ, Chamberlin DG (1980). *Colour: Its Measurement, Computation, and Application*. John Wiley & Sons, Ltd., Chichester, West Sussex, United Kingdom.

Cohen JB (1964). Dependency of the spectral reflectance curves of the Munsell color chips. Psychon. Sci. **1**:369–370.

Dartnall HJA, Bowmaker JK, Mollon JD (1983). Human visual pigments: Microspectrophotometric results from the eyes of seven persons. Proc. R. Soc. London B **220**:115–130.

Derrington AM, Krauskopf J, Lennie P (1984). Chromatic mechanisms in lateral geniculate nucleus of macaque. J. Physio. **357**:241–265.

DeValois RL, DeValois KK (1975). The neural basis of seeing. In *Handbook of Perception* EC Carterette, MP Friedman, ed., Vol. V:117–166. Academic Press, New York.

D'Zmura M, Lennie P (1986). Mechanisms of color constancy. J. Opt. Soc. Am. **A3**:1662–1672.

Enoch JM (1963). Optical properties of the retinal receptors. J. Opt. Soc. Am. **53**:71–85.

Frome FS, Buck SL, Boynton RM (1981). Visibility of borders: Separate and combined effects of color differences, luminance contrast, and luminance level. J. Opt. Soc. Am. **71**:145–150.

Grinberg DL, Williams DR (1985). Stereopsis with chromatic signals from the blue-sensitive mechanism. Vision Res. **25**:531–537.

Gregory RL (1977). Vision with isoluminant colour contrast: I. A projection technique and observations. Perception **6**:113–119.

Grum F (1980). Colorimetry of fluorescent materials. In Vol. 2: *Color Measurement*, pp. 236–288. F Grum, CJ Bartleson, ed., *Optical Radiation Measurements*: Academic Press, New York.

Helmholtz H (1924). *Physiological Optics* (edited by JPC Southall, ed., 3 volumes). Optical Society of America, Rochester, NY.

Henderson ST (1977). *Daylight and Its Spectrum*, 2nd ed. Adam Hilger, Bristol, England.

Hering E (1964). *Outlines of a Theory of the Light Sense* (LM Hurvich, D Jameson, trans.), Harvard University Press, Cambridge, MA.

Hunter RS, Harold RW (1987). *The Measurement of Color Appearance*. Wiley, New York.

Hurvich LM, Jameson D (1955). Some quantitative aspects of an opponent-colors theory: II. Brightness, saturation, and hue in normal and dichromatic vision. J. Opt. Soc. Am. **45**:602–616.

Ingling CR Jr, Martinez-Uriegas E (1985). The spatiotemporal properties of the r-g X cell channel. Vision Res. **25**:33–38.

Judd DB (1960). Appraisal of Land's work on two-primary color projections. J. Opt. Soc. Am. **50**:254–268.

Judd DB, MacAdam DL, Wyszecki G (1964). Spectral distribution of typical daylight as a function of correlated color temperature. J. Opt. Soc. Am. **54**:1031–1040.

Kaiser PK (1984). Physiological response to color: A critical review. Color Res. Appl. **9**:29–36.

Kaufman JE (ed.) (1981). *IES Lighting Handbook* (Reference Volume). Illuminating Engineering Society of North America, New York.

Kelly KL, Judd DB (1976). The ISCC-NBS method of designating colors and a dictionary of color names. NBS Special Publication 440, U.S. Department of Commerce, U.S. Superintendent of Documents, Washington DC.

Krauskopf J, Williams DR, Heeley DW (1982). Cardinal directions of color space. Vision Res. **22**:1123–1131.

Land EH (1959a). Color vision and the natural image: Part I. Proc. Natl. Acad. Sci. U.S.A. **45**:115–129.

Land EH (1959b). Color vision and the natural image: Part II. Proc. Natl. Acad. Sci. U.S.A. **45**:636–644.

Land EH (1986). Recent advances in retinex theory. Vision Res. **26**:7–21.

Lennie P, D'Zmura M (1988). Mechanisms of color vision. CRC Crit. Rev. Neurobiol. **3**:333–400.

MacAdam DL (1942). Visual sensitivities to color differences in daylight. J. Opt. Soc. Am. **32**:247–274.

MacAdam DL (ed.) (1970). *Sources of Color Science*. MIT Press, Cambridge, MA.

MacAdam DL (1981). *Color Measurement: Theme and Variations*. Springer-Verlag, New York.

Maloney LT, Wandell BA (1986). Color constancy: A method for recording surface spectral reflectance. J. Opt. Soc. Am. **A3**:29–33.

Mullen KT, Baker CL Jr (1985). A motion aftereffect from an isoluminant stimulus. Vision Res. **25**:685–688.

Nagy AL, Eskew RT Jr, Boynton RM (1987). Analysis of color discrimination ellipses in a cone excitation space. J. Opt. Soc. Am. **A4**:756–768.

Naka KI, Rushton WAH (1966). S-cone potentials from colour units in the retina of fish (Cyprinidae). J. Physiol. **185**:536–555.

Nassau K (1983). *The Physics and Chemistry of Color: The Fifteen Causes of Color.* Wiley-Interscience, New York.

Perrin FH (1953). Physical concepts: Radiant energy and its measurement. *The Science of Color*, Chap. 6 (Committee on Colorimetry, Optical Society of America). Thomas Y. Crowell, New York.

Pokorny J, Smith VC, Verriest G, Pinckers AJL (1979). *Congenital and Acquired Color Vision Defects*. Grune & Stratton, New York.

Schnapf JL, Kraft TW, Baylor DA (1987). Spectral sensitivity of human cone photoreceptors. Nature (London) **325**:439–441.

Smith VC, Pokorny J (1975). Spectral sensitivity of the foveal cone photopigments between 400 and 500 nm. Vision Res. **15**:161–171.

Stefurak DL, Boynton RM (1986). Independence of memory for categorically-different colors and shapes. Percept. Psychophys. **39**:164–174.

Tansley BW, Boynton RM (1978). Chromatic border distinctness perception: The role of red-green-sensitive cones. Vision Res. **18**:683–697.

Trezona PW (1976). Aspects of peripheral color vision. Modern Problems Ophthalmol. **17**:52–70.

Van Essen DC, Newsome WT, Maunsell JHR (1984). The visual field representation in striate cortex of the macaque monkey: Asymmetries, anisotropies, and individual variability. Vision Res. **24**:429–448.

Walls GL (1960). Land! Land! Psychol. Bull. **57**:29–48.

Ware C, Cowan WB (1983). Specification of heterochromatic brightness matches: A conversion factor for calculating luminances of stimuli that are equal in brightness. NRC Tech. Rep. 26055, National Research Council, Ottawa, Canada.

Webster MA, MacLeod DIA (1988). Factors underlying individual differences in the color matches of normal observers. J. Opt. Soc. Am. **A5**:1722–1735.

Weisskopf V (1968). How light interacts with matter. Sci. Am. **219**(3):60–71.

Wiesel TN, Hubel DH (1966). Spatial and chromatic interactions in the lateral geniculate body of the rhesus monkey. J. Neurophysiol. **29**:1115–1156.

Worthey JA, Brill MH (1966). Heuristic analysis of Von Kries color constancy. J. Opt. Soc. Am. **A3**:1708–1712.

Wyszecki G, Stiles WS (1982). *Color Science*, 2nd ed. Wiley, New York.

Zeki S (1985). Colour pathways and hierarchies in the cerebral cortex. In *Central and Peripheral Mechanisms of Colour Vision* (D Ottoson, S Zecki, eds.), pp. 19–44.

9
From Visual Structure to Perceptual Function

William Maguire, Naomi Weisstein, and Victor Klymenko

Introduction

We are bombarded by light. It caroms off the objects around us in an apparent chaos of changing wavelengths and intensities. Every surface that deflects its path leaves its stamp, however, and so organisms have evolved that form and process images, the tracks of those distant objects. Vision experienced is an apprehension of those objects within a structure of sensory qualities. We describe our experienced images in terms of size, shape, color, brightness, direction, velocity, depth. These qualities arise in the brain's image processing circuitry and resonate to the world's intrinsic structure.

In this chapter we will explore how the visual system constructs object and event representations from the retinal image, first by analyzing all small (local) regions of the image in terms of specific values along a set of separate image parameters or dimensions and then by integrating that information into global object-based perceptual representations. We will focus our attention on the methods of perceptual psychology and psychophysics. We believe, however, that the goals of the psychologist attempting to describe the functional architecture of visual perception and the neurophysiologist trying to reveal the mechanics of the visual system's neural circuitry are finally the same and that eventually they will speak a common language. (See Teller, 1984 for an analysis of current bridging or linking assumptions between the two disciplines.)

Early on, those interested in psychophysical investigations of basic visual sensitivity attempted to integrate their results with what was known or believed about physiological structures with the goal of creating a complete psychobiology. Color vision presents an instructive example of particular success in this enterprise with the psychophysics of color leading the known physiology, and significantly predicting the later findings. On the basis of color matching experiments, the three receptor theory of color vision was born, following Newton's and Young's theories. With data from color-blind individuals, and the assumption that their problem was due to the absence of one or another photoreceptor, the basic absorption spectra of the three human cone photopigments were predicted. These predictions were surprisingly accurate as revealed by later direct measurement with microspectrophotometry (Pokorny and Smith, 1986).

Investigating color appearance, Hurvich and Jameson (1955) built a model that functioned by opponent processes, a mechanism first advanced by Hering, to account for their own hue cancellation data. They posited three color channels: one achromatic, combining information from different cone types additively, and two color opponent, combining inputs from the different cone types subtractively. Subsequent neurophysiology has established the existence of color opponent cells that combine cone inputs subtractively as well as broad band cells that combine cone inputs additively in primate retina, thalamus, and cortex (Gouras, 1974; DeMonasterio and Gouras, 1975).

Although this psychophysically based theorizing cannot explain all the facts of color vision, nor has it been particularly useful in discovering important details of physiology such as the great variety of color-coded cells or the specific locations of the color channels, still it has been particularly successful in anticipating the most important neurophysiological data in color vision.

The analysis of how we register and represent the multitude of forms that we encounter, both rigid and deformable objects moving in a three-dimensional space, appears to be relatively more complex than the analysis of their surface color appearances, and the accomplishments of theory and psychophysical investigation less dramatic. Nonetheless there has been progress in understanding how we register the form, motion, and spatial position of objects. Likewise, a tradition of close cooperation between physiological and psychophysical laboratory is evident in this research.

Building Blocks

Some Definitions

Our concern is with the role of specialized analyzers, channels, and pathways in the analysis and integration of information about object dimensions. By analyzers we mean psychophysically identified mechanisms that are sensitive to a limited range of values on a specific stimulus dimension, and respond to a stimulus in a local region of the retinal image. Physiologically, the equivalent level of analysis is at the level of individual neurons and their receptive fields. Channel refers to a set of otherwise identical analyzers that differs in response along one or more dimensions. The variant dimension might typically be spatial position, in which case the channel would be a spatial distribution of identical analyzers. One might also speak of a channel responsive to vertical orientation that contained analyzers with identical orientation tuning, but different size and/or spatial position tuning. A pathway is a group of channels that shares response characteristics on a limited set of dimensions. Thus we might speak of a motion pathway, consisting of channels sensitive to different speeds an directions of motion, which in turn are composed of spatially distributed sets of analyzers, tuned to particular speeds and directions. The term pathway is used as well to refer to the physiological structure underlying the psychophysically defined pathway.

Early Processing Stages

Many current theories of form perception include an initial stage of image processing in which a dimensional analysis of the image takes place. That is the local components of the image are analyzed and values determined along a number of image dimensions. These include dimensions of extension such as size or spatial frequency, orientation of edge components, binocular disparity, direction and magnitude or temporal frequency of motion vectors, as well as intensive dimensions such as brightness, saturation, and hue. We will concentrate on the extensive dimensions. The intensive dimensions are treated by Boynton in Chapter 8.

At the present time, evidence from psychophysics and neurophysiology converges on a characterization of the primitive analyzers that are involved in the initial local analysis of the stimulus. The term local refers to the fact that a small neighborhood of the image is analyzed in relative isolation from adjacent parts of the image. Global analysis occurs when relatively large neighborhoods of the image are analyzed interactively. The image of a circle provides an example. It consists locally of curved line segments, and the set of local properties includes all orientations. Globally it is round, convex, and closed.

Orientation provides an example of how one image dimension is locally analyzed. The psychophysical evidence is consistent with an array of analyzers localized in a space with retinal coordinates. Each individual analyzer is most sensitive to a particular edge orientation and sensitivity declines to zero for edge orientations 10–15° from the optimum. There are sufficient analyzers at each location that the full set of edge orientations can be represented (Kulikowski et al., 1973).

Neurophysiology and neuroanatomy lead to a similar conclusion. Area V1 of the visual cortex appears to be organized into orientation columns. Within a column, cells are responsive to the same orientation (though they will differ in other properties such as optimal stimulus width or spatial frequency). Nearby columns are responsive to the same retinal coordinates, but to slightly different orientations, an arrangement called a hypercolumn (Hubel and Wiesel, 1974). The full set of edge orientations at each location is thus represented by adjacent columns coding for the same retinal coordinates (see Dow, Chapter 4).

A critical question that arises in the investigation of the analyzers is the extent to which they can be said to function independently. The question can be asked at a number of levels. We can ask: Do the individual analyzers whose sensitivities are distributed along a particular stimulus dimension work independently or in concert with each other? Is the output of analyzers at the early stages of visual processing subject to influences from stages higher in the processing hierarchy. We can also examine independence with regard to clusters of analyzers, rather than individual analyzers. We might ask: Is local edge orientation analyzed independently of the size or spatial frequency characteristics of the edge? Is information about stimulus motion, and its associated dimensions isolated from information about pattern characteristics? Are the extensive dimensions, in general, analyzed separately from the intensive dimensions?

All of these questions will be addressed at different points in the chapter and current knowledge discussed. In general the empirical evidence does not support simple yes or no answers about independence. Independence as characterized above is relative. That is, it is likely to be found under some circumstances, while interaction is the rule under other circumstances. The great challenge for theory is discerning the functional and evolutionary purpose behind the specific organization that emerges from these investigations.

It is also clear that our knowledge of how interacting analyzers and channels contribute to coherent object representations is at best rudimentary. Suprathreshold psychophysics has produced many results that indicate a variety of interactions between analyzers but a clear picture of the overall organization of these interactions has not yet emerged. In this respect, the brain physiology is even more rudimentary. Advanced anatomical techniques allow us to see the rich interconnectedness within and between the numerous visual field representations of the cortex. It can be quite difficult, however, for a variety of technical and methodological reasons to observe simultaneously the responses of the neurons so interconnected.

Analyzers, Sets of Analyzers, and Pathways

Individual Analyzers

At the level of individual dimensions, there is much agreement about the existence of sets of relatively independent analyzers differing from each other in the range of dimensional values to which they are tuned. Graham (1985, 1989) has proposed a number of candidate dimensions for which analyzers might exist whose responses would provide a complete description of temporally modulated patches with one-dimensional sinusoidal luminance profiles. Such stimuli are typically used in a variety of psychophysical investigations. Excluding color information the dimensions include spatial frequency, orientation, spatial position, spatial extent, spatial phase, temporal frequency, temporal position, temporal extent, temporal phase, direction of motion, contrast, mean luminance, eye of origin, and binocular disparity. (Schematic illustration of some of these dimensions is provided in Figure 9.1.)

Throughout this chapter we will discuss spatial and temporal frequency sensitivity. It is common to speak of low and high temporal and spatial frequencies within the range of human sensitivity. Typically, low spatial frequencies are those less than 2 cycles per degree, and people show some sensitivity to contrast modulation at least as low as 0.2 cycles per degree. High spatial frequencies are typically at least 6 cycles per degree. Human contrast sensitivity extends up to 40

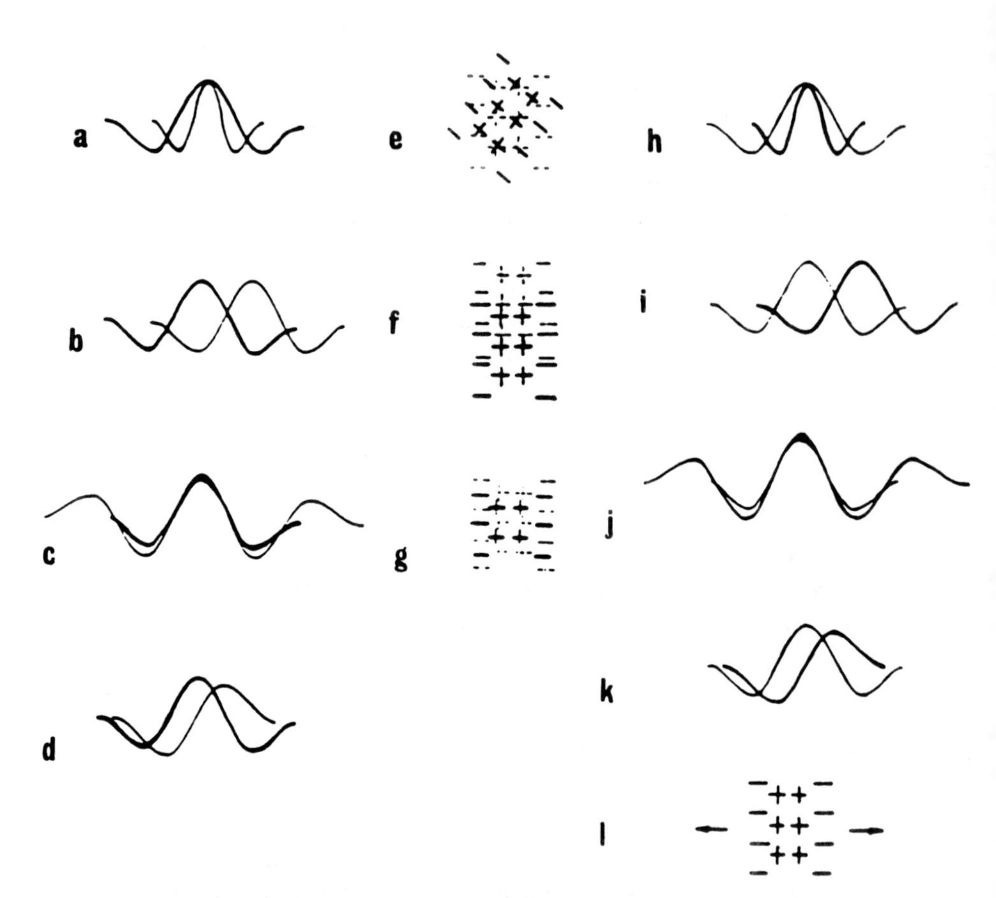

FIGURE 9.1. Illustration of a number of the spatial and temporal dimensions along which sine wave patterns can differ. Illustrated from top to bottom and then from left to right are (a) spatial frequency, (b,f) spatial position, which can vary in two dimensions for flat patterns, as can (c,g) spatial extent or spatial bandwidth, (d) spatial phase. An additional spatial dimension is (e) orientation. Temporal dimensions are (h) temporal frequency, (i) temporal position, (j) temporal extent or duration, (k) temporal phase, and (l) direction of motion. (From Graham, 1985.)

or more cycles per degree under some conditions (Van Nes and Bouman, 1967), but typically stimuli above 15 cycles per degree are seldom used. Correspondingly low temporal frequencies are typically below 1 Hz. (A 0 Hz stimulus is one that is stationary; a 0 spatial frequency stimulus is one of uniform luminance.) High temporal frequency stimuli are generally above 5 Hz and contrast sensitivity limits can be above 50 Hz (Kelly, 1961), but stimuli are typically not higher than 20 Hz.

Not all of the dimensions described in Figure 9.1 have been extensively studied, but strong evidence exists for tuned analyzers each covering part of the range of human sensitivity for spatial frequency, spatial position, temporal frequency, direction of motion, and binocular disparity. Using stimulus patterns near the threshold of detection, psychophysical evidence is consistent with substantial independence along these stimulus dimensions between individual analyzers (Graham, 1989). Outside this near-threshold range, however, cooperative and inhibitory interactions are the rule rather than the exception (Breitmeyer and Ganz, 1976; Greenlee and Magnussen, 1988; Grossberg and Mingolla, 1985a,b).

Analyzers are either labeled or unlabeled (Graham, 1989; Watson, 1986; Treisman, 1986). A labeled analyzer is not only sensitive to a limited range of stimulus values but mechanisms above in the processing hierarchy have information about the sensitivity range of this analyzer on the particular dimension so that even at the analyzer's threshold of responding, the stimulus may be identified and discriminated from other patterns. An example of labeling is spatial position. A light flashed at one location is discriminable from an identical light flashed at another spatial location at threshold. Thus the analyzers responsive at threshold are labeled for spatial position. On the other hand, human observers are unable to determine to which eye a monocularly presented stimulus has been presented from the visual information. Given evidence from neurophysiology for cells in the visual system that respond only to patterns presented to one eye, we would say that the corresponding analyzers are sensitive to eye of origin, but they are unlabeled (Graham, 1989).

Sets of Analyzers: Transient and Sustained Pathways

Groups of channels responsive to different stimulus characteristics may be gathered into streams or pathways with significant suprathreshold interactions within the pathway, and relative independence between pathways. These pathways may in turn have physiological reality in relatively independent groups of interconnected neurons found in visual areas of the brain. A number of investigators have speculated about a pathway specialized to process motion and possibly spatial position information relatively independently from another pathway specialized to process pattern information (see Ikeda and Wright, 1975; Enroth-Cugell and Robson, 1966 for neurophysiological evidence; Weisstein et al., 1975 and Breitmeyer and Ganz, 1976 for psychophysical evidence). There are a number of convergent findings that point to the existence of a set of fast responding channels with relatively transient responses to stimuli. These prefer low spatial frequency and high temporal frequency stimuli. Another set of channels show slower, relatively sustained responses. These prefer high spatial and low temporal frequencies. These channels have been called transient and sustained, respectively (Ikeda and Wright, 1975; Weisstein et al., 1975; Breitmeyer and Ganz, 1976; Meyer and Maguire, 1977), though other terms have been used. (An example of sustained and transient responses at a neurophysiological level is shown in Fig. 9.2 taken from Ikeda and Wright, 1975.) At the same time there is

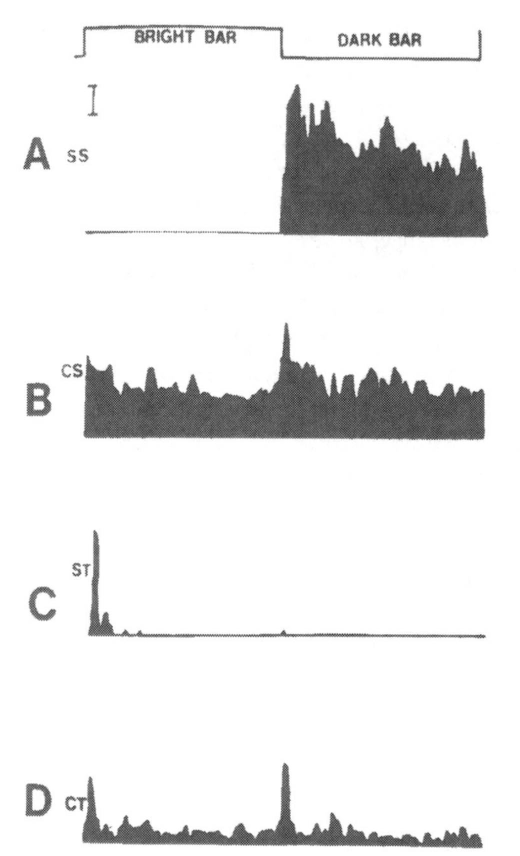

FIGURE 9.2. Poststimulus histograms showing the response of sustained and transient cortical cells in V1 of the cat. The left half is response to a light bar, the right half, a dark bar. Both simple (s) and complex (c), sustained (s) and transient (T) cells are illustrated. (A) Simple sustained cell; (B) complex sustained cell; (C) simple transient cell; (D) complex transient cell. (Data and figure from Ikeda and Wright, 1975.)

abundant evidence to indicate that there is extensive interaction between these channels. It is likely that these sustained and transient channels overlap substantially in the patterns that they respond to, although due to differential sensitivity only one or the other type will be active at threshold.

Physiologically it is known that different groups of neurons process and transmit information at different rates. There has been a good deal of interest in this regard centering on the X and Y cells first described in the cat retina (Enroth-Cugell and Robson, 1966). This classification was initially based on the linear summation within the receptive fields of X cells and the nonlinear summation within the receptive fields of Y cells. Later a number of other characteristics were described. X cells have smaller axons and thus respond more slowly. They also have more sustained responses and show sensitivity to higher spatial frequencies and lower temporal frequencies of stimulation. Y cells respond faster and more transiently and are sensitive to lower spatial frequencies and higher temporal frequencies (see also Slaughter, Chapter 3).

It has also been suggested that the segregated output from these cells constitute the sustained and transient pathways evident in the LGN and cortex of the cat (Ikeda and Wright, 1975). The significance of these results for human vision is largely conjectural. It is clear that the distribution, properties, and targets of X and Y cells differ for the cat and monkey (Shapley et al., 1981).

There is, however, a body of psychophysical work that supports models of visual processing where pathways with X-like and Y-like properties play a role. These studies show correlations between the properties of spatial frequency tuning, temporal frequency tuning, response time, and the sustained or transient nature of the response. Simple reaction time to the presentation of a sinusoidal grating increases with grating spatial frequency (Breitmeyer, 1975). Visual persistence, the phenomenal presence of the stimulus after physical offset, also increases with spatial frequency (Meyer and Maguire, 1977). Experiments that attempt to describe the temporal impulse response of the visual system near threshold find the response to high spatial frequencies to be sustained, while the response to low spatial frequencies is transient. The impulse response is a convenient way of describing the response properties of a system if that system is a linear system. Visual analyzers are thought to be approximately linear at threshold (Watson and Nachmias, 1977). Figure 9.3 illustrates data from Watson and Nachmias (1977) for a number of spatial frequencies. It shows that the impulse response to high spatial frequencies is monophasic, while the impulse response to low spatial frequencies is biphasic. A stimulus of more than brief duration will generate a steady response from a channel with a monophasic impulse response, while a channel with a biphasic impulse response will mainly be responsive at stimulus onset and offset.

The term transient means that the analyzers have a biphasic impulse response and thus are most responsive to stimulus change. The term sustained means that the analyzers display a monophasic impulse response and thus respond steadily throughout stimulus presentation. However, because high temporal frequency sensitivity, fast response latency, and low spatial frequency sensitivity are empirically associated with transient response, and low temporal frequency sensitivity, slow response latency, and high spatial frequency sensitivity are empirically associated with sustained response, the terms transient and sustained are sometimes used to describe pathways with all such characteristics. We will use the term pathway to describe the full set of channels with sustained or transient characteristics.

A number of investigators have reported the existence of two discriminable thresholds for flickering sine wave gratings, a pattern and flicker threshold. Flickering high spatial frequency stimuli, when presented at a contrast where they are just detectable, appear stationary, and flicker can be discerned only at a higher contrast. Flickering low spatial frequency stimuli appear as spatially uniform flicker at contrasts where they are just detectable and higher contrast is required for the grating pattern to be seen (Kulikowski and Tolhurst, 1973). This evidence is consistent with a description of sustained and transient pathways overlapping in sensitivity above threshold. It is unclear, however, how segregated pattern and motion or flicker information is, even at threshold. It has been shown that at least some pattern and orientation information is available for flickering low-frequency gratings at threshold using identification at threshold methodologies (Watson, 1986) (see Spatial Analyzers Revealed by Identification Experiments).

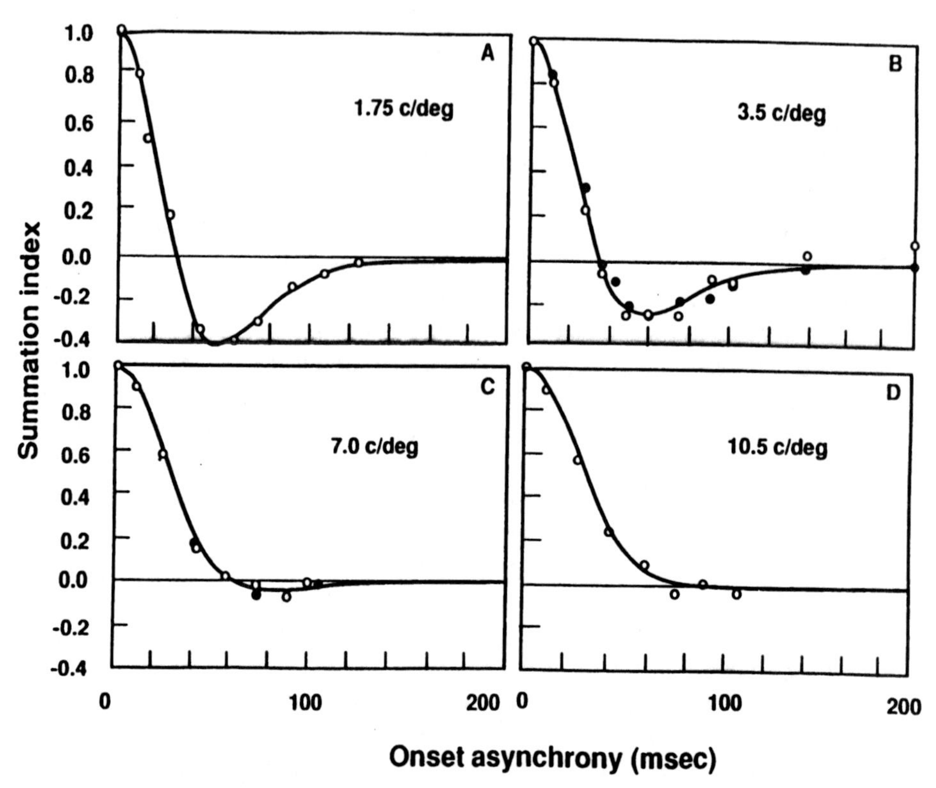

FIGURE 9.3. Data from Watson & Nachmias (1977). Copyright (1977), Pergamon Press plc. Reprinted with permission. Each curve describes the impulse response to a different spatial frequency. The response is constructed by using a subthreshold summation procedure where the gratings to be summed have identical spatial frequencies and are presented extremely briefly. They are presented with small asynchronies in time of presentation. This stimulus onset asynchrony varies in different experimental conditions. The x axis plots the asynchrony in stimulus onset. The y axis is a summation index. It measures the interactions of the two gratings at threshold. Positive values of the index indicate that the energy of the two gratings sums to produce a lower threshold than either presented alone. Negative values mean that the two gratings actually interact in an inhibitory fashion. Negative values are associated with a biphasic impulse response and are much more prominent for low spatial frequencies. (From Graham, 1989.)

We discuss later the possibility that there are multiple channels of the sustained and transient type. The transient analyzers individually may show directional selectivity. The two groups of channels overlap in their sensitivity over the spatial and temporal frequency domain, with each one dominating its respective part of the domain. Most suprathreshold stimuli would be expected to activate both sustained and transient mechanisms, each one carrying somewhat different labeled information about the stimulus. (See Graham, 1989, chapter 12, for an illustration of this configuration.)

Pathways Defined by Response at Equiluminance

Analysis of the intensive and extensive dimensions of images does appear to take place in relative isolation. The most compelling evidence for this is from experiments that use images that contain only isoluminant edges. These are edges that are defined by differences in hue, but have no brightness differences across the color discontinuity. Perception of motion, depth, as well as some aspects of form is remarkably impaired with these images (Cavanaugh et al., 1984; Livingstone and Hubel, 1987). These results have been interpreted in terms of a relatively isolated system specialized for the analysis of color and another system blind to all intensive differences except differences in overall brightness to analyze motion and spatial position. The presumed physiological basis of these separate systems is to be found in sets of neurons in LGN, V1, and visual associative cortex, grouped into separate nuclei within these structures and having their own unique input–output relations with surrounding visual structures. The lateral geniculate nucleus provides an example. It consists of magnocellular and parvocellular layers that contain cells with different response properties, the cells in parvocellular layers showing more color opponent properties and more sustained responses in the rhesus monkey (Derrington and Lennie, 1984). These cells, in turn, have different targets in V1. These orderly sets of independent connections appear to point to the existence of at least two, more likely three separate pathways running through LGN, and V1, each with unique connections in associative cortex. One of these pathways, the parvocellular pathway, processes information about intensive characteristics of the image. Another, the magnocellular pathway, processes gross form, spatial position, and motion. The third, also parvocellular, processes pattern detail carried by the high spatial frequencies of the image (Livingstone and Hubel, 1987). Whether the latter two pathways can be characterized as transient and sustained pathways is at present unclear and will be discussed more fully in Equiluminant Contours.

Independent Analyzers Tuned to Specific Stimulus Dimensions

The following sections examine evidence for multiple independent analyzers tuned to particular stimulus dimensions as revealed by experiments that measure threshold visual response.

Spatial and Temporal Frequency Analyzers Revealed by Subthreshold Summation Experiments

The most compelling evidence for multiple analyzers in human vision comes from experiments with near-threshold stimuli, that are of such low intensity or contrast that they are barely detectable by a human observer. There are a number of experimental paradigms.

For example, in subthreshold summation experiments, the detection threshold of a stimulus is examined in the presence of another stimulus that is too weak to be itself detected. If the presence of the subthreshold stimulus does not affect the detection of the second stimulus, then it is concluded that the stimuli activate different analyzers. Since the stimuli typically differ only in their values along a single dimension, independent analyzers are demonstrated for that dimension. Sachs et al. (1971) presented pairs of vertical sine wave gratings, superimposed and differing only in spatial frequency, to investigate whether there were multiple channels for spatial frequency. With one vertical grating kept just below its threshold, they found that the threshold of the second grating was unaffected if it differed in spatial frequency by more than one octave.

Graham and Nachmias (1971) compared thresholds for complex grating patterns. These compound gratings consist of a single spatial frequency and its third harmonic. The relative position of the two gratings could be varied so that their luminance peaks and troughs were coincident, a condition they called "peaks add," or they were aligned peak to trough, a condition they called "peaks subtract." This manipulation did not affect the threshold visibility of the compound, which was visible when one of the component simple gratings reached its independent threshold. A profile of the stimuli and data are shown in Figure 9.4. A perfect corner would indicate perfect independence. The slightly rounded corner they obtained indicates almost complete independence. Spatial frequency analyzers have been investigated perhaps more than analyzers on any other simple extensive dimension.

The absence of subthreshold summation has been shown for pairs of stimuli differing along a number of significant visual dimensions. Sine wave stimuli that differ only in spatial frequency appear to be detected independently when they differ in spatial frequency by more than an octave (Graham and Nachmias, 1971; Sachs et al., 1971). Similarly, line or grating stimuli that differ by more than 10° in orientation on the retina show independence in subthreshold summation experiments (Kulikowski et al., 1973). Not generally investigated with the same interest but generally known for some time has been the independent detection for two stimuli that differ in spatial position on the retina (Olzak and Thomas, 1986).

Spatial Analyzers Revealed by Identification Experiments

Another paradigm involves identification. These experiments require the subject to identify which of two near-threshold stimuli is presented on any trial. The two stimuli differ only in their value on the dimension being studied. If discrimination of the stimuli is possible at threshold, the analyzers underlying detection of

►

FIGURE 9.4. Subthreshold summation experiment of Graham and Nachmias (1971). Copyright (1971), Pergamon Press plc. Reprinted with permission. (a,b) Luminance profiles of gratings with spatial frequency f and $3f$. (c) Luminance profile of "peaks subtract" compound. (d) Luminance profile of "peaks add" compound. Data figure on right. Axes represent "contrast of f" $\equiv c_f$, (x axis) and "contrast of $3f$" $\equiv c_{3f}$, (y axis) components at threshold in compound divided by the component's threshold when presented alone (c_{f0}, c_{3f0}). Note that data points, which represent thresholds for compounds with various contrast ratios of the components, tend to cluster around solid lines that represent x or y values of 1.0. Thus the compound, whatever the contrast ratio, tends to reach threshold when one of the component gratings reaches threshold. The data points are represented by letters because different frequency gratings were used on different trials. (See Graham and Nachmias, 1971 for details.)

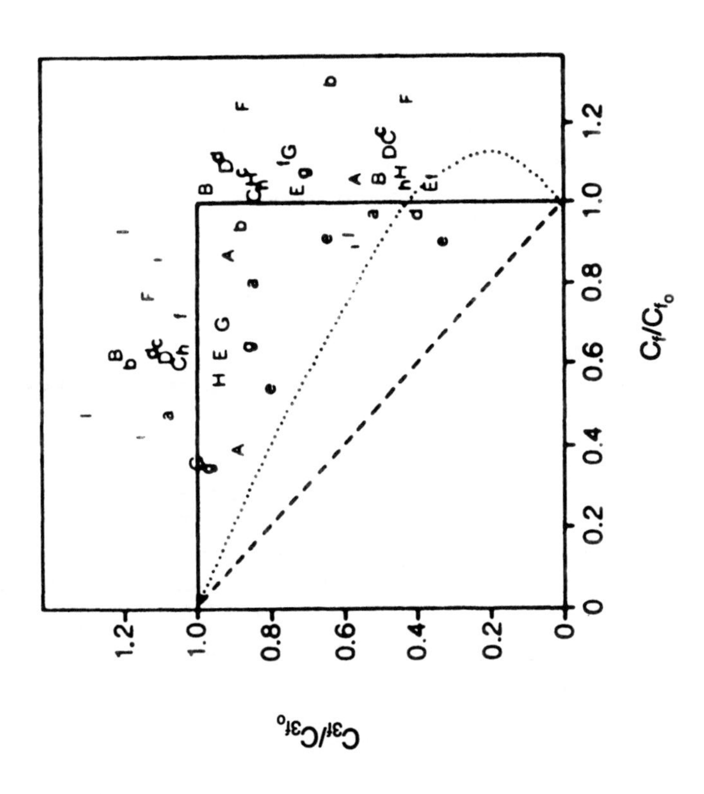
C_{3f}/C_{3f_o}
C_f/C_{f_o}
0
0.2
0.4
0.6
0.8
1.0
1.2

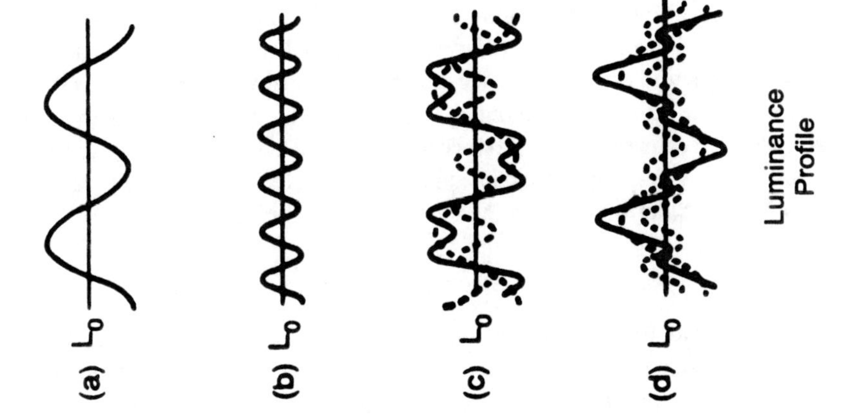
(a) L_o
(b) L_o
(c) L_o
(d) L_o
Luminance
Profile

each stimulus are assumed to be different. They also must be labeled, that is, carry information about the properties of the stimuli to which they respond (Graham, 1989).

Some discrimination of spatial frequency, orientation, and spatial position is present at detection threshold. Identification experiments generally yield results that are consistent with summation data. One discrepancy between summation and identification experiments occurs when the stimuli used vary in spatial frequency and are flickering at high temporal frequencies. For these stimuli the apparent spatial frequency bandwidth estimates in identification are much broader than summation (Graham, 1989).

Analyzers Revealed by Uncertainty Experiments

Uncertainty experiments compare thresholds in blocks of trials where the identity of the stimulus to be detected is known beforehand, with blocks of trials where the observer knows only that the stimulus to be presented on each trial is one of several stimuli, identical except for their values on the dimension of interest. Uncertainty about spatial frequency or spatial position of the stimulus leads to decrements in performance in such tasks (Davis et al., 1983). These decrements are attributed to the requirement that the observer monitor the output of more than one analyzer under conditions (near-threshold) in which noise in the analyzer output affects performance. Such effects imply multiple analyzers. In the spatial frequency domain, these effects are found with spatial frequencies one octave apart (Davis et al., 1983).

Temporal Frequency Analyzers

Dimensions associated with stimulus movement produce a somewhat different set of results from the spatial analyzers previously discussed. The analyzers that underlie detection in subthreshold summation experiments generally do not show narrow tuning when temporal frequency is the variable of interest (Watson, 1986). When low to middle spatial frequency stimuli are used, temporal frequency tuning width estimates are narrower for identification experiments than subthreshold summation. These tuning widths are still quite large. Stimuli must be at least three octaves apart to produce near perfect discrimination (Watson and Robson, 1981). It is apparent that the analyzers that detect stimulus temporal frequency have broader tuning than the spatial analyzers discussed.

Velocity Analyzers Sensitive to Direction of Motion

When subthreshold summation of two gratings moving in opposite directions is examined, evidence of independence when stimulus velocity is greater than 1 degree per second is obtained. When stimulus velocity is below 1 degree per second, the two gratings appear to be detected by the same mechanism (Graham, 1989).

Binocular Disparity Analyzers

When threshold stimuli fall onto correspondent locations on the two retinas they show interactions consistent with the operation of binocular disparity analyzers at threshold. These analyzers appear to show selectivity for both spatial frequency and orientation of the stimulus as well (Arditi, 1986).

Threshold Interactions of Analyzers Revealed by Adaptation

A methodology commonly used to explore the properties of analyzers is adaptation (see also Frumkes, Chapter 7). The observer is exposed to a stimulus well above threshold for a reasonably prolonged period (>30 sec). This appears to make analyzers that respond to that stimulus less sensitive for some time after. A second stimulus, the test stimulus, is then presented and its threshold measured. If the threshold of the second stimulus is unaffected following adaptation, it is assumed that it is not detected by the adapted analyzer and the hypothesis of separate analyzers on the dimension where the two stimuli differ is supported.

Adaptation experiments, however, can reveal interactions between the separate analyzers found in subthreshold summation and identification experiments. Consistent with these methodologies spatial frequency thresholds generally are raised when the adaptation stimulus is within one octave of the test stimulus. Between one and two octaves, however, sensitivity to a grating is actually increased (De Valois, 1977). Orientation-specific adaptation also shows facilitation effects between stimuli that are independent in subthreshold summation experiments (Olzak and Thomas, 1986).

An experiment showing interaction between analyzers for both orientation and spatial frequency is that of Greenlee and Magnussen (1988). They presented two alternating gratings during the adaptation period, as a means of producing simultaneous adaptation to both patterns. These gratings differed either in orientation or in spatial frequency. Threshold was then tested for one of the gratings and compared to an adaptation control condition where that grating was presented alternating with a neutral field, to isolate the effects on adaptation of the second stimulus. When the two adapting spatial frequencies or orientations were similar, then thresholds were raised relative to the control condition implying both stimuli were adapting the same analyzer. However, when the disparity in orientation or spatial frequency was greater, then threshold was lowered relative to the control condition, implying facilitation of detection by the presence of the second adaptation grating. Facilitation occurred when the two gratings differed by more than 15° in orientation or one octave in spatial frequency. Facilitation was greatest when adapting gratings differed in spatial frequency by one and a half octaves. In another experiment where orientation varied they found lowest thresholds with a 45° difference in orientation of the two gratings.

Broader bandwidth interactions are to be expected given the suprathreshold contrast of the adapting stimulus. Above threshold, strict independence is the exception rather than the rule. Analyzers, silent at threshold, may begin to respond when stimuli are above threshold and channel interaction processes may be triggered that are not evident with threshold stimuli. One explanation for the facilitation effects noted in the adaptation paradigms is that analyzers tuned to similar values tonically inhibit each other. When the responsiveness of an analyzer is reduced through adaptation, the inhibition it produces is reduced, producing the observed facilitation in other analyzers (Greenlee and Magnussen, 1988).

Adaptation of Direction and Temporal Frequency

When grating stimuli that flicker or drift are used as adaptation and test stimuli, adaptation selective for temporal frequency is found (Watson, 1986). Green (1981) using flickering, spatially uniform fields found some selectivity for different flicker rates when measuring the threshold following adaptation for a 0.5 cycle per degree sine wave that could drift at different rates. The tuning, however, was very broad, with full bandwidth at half amplitude tuning exceeding 4 octaves.

Directionally selective sensitivity changes have been demonstrated. If a vertical grating pattern moving from left to right is watched for some time, the observer shows raised thresholds for patterns moving in that direction. However, patterns moving right to left also show raised thresholds, only less so (Pantle and Sekuler, 1968). This is true even for pattern velocities at which independence of directional analyzers is found in subthreshold summation experiments.

Adaptation of Global Stimulus Properties

Adaptation and threshold measurement techniques have been used to probe for the existence of analyzers that are sensitive to more global properties of images. In the domain of motion, Regan and Beverly report the existence of analyzers for image transformations such as changing size along a single axis, changing size along all axes in the image plane (looming), and global rotation of a stimulus around its own axis (Regan, 1986). In their research they use outline squares that may loom and recede, or rotate. Exposure to these complex patterns should produce adaptation within local unidirectional motion analyzers. They control for this adaptation by comparing patterns where the local components of motion are globally organized into one pattern (a looming square, which oscillates back and forth in depth) with those where the local components are identical but their relative temporal phase is different, forming a different global pattern (a square that translates back and forth along an oblique axis in the image plane). In the first pattern, the pattern appears to loom and recede because the sides that face each other move apart and then together again synchronously. In the second pattern, individual sides exhibit identical patterns of motion, but now the sides that face each other move back and forth synchronously always moving in the same direction. They find that only the first pattern is an effective adaptation stimulus, if one measures displacement needed to detect the looming and receding of an oscillatory stimulus (Regan and Beverly, 1978). Since both adapting stimuli have the same local motion components, the first stimulus must adapt an additional global analyzer, sensitive to looming and receding stimuli. They have produced similar results for global rotation (Regan, 1986). Although there are a number of methodological differences between these studies and the other threshold after adaptation studies we have discussed (displacement thresholds rather than contrast thresholds, identification rather than detection), nonetheless they are provocative and provide some evidence for analyzers tuned to more global relational features of the image. These analyzers are nonetheless retinally localized in that threshold changes are limited to the area of the retina in which the adapting stimulus was placed (Regan and Beverly, 1978; Regan, 1986).

Comparable Findings with Neurons in Visual Cortex

Neurophysiological investigations appear generally consistent with a psychophysical model of visual function that involves an analysis of the image into local properties defined on a set of stimulus dimensions. Striate cortex or V1 is known to have large numbers of cells selective for the orientation of local bar or edge components (see Dow, Chapter 4). This orientation selectivity in the form of orientation columns appears to be a strong organizational principle for this cortical area. In rhesus monkey, these orientation-selective units vary in their degree of specificity. Simple cells show fairly narrow tuning, most having full bandwidths of 30–40° at half amplitude (a 50% reduction in response 15–20° to either side of the optimal orientation) when stimulated with a suprathreshold moving bar stimulus. The tuning of complex cells appears to be broader (Schiller et al., 1976). It is sometimes difficult to compare quantitative estimates of analyzer tuning width psychophysically derived from threshold studies, with physiological estimates of cell tuning width, typically derived from response profiles to suprathreshold stimuli. The mean tuning width of simple cells does nonetheless imply multiple analyzers.

It has been demonstrated that thresholds are raised and responsiveness reduced in orientation-specific cells that have been exposed to their optimal stimuli for a prolonged period of time and that the time course of adaptation and recovery is similar in studies of neurons of the striate cortex of the cat and human psychophysical experiments (Albrecht et al., 1984).

The most extensive study of both spatial and temporal sensitivity in rhesus monkey visual cortex is that of Foster et al. (1985). They examined the sensitivity of cells in both V1 and V2 to moving sine wave gratings of differing spatial frequencies. The full bandwidths for spatial frequency tuning of individual cells averaged 1.4 octaves within foveal V1 and 1.6 octaves within V2. The range of spatial frequency bandwidths was 0.8–3.0 octaves. V2 neurons were generally tuned to lower spatial frequencies with peak sensitivity of neurons generally 2 octaves lower than V1. Most neurons in V1 had low pass temporal frequency tuning and a mean full bandwidth of 2.9 octaves. V2 neurons generally had bandpass characteristics and full bandwidth characteristics of 2.1 octaves on average. The cells in V1 with narrow spatial frequency and broad temporal frequency tuning seem roughly comparable with the psychophysics, though 1.4 octaves spatial frequency bandwidth is somewhat broader than most psychophysical estimates and

2.9 octaves temporal frequency bandwidth somewhat narrower. The temporal frequency bandwidths of V2 cells are clearly narrower than psychophysical estimates. It is worth noting that V1 can be roughly characterized by higher spatial frequency tuning and low pass temporal frequency tuning, while V2 is characterized by low spatial frequency tuning and bandpass temporal frequency tuning. This structural feature of the cortex invites further investigation in the search for neural analogues of sustained and transient analyzers.

There have been a number of studies of adaptation of neurons exposed to unidirectional motion. Vautin and Berkley (1977) found that the time course of reduced responsiveness following prolonged exposure to optimal moving stimuli of cat V1 neurons was comparable to the time course of human adaptation. There are cells in the visual associative cortex of the rhesus monkey that are responsive specifically to global rotation or motion in depth (looming) of a stimulus. These cells have been described in area MST, an area that receives input from MT (Saito et al., 1986). MT is an area in which a majority of cells have a preferred direction of motion.

A Current Picture of Analyzers on the Spatial and Temporal Frequency Dimensions

The existence of independent analyzers at threshold has been demonstrated for the major pattern dimensions that concern us here. Given this a number of questions arise. How many analyzers are there on each dimension? Are there distinct analyzers on each dimension or are individual analyzers narrowly tuned on multiple dimensions? The neurophysiology reveals that cells that show tuning on a dimension such as orientation or spatial frequency will show a wide variety of peak sensitivities virtually continuously covering the dimension. The psychophysical evidence, however, while specific about the bandwidth of the analyzers and thus the minimum number of analyzers needed to encode the dimension does not set a maximum limit on the number of analyzers, if one assumes overlapping tuning. It is thus consistent with a large number of appropriately tuned analyzers (see Graham, 1989 for discussion).

In another relevant comparison, most cells will show tuning on more than one dimension. For example, orientation-selective cells are often spatial and temporal frequency selective as well. Subthreshold summation studies of analyzers at threshold keep all "irrelevant" dimensions constant. When the two stimuli differ by more than the tuning width on the studied dimension, the subthreshold stimulus generally has no effect on threshold, despite the fact that it shares identical values with the threshold stimulus on the irrelevant dimensions. This is at least consistent with multiple selectivities in the psychophysically defined analyzers (Olzak and Thomas, 1986; Graham, 1989).

Near-threshold psychophysics has revealed the existence of sets of narrowly tuned, independent analyzers. These provide a set of basic building blocks from which perceptual representations can be formed. In the following sections, we will consider psychophysical experiments using suprathreshold stimuli. These reveal extensive interactions between the analyzers revealed at threshold in their normal suprathreshold operation.

Perceptual Functions

We assume that integrated groups of analyzers identified at threshold underlie our perception of objects and events. To learn just how they do so would seem a complex task. The simplest possible model would treat such analyzers as labeled detectors. When the unit is responding, then a particular stimulus feature specified by the unit's label would be perceived. Models of that sort have been proposed. A serious shortcoming of the simple detector model is that it is physiologically unlikely. (There are other reasons to question the validity of such models. See Leibovic, 1969, 1971; Leibovic et al., 1971; Weisstein, 1973.) Visual system neurons generally show sensitivity in their superthreshold responding to multiple stimulus dimensions. As a result the response of each individual cell is ambiguous as to the presence of a particular stimulus feature. For example, a cell that shows maximal sensitivity at threshold to a high spatial frequency, vertically orientated stimulus may show an identical response to a low contrast, high spatial frequency, vertically orientated stimulus and a high contrast, high spatial frequency stimulus, orientated 10° off the vertical axis, and a high contrast, medium spatial frequency, vertically orientated stimulus. The individual cell's response is ambiguous with regard to the actual stimulus profile across dimensions (Weisstein, 1973; Teller, 1984).

Alternative models of the relationship of analyzer response to pattern appearance involve considering sets of analyzers as components in networks. These networks may be serial or parallel, distributed or nondistributed, competitive or cooperative. Object qualities in this view are related to network responses. These network responses will be the result of inhibitory and facilitative interactions between the sets of analyzers identified near threshold. There is a body of psychophysical work that supports the existence of interactions between analyzers.

At a more complex level of analysis we wish to know what the physiological substrate is for our recognition of familiar objects and events. We might posit a model in which individual neurons, very selective in their responding, represent complex percepts. The increase in selectivity implies that a very large number of such units are needed to represent the full range of possible percepts. In the extreme, the model is reduced to an absurdity, with sets of grandmother cells to represent grandmother from all possible viewpoints, and equivalently large sets for all other possible persons and objects in the environment. Given the unlikelihood of a one-to-one mapping between complexly labeled cells and unique percepts, it is likely that percepts have a physiological realization in multiple responses from arrays of cells. We might think of unique percepts as corresponding to unique states in networks of interacting neurons labeled for object features. What the specific organization of such a network would be, and the hierarchy of labels would be, constitutes a knotty challenge for those who would understand the relationship of brain states to perceptual states. Answers at the present time are largely speculative.

A number of simpler perceptual phenomena can be understood by reference to mechanisms that involve sets of dimensionally coded analyzers. These include aftereffects and simultaneous contrast effects, masking and metacontrast phenomena, texture segregation and search phenomena, persistence phenomena and deadline procedures. Investigation of these and related paradigms provides information about the function of local analyzers in global perception, as well as revealing the mechanisms by which that function is realized. Some of these investigations are discussed below.

Simple Suprathreshold Effects

Visual Aftereffects

We previously discussed how adaptation leads to specific increases in threshold, consistent with the existence of analyzers whose sensitivity is decreased following prolonged stimulation. Such adaptation also has dramatic effects on the appearance of subsequently viewed suprathreshold patterns. For example, if a pattern of bars or texture moving in one direction is observed for a relatively short time (30 sec), subsequently viewed stationary patterns will appear to be moving in a direction opposite to that of the adapting motion (the motion aftereffect). In another example, adaptation to stationary grating patterns leads to alteration in the apparent spatial frequency of subsequently viewed grating patterns (Blakemore and Sutton, 1969). Similarly, the tilt aftereffect is a well-known effect of adapting to orientation. One can also produce apparent shifts in depth from prolonged exposure to binocular disparity cues (Blakemore and Julesz, 1971).

There is reason to believe that these alterations in pattern appearance arise from changes at the level of dimensional analyzers. The motion aftereffect provides an illustrative case. The aftereffect is limited to the retinal area stimulated (Wohlgemuth, 1911; Sekuker and Pantle, 1967). Perceived direction of illusory motion is also mapped in retinal coordinates, and changes its three space coordinates with tilt of the head. The aftereffect has both monocular and binocular components, and so is not confined to retinal mechanisms (Wohlgemuth, 1911). Adaptation is generally complete after 60 sec exposure to the moving stimulus. The time course of recovery from adaptation is affected by a number of factors, but is typically around 20–30 sec. As noted below, such characteristics are consistent with observed changes in sensitivity of directionally selective cortical neurons.

A major feature of the data on changes in suprathreshold appearance following adaptation is that interactions above threshold are found for stimuli that appear to stimulate independent channels at threshold. Adaptation to a grating moving at an intermediate velocity provides a case in point. Subsequently viewed stationary patterns produce a motion aftereffect. Additionally, subsequently viewed moving patterns, moving in both the same and opposite directions, will alter perceived velocity. Generally they are slowed, unless they are slow moving and moving in the opposite direction, in which case they appear speeded up (Thompson, 1981; Smith, 1985). These interactions of patterns across the full spectrum of velocities in both directions contrast with threshold results, which have generally demonstrated independence of channels for opposite directions of motion when the patterns are moving above 1 degree per second (Graham, 1989).

Adaptation and recovery times described above are comparable to the physiologically recorded time course of adaptation and recovery from adaptation to unidirectional motion observed with neurons in V1 (Vautin and Berkley, 1977). The effects of such adaptation on neurons can be complex, however. Neurons tuned to opposite directions of motion can be affected by exposure to stimuli that are poor in eliciting a response. Some neurons show enhanced responding and others show reduced responding. In this way exposure to a moving grating will affect the postexposure sensitivity of a large number of cells (Marlin et al., 1988). Thus different classes of neurons may be affected by threshold and suprathreshold investigations and this in turn could contribute to differences found between threshold and suprathreshold methods.

Related to the simple aftereffects discussed above, but having somewhat different characteristics, are what are known as contingent aftereffects. In the contingent aftereffect procedure, the observer adapts simultaneously to two stimuli by viewing them in a repeating sequence of brief alternate presentations. The stimuli are typically gratings that differ along two dimensions. The McCollough effect, the most studied contingent aftereffect, provides an illustrative case. An observer is simultaneously adapted to red and black vertical square wave gratings, and green and black horizontal square wave gratings by alternating brief presentations of each.

The test stimulus in a contingent aftereffect procedure will be identical to one of the adapting stimuli except along one of the dimensions where the two adapting stimuli differ. The appearance of the test stimulus is altered in a manner appropriate to exposure to the adapting stimulus that it most resembles. In the case of the McCollough effect, following adaptation, the observer will see a green aftereffect when the test stimulus is a black and white vertical square wave grating, and a red aftereffect on a black and white horizontal square wave grating (McCollough, 1965).

The McCollough effect has certain characteristics that resemble characteristics of simple aftereffects. The aftereffect is limited to the retinal area that was adapted. The orientation specificity is measured in retinal coordinates. Turning the head sideways while viewing a test grating following adaptation causes the apparent color of the grating to change consistent with the retinal orientation of the grating (McCollough, 1965; Strohmeyer, 1978). Other characteristics seem unique to contingent aftereffects. The aftereffect has a very slow decay constant in the absence of test stimuli that elicit the aftereffect. It can be elicited days later if the adaptation period is long enough, generally greater than 15 min (Strohmeyer, 1978). Although orientation-contingent color aftereffects are easily the most studied, other types of contingent aftereffects have been demonstrated. These include movement aftereffects contingent on either color or spatial frequency, and color aftereffects contingent on direction of motion, or spatial frequency to name just a few (see Treisman, 1986 for a review).

There is less unanimity about the mechanisms responsible for contingent aftereffects than there is for simple aftereffects. Some ascribe the effect to selective adaptation of neurons doubly selective for the relevant dimensions (McCollough, 1965). A neuron is doubly selective if it shows narrow tuning on two stimulus dimensions. But such explanations must account for the long decay constants. This appears to make an explanation based solely on adaptation of tuned neurons unlikely. Others have adapted an associative learning, or classical conditioning model for the effect (Murch, 1976). Such models, however, do not really explain the time course of decay of the McCollough effect since conditioned responses generally show longer decay times than the McCollough effect (Skowbo, 1984). Neural units with the proper double selectivities are known to exist in macaque striate cortex (Michael, 1978a,b). However, it is currently unclear whether they are carrying doubly labeled information. Moreover, recent research indicates that there is a rather strong independence of color dimensions and extensive dimensions such as motion and form in early visual processing. This is discussed in Equiluminant Contours.

Simultaneous Contrast

In simultaneous contrast, the dimensional value of a surrounding context stimulus affects the appearance of an embedded stimulus having a different value on the same dimension. Simultaneous tilt provides an example, and is illustrated in Figure 9.5. The two small central grating patches are both vertically oriented, yet each is slightly shifted in apparent orientation due to the presence of the oblique context gratings. Similar contrast effects have been demonstrated with direction, velocity of motion (Tynan and Sekuler, 1975a; Walker and Powell, 1974), and size or spatial frequency (Blakemore and Sutton, 1969). Simultaneous tilt occurs over a range of orientation disparities between the context figure and the figure that appears apparently tilted. Tuning is similar to the tilt aftereffect. The effect is maximal with an orientation disparity of 10–20°. For large orientation disparities between 60 and 90°, assimilation, a shift in apparent orientation toward the inducing orientation, rather than contrast has been reported (Gibson and Radner, 1937).

A possible basis for these contrast effects of size, orientation, direction, and velocity is the surrounds found beyond the classical receptive field of cells in a number of visual cortical areas including V1 (Allman et al., 1985). The orientation-selective surrounds reported by Fries et al. (1977) in cat striate cortex provide a case in point. These are regions that surround the classical receptive field that are unresponsive to an optimally oriented stimulus when it is restricted to the surround region. When a stimulus of optimal orientation stimulates the receptive field, presentation of a pattern of similar orientation in the apparently unresponsive surround region now reduces the response to the optimal stimulus,

FIGURE 9.5. Simultaneous tilt. The circular grating patches actually are in the same orientation. This is easily demonstrated by laying a straight edge parallel to the pair.

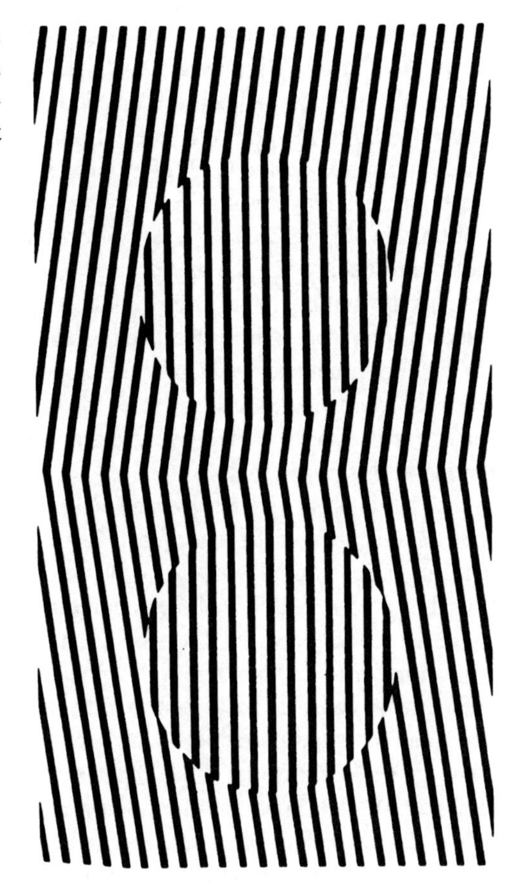

revealing the presence of lateral orientation-selective inhibition. Stimulating the surround affects the pattern of response across orientation channels in a manner quite similar to orientation-selective adaptation with an optimal stimulus within the receptive field. This provides a physiological basis for the similar percepts arising from simultaneous tilt, and orientation-selective adaptation.

Ambiguous Figures

Direction of motion of gratings within an aperture is ambiguous, in that the proximal motion stimulus can arise from distal motion with a variety of directions and velocities. This ambiguity arises because the grating pattern has only contours in one orientation, and uniform texture. Within an aperture its outer boundaries are occluded. Whatever direction and velocity the pattern is moved in, the image within the aperture is indistinguishable from motion perpendicular to the grating orientation at some slower velocity. A large grating tilted 45° from the vertical provides an illustration. Imagine the grating moving horizontally behind an aperture and the same grating moving vertically behind the aperture. The two cases cannot be discriminated while viewing the grating within the aperture. The grating will always appear to be moving in a direction perpendicular to the angle of tilt. A cell's receptive field also constitutes an aperture. When direction of motion of a moving grating is discussed in that context it is always assumed to be perpendicular to the grating's orientation.

Adelson and Movshon (1982) investigated the appearance of pairs of superimposed gratings moving within an aperture, differing in speed and direction of motion. For all possible speeds and directions, the pattern within the aperture is indistinguishable from the appearance of the checklike pattern (seen when the gratings are stationary) moving rigidly in a particular direction at a particular velocity. Invariably, this is what the observer sees. This percept (a global motion percept, not simply derived from the component grating motions) is most salient when the component gratings are of the same or similar frequencies, implying some size or spatial frequency tuning in the mechanism that computes global motion in this case.

Such patterns have been presented while observing the responses of visual neurons. In the cortex of the macaque, cells in V1 respond only if they are sensitive to the direction of motion of one of the gratings. In MT, on the other hand, directionally selective cells have been found that respond to the checklike plaid moving in the globally derived direction just as they do to a single grating orientated perpendicular to and moving in the globally defined direction (Movshon et al., 1986). If we consider these results with those considered earlier for motion in depth and for rotation, we see that information from local motion channels may converge in MT allowing the computation of more global patterns of motion. However, there is no global rotation detection in MT. This implies that convergence may take place in stages, and that MST with cells sensitive to motion in depth, rotation in the frontal plane, and rotation in depth may represent a stage beyond MT (Maunsell and Newsome, 1987).

Illusory Contours

Simultaneous contrast and metacontrast phenomena (see Metacontrast) provide support for models of feature interaction that posit a group of specific inhibitory interactions that take place between analyzers tuned to slightly different spatial positions. If one assumes the functional analyzers correspond to receptive fields of identified cells in visual cortex, then these inhibitory interactions may correspond to response modulation, outside the traditional receptive field, such as previously described.

Not all local interactions between analyzers are inhibitory however. There are a number of phenomena that would appear to indicate that spatially adjacent analyzer interactions can be cooperative rather than inhibitory. This appears to be the case for stimulus configurations that produce illusory contours. Illusory contours are appearances of edges in physical areas of uniform intensity and texture. The illusion is produced by the contours that immediately surround the homogeneous region. Some examples are provided in Figure 9.6. (See Klymenko et al., 1987 for a moving illusory contour produced by higher order spatiotem-

FIGURE 9.6. Two displays that produce illusory contours. The top figure is a Kanisza triangle. Covering various parts of the figure shows that contours of the triangle lying above, while perceptually present in the uncovered figure, are physically absent. The bottom figure is a display used to produce phantom contours. A vertical grating is occluded by a horizontal strip. When the vertical bars are moved continuously from left to right, they are clearly visible on the surface of the stationary occluding horizontal surface.

poral context.) An interesting property of these illusions is that they produce many basic perceptual effects that normally are produced by real contours. Illusory contours are effective masks in masking paradigms (Weisstein et al., 1974). When phantom contours, moving illusory contours, which look like illusory square or sine wave gratings in motion (see Fig. 9.6) are viewed for some time, a retinolocal motion aftereffect is seen with subsequent stationary contours in the area corresponding to the retinal area where the contours were seen (Weisstein et al. 1977; Maguire and Brown, 1987). Tilt aftereffects after exposure to illusory contours have also been reported (Smith and Over, 1975).

Illusory contours not only give the appearance of real brightness discontinuities in the image, but they produce effects in corresponding retinal locations that one associates with real brightness discontinuities. An intriguing hypothesis is that illusory contour configurations produce activation of local analyzers like an actual contour would. The inducing contours, the physically present contours that are necessary for the illusion to be seen may be producing activation of the relevant channel by cooperative interactions that act across limited regions of the visual field (Weissten and Maguire, 1978). There is evidence that illusory contours are produced by lateral interactions between analyzers with similar properties. Phantom contours will not be seen when inducing figures exceed 5° of visual angle, and the maximum separation appears to get smaller as the spatial frequency of the inducing contours increases (Genter and Weisstein, 1981; Tynan and Sekuler, 1975b; Gyoba, 1983; Maguire and Brown, 1987). Likewise, illusory contours are reduced in vividness with increased physical separation of inducing contours (Shapley and Gordon, 1985).

There has been recent interest in the visual properties of amodal contours. Amodal contours are contours of objects partially blocked from view that the visual system infers but does not modally represent (Kanisza and Gerbino, 1982). These have been shown to have a number of interesting properties. Such contours can give rise to orientation specific adaptation (Weisstein, 1970). They also have been recently shown to affect perceptual grouping, depth perception, and the perception of apparent motion (Nakayama, Shimojo, and Silverman, 1989; Simojo, Silverman, and Nakayama, 1988; Shimojo and Nakayama, 1990). Such results imply that amodal contours may share a number of representative functions with illusory contours, including representation at a relatively early stage of visual processing.

Some recent models of the cortical neural network have explicitly postulated mechanisms by which illusory contours would be produced by activation of dimensional analyzers in a cooperative network (Grossberg and Mingolla, 1985b). There is currently evidence for the existence of such mechanisms in V2 of the rhesus monkey. von der Heydt et al. (1984) found that they were able to activate cells in V2 (but not V1) by presenting patterns that produced illusory edges in retinal locations that corresponded to the cell's receptive field. These responses were not simply explained by activation of the receptive field by the inducing contours, as all the inducing contours fell outside the traditional receptive field. Furthermore these responses showed the same specificity to the orientation of the nonexistent edge that the cell's responses to standard bar stimuli showed. As Allman et al. (1985) have pointed out, this is one of the few demonstrations of activation rather than inhibition of a localized receptive field by specific stimulation outside the receptive field.

Illusory contours have other characteristics that are particularly relevant given their possible neural basis. Both depth, motion, and figure/ground ordering information are important determinants of the vividness of illusory contours seen with a particular stimulus configuration. Phantom contours are reduced when cues such as interposition, kinetic occlusion, and binocular disparity indicate that the inducing region is behind the region where the phantoms are seen (Weisstein et al., 1982a; Brown and Weisstein, 1986; Maguire and Brown, 1987). Similarly, the appearance of illusory contours is enhanced by congruent disparity cues and suppressed by contradictory cues (Lawson et al., 1974). Phantom contours are also influenced by variables that influence the apparent figure/ground organization of the display. Phantom contour completion is more likely for nonflickering regions when these are adjacent to flickering regions (Brown and Weisstein, 1985; Maguire and Brown, 1987). In general when flickering and nonflickering regions are juxtaposed, the flickering regions are seen as ground and recede in depth (Wong and Weisstein, 1984). Similar effects of flicker have been demonstrated with a modified Kanisza triangle (Meyer and Dougherty, 1987).

Movement can play a similar role in enhancing or disrupting the appearance of illusory contours. Phantom contours can be seen when the inducing display is stationary (Gyoba, 1983), but generally their appearance is more vivid when the display is in motion (Tynan and Sekuler, 1975b; Maguire and Brown, 1987) or is flickering (Genter and Weisstein, 1981).

All these characteristics suggest interactions among dimensional analyzers. Combined with information from other paradigms, they help reveal the characteristics of independent channels or pathways carrying different types of dimensional information within the visual system.

Pathways

Equiluminant Contours

As was previously noted, color and brightness information seem to be extracted relatively independently from information about motion, depth, and form. At the level of V1 in the macaque, nonoriented cells are more likely to be color coded than orientation-specific cells (Dow and Gouras, 1973; Dow, Chapter 4). They also respond differentially over a greater range of stimulus luminances and contrasts (Maguire and Baizer, 1982).

Some of the most compelling evidence for separate analysis of extensive and intensive dimensions has been gathered recently, using psychophysical techniques. The stimuli are colored displays where regions may differ in hue, but they are all of equivalent brightness. These displays form equiluminant edges at the boundaries between different colored regions, and the appearance of these edges and various figures composed of these edges has drawn a good deal of interest. A number of distortions of perception have been reported with these stimuli. Moving stimuli appear to be moving much more slowly than normal (Cavanaugh et al., 1984). Phi or apparent movement appears to break down (Ramachandran and Gregory, 1978). Depth perception is impaired with equiluminant random dot stereograms (Lu and Fender, 1972; Livingstone and Hubel, 1987). Equiluminant depictions of three-dimensional objects have a flat appearance. Stimulus configurations that produce illusory contours under conditions of luminance contrast fail to do so with equiluminant edges (Livingstone and Hubel, 1987). This evidence is generally consistent with the existence of a pathway that relays information about stimulus depth and motion but is insensitive to chromatic signals. These chromatic signals are presumably analyzed by another pathway that codes color but does not register detailed information about motion and depth. There is also some evidence for a third pathway carrying detailed form information (Livingstone and Hubel, 1987). Unlike the dimensional analyzers previously discussed, the color and motion/depth pathways display considerable independence even with suprathreshold stimuli.

It has been suggested that the separation of color and motion/depth processing has its neural basis in the existence of separate neural pathways running through the LGN and visual cortex for color and motion/depth information. Livingstone and Hubel (1987) associate the motion/depth pathway with the magnocellular layers of the LGN that project to V1, layers 4C and 4B. From V1 projections have been traced to the thick stripes of V2 and MT. The color pathway is associated with the cytochrome oxidase staining "blob" areas in layers 2 and 3 of V1, which in turn project to the thin stripes of V2, which in turn projects to V4. The pathway concerned with detailed form information has the following neural basis. Cells in the parvocellular layers of the LGN project to V1 layer 4C and in turn to the interblob regions of layers 2 and 3, which in turn project to the pale stripes of V2.

These different pathways have been associated with different classes of cells, whose receptive field organization appears to indicate specialization for different types of stimulus information. An important question is whether these pathways can be identified with the sustained and transient pathways psychophysically investigated. A reasonable hypothesis is that the magnocellular pathway is the transient pathway, and the parvocellular interblob pathway is the sustained pathway. There is a general tendency for cells in the magnocellular layer of the LGN to be responsive to lower spatial and higher temporal frequencies than parvocellular units (Derrington and Lennie, 1984). A recent study provides some further support for this hypothesis. Merigan (1989) found that macaques treated with acrylamide, which severely damages the parvocellular pathways while sparing for the most part the magnocellular pathway, showed reduced contrast sensitivity for the full range of stationary spatial frequencies, while having normal contrast sensitivity for a 0.7°, 10-Hz flickering grating. The animals showed reduced chromatic sensitivity throughout the spatial and temporal frequency range. Further research at cortical levels of these pathways should give a clearer picture of the extent these pathways can be labeled sustained or transient.

Texture Grouping and Segregation Effects

Many theories of pattern perception require that a segmentation of the visual image must take place before pattern recognition takes place. Segmentation is the parsing of local image elements into meaningful units, such as line segments and borders. These segregated units are then assigned to different surfaces, objects etc. Texture segregation studies investigate the conditions under which segmentation occurs. The results generally show that the characteristics of the image considered in terms of simple dimensions (precisely those that undergo independent analysis near threshold) are critical in predicting the segmentation of the image.

A number of investigators have put forward theories relating output of early neural channels to some aspects of the segmentation problem (Beck, 1982; Grossberg and Mingolla, 1985a; Julesz, 1975; Graham et al., 1989; Treisman and Gormican, 1988). Texture segregation is one paradigm that has been used extensively. In these experiments two classes of texture elements are intermixed or placed adjacent to each other (see Fig. 9.7). For some pairs texture segregation succeeds. That is they are perceived as two distinct groupings. For others segregation fails. Texture segregation is easily demonstrable for groups of elements that differ in brightness, direction of movement, binocular disparity (random dot stereograms), velocity of movement, size, line orientation, line length (Beck, 1982), as well as number of line terminators and line crossings (Julesz, 1987). Texture segregation always appears to succeed when the discrimination of texture

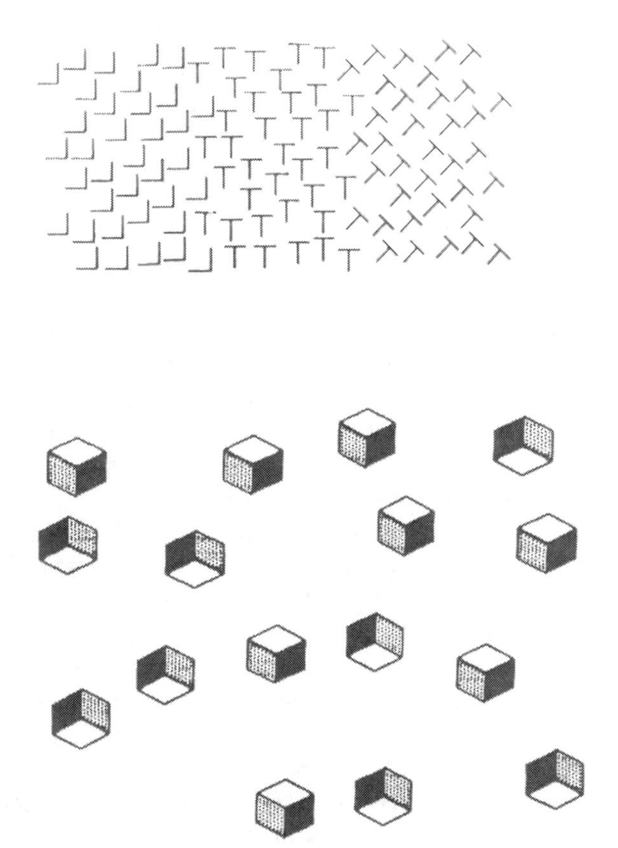

FIGURE 9.7. The figure above consists of three regions of texture elements. The tilted T's on the right segregate easily from the central upright T's. The left side is more difficult to segregate because the texture elements share the basic features of the upright T's, vertical and horizontal lines. Below, shaded cubes in different orientations can support texture segregation, although they do not differ in basic features. From Beck J (1966). Effect of orientation and of shape similarity on perceptual grouping. *Perception and Psychophysics* **1**:300–302. Reprinted by permission of Psychonomic Society, Inc.

can take place along a single stimulus dimension. Thus vertical and horizontal line segments are easily seen as two distinct groups because of their orientation difference. The formation of discriminable textures appears logically to be the result of two processes. Different analyzers arrayed along a single dimension are activated by the textures. Then a global grouping process operates by which the outputs of similarly tuned analyzers at different spatial locations would form a unified grouping (see also Grossberg and Mingolla, 1985a).

Treisman expanded these ideas in the following way. She hypothesizes that grouping is performed in neural structures each dedicated to processing a single stimulus dimension (V1 orientation, MT motion, V4 color). Presumably there are other structures not yet discovered, or not yet adequately characterized, dedicated to other stimulus dimensions. This idea is part of a larger theory of perception that claims certain perceptual operations take place effortlessly without need of attention. This implies that these operations are automatic, unconscious, and are performed on all parts of the image. These include segmentation on the basis of differences on a single stimulus dimension, and location of a target with a unique dimensional value in a field of other elements. The theory then posits multiple independent representations of the visual image, each restricted to information about dimensional values within the array. When a texture is defined by a unique conjunction of dimensions, or a search target by such a conjunction (such as a black vertical line in a field of black oblique and red vertical lines) it will not have a unique representation within any of these feature maps, and by her theory texture segregation will fail and visual search will be effortful. The theory posits that putting together the separate representations of specific dimensions requires attention. This in turn implies that objects identified by combinations of features can be located only with attentive search.

This is an ambitious theory that brings together data from a number of different paradigms. It posits that texture segregation and visual search paradigms can be used as converging operations to discover the full range of elementary dimensional representations and describe them.

There are data, however, that are not entirely compatible with the theory. Graham et al. (1989) have shown that texture segregation is not simply predicted by the differences in output in a set of physiologically plausible elementary analyzers (oriented masks with one octave spatial frequency bandwidths). They propose a nonlinear transformation of the output, and a second stage of spatial filtering to explain their texture segregation results. Also some features that show easy texture discriminability, like variations in line terminators and line crossings (Julesz, 1987), have characteristics not generally thought of in terms of the initial pattern analysis of the visual stimulus. They have not been extensively studied in either psychophysical or physiological paradigms outside the texture segmentation literature, however.

In visual search tasks, a number of elements are presented simultaneously. One or more elements have been designated as "targets." The observer's task is to determine as fast and accurately as possible whether a target is present in the array. Other nontarget elements are called distractors. If the speed or accuracy of an observer's performance when a target is present is not affected by the number of distractors that are present, then it is considered to be effortless. The target is said to pop out. With repeated practice, on a small set of target letters, which never function as distractors in the search task (another set of letters is used as distractors), the targets come to be located without effort (Shiffrin and Schneider, 1977). A related search phenomenon is the effortless location of numbers among letter distractors (Brand, 1971). Despite the effortless search process, these target/distractor pairs do not produce discriminable textures in a texture segregation task (Maguire and Sitkowkski, 1984). This would appear to imply that search and texture segregation processes do not involve identical mechanisms.

Recently some phenomena of texture segregation and visual search have been discovered that further cloud the issue. A texture that consists of egglike shapes defined by shading cues so that some of the elements of the display look like eggs and some look like eggcups will show texture segregation of the eggs from the cups (Ramachandron, 1988). These figures are in fact ambiguous, the shading cues indicating a concave or convex surface only after a clear interpretation of the direction of the putative light source has been made. Manipulations that give independent information about the direction of the light source, such as the presence of a clearly illuminated hand, affect the perception of shape that results from the ambiguous cues (Berbaum et al., 1983, 1984). Altering the shading cues so that a sharp black/white edge is present would appear to be a relatively minor alteration of a gradual to an abrupt edge at the feature level. Under these altered conditions no depth is present in the display and no texture segregation takes place (Ramachandran, 1988). (A related display where shaded cubes differing in orientation are segregated is illustrated in Fig. 9.7.) Does this imply that shapes from shading cues act as elementary features in the visual system? One might grant the possibility, however unlikely, but it is difficult to see how such features could be elementary, depending as they do critically on information at a perceptual level about the direction of the light source.

Some analogous results have been found with the visual search paradigm. Higher order features such as orientation in three-dimensional space, and differences in the apparent direction of illumination for target and distractor items (see cubes in Fig. 9.7) produce effortless search results (Enns, 1988). Other recent studies report effortless search for targets composed of feature conjunctions when the feature differences among the distractors promote separation of the display into different depth planes, producing figure/ground differences between the elements (Kaufman, 1989). Once again one can speculate that perhaps information about the spatial layout and three dimensionality of objects functions as elementary features, and perhaps there are brain structures specialized to extract this information relatively early in the perceptual process. This seems unlikely though given the difficulty in specifying these features, in the absence of relatively complete processing of the visual array. The principles governing texture segmentation and visual search may be more complex than they originally appeared. Higher order perceptual processes may enter into these paradigms in ways not yet well specified.

Figure-Ground Effects

We have seen from the discussion of equiluminant contours and texture segregation that the set of elementary dimensional analyzers may be responsible for perceptions more global and complex than might be expected from their role as analyzers over a certain small range of values on various pattern dimensions. For example, color pathways carry little information about three dimensionality from monocular cues. On the other hand, achromatic pathways register three dimensionality quite well.

Another example is figure/ground perception. The boundary contour that results from the apposition of two adjacent surfaces is generally determined by the intrinsic shape of one of the surfaces, and not the other. This is because such boundary contours are generally due to occlusion of one surface by the other. Assignment of the boundary contour to the appropriate surface is an important function of visual processing. The assignment leads to perception of foreground and background surfaces, a segmentation into two depth planes. Such perceptions also arise from the apposition of two discriminable textures.

The gestalt psychologists were interested in this process, which they called the figure/ground problem. They proposed certain principles to determine under what conditions a region would be perceived as figure or ground. For example, a surrounded surface is preferentially seen as a figure, relative to the surface that surrounds it. Symmetrical or convex surfaces are likewise seen as a figure relative to asymmetrical or concave ones. A number of other factors also contributed to figural preference (see Pomerantz and Kubovy, 1986 for a review).

Such figure/ground segregation is seen also with intermixed textures. Longer lines form stronger groups and are perceived as foreground in a mixed field of long and short lines (Beck, 1982). Triangles form stronger groupings in a field of complete triangles and triangle fragments. In shape from shading grouping, the eggs appear as foreground in a background of eggcups (Ramachandran, 1988).

We have recently found that high-spatial, low temporal frequency channels are involved in figure perception, while low-spatial, high temporal frequency channels are involved in ground perception. In particular, we have found that figure and background regions generated by ambiguous figures (Fig. 9.8) exhibit differential sensitivities to basic stimulus parameters. Gestalt psychologists noted that the figure tends to be more structured (Koffka, 1935) and to have a more thing-like character (Rubin, 1921/1958). The region that is more articulated and detailed tends to be seen as the figure. Greater detail is more or less correlated with the presence of higher spatial frequencies in the stimulus (see Ginsburg, 1978; Graham, 1980; Weisstein and Harris, 1980, for reviews).

We reasoned that if detail demands higher spatial frequencies for analysis, then perhaps figure perception also involves those higher spatial frequencies. Likewise, if ground appears more global and less articulated then perhaps ground perception involves lower spatial frequencies (Julesz, 1978; Wong and Weisstein 1982, 1983; Weisstein and Wong, 1986, 1987). These links between spatial frequency and figure perception indeed were found to exist.

Wong and Weisstein (1982, 1983) examined orientation discrimination of a thin target line projected on the figure shown in Figure 9.8. This is the well known faces/goblet figure first devised by Rubin. (Their figure was the more common outline faces/goblet. The regions defining faces and goblet were uniform and an outline defined the border between the regions.) The target line was placed on the goblet surface. Subjects viewed the display indicating continuously whether the faces or goblet percept was dominant. Orientation discrimination was better when the target line was presented during the goblet percept than when it was presented during the faces percept. Increased orientation sensitivity is associated with

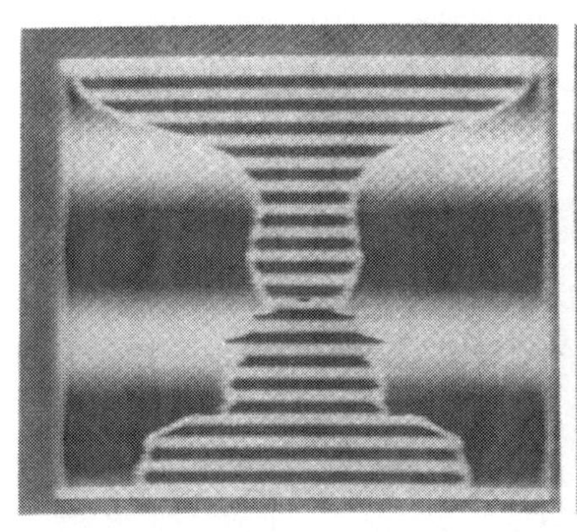
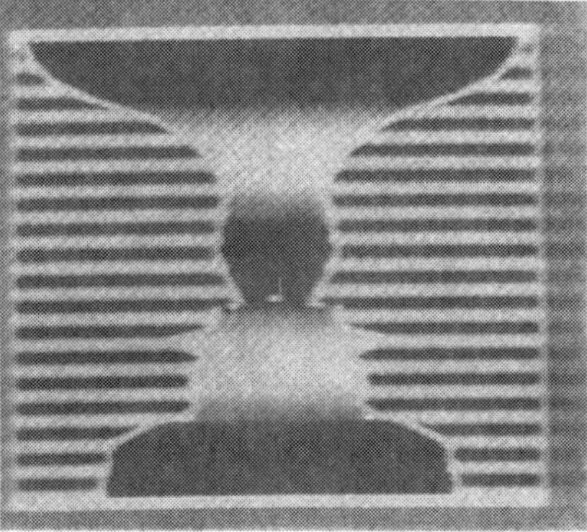

FIGURE 9.8. Modifications of Rubin's faces/goblet figure. The figure is inherently ambiguous; either a pair of faces or a goblet may be seen. The percept tends to alternate between the two. When the regions are filled with sine waves as above, the region that contains the higher frequency sine waves tends to be seen as the figure. This can be seen by comparing the two panels.

analyzers tuned to higher spatial frequencies and lower temporal frequencies (Graham, 1989). In another experiment they found that the spatial region perceived as figure produced better detection of the target lines when they were sharp with many high spatial frequency components, but the ground percept led to better detection of blurred target lines, lacking these high spatial frequencies.

It is interesting to note that when sharp or blurred lines are presented in a discrimination task, figure enhances the discrimination accuracy of sharp but not blurred lines (as expected) but ground does not enhance the ability to discriminate blurred lines even though line detection is enhanced (Weisstein and Wong, 1986, 1987). As we shall discuss in the sections on Object Superiority and Metacontrast, this points to another possible characteristic of high spatial frequency channels not shared by low ones, namely the ability of certain global perceptions to enhance their functioning.

Klymenko and Weisstein (1986) directly assessed the influence of spatial frequency on figure/ground organization by replacing the uniform surfaces of ambiguous patterns with sine wave gratings differing in spatial frequency. An example of two of the patterns from Experiment 1 in Klymenko and Weisstein (1986) is shown in Figure 9.9. Observers viewed all combinations of spatial frequency differences for the spatial frequencies of 0.5 to 8 cycles per degree. The response measure was the percentage of the 30 sec of stimulus presentation time that observers saw the goblet as figure. Observers saw the goblet as figure a higher percentage of the time when the stimulus was one like the pattern on the left of Figure 9.8 where the goblet contained the relatively higher spatial frequency grating. They replicated this result for several patterns and found that, in general, the greater the separation in spatial frequency between the two sine wave gratings in a pattern, the greater the percentage of viewing time the lower

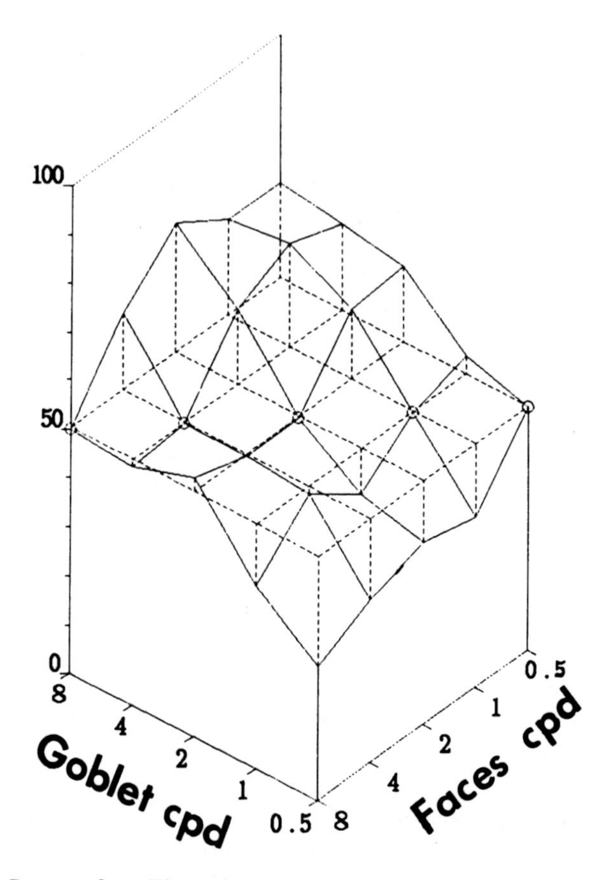

FIGURE 9.9. Data are from Klymenko and Weisstein (1986). The vertical axis represents the percentage of time that the goblet was seen as the figure for the various combinations of goblet region and face region spatial frequency. In general, the region containing the higher spatial frequency is seen as figure and the bias increases with the magnitude of the differences in spatial frequency between the two regions. cpd, cycles per degree.

spatial frequency region will be seen as the ground (also see Wong and Weisstein 1985). Figure 9.8 shows the percentage response times that the goblet was seen as figure for all combinations of spatial frequency differences. When high and low spatial frequency patterns meet at a boundary contour, it appears that there is a perceptual bias to assign figure to the high spatial frequency region and assign the low frequency region to the ground.

Figure/ground perception is related to the perception of depth. Brown and Weisstein (1988) found that a column of rows of alternating high and low spatial frequencies viewed stereoscopically produce vivid impressions of depth that disappear only after adjusting the gratings so that a difference in binocular disparity between them cancels the depth induced by the spatial frequency disparities in the regions. Wong and Weisstein (1990a) further showed that perceived depth appears to mediate figure/ground perception. Regions to one or the other side of the boundaries in ambiguous figures are filled with high or low spatial frequencies and stereoscopic depth is introduced to counter the perceived depth that is generated. This manipulation causes the impression of figure vs ground to disappear. Interestingly, when spatial frequency is held constant and stereoscopic depth alone is used in the adjacent regions, figure and ground perceptions again appear. These perceptions resemble the effects of spatial frequency, in that they are bistable, with binocular disparity biasing the bistable perception.

So far, we have discussed figure and ground perception only in terms of spatial frequencies. Temporal frequencies might also be expected to influence figure and ground perception with low temporal frequencies favoring a figure analysis and high temporal frequencies favoring a ground analysis. Julesz (1978) suggested that the speed of response to backgrounds and to the information contained therein may be faster than the speed of response to information contained in figures. He labeled the visual process that analyzes ground an "early warning system" and proposed that its function is to detect the presence of things without identification of details (see also Breitmeyer, 1984). Speed of response can be separated into latency and rise time. Latency of course is logically separate from temporal frequency, but faster rise times as well as faster latencies may be involved in ground processing. We should note that this "early warning" conjecture converges with the ideas about spatial frequency discussed above, since a high temporal frequency response has been associated with a low spatial frequency response and vice versa. This coupling would result in a ground analysis that produces not only a rough (low spatial frequency) initial scan, but a rapid one as well.

Using a metacontrast paradigm (see Metacontrast) Weisstein and Wong found evidence that the visual response to ground is twice as fast as that to figure (Weisstein and Wong, 1986, 1987).

They also investigated the direct effects of temporal frequency on the perception of figure and ground. Wong and Weisstein (1984, 1985, 1987) divided a rectangular array of random dots into flickering and nonflickering regions. These differences in temporal modulation were sufficient for the field to segregate into regions. They found that the regions with flickering dots are generally seen as a flickering background and the stationary regions as foreground. The bias toward a perception of the stationary region as foreground figure was greatest at flicker rates around 6–8 Hz. These rates of flicker are within the range where the visual system is maximally sensitive when measured by threshold contrast of flickering gratings (Keesey, 1972; Kelly, 1961; King-Smith and Kulikowski, 1975). Recently Weisstein and Wong (1990) found similar effects using sine wave gratings flickering in bounded regions of an ambiguous figure.

As with the early investigations of visual responses to spatial and temporal frequency, these initial investigations of the visual system's responses to figure and ground varied only spatial or only temporal frequency while holding the other dimension constant. More recently, more complex variations of spatial and temporal frequency have been investigated (Klymenko et al., 1989; Klymenko and Weisstein, 1989a,b). Klymenko et al. (1989) systematically examined the combined influence of spatial and temporal frequency on figure/ground organization. An ambiguous pattern, a circle sectored into eight parts of which alternate sectors formed two Maltest crosses was used. The crosses were modulated in both space and time. As in the Klymenko and Weisstein (1986) study, the two regions of the patterns (the two crosses) were filled with sine wave gratings of 0.5, 1, 4, or 8 cpd. One or both regions underwent on–off flicker at the rates of 0, 3.75, 7.5, or 15 Hz. In the first experiment all combinations of temporal frequency differences were tested for each spatial frequency. When a region contained the relatively higher temporal frequency grating, it was seen as the ground more often than when it contained the relatively lower temporal frequency grating. This temporal frequency effect was more pronounced in the patterns containing higher spatial frequency gratings indicating that the figure and ground system is more sensitive to temporal frequency differences for higher spatial frequencies.

In their second experiment, Klymenko et al. (1989) tested patterns in which the two regions differed in spatial frequency rather than temporal frequency. Both regions flickered at the same rate. As found by Klymenko and Weisstein (1986) with stationary patterns, the region with the lower spatial frequency tended to be seen as the ground; however, this figure and ground separation was attenuated for the flickering patterns for some temporal frequencies.

The combined results of these two experiments suggest that if one of the regions of the ambiguous pattern has both a lower spatial frequency and a higher temporal frequency, then both factors should bias this region to be perceived as the background. Observation indicated that this is the case.

In their final experiment, Klymenko et al. (1989) tested the perceptual organization of patterns when spatial and temporal frequency cues to figure and ground organization were placed in opposition. A tradeoff was evident when the temporal and spatial frequencies were set in opposition. Regions with relatively low spatial frequency and low temporal frequency were seen as ground as often as were regions with relatively high spatial frequency and high temporal frequency.

The depth effects that accompany figure and ground perception may have a functional basis in the following. It has been suggested for images with limited depth of field (see Frisby and Mayhew, 1978) that high and low spatial frequencies, respectively, simulate sharp and blurred focus supporting an interpretation of surfaces lying in different depth planes. (A consequence of blurring an image is attenuation of high spatial frequencies in the image.) Pentland (1985) geometrically demonstrated that the gradient of blur is a valid indicator of relative distance in depth. Correspondingly, he showed that increasing the blur of a partially occluded object increases the perceived depth separation between the occluded and occluding objects. However, under natural conditions surfaces in front of as well as behind the point of fixation will be out of focus (see Pentland, 1985).

There are countless real-world situations consistent with the pattern of temporal frequency results. Consider an observer who visually pursues a small, laterally moving object in a textured environment (see Hochberg, 1971). More background than foreground texture elements will sweep over the retina, producing a higher rate of transients for the background than the foreground. In addition, transients are produced by the occlusion and disocclusion of background texture by the moving object, so called kinetic occlusion (see Gibson et al., 1969; see also Ramachandran and Anstis, 1986).

It has been argued (Calis and Leeuwenberg, 1981; Breitmeyer and Ganz, 1976; Julesz, 1978) that the ground system, which is faster, acts as an early warning system signaling global information and motion and thus regions of potential interest, whereas the figure system is specialized for finer spatial resolution and detailed analysis of the region of interest itself (see discussions in Breitmeyer, 1980, 1984; Breitmeyer and Ganz, 1976; Julesz, 1978; Weisstein and Wong, 1986, 1987). For example, moving low spatial frequency stimuli are more likely to produce involuntary optokinetic eye movements than high spatial frequency stimuli (see discussion in Wertheim, 1990). The systematic effect of spatial and temporal frequency on figure/ground organization suggests a critical role for the interaction of sustained and transient pathways in the segmentation process. Thus the building blocks previously discussed are implicated not only in the registration of simple object properties, but in such higher order perceptual computations as the assignment of contours and regions to figure and ground.

Complex Spatiotemporal Interactions

With the large body of evidence indicating a stage of visual information processing where local, interacting analyzers process information about specific stimulus qualities in parallel and perhaps in separate feature specific pathways, an important question is how this information is integrated globally into perceptual representations of object and events. Current understanding of how isolated pattern information might be used to construct and recognize the myriad forms that humans and animals confront in their natural environment is still rudimentary.

Texture segregation studies have been generally used to further support the idea of independent "modular" analyzers of patterns. But analyzers interact in space and in time. Most theories of pattern recognition assume different stages of processing, with the basic building blocks of pattern recognition, analyzers, first registering information. Successive stages integrate this information to produce representations of objects and events.

Masking

A way to study the temporal development of object representations is to study the effect of one stimulus, relatively briefly flashed, on the visibility of another. Visual masking refers to the reduction of the visibility of one stimulus called the target by a spatiotemporally overlapping or contiguous second stimulus called the mask. Unlike subthreshold summation experiments in which the subthreshold stimulus adds to target detectability, most masks are above threshold and like adaptation stimuli reduce target detectability and discriminability. The mask can be presented before, after, or simultaneously with the presentation of the target. Simultaneous or prior presentations of the mask are called forward masking: the mask comes first. Subsequent presentations of the mask are called backward masking: the mask comes second. The terms target and mask are arbitrary and depend on which stimulus the observer is responding to. The term target designates the stimulus whose visibility is being measured.

There is a huge body of literature on masking too extensive to be reviewed here. A prototypical masking study is Crawford's (1947) designed to study the time course of light adaptation. The target's threshold is measured at various times before, during, and after the presentation of a full field of light. Unlike steady-state conditions where the increment necessary to just detect a difference in brightness is a constant fraction of the brightness of the field (see Weber fraction, Fromkes, Chapter 7), a Crawford function has an appearance as shown in Figure 9.10. Target threshold begins to rise when it is presented 100 msec before the mask, and it reaches a peak at simultaneous onset of target and mask. It then gradually declines to a steady-state value when the mask has been on a while. Just before the mask is turned off target presentation leads to a threshold rise again. The threshold peaks for target onset immediately after mask offset, and declines as the offset–onset interval lengthens. Crawford functions illustrate key characteristics of masking of light by light. Backward masking (the front of the function in Fig. 9.10) is generally greater than forward masking (see also Frumkes, Chapter 7).

Masking is used here to study the response in frozen moments to a prolonged stimulus. That is, the target, localized in time, is used as a probe to reveal the changing response to the prolonged full field mask. The results show that temporal change (the beginning and end of the mask presentation) is signaled by increased visual system activity. The result is raised target thresholds at these locations in time. Why is target appearance affected when it is flashed 100 msec before the mask onset? Certain components of the response to the mask travel faster in the brain than do certain components of the response to the target (see Metacontrast) thereby "catching-up" with the target response and raising the threshold for detection. The reason why the target is affected before the mask is turned off is comparable. The gradual rather than immediate decline of threshold after the light is turned off indicates that response to the light persists in the visual system (but see Frumkes, Chapter 7).

Conceptually we can categorize masking studies into three types: masking of light by light, masking of pattern by light, and masking of pattern by pattern. Although masking of light by light provides a useful introduction to some of the salient characteristics of masking, masking of pattern by pattern is of more concern to us here. Masking of pattern by pattern can provide a great deal of information about the spatial and temporal properties of visual perception, and hence about the function and circuitry of channels selectively sensitive to these properties.

Object Superiority

Forward or simultaneous masking occurs when the mask is presented before or simultaneously with the target. The mask generally produces effects on threshold similar to an adaptation stimulus. A sine wave grating's visibility is maximally

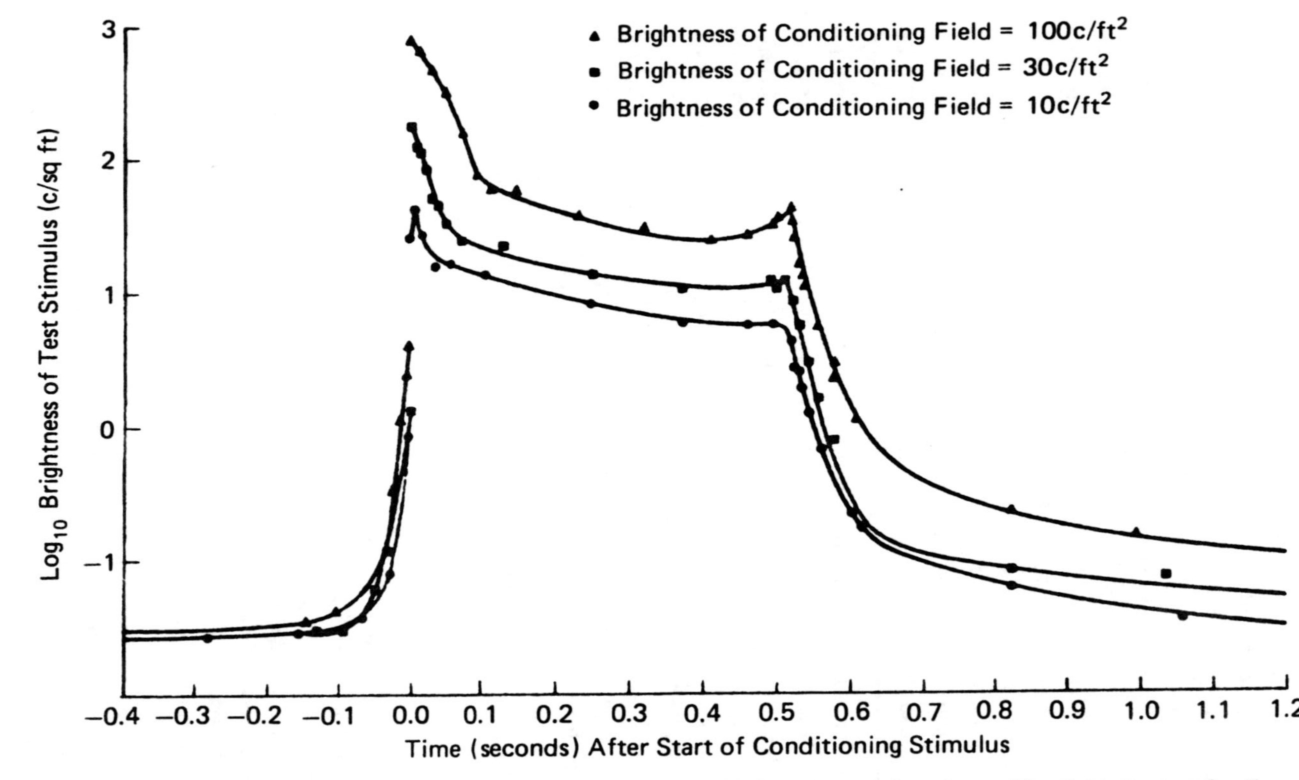

FIGURE 9.10. Threshold for detecting a target light before, during, and after presentation of a masking light. See text for discussion. (From Crawford, 1947.)

degraded when the mask has a similar orientation (Campbell and Kulikowski, 1966) and spatial frequency (Tolhurst and Barfield, 1978) to the target. The orientation and spatial frequency range over which masks can affect targets is similar to that found with adaptation studies.

An interesting exception to the general rule that masks impede target visibility are studies that have examined the discriminability of a thin line that is flashed simultaneously with a context that consists of arrangements of connected line segments. Some of these contexts are outlines of apparently three-dimensional objects (see Fig. 9.11). Such three-dimensional contexts enhance visibility in comparison to other nonobject contexts matched for number and orientation of component lines (Fig. 9.11). This phenomenon is known as the object-superiority effect (Weisstein and Harris, 1974). When visibility is enhanced relative to the target lines presented alone, this is known as the object-line effect (Williams and Weisstein, 1978).

Object-superiority and related effects indicate a greater complexity in analyzer, channel, and pathway functions than we have discussed up to this point. Normal perception involves integration of information to form the coherent representations of objects and events that are present at the level of awareness. Object contexts tap into these integration processes in a number of ways according to our present understanding. These contexts can form local configurations with the target lines that enhance discriminability. A clear example is the formation of either "arrow" or "triangle" forms when a right angle context is added to one or another oblique line that are reflections of each other across the vertical meridian. Despite the fact that the right angle is added to both targets to be discriminated, and hence does not contribute to discrimination in and of itself, the resulting forms are more easily discriminated than the oblique lines alone (Enns and Printzmetal, 1984; Lanze et al., 1985). The effects of these configurations can be seen by comparing accuracy for parts C and D on the right of Figure 9.11.

▶

FIGURE 9.11. The types of stimuli used in object-superiority experiments. On the left, the stimuli used by Weisstein and Harris (1974). (Copyright 1974 by the AAAS.) At the far left, the contexts are presented alone. When one of the four target contexts was added, one of the four figures to the right of the solid line were seen. The subject was to decide which target line was presented. The numbers to the right of the targets show relative performance in identifying target lines in each context. The numbers represent the difference in percentage correct between the context in that row and the most objectlike context shown at the top (a). All values are negative, showing that the object context led to most accurate target identification. On the right are figures from another experiment (From Lanze et al., 1985. Reprinted by permission of the Psychonomic Society.) In this case object depth was manipulated independently from the local arrow/triangle configuration, which helps distinguish the two alternatives in context. Relative ratings of apparent depth are in column 3, and accuracy in the two alternative forced choice task in column 4. The data indicate that both the formation of salient local configurations like arrow/triangle and the global configuration of depth contribute to increased accuracy in the object-superiority effect.

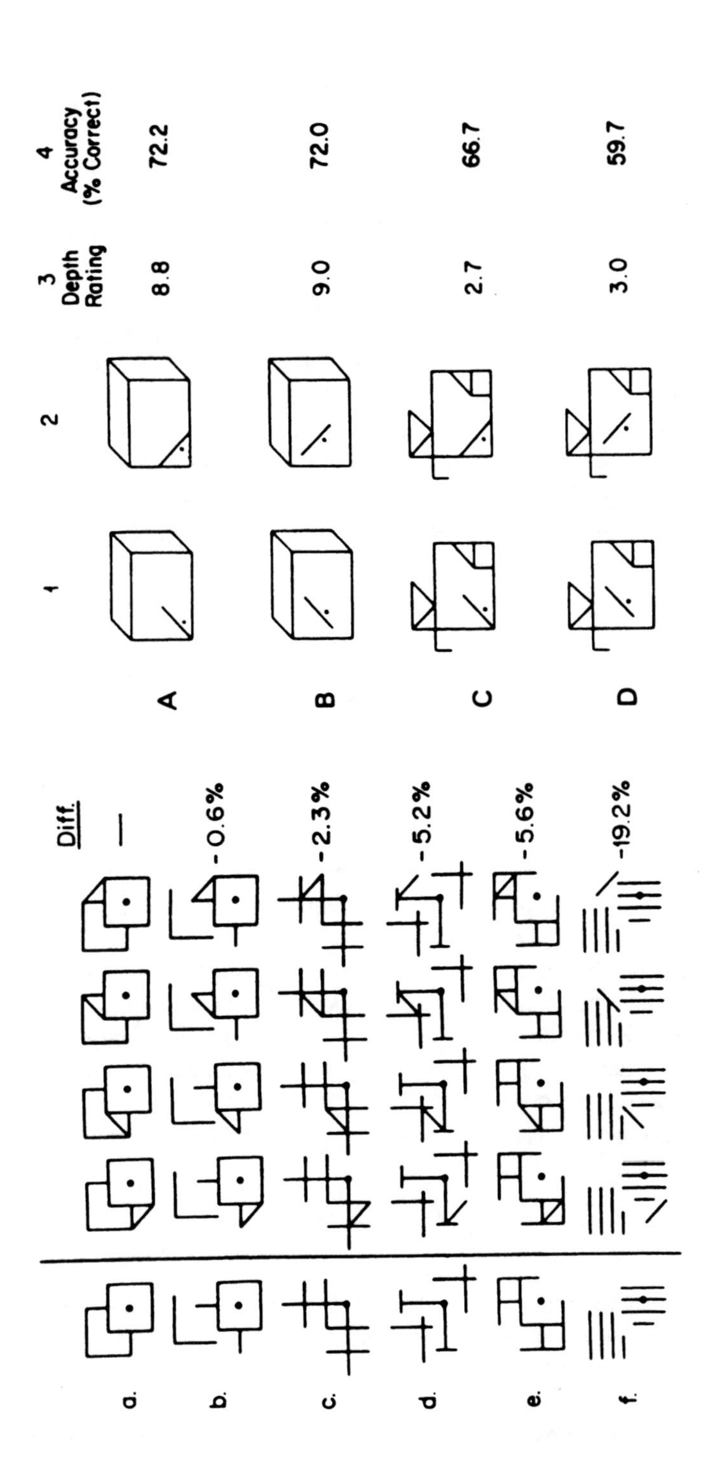
a.
b.
c.
d.
e.
f.
Diff.
—
-0.6%
-2.3%
-5.2%
-5.6%
-19.2%
A
B
C
D
1
2
3
Depth
Rating
8.8
9.0
2.7
3.0
4
Accuracy
(% Correct)
72.2
72.0
66.7
59.7

The target lines in the bottom left corner of C form arrow or triangle configurations with the sides of the square; those in D do not. Object superiority involves more than the heightened discriminability of local configurations, however. There is enhanced detectability with three-dimensional object contexts, even when local configuration does not differ for three-dimensional and two-dimensional contexts (Lanze et al., 1985; Enns and Gilani, 1988). Three dimensionality appears to be the most important global characteristic. It has been shown to be more related to discrimination performance than such generally correlated figural qualities as relevance of the target to the overall structure (Lanze et al., 1982) and the overall connectedness of the figure (Weisstein et al., 1982b).

The facilitation of target discrimination by the presence of global object like masks fits in nicely with a number of theories of pattern perception. These assume that in response to the representation of a visual pattern, a low spatial frequency sensitive global system extracting overall shape and three dimensionality has temporal priority over a high spatial frequency system, computing local detail (Calis and Leeuwenberg, 1981; Julesz, 1978; Alwitt, 1981; Breitmeyer and Ganz, 1976; Kinchla and Wolfe, 1979). In some way, the global information, extracted first, affects later processing of target information.

The exact mechanism by which a pathway extracting global pattern information influences responding in a pathway concerned with pattern detail is unknown. Some possibilities are that for object superiority the specific combinations of local monocular cues to depth may selectively open a gate by which higher order pattern information feeds back information to the level of the basic analyzers, or alternatively filters such information before it reaches a level at which decision processes operate. A third possibility is that line segments arranged in perspective images may trigger cooperation among channels such that the signal associated with each line segment in the array is augmented producing the observed facilitation (see discussion Weisstein and Maguire, 1978; Williams and Weisstein, 1984).

Physiologically the existence of reciprocal connections between higher and lower order visual representations in the cortex is now well established (Maunsell and Van Essen, 1983). The circuitry certainly exists for feedback loops by which more complex visual patterns could influence the basic dimensional encoding to produce effects like the object-superiority effect, but compelling evidence for picking feedback over filtering or cooperative networks does not appear to exist at this time.

Metacontrast

Studies of metacontrast masking also reveal the role of sustained and transient pathways in pattern perception and in the perception of apparent depth. A typical metacontrast study consists of a line segment presented asynchronously with two flanking mask lines. Masking is generally weak or nonexistent at simultaneous presentation and increases steadily to a maximal value when mask onset follows target onset by 50–150 msec. How is it that maximal interaction in the visual system occurs at a nonzero stimulus onset asynchrony? The explanation is that although the mask signal comes later, it travels faster, catching up with the target signal somewhere in the processing sequence. Figure 9.12 (Williams and Weisstein, 1984) illustrates this logic.

Masking here appears to be due to inhibitory interactions between channels that have different response latencies. Specifically a channel (or channels) with a short response latency, a fast channel, appears to inhibit the output of a channel (or channels) with a longer response latency (Weisstein, 1968, 1972; Weisstein et al., 1975; Breitmeyer and Ganz, 1976; see Breitmeyer, 1984 for a review). Masking interactions between channels with different temporal characteristics again suggest possible interactions between the sustained and transient pathways, which differ precisely in rise times and latencies of channels.

Converging evidence for a model of sustained and transient inhibitory interactions comes from a variety of metacontrast results. Because transient channels respond more quickly than sustained channels, a mask stimulus that follows the target stimulus will produce maximal overlap of sustained and transient response in the visual system. Under these conditions inhibition is most pronounced and masking maximal (Fig. 9.12 shows the way this would work). This inhibition is spatially localized, with metacontrast decreasing with target mask separation, and being largely absent for target mask separations greater than 2° outside the fovea and being even more localized within the fovea (Growney et al., 1977; Breitmeyer, 1984).

A number of results are consistent with the characterization of the fast masking process as transient, and the slower channels underlying target appearance as sustained. Masking is little affected by blurring the mask, suggesting that it is low spatial frequency components that are responsible for the masking effect (Growney, 1976). When the target consists of a central sine wave grating of variable spatial frequency and the mask consists of two flanking gratings also of variable spatial frequency, masking is produced at shorter stimulus onset asynchronies and is weaker when high spatial frequency masks are used, consistent with weaker longer latency responses to higher spatial frequencies in the mask channel (Rogowitz, 1983).

Metacontrast functions extend our knowledge of mechanisms involved in object superiority. When the object contexts illustrated in Figure 9.11 are used as masks in metacontrast experiments they produce some interesting results. In these experiments the target line is presented asynchronously with the surrounding context. The zero delay condition corresponds to a standard object superiority condition. When context is delayed relative to the target line, line identification suffers and typical U-shaped metacontrast functions result. An interesting feature of these metacontrast functions is that the delay that produces maximal masking is longer for the object-like context than it is for the flat contexts, as shown in Figure 9.13 (Williams and Weisstein, 1977, 1984). Recall that longer delays in metacontrast are associated either with faster inhibitory responses from the mask or slower responses to the target. This in turn implies that the three-dimensional object context either reduces the latency or increases

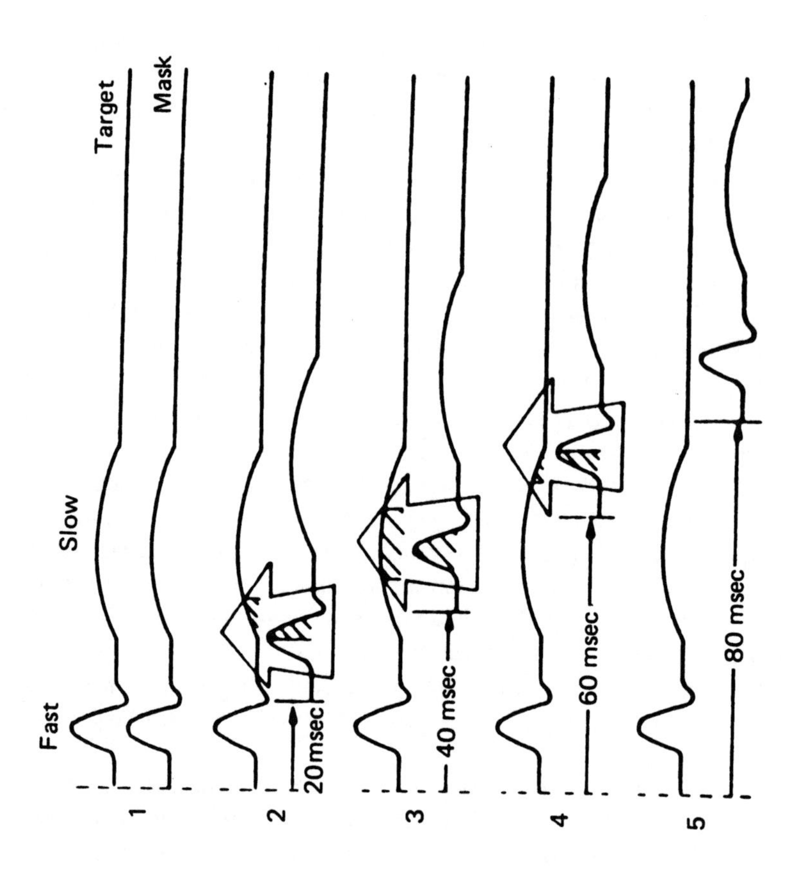
Target
Mask
Fast
Slow
1
2
3
4
5
20msec
40 msec
60 msec
80 msec

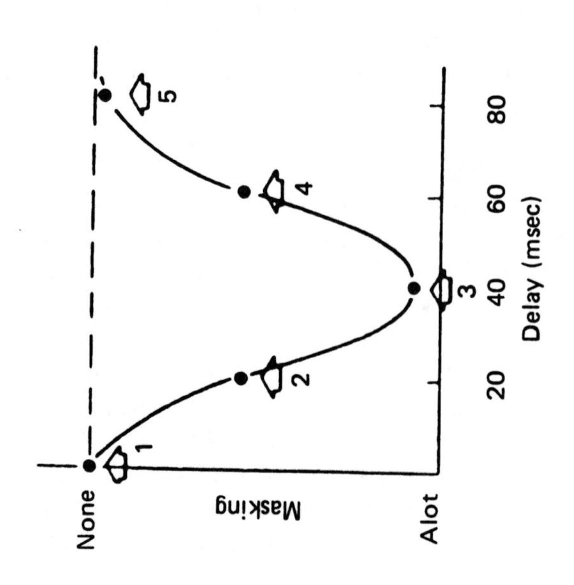
None
Masking
Alot
1
2
3
4
5
20
40
60
80
Delay (msec)

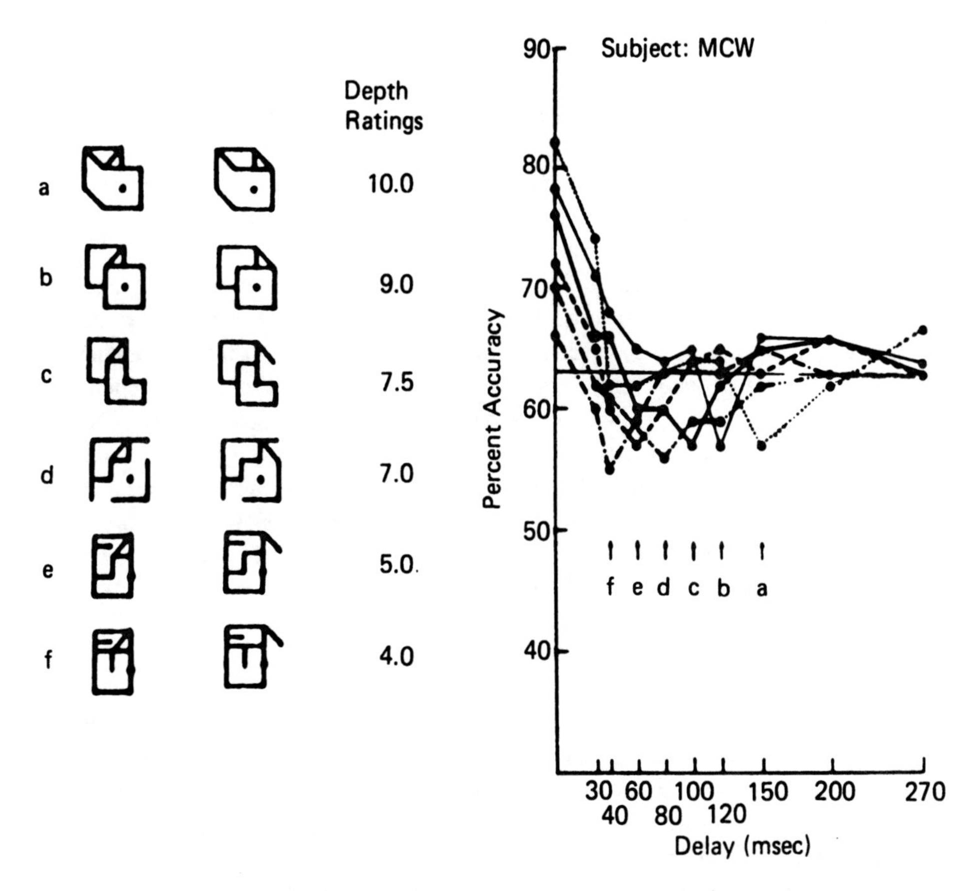

FIGURE 9.13. Data showing the shift in metacontrast masking function asymptote with changes in apparent depth of context. The contexts used in the experiment are shown at the left. Subjects rated these contexts for apparent depth and the relative ratings are shown in the column to the right of the contexts. Representative data for one subject (MCW) is presented on the right, with the delay at which each function reached its minimum (most masking) displayed below. It can be seen that for the context with the greatest perceived depth (a) the minimum shows the longest delay. Reprinted with permission from Williams and Weisstein, 1984, Copyright (1984), Pergamon Press plc.

◄

FIGURE 9.12. On the left a hypothetical U-shaped masking function (solid line) is presented plotted as a function of mask delay relative to the target. Dotted line represents the target alone. y axis plots visibility. On the right, a model shows how the interaction of a fast masking response and a slow response can produce these functions. (1) The response to both target and mask over time when they are presented simultaneously. Both give rise to both a fast and slow response. (2–5) As the mask is delayed there is increasing overlap between the slow target response and the fast masking that inhibits it. In this panel overlap is maximal at 40 msec (trace 3) and the U minimum is seen at 40 msec. (Reprinted with permission from Williams and Weisstein, 1984,Copyright (1984), Pergamon Press plc.)

the rise time in the mask activated inhibitory channel, or that the output of the target activated sustained channel is increasingly dominated by analyzers responding to the higher spatial frequencies as apparent depth of the mask and target increases. Other research, particularly the use of equiluminant stimuli discussed above, converges on the hypothesis that the transient, mask-activated channels may be sensitive to information about pattern depth even when this information is essentially figural.

Research into the effects of removing high spatial and temporal frequency information yields more specific information about the role of sustained and transient pathways in the object-superiority effect and metacontrast masking. High temporal frequency information can be removed from the masking stimulus by presenting it with a gradual onset and offset (ramped temporal window). This manipulation reduces metacontrast masking with object superiority effect stimuli, and produces maximal masking at shorter stimulus onset asynchronies. This implies that masking channel responses are reduced and slowed (possibly by a shift to more slowly responding analyzers) by removing the high temporal frequency transient information (Williams, 1980).

Faster Processing of Three-Dimensional Images

When the contexts surrounding the perspective target line in Figure 9.11 are delayed in a metacontrast paradigm, the stimulus onset asynchronies at which maximum masking occurs systematically increase with increasing apparent depth. Above we discussed the factors that could account for such a result. The mask-activated analyzers could respond faster, or faster analyzers could be activated as the pattern contains more and more cues to three dimensionality. Alternatively the mask-activated analyzers could interact with higher and therefore slower acting target frequencies (see Breitmeyer, 1984 discussed above).

Converging evidence points to faster processing of three-dimensional images. Flickering object-like contexts yield better discrimination of the target lines at high temporal frequencies than do flickering flatter ones (Genter, 1981). Recall that sensitivity to high temporal frequencies implies a faster rise time in the temporal impulse response (see Figs. 9.3 and 9.12). Object-like contexts also have a faster decay time, that is the flatter contexts have a longer visual persistence (Genter, 1981). Longer persistence is generally associated with the slower sustained channels (Meyer and Maguire, 1977, 1981). Reaction time to targets in a speeded discrimination task is faster for object contexts than for flatter ones (Wong and Weisstein, 1990b). Although the perspective line of the target provides much of the depth in the object-like contexts, flickering the contexts without the target lines yields higher critical flicker fusion values for the more three-dimensional contexts. This is consistent with faster processing of the more three-dimensional contexts even without the target line.

Perhaps the most thorough study of speed of processing for these contexts is by Wong and Weisstein (1990b). This study utilized a deadline procedure in which observers respond to the target as quickly as possible after they hear a tone. Their task is as usual to discriminate which of two target lines was presented. The time at which the tone is presented varies from simultaneous with to 500 msec after the target plus context are presented. Target and context are always presented simultaneously. The accuracy of response at various time intervals is thus recorded. In this way a signal integration function is obtained. This traces the sensitivity to the target line as a function of time since target and context onset. A type of temporal impulse response to the target line in a particular context is obtained. Comparing three-dimensional, flat contexts and target alone yielded three significant findings. The initial response latency is shortest for the target alone and slowest for the target in the flat context. Asymptotic accuracy is in agreement with previous object superiority results in that it is highest for the object context, intermediate for the target alone, and lowest for the flat pattern. The rise time (duration from initial response to asymptote) is fastest for the target in the three-dimensional context, intermediate for the target presented alone, and slowest for the target in the flat context.

These results suggest a pattern of interactions, which is also consistent with object superiority and metacontrast results discussed above. Each line in the context is at first registered independently by analyzers at that location. (It is expected that a large number of spatial and temporal frequency analyzers will respond to each line, because the lines are all aperiodic broadband stimuli, which are composed of a wide range of spatial frequencies.) Some of these will be sustained analyzers sensitive to low temporal and high spatial frequencies. Others will be transient analyzers with their typical sensitivities. Quickly the transient "early warning system" that registers the global three dimensionality of the patterns feeds back on the figural detail sustained system telling it where to look and what to look for. This transient response is faster for the three-dimensional patterns and thus allows the sustained system to respond faster.

Obviously the analyzers do not speed up internally; one assumes that they have relatively fixed latencies. But in a cooperative network in which each analyzer mutually enhances the response of the other, a rapid and amplified response can result from analyzers with fixed latencies. Recall that enhanced target discrimination in the object superiority effect depends on the target containing high spatial frequencies. Metacontrast data suggest inhibition of sustained by transient channels when the mask is delayed by 100–200 msec. We assume that when both channels are simultaneously activated, the advance information feeding back from the faster responding transient channels (and the faster responding signal from a cooperative network of these channels, when an object-like context is presented) leads to faster and greater sustained activity. Thus the interaction of sustained and transient pathways in a global cooperative network presents a plausible model of the object-superiority effect. Further research is needed to determine the accuracy of this model.

Summary of Suprathreshold Interactions

A preponderance of psychophysical evidence indicates that there are basic visual dimensions that are analyzed in parallel by sets of tuned analyzers. Although the existence of these analyzers is demonstrated by their independent function at threshold, under conditions of suprathreshold perception, they show a number of distinct forms of interaction. Within the set of analyzers tuned to a single stimulus dimension, the following inhibitory interactions have been described. Spatially adjacent analyzers tuned to the same dimensional value inhibit each other. Presumably this mechanism enhances boundaries that are defined by texture differences, in the way that lateral inhibition enhances boundaries that are defined by luminance differences (Ratliff, 1965). This is because in a field of uniform texture, the analyzers activated within the field will be inhibiting each other. At the texture boundary, however, where different sets of analyzers will be activated across the boundary, inhibition will be reduced. Thus the most vigorous responses will be found at the texture boundaries. In another form of inhibitory interaction spatially superimposed channels tuned to different values on the same dimension inhibit each other. This mechanism may serve to narrow the response of the individual analyzers to stimuli along the dimension.

Additionally, cooperative interactions appear at this level. Illusory contours appear to be the result of cooperative interactions between analyzers sensitive to spatial characteristics of the physically present contours. Orientation-selective analyzers activated when two contours are aligned may under certain circumstances activate analyzers that normally respond to contours of the same orientation that lie between them. These interactions are spatially limited to distances under 5° around the fovea. Such mechanisms may play an important role in filling in information missing from the visual image. Such missing information may arise from retinal or nervous system damage (scotomata), from physiological structure (blind spot, retinal blood vessels), or from properties of the image itself, such as glare.

Analysis of the visual system into sustained and transient pathways has revealed a number of other interactions within and between these pathways. Transient responses inhibit sustained responses in a limited spatial neighborhood. This mechanism appears to play a role in the erasure of the persisting image from the previous fixation following a saccadic eye movement (saccadic suppression) (Breitmeyer and Ganz, 1976).

Activity within sustained and transient pathways may be modulated by higher order information that affects the segmentation of the image and the assignment of relative depth to surfaces. Figural qualities that contribute to three dimensionality enhance activity in the sustained pathway. The assignment of a region of the image to ground enhances transient processing within that region. The modulation of analyzer output, by the information involved in the global segmentation of the image, may be a rather general phenomenon. We see it in the cooperative interactions that produce illusory contours.

The segmentation of the visual image and the assignment of relative depth to surfaces are reciprocally affected by the relative magnitude of activity in sustained and transient pathways. Regions of the image where sustained activity dominates tend to be seen as forward and boundary contours are assigned to the region giving it figural qualities, particularly where adjacent regions are rich in transient information.

The integration of local information into global percepts capable of interacting with memory processes to produce pattern recognition is the critical function of the visual system and the discovery of the mechanisms by which it occurs is the great challenge for visual science. The investigation of early visual mechanisms already reveals some complex processes. We see global grouping processes working to turn feature output into global surfaces in texture segregation. We see evidence for detectors that process global information such as motion in depth, but which show adaptation effects like elementary analyzers. We see context apparently modulating local feature signals in complex ways. These mechanisms remain in the shadows as it were, with only the vaguest outlines apparent. It remains for the methods of visual science, including the psychophysical methods discussed here, to move them into the light. We discuss the relationship of psychophysics to the other tools of visual science below.

Psychophysics, Cortical Organization, and Formal Theories of Vision

The visual system is a component in a complex information processing system, whose purpose is to direct the behavior of the organism so that it and its offspring will prosper despite a forever changing and often threatening environment. Because the optical imaging of distant objects and events serves an adaptive purpose eyes are to be found on a plethora of animal forms of varying complexity and requirements. These variant visual systems evolved to the different needs of different species in different ecological niches. A clear example is the presence of frontally directed eyes in both mammalian and avian predators, compared with more laterally directed eyes providing panoramic vision to those species that as often find themselves to be dinner as diners.

It follows that to understand the visual system, one must understand its role in the behavior of the organism. To understand the human visual system requires then a coming together of a variety of disciplines. On the one hand, the contributions of the neurosciences are required to delineate the anatomy, physiology, neurochemistry, and biophysics of the visual system. On the other hand, one needs perceptual psychology to specify what information we can use and do use under varying environmental conditions. As we have seen, perceptual psychology and psychophysics can play another important role as a testing ground for specific theories of how perception works. Whether these theories are physiological in spirit or more formal, they have important implications for the neuroscientists and their enterprise.

At our present level of knowledge, it becomes practically important to do research in comparative psychology to fully understand human vision. This is because the best work in neural function uses animal subjects. When we apply this work to human perception, we implicitly assumes that the perceptual world of the animal is similar to our own. It is good to test that assumption at times, for certainly it is not always true. A number of the phenomena discussed in this chapter have been substantiated with animal subjects. For example, rhesus monkeys experience both motion aftereffects (Scott and Powell, 1963) and contingent aftereffects (Maguire et al., 1980). It would be comforting to have a larger body of animal psychophysics to consult when making assertions about human brains while staring into monkey skulls, but such research can be time consuming and presents many methodological challenges, so knowledge in this critical area trails relative to the others.

Some have concentrated their interests in vision, in the formal analysis of the properties of images, and have sought computer algorithms and heuristics that would successfully solve perceptual problems analogous to the types that the human visual system solves. The formal analysis of visual information processing is an exciting field in its own right, and one that promises many important applications. It cannot subsume the empirical questions of perceptual psychology, however. If our interest were the movement of animals, a formal analysis of the physics of locomotion on land might lead us to expect to find wheels. Nature, however, has failed to provide them, relying for most creatures on rather more cumbersome means.

By analogy, it is unclear that the visual system will be found to work by the most elegant formal solutions that we can devise. In an area where computer abilities have fair parity with human abilities, chess, the computers use strategies that are really beyond the capabilities of human computation to attain comparable play, while the exact heuristics that humans use remain to be described (Best, 1989).

Nevertheless, the formal analysis of images and image processing has been a fruitful source of ideas in vision. The concept of spatial frequency channels as a component of the visual system arose from the formal analysis of optical information channels, in which spatial frequency analysis can lead to some useful properties in the frequency domain representation. This led, in turn, to the psychophysical search for the existence of such channels, followed by the search for a cellular basis for these channels. This search did not yield the very narrowband, spatially distributed spatial frequency channels that the formal analysis might require. Nonetheless, the psychophysics and neurophysiology did show selectivity in the spatial and even the temporal frequency domain. So the formal analysis was fruitful, even though it could hardly substitute for the empirical investigations.

Formal analysis and computer modelling remain powerful tools in the investigation of human vision (see the following chapters). If one begins with an image processing stage where the image is represented as the output of relatively segregated spatially localized orientation specific analyzers covering the spatial frequency spectrum, with similar coverage of the temporal frequency spectrum for dynamic images, one can test segmentation, object finding, and recognition algorithms and heuristics. Such an approach can yield useful theory about the higher levels of information processing within the visual system.

Multiscale, multiresolution techniques are being intensively investigated within the computer vision community (Pizer, 1988). In these methods, the image is sampled at a number of spatial scales. These methods have obvious analogues in the sets of spatial and temporal frequency channels in human vision.

The great interest that the problems of vision have generated causes us to anticipate more detailed and successful theories of segmentation, object formation, and object recognition in the years to come. The psychophysical techniques that have been successful in characterizing the earliest stages of vision should also serve in investigation of more complex processes. Threshold measurements and adaptation effects have been successfully applied to more global analyzers. Masking studies allow one to interrupt processing at various points producing optimally a record of visual processing in a series of still frames. Illusions and ambiguous displays reveal the boundary conditions under which perceptual mechanisms work, and thus serve to reveal their nature. We expect the great progress of the last few decades born of a cooperative enterprise between formal, psychological, and physiological approaches to vision will continue.

References

Adelson EH, Movshon JA (1982). Phenomenal coherence of moving visual patterns. Nature (London) **300**:523–525.

Albrecht DG, Farrar SB, Hamilton DB (1984). Spatial contrast adaptation characteristics of neurones recorded in the cat's visual cortex. J. Physiol. **347**:713–739.

Allman J, Meizin F, McGuiness E (1985). Stimulus specific responses from beyond the classical receptive field: Neurophysiological mechanisms for local-global comparisons in visual neurons. Annu. Rev. Neurosci. **8**:407–430.

Alwitt LF (1981). Two neural mechanisms related to modes of selective attention. J. Exp. Psychol.: Human Percept. Perform. **7**:324–332.

Arditi A (1986). Binocular vision. In Chap. 23. *Handbook of Perception and Human Performance*, (K Boff, L Kaufman, JP Thomas, eds.), Wiley, New York.

Beck J (1966). Effect of orientation and of shape similarity on perceptual grouping. Percept. Psychophys. **1**:300–302.

Beck J (1982). Textural segmentation. In *Organization and Representation in Perception* (J Beck, ed.), pp. 285–317. Erlbaum, Hillsdale, NJ.

Berbaum K, Bever T, Chung CS (1983). Light source position in the perception of object shape. Perception **12**:411–416.

Berbaum K, Bever T, Chung CS (1984). Extending the perception of shape from known to unknown shading. Perception **13**:479–488.

Best J (1989). *Cognitive Psychology*, 2nd ed. West Publishing.

Blakemore C, Julesz B (1971). Stereoscopic depth aftereffect produced without monocular cues. Science **171**:286–288.

Blakemore C, Sutton P (1969). Size adaptation: A new aftereffect. Science **166**:245–247.

Brand J (1971). Classification without identification in visual search. J. Exp. Psychol. **23**:178–186.

Breitmeyer BG (1975). Simple reaction time as a measure of the temporal response properties of transient and sustained channels. Vision Res. **15**:1411–1412.

Breitmeyer BG (1984). *Visual Masking: An Integrative Approach*. Oxford University Press, New York.

Breitmeyer BG, Ganz L (1976). Implications of sustained and transient channels for theories of visual pattern masking, saccadic suppression and information processing. Psychol. Rev. **83**:1–35.

Brown J, Weisstein N (1985). Flickering phantoms: A figure/ground approach. Paper presented at the *57th Annual Meeting of the Eastern Psychological Association*, Baltimore, MD.

Brown J, Weisstein N (1986). Depth information within phantom inducing regions can influence phantom visibility. Paper presented at the *58th Annual Meeting of the Eastern Psychological Association*, Baltimore, MD.

Brown J, Weisstein N (1988). A spatial frequency effect on perceived depth. Percept. Psychophys. **44**:157–166.

Calis G, Leeuwenberg E (1981). Grounding the figure. J. Exp. Psychol.: Human Percept. Perform. **7**:1386–1397.

Campbell FW, Kulikowski JJ (1966). Orientation selectivity of the human visual system. J. Physiol. **187**:437–445.

Cavanaugh P, Tyler CW, Favreau OE (1984). Perceived velocity of moving chromatic gratings. J. Opt. Soc. Am. Sect. A:Opt. Image Sci. **1**:893–899.

Crawford BH (1947). Visual adaptation in relation to brief conditioning stimuli. Proc. R. Soc. London **134**:283–302.

Davis ET, Kramer P, Graham N (1983). Uncertainty about spatial frequency, spatial position, or contrast of visual patterns. Percept. Psychophys. **33**:20–28.

DeMonasterio FM, Gouras P (1975). Functional properties of ganglion cells of the rhesus monkey retina. J. Physiol. **251**:167–195.

Derrington AM, Lennie P (1984). Spatial and temporal contrast sensitivities of neurons in lateral geniculate nucleus of macaque. J. Physiol. (London) **357**:219–240.

De Valois K (1977). Spatial frequency adaptation can enhance contrast sensitivity. Vision Res. **17**:1057–1065.

Dow B, Gouras P (1973). Color and spatial specificity of single units in rhesus monkey foveal striate cortex. J. Neurophysiol. **36**:79–100.

Enns JT (1988). Three dimensional figures that pop out in visual search. Paper presented at First International Conference on Visual Search. Durham, U.K.

Enns JB, Gilani AB (1988). Three dimensionality and discriminability in the object superiority effect. Percept. Psychophys. **44**:243–256.

Enns JT, Printzmetal W (1984). The role of redundancy in the object-line effect. Percept. Psychophys. **35**:22–32.

Enroth-Cugell C, Robson JG (1966). The contrast sensitivity of retinal ganglion cells of the cat. J. Physiol. **187**:517–552.

Foster KH, Gaska JP, Nagler M, Pollen D (1985). Spatial and temporal frequency selectivity of neurones in visual cortical areas V1 and V2 of the macaque monkey. J. Physiol. (London) **365**:331–363.

Fries W, Albus K, Creutzfeldt O (1977). Effects of interacting visual patterns on single cell responses in cat's striate cortex. Vision Res. **17**:1001–1008.

Frisby JP, Mayhew JE (1978). The relationship between apparent depth and disparity in revalrous-texture stereograms. Perception **7**:661–678.

Genter CR (1981). Temporal factors in the perception of complex imagery. Unpublished Ph.D. thesis, State University of New York, Buffalo.

Genter CR, Weisstein N (1981). Flickering phantoms: A motion illusion without motion. Vision Res. **21**:963–966.

Gibson JJ, Radner M (1937). Adaptation, after-effect, and contrast in the perception of tilted lines. I. Quantitative studies. J. Exp. Psychol. **20**:453–467.

Gibson JJ, Kaplan GA, Reynolds HN, Wheeler K (1969). The change from visible to invisible: A study of optical transitions. Percept. Psychophys. **5**:113–116.

Ginsburg A (1978). *Visual Information Processing Based on Spatial Filters Constrained By Biological Data*. Aerospace Medical Research Laboratory, Wright Patterson Air Force Base, Ohio.

Gouras P (1974). Opponent-color cells in different layers of foveal striate cortex. J. Physiol. **238**:583–602.

Graham N (1985). Detection and identification of near-threshold visual patterns. J. Opt. Soc. Am. A **2**:1468–1482.

Graham N (1989). *Visual Pattern Analyzers*. Oxford University Press, New York, in press.

Graham N, Nachmias J (1971). Detection of grating patterns containing two spatial frequencies: A comparison of single channel and multiple-channels models. Vision Res. **11**:251–259.

Graham N, Beck J, Sutter A (1989). Two nonlinearities in texture segregation. Invest. Ophthalmol. Visual Sci. **30**(Suppl):161.

Green M (1981). Psychophysical relationships among mechanisms sensitive to pattern, motion, and flicker. Vision Res. **21**:971–983.

Greenlee MG, Magnussen S (1988). Interaction among spatial frequency and orientation channels adapted concurrently. Vision Res. **28**:1303–1310.

Grossberg S, Mingolla E (1985a). Neural dynamics of perceptual grouping: Textures, boundaries, and emergent features. Percept. Psychophys. **38**:141–171.

Grossberg S, Mingolla E (1985b). Neural dynamics of form perception: Boundary completion, illusory figures, and neon color spreading. Psychol. Rev. **92**:173–211.

Growney R (1976). The function of contour in metacontrast. *Vision Res*. **16**:253–261.

Growney R, Weisstein N, Cox S (1977). Metacontrast as a function of spatial separation with narrow line targets and masks. Vision Res. **17**:1205–1210.

Gyoba J (1983). Stationary phantoms: A completion effect without motion or flicker. Vision Res. **22**:119–134.

Hochberg J (1971). Perception: Space and movement. In *Woodworth and Schlosberg's Experimental Psychology*, (JA Kling, LA Riggs, eds.), pp. 475–550. Holt, Rinehart, & Winston, New York.

Hubel DH, Wiesel TN (1974). Sequence regularity and geometry of orientation columns in the monkey striate cortex. J. Comp. Neurol. **158**:267–294.

Hurvich LM, Jameson D (1955). Some quantitative aspects of an opponent-colors theory. II. Brightness, saturation and hue in normal and dichromatic vision. J. Opt. Soc. Am. **45**:602–616.

Ikeda H, Wright MJ (1975). Spatial and temporal properties of "sustained" and "transient" neurones in area 17 of the cat's visual cortex. Exp. Brain Res. **22**:363–383.

Julesz B (1975). Experiments in the visual perception of texture. Sci. Am. **232**:34–43.

Julesz B (1978). Perceptual limits of texture discrimination and their implications for figure-ground separation. In *Formal Theories of Perception* (E Leeuwenberg, ed.), pp. 205–216. Wiley, New York.

Julesz B (1987). Preattentive human vision: Link between neurophysiology and psychophysics. In *Handbook of Physiology Section 1-Nervous System. Vol. 5, Higher Functions of the Brain, Pt 2* (V B Mountcastle, ed.). American Physiological Society, Bethesda, MD.

Kanisza G, Gerbino W (1982). Amodal completion: Seeing or thinking? in *Organization and Representation in Perception* (J. Beck, ed.), pp. 167–190. Lawrence Erlbaum, Hillsdale, N.J.

Kaufman D (1989). Visual search for conjunction of motion and orientation: The effects of varying distractor-type ratio and relative temporal frequencies. Invest. Ophthalmol. Visual Sci. **30**:456.

Keesey UT (1972). Flicker and pattern detection: A comparison of thresholds. J. Opt. Soc. Am. **56**:446–448.

Kelly DH (1961). Visual response to time-dependent stimuli. I. Amplitude sensitivity measurements. J. Opt. Soc. Am. **51**:422–429.

Kinchla HA, Wolfe JM (1979). The order of visual processing: "Top-down," "bottom-up," or "middle-out." Percept. Psychophys. **25**:225–231.

King-Smith PE, Kulikowski JJ (1975). Pattern and flicker detection analysed by subthreshold summation. J. Physiol. **249**:519–548.

Klymenko V, Weisstein N (1986). Spatial frequency differences can determine figure-ground organization. J. Exp. Psychol. Human Percept. Perform. **12**:324–330.

Klymenko V, Weisstein N (1987). The resonance theory of kinetic shape perception and the motion-induced contour. In *The Perception of Illusory Contours* (S Petry, GE Meyer, eds.), pp 143–148. Springer-Verlag, New York.

Klymenko V, Weisstein N (1989a). Figure and ground in space and time: 1. Temporal response surfaces of perceptual organization. Perception **18**:627–637.

Klymenko V, Weisstein N (1989b). Figure and ground in space and time: 2. Frequency velocity and perceptual organization. Perception **18**:639–648.

Klymenko V, Weisstein N, Ralston J (1987). Illusory contours, projective transformations, and kinetic shape perception. Acta Psychol. **64**:229–243.

Klymenko V, Weisstein N, Topolski R, Hsieh CH (1989) Spatial and temporal frequency in figure-ground organization. Percep. Psychophys. **45**:395–403.

offka K (1935). *Principles of Gestalt Psychology* Harcourt Brace, New York.

Kulikowski JJ, Tolhurst DJ (1973). Psychophysical evidence for sustained and transient detectors in human vision. J. Physiol. **232**:149–162.

Kulikowski JJ, Abadi R, King-Smith PE (1973). Orientation selectivity of grating and line detectors in human vision. Vision Res. **13**:1479–1486.

Lanze M, Weisstein N, Harris J (1982). Perceived depth vs. structural relevance in the object superiority effect. Percept. Psychophys. **31**:376–382.

Lanze M, Maguire W, Weisstein N (1985). Emergent features: A new factor in the object-superiority effect? Percept. Psychophys. **38**:438–442.

Lawson RB, Cowan E, Gibbs TD, Whitmore CG (1974). Stereoscopic enhancement and erasure of subjective contours. J. Exp. Psychol. **103**:1142–1146.

Leibovic KN (1969). Some problems of information processing and models of the visual pathway. J. Theoret. Biol. **22**:62–79.

Leibovic KN (1972). *Nervous System Theory*. Academic Press, New York.

Leibovic KN, Balsleve E, Mathieson TA (1971). Binocular vision and pattern recognition. Kybernetic **8**:14–23.

Livingstone MS, Hubel DH (1987). Psychophysical evidence for separate channels for the perception of form, color, movement, and depth. J. Neurosci. **7**:3416–3468.

Lu C, Fender DH (1972). The interaction of color and luminance in stereoscopic vision. Invest. Ophthalmol. Visual Sci. **11**:482–490.

Maguire W, Baizer J (1982). Luminance coding of briefly presented stimuli in area 17 of the rhesus monkey. J. Neurophysiol. **47**:128–137.

Maguire W, Brown J (1987). The current state of research into visual phantoms. In *The Perception of Illusory Contours* (S Petry, GE Meyer, eds.), pp 213–219. Springer-Verlag, New York.

Maguire W, Sitkowski S (1984). The role of target discriminability and distractor redundancy in discrimination of numbers among letters. Paper presented at the 57th Annual Meeting of the Eastern Psychological Association, Baltimore, MD.

Maguire W, Meyer GE, Baizer JS (1980). The McCollough effect in the rhesus monkey. Invest. Ophthalmol. Visual Sci. **19**:312–324.

Marlin SG, Hasan SS, Cynander MS (1988). Direction-selective adaptation in simple and complex cells in cat striate cortex. J. Neurophysiol. **59**:1314–1330.

Maunsell JHR, Newsome WT (1987). Visual processing in monkey extrastriate cortex. Annu. Rev. Neurosci. **10**:363–401.

Maunsell JHR, Van Essen DC (1983). The connections of the middle temporal visual area (MT) and their relationship to a cortical hierarchy in the macaque monkey. J. Neurophysiol. **3**:2563–2586.

McCollough C (1965). Color adaptation of edge detectors in the human visual system. Science **149**:1115–1116.

Merigan WH (1989). Chromatic and achromatic vision of macaques: Role of the p pathway. J. Neurosci. **9**:776–783.

Meyer GE, Dougherty T (1987). Effects of flicker-induced depth on chromatic subjective contours. J. Exp. Psychol.: Human Percept. Perform. **13**:355–360.

Meyer GE, Maguire W (1977). Spatial frequency and the mediation of short term visual storage. Science **198**:524–525.

Meyer GE, Maguire W (1981). Effects of spatial-frequency specific adaptation and target duration on visual persistence. J. Exp. Psychol.: Human Percept. Perform. **7**:151–156.

Michael CR (1978a). Color vision mechanisms in monkey striate cortex: Simple cells with dual opponent-color receptive fields. J. Neurophysiol. **41**:1233–1249.

Michael CR (1978b). Color sensitive complex cells in monkey striate cortex. J. Neurophysiol. **41**:1250–1266.

Movshon JA, Adelson EH, Gizzi SM, Newsome WT (1986) The analysis of moving visual patterns. In *Pattern Recognition Mechanisms*, (C Chagas, R Gatass, C Gross, eds.), pp 117–151. Springer-Verlag, New York.

Murch GM (1976). Classical conditioning of the McCollough effect: Temporal parameters. Vision Res. **19**:939–942.

Nakayama K, Shimojo S, Silverman GH (1989). Stereoscopic depth: Its relation to image segmentation, grouping, and the recognition of occluded objects. Perception, in press.

Olzak LA, Thomas JP (1986). Seeing spatial patterns. In *Handbook of Perception and Human Performance* (K Boff, L Kaufman, JP Thomas, eds.), Chapt. 7. Wiley, New York.

Pantle A, Sekuler R (1968). Velocity-sensitive elements in human vision: Initial psychophysical evidence. Vision Res. **8**:445–450.

Pentland AP (1985) The focal gradient: Optics ecologically salient. Invest. Ophthalmol. Visual Sci. **26**:243.

Pizer S (1988). Multiscale methods and the segmentation of medical images. TR88-051 Department of Computer Science, University of North Carolina at Chapel Hill.

Pokorny J, Smith VC (1986). Colorimetry and color discrimination. In *Handbook of Perception and Human Performance* (K Boff, L Kaufman, JP Thomas, eds.), Chapt. 8. Wiley, New York.

Pomerantz JR, Kubovy M (1986). Theoretical approaches to perceptual organization. In *Handbook of Perception and Human Performance Vol. 2: Cognitive Processes and Performance* (KR Boff, L Kaufman, JP Thomas, eds.), Chapt. 36. Wiley, New York.

Ramachandron VS (1988). Perception of shape from shading. Nature (London) **331**:133–136.

Ramachandron VS, Anstis S (1986). Figure-ground segregation modulates apparent motion. Vision Res. **26**:1969–1975.

Ramachandran VS, Gregory RL (1978). Does colour provide an input to human motion perception? Nature (London) **275**:55–56.

Ratliff F (1965). *Mach Bands: Quantitative Studies on Neural Networks in the Retina*. Holden-Day, San Francisco.

Regan D (1986). Visual processing of four kinds of relative motion. Vision Res. **26**: 127–145.

Regan D, Beverly KI (1978). Looming detectors in the human visual pathway. Vision Res. **18**:415–421.

Rogowitz BE (1983). Spatial temporal interactions– Backward and forward metacontrast masking with sine-wave gratings. Vision Res. **23**:1057–1093.

Rubin E (1958). Figure and ground. In *Readings in Perception*. (DC Beardslee, M Wertheimer, eds.). Van Nostrand, Princeton, NJ. (Original work published 1921.)

Sachs MB, Nachmias J, Robson J (1971). Spatial-frequency channels in human vision. J. Opt. Soc. Am. **61**:1176–1186.

Saito H, Yukie M, Tanaka K, Hikosaka K, Fukada Y, Iwai E (1986). Integration of direction signals of image motion in the superior temporal sulcus of the macaque monkey. J. Neurosci. **6**:145–157.

Schiller P, Finlay BL, Volman SF (1976). Quantitative studies of single-cell properties in monkey striate cortex II. Orientation specificity and ocular dominance. J. Neurophysiol. **39**:1320–1333.

Scott TR, Powell DA (1963). Measurement of a visual motion aftereffect in the rhesus monkey. Science **140**:57–59.

Sekuler R, Pantle A (1967). A model for the aftereffects of seen movement. Vision Res. **7**:427–439.

Shapley R, Gordon J (1985). Nonlinearity in the perception of form. Percept. Psychophys. **37**:84–88.

Shapley R, Kaplan E, Soodak R (1981). Spatial summation and contrast sensitivity of X and Y cells in the lateral geniculate nucleus of the macaque. Nature (London) **292**: 543–545.

Shiffrin RM, Schneider W (1977). Controlled and automatic information processing II. Perceptual learning, automatic attending, and a general theory. Psychol. Rev. **84**: 127–190.

Shimojo S, Nakayama K (1990). Amodal representation of occluded surfaces: Role of invisible stimuli in apparent motion correspondence. *Perception* (In Press).

Shimojo S, Silverman GH, Nakayama K (1988). An occlusion-related mechanism of depth perception based on motion and interocular sequence. *Nature*, **333**:265–268.

Skowbo D (1984). Are McCollough effects conditioned responses. Psychol. Bull. **96**: 215–226.

Smith AT (1985). Velocity coding: Evidence from perceived velocity shifts. Vision Res. **25**:1969–1976.

Smith AT, Over R (1975). Tilt aftereffects with subjective contours. Nature **257**:581–582.

Strohmeyer CF (1978). Form-color aftereffects in human vision. *Handbook of Sensory Physiology: Vol 8. Perception* (R Held, H Leibowitz, HL Teuber, eds.), pp 97–142. Springer-Verlag, New York.

Teller D (1984). Linking propositions. Vision Res. **10**:1233–1246.

Thompson P (1981). Velocity aftereffects: The effects of adaptation to moving stimuli on the perception of subsequently seen moving stimuli. Vision Res. **21**:337–345.

Tolhurst DJ, Barfield LP (1978). Interactions between spatial frequency channels. Vision Res. **18**:951–958.

Treisman A (1986). Properties, parts, and objects. In *Handbook of Perception and Human Performance* (K Boff, L Kaufman, JP Thomas, eds.), Chapt. 35. Wiley, New York.

Treisman A, Gormican S (1988). Feature analysis in early vision: Evidence from search asymmetries. Psychol. Rev. **95**:15–48.

Tynan P, Sekuler R (1975a). Simultaneous motion contrast: Velocity, sensitivity, and depth response. Vision Res. **15**:1231–1238.

Tynan P, Sekuler R (1975b). Moving visual phantoms: A new completion effect. Science **188**:951–952.

Van Nes FL, Bowman MA (1967). Spatial modulation transfer in the human eye. J. Opt. Soc. Am. **57**:401–406.

Vautin RG, Berkley MA (1977). Responses of single cells in cat visual cortex to prolonged stimulus movement: Neural correlates of visual aftereffects. J. Neurophysiol. **40**:1051–1065.

von der Heydt R, Peterhans E, Baumgartner G (1984). Illusory contours and cortical neuron responses. Science **224**:1260–1261.

Walker P, Powell DJ (1974). Lateral interaction between neural channels sensitive to velocity in human visual system. Nature (London) **252**:732–733.

Watson AB (1986). Temporal Sensitivity. In *Handbook of Perception and Human Performance* (K Boff, L Kaufman, JP Thomas, eds.), Chapt. 6. Wiley, New York.

Watson AB, Nachmias J (1977). Patterns of temporal interaction in the detection of gratings. Vision Res. **17**:893–902.

Watson AB, Robson JG (1981). Discrimination at threshold: Labelled detectors in human vision. Vision Res. **21**:1115–1122.

Weisstein N (1968). A Rashevsky-Landahl neural net: Simulation of metacontrast. Psychol. Rev. **75**:494–521.

Weisstein N (1970). Neural symbolic activity: a psychophysical measure. *Science*, **168**: 1489–1491.

Weisstein N (1972). Metacontrast. In *Handbook of Sensory Physiology* (D. Jameson & L. Hurvich, eds.), Vol. 7:233–272. Springer-Verlag, Berlin.

Weisstein N (1973). Beyond the yellow Volkswagen detector and the grandmother cell: A general strategy for the exploration of operations in human pattern recognition. In *Contemporary Issues in Cognitive Psychology: The Loyola Symposium* (R Solso, ed.). Winston & Sons, Washington, DC.:17–51.

Weisstein N, Harris CS (1974). Visual detection of line segments: An object-superiority effect. Science **186**:752–755.

Weisstein N, Harris CS (1980). Masking and unmasking of distributed representations in the visual system. In *Visual Coding and Adaptability* (CS Harris, ed.). Erlbaum, Hillsdale, NJ:317–364.

Weisstein N, Maguire W (1978). Computing the next step: Psychophysical measures of representation and interpretation. In *Computer Vision Systems* (AR Hanson, EM Riseman, eds.), pp 243–260. Academic Press, New York.

Weisstein N, Wong E (1986). Figure-ground organization and the spatial and temporal responses of the visual system. In *Pattern Recognition by Humans and Machines Vol. 2.* (E Schwab, HC Nusbaum, eds.), Academic Press, New York.

Weisstein N, Wong E (1987). Figure-ground organization affects the early visual processing of information. In *Vision, Brain, and Cooperative Computation* (MA Arbib, AR Hanson, eds.). MIT Press, Cambridge MA.

Weisstein N, Matthews M, Berbaum K (1974). Illusory contours can mask real contours. Bull. Psychon. Soc. **4**:266.

Weisstein N, Ozog G, Szoc R (1975). A comparison and elaboration of two models of metacontrast. Psychol. Rev. **82**:375–343.

Weisstein N, Maguire W, Berbaum K (1977). A phantom motion aftereffect. Science **189**:955–958.

Weisstein N, Maguire W, Williams MC (1982a). The effect of perceived depth of phantoms and the phantom motion aftereffect. *Organization and Representation in Perception*. Erlbaum, Hillsdale, NJ.:235–249.

Weisstein N, Williams MC, Harris CS (1982b). Depth, connectedness, and structural relevance in the object-superiority effect: Line segments are harder to see in flatter patterns. Perception **11**:5–17.

Wertheim AH (1990). Visual vestibular and oculomotor interactions in the perception of object motion during egomotion. In *The Perception and Control of Egomotion* (R Warren, AH Wertheim, eds.). Erlbaum, Hillsdale, NJ.

Williams A, Weisstein N (1977). The time course of object-superiority with contexts whose local environments are similar. Bull. Psychon. Soc. Abstr. **10**:9243.

Williams A, Weisstein N (1978). Line segments are perceived better in a coherent contest than alone: An object-line effect. Memory Cognit. **6**:85–90.

Williams MC (1980). Fast and slow response to configurational factors in "object superiority" stimuli. Unpublished Ph.D. thesis, State University of New York, Buffalo, NY.

Williams MC, Weisstein N (1984). The effect of perceived depth and connectedness on metacontrast functions. Vision Res. **24**:1279–1288.

Wohlgemuth A (1911). On the aftereffect of seen movement. Br. J. Psychol. Monogr. Suppl. 1.

Wong E, Weisstein N (1982). A new perceptual contest-superiority effect: Line segments are more visible against a figure than against a ground. Science **218**:587–589.

Wong E, Weisstein N (1983). Sharp targets are detected better against a figure, and blurred targets are detected better against a background. J. Exp. Psychol: Human Percept. Perform. **9**:194–202.

Wong E, Weisstein N (1984). Flicker induces depth: Spatial and temporal factors in the perceptual segregation of flickering and nonflickering regions in depth. Percept. Psychophys. **35**:229–236.

Wong E, Weisstein N (1985). A new visual illusion: Flickering fields are localized in a depth plane behind nonflickering fields. Perception **14**:13–17.

Wong E, Weisstein N (1987). The effects of flicker on the perception of figure and ground. Percept. Psychophys. **41**:440–448.

Wong E, Weisstein N (1990a). Spatial frequency, perceived depth, and figure/ground perception. Vision Res. (in press).

Wong E, Weisstein N (1990b). Time course of context effects on target discrimination: Studies of the object superiority and object-line effect in reaction time, speed/accuracy tradeoff and critical flicker fusion. (manuscript in prep.)

Part 3
Theory and Computation

10
Visual Information: Structure and Function

K.N. LEIBOVIC

Introduction

An information system, whether biological or man made, acquires, processes, transmits, and uses information. It is designed to detect the raw physical signal and it employs a representation that is matched to its use. Signal processing is involved at each stage. In this chapter we make a transition from an emphasis on biology to theory and computation. We consider examples of information acquisition in photoreceptors and of its transmission in neural nets. We inquire how structure and function are matched to each other and what this implies for information processing in the nervous system. Other aspects of information processing and computation are presented in the following chapters.

The physical signal in vision is the light that passes through the optical system of the eye and is projected onto the retina. Here, as described in Chapter 3, it is captured by the rhodopsin contained in the rods and cones. The latter are more or less cylindrical cells that present the mosaic of their circular cross-sections to the photon flux. In spite of the discrete nature of both photons and visual cells the world does not appear punctate but continuous. Of course, a perceived discontinuity implies the existence of gaps or changes. If the same signal is received by adjacent cells then no matter what signal there may be in between, no gap will be detected. Interphotoreceptor spacing therefore limits resolution. But there is another limit to resolution in any optical system due to the nature of light. This arises from diffraction. As we shall see in the following paragraphs, photoreceptor diameter and spacing are matched to the optical limits of resolution. This is one example of structure subserving function in visual information.

Although there exists an extensive literature on the significance of photoreceptor diameter and spacing (for reviews see e.g. Snyder, 1979; Hughes, 1987; Pugh, 1988), no comparable attention has been paid to photoreceptor length. In spite of the 10^5-fold amplification of the photon energy into a membrane response, in spite of the many rhodopsin molecules with their associated transduction machinery and the thermal noise that must be inherent in all this, even the single photon response is highly reliable and reproducible (Baylor et al., 1979). Absorption probability and noise in the transduction machinery both depend on outer segment length. Perhaps an optimality principle operates with respect to this length in photoreceptor design and evolution. We shall suggest such a principle based on maximizing photon absorptions and minimizing noise. It turns out that outer segment length is of the right order to be consistent with this principle, another example of how structure may subserve function.

Finally, we consider the transmission of signals through successive stages in the nervous system. It appears that nature has gone to great lengths in matching light detection to the physical limits of resolution and generating a high fidelity signal in rods and cones. But transmission of the signal involves converging and diverging pathways. Does not convergence and divergence degrade the signal? Is not resolution lost? We know from experiment and our perceptual experience that our visual judgments can be very precise. Based on an analysis of converging and diverging pathways we shall show that fidelity and resolution can, in fact, be preserved. This has implications for the representation of information in terms of structure and function.

Spatial Resolution of the Photoreceptor Array

Due to the wave properties of light the image of a point source is not a replica of the original. Waves emanating from the point source interfere with each other in passing through an aperture such as the pupil and produce a diffraction pattern of alternating light and dark rings. The central area of the diffraction pattern from a small circular pinhole is a circular patch of light, the so called Airy disc, of radius (see, e.g., Born and Wolf, 1987)

$$\Delta\theta = 1.22\lambda/d \tag{1}$$

Here λ is the wavelength of a monochromatic light passing through the pinhole of diameter d and $\Delta\theta$ is measured in radians and is the angle subtended by the circular patch at the pinhole. The intensity distribution in the diffraction pattern is sketched in Figure 10.1A. When two such overlapping patterns arise from adjacent point sources, there is a minimum separation where the two intensity peaks have sufficient contrast with the trough in between to be seen as distinct. This is sketched in Figure 10.1B. A generally accepted rule is the Raleigh criterion for minimum separation in which the diffraction patterns are a distance apart given by the radius of the Airy disc of Eq. (1). It is a limit of resolution of any optical system.

To translate this to retinal dimensions we make a brief excursion to the schematic eye (see, e.g., Linksz, 1950). This is a standard representation of the many individual variations of human ocular optics. A simplified version is sketched in Figure 10.1C. To obtain the size of the retinal image i of a distant object o, we draw a ray from o through the front focus to the principal plane. From there the ray is refracted parallel to the optic axis to meet the retina as shown. Clearly, the size of the retinal image is $PF_1 \tan \Delta\theta = PF_1\Delta\theta$ approximately when $\Delta\theta$ is small. Now, let the distant object o be two point sources at the limit of resolution of Eq. (1). Let the pupil diameter $d = 2$ mm, which is near the

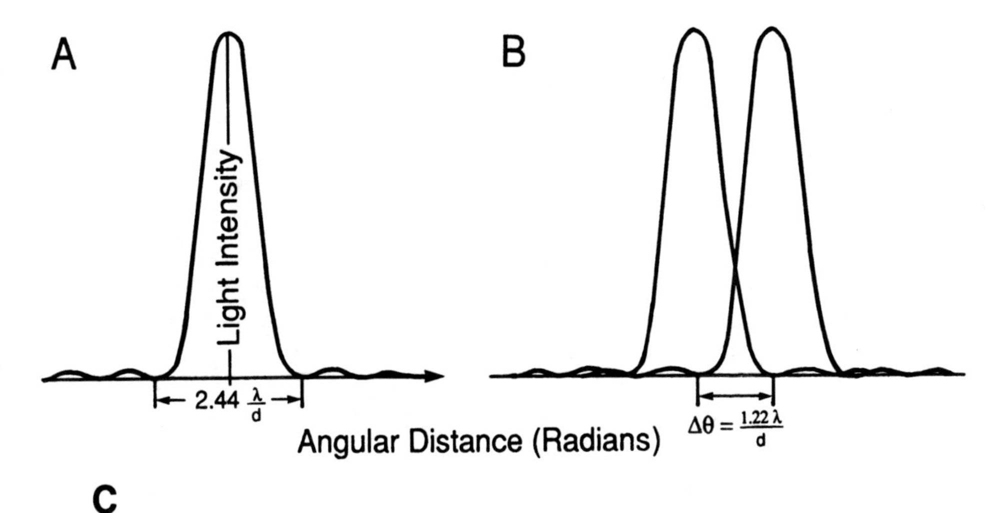

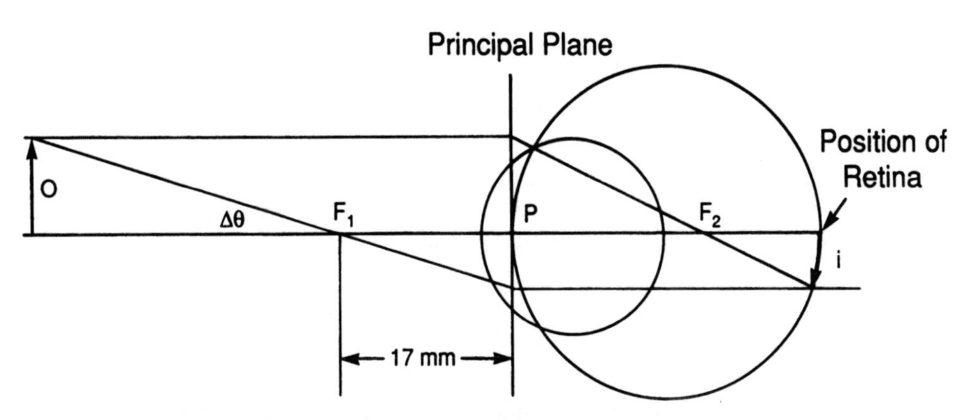

FIGURE 10.1. (A) Intensity distribution in the diffraction pattern from a circular pinhole of diameter d and for light of wavelength λ. (B) Illustration of minimum angle of resolution for two superposed Airy discs $\Delta\theta$ radians apart as in Eq. (1). (C) Simplified schematic eye. In this simplified version the two principal planes for the object and image space respectively are combined into one passing through P. F_1 is the front focus. F_2 is the back focus. $PF_1 = 17$ mm corresponding to the unaccommodated eye viewing the object at a large distance. o and i are object and image, respectively.

optimum for this task, and let $\lambda = 500$ nm. Then the separation of the two point sources on the retina is

$$17 \times 10^{-3} \times 1.22 \times 500 \times 10^{-9}/2 \times 10^{-3} \text{ m} = 5.2 \times 10^{-6} \text{ m}$$

or approximately 5 μm. In the human fovea the distance between cones varies from 3 μm in the center to 5 or 6 μm in the macula. Thus intercone spacing is not a limitation on the physically resolvable stimulus, at least not in the central fovea where acuity is highest.

Instead of point sources one may use gratings. These have advantages in the study of the optical properties of the eye with respect to contrast (cf. Campbell, 1968). In an array of photoreceptors, the detection of contrast depends on diameter and spacing. It is found that the highest spatial frequency that can be detected by the human eye is about 60 cycles/degree. This corresponds to about 5 μm on the retina, the same as for two point sources.

In a more extensive treatment additional factors must be included, such as waveguide properties and the effective entrance aperture of photoreceptors (for a review see Snyder, 1979). Due to their shape and slight differences in the refractive index inside and outside the cell, photoreceptors act as light guides. Consequently, light rays entering at a small angle to the axis can be confined within a rod or cone by internal reflection. As a result photoreceptors are directionally selective for light, their effective entrance aperture is larger than the diameter of the cell, and they capture more than the light directed along the axis. These effects, however, are not as large as those arising from diffraction and we shall not consider them further, except to note that in the adaptation to different functions and due to variations of function across the retina these effects will influence optimal design. In general, a photoreceptor with a larger diameter is exposed to more of the incident photons, while one with a smaller diameter is better for high acuity.

We now assume that the photoreceptor diameter has been determined, based on its functional role as discussed above, and we turn next to an analysis of outer segment length, the other important dimension of rods and cones.

Outer Segment Length and Optimal Detection

In Chapter 2 we saw the need for numerous layers of rhodopsin molecules in the photon path to increase the probability of absorption. We also saw that the light response involves a transduction biochemistry and membrane conductance changes that are restricted to a region within the outer segment where a photon is absorbed. The transduction and membrane processes inevitably produce noise and the longer the outer segment, the more noise there is. We therefore have two competing demands: on the one hand, the outer segment should be long to maximize absorption; on the other hand it should be short to minimize noise. What is the optimum length to satisfy these conditions? We shall consider this question for the red rods of the toad for which the best data are available. Our premise will be that these rods are optimized for the detection of single photons.

A simple expression for the photon absorption probability in an outer segment is given by the Beer–Lambert law

$$p_a = 1 - \exp(-\alpha s) \tag{2}$$

where p_a = absorption probability, α = optical density per micron (OD/μm) of an outer segment, and s = length of outer segment.

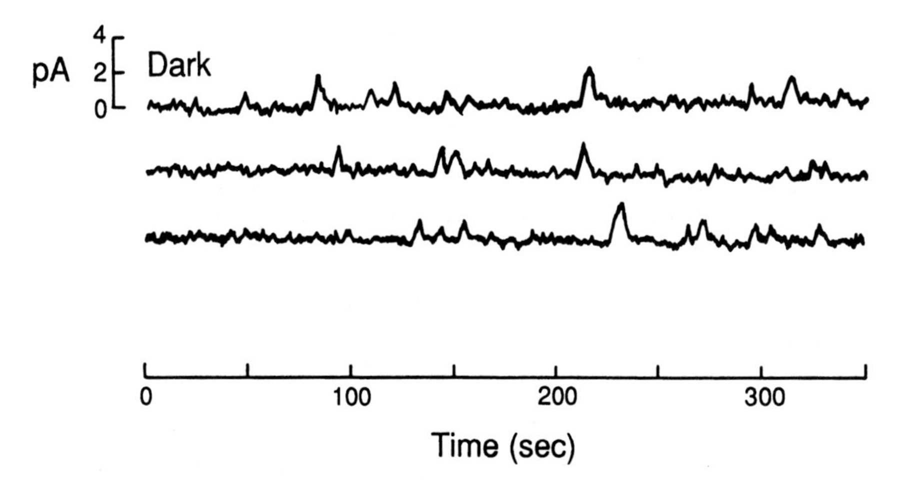

FIGURE 10.2. A sample of the current recorded in the dark from a toad rod, showing the continuous noise from the transduction cycle interrupted by the larger, spontaneous events ascribed to rhodopsin isomerizations of thermal origin. (From Baylor et al. 1980. © The Physiological Society Hon. Treasurer, by permission.)

Photoreceptor noise can be separated into three components (Baylor et al., 1980; Bodoia and Detwiler, 1984; Matthews, 1986). One arises from the transduction biochemistry, another from the spontaneous, thermal isomerizations of rhodopsin, and a third from fluctuations in the states of the light-sensitive channels in the plasma membrane. The third is small by comparison with the other two and we shall neglect it. The biochemical steps in transduction give rise to a continuous noise in the current flowing through a rod in the dark, while the spontaneous isomerizations give rise to discrete events. Figure 10.2 shows a sample record of the dark current obtained with a suction electrode (Baylor et al., 1980). The continuous component has a zero mean and a standard deviation of 0.19 pA, while the discrete component has a mean of 0.96 pA and a standard deviation of 0.2 pA (Baylor et al., 1979, Table 1). The single photon response has the same characteristics as the discrete component since it, too, is due to the isomerization of a rhodopsin molecule.

Now, for a current $i(t)$ fluctuating in time and characterized by a stationary probability distribution $p(i)$ the variance σ^2 is given by

$$\sigma^2 = \int p(i-\bar{i})^2 \, dt$$

where $\bar{i}$ is the mean current and $p \equiv p(i)$, $i \equiv i(t)$. From this it follows that the root mean square (RMS) current is given by

$$\sqrt{\bar{i^2}} \equiv i_{\mathrm{RMS}} = \sqrt{\bar{i}^2 + \sigma^2} \tag{3}$$

Using the figures for the single photon response above we calculate from (3) the

$$\text{RMS signal} = \sqrt{0.96^2 + 0.2^2}\ \text{pA}{=}0.98\ \text{pA} \tag{3a}$$

Similarly, for the continuous noise component,

$$\text{RMS noise}_{\text{cont}} = \sqrt{0.0 + 0.19^2}\ \text{pA}{=}0.19\ \text{pA}$$

But this latter term is for recordings from a 60-μm-long outer segment. We assume that transduction noise is proportional to outer segment length. Therefore for a length s

$$\text{RMS noise}_{\text{cont}} = (0.19s/60)\ \text{pA} \tag{3b}$$

When a photon is incident on a rod, the expected signal size is $p_a \times$ RMS signal.

Within a 50-sec period there is a mean probability that one spontaneous thermal isomerization occurs in a 60-μm-long outer segment. It produces a signal of the same size as the single photon response and occupies about 7 sec out of the 50 sec (cf. Fig. 4 of Baylor et al., 1979). Thus, the probability that a photon is incident when it cannot be confused with such a "false alarm" is

$$(1 - 7s/50 \times 60) \tag{4}$$

Hence, the expected signal size should be corrected to

$$p_a \times (1 - 7s/3000) \times \text{RMS signal} \tag{5}$$

Subtracting the continuous noise from this, we get for the "useful signal size"

$$R \equiv \text{useful signal size}$$
$$= [1 - \exp(-0.016\, s)][1 - 7s/3000] \times 0.98 - 0.19s/60 \tag{6}$$

using (2), (3a), and (3b) and substituting $\alpha = 0.016$ (Harosi, 1975). When (6) is optimized by differentiating R with respect to s and equating to zero, we find that, approximately,

$$s = 65\ \mu\text{m}$$

This is close to the actual outer segment length of between 60 and 70 μm of the red rods of *Bufo marinus* (Baylor et al., 1980; Fain, 1976).

It should be noted that as is true of photoreceptor diameter so the outer segment length may vary between different rods and across the retina, not to mention species differences. In *Bufo*, the green rods have outer segments between 30 and 50 μm long (Yau, personal communication; Fain, 1976). According to Walls (1963) the green rods may be more like cones. They may be involved in some form of color vision and ultimate detection sensitivity may not be as important. Therefore, false alarm and noise might be given a different emphasis in relation to absorption of photons. If the green rods are designed to operate at higher light intensities the continuous noise and false alarm probabilities can also be expected to be different. Although exact data for a quantitative analysis are not available, a shorter outer segment would be consistent with these factors. One or another of the criteria we have considered may also have priority in different species and different regions of the eye. In line with this, as we go out from the fovea, human cones become more sparsely distributed and their outer segments get shorter and more conical. Their shorter length is only partly compensated by their larger aperture, which waveguides the light through the outer segment (Snyder, 1979). These cones are not involved in high acuity. It has been proposed that the conical shape of their outer segments helps radiate those photons that are not absorbed to surrounding rods, thus enabling the latter to absorb more light (Miller and Snyder, 1973). Foveal cones, on the other hand, subserve high acuity. They have smaller diameters well compensated by greater outer segment length to maximize their sensitivity under their operating conditions. In addition to the criteria we have considered other factors, such as metabolic ones, may enter into the design and evolution of rods and cones. Not having included all these we have no right to expect too close agreement between the actual values and those calculated on the basis of our few criteria. On the other hand, the function of a theoretical model is to describe and conceptualize a physical phenomenon. It is neither necessary nor desirable to include all the parameters that may affect the operations of the physical system. A model should highlight selected principles that govern these operations. In this sense our model achieves its purpose. For it is safe to say that photoreceptors are well matched to the optics and signal processing requirements of the eye; and the criteria we have invoked yield dimensions of the order that are found in nature. It supports the notion that these are important and relevant criteria.

Convergence and Divergence of Transmitted Information

General Considerations

We have seen in the foregoing paragraphs that retinal photoreceptors are very sensitive and reliable elements and are capable of high resolution. To preserve these in transmission a human designer might connect each rod or cone to its intended target. But this is not what we find in the retina, nor in the nervous system in general. Instead there is a great convergence and divergence of fiber tracts. In the human retina some 125 million rods and cones converge upon 1 million ganglion cells (Polyak, 1941), and at the same time the signals from one rod or cone are transmitted to more than one ganglion cell. In the macaque monkey 102.6 million rods and cones project to 1.1 million ganglion cells, these in turn to 1.1 to 2.3 million principal cells in LGNd, which in turn are connected to between 50 and 84 million cells in layer IV of the primary visual cortex V1. These connect with 90 to 150 million cells in the other layers of V1. (Estimates are quoted by Orban, 1984 from various sources.) Similarly in other parts of the nervous system, for example, in the cat cerebellum, one parallel fiber is estimated to diverge to 300 Purkinje cells, while between 100,000 and 200,000 parallel fibers converge on one Purkinje cell (Eccles, 1969). These overall numbers do not tell the whole story of convergence and divergence, which exists between any two connected cell populations. It may seem like a smearing out of information and one may well ask how it is possible that, for example, acuity judgments are so precise (see also Leibovic, 1972, Ch. 26).

Of course, the nervous system cannot be concerned with point to point transmission, since this would imply some homunculus sitting somewhere in the brain and reading the information on the retinal screen. Instead we have numerous computations carried out between successive layers in the nervous system. These computations extract behaviorally and functionally relevant parameters and they take place in neural networks. One such network is represented by a target cell and its receptive field. Another such network is represented by a set of target cells with overlapping receptive fields. Experimentally, only the first type has been investigated extensively. For it is easier to record from a single cell while stimulating its receptive field than from a collection of cells. Based on target cell responses, it has been shown that there are different kinds of network: some are tuned to contrast, others to movement, and still others to orientation. There have even been reports of cells which respond with some preference to hands and faces (Gross et al., 1972; Bruce et al., 1981; Kendrick and Baldwin, 1987). The human experimenter cannot be certain, however, that the stimulus he uses is not merely sharing some chords with nature's song to which the cell is tuned. There has been a tendency to ascribe a high specificity to target cell responses (Barlow, 1972). But in fact they are not sharply tuned to their stimuli: a cell responding to contrast and orientation can also respond to movement. The responses are multivariable. The value of one variable can be traded for another and for any one variable, for example, the orientation of an edge, the response may be strongest for one orientation, but the cell will also respond, though less strongly, to a range of orientations around the optimal one. These facts can easily be forgotten, especially in the excitement of having found a meaningful stimulus to which a cell talks back to the experimenter.

Representation and Preservation of Information

To consider a simple example, suppose we have a target cell with an ON center, OFF surround receptive field. It responds to a spot of light with excitation when presented in the center and with inhibition when presented in the surround. When the spot is moved across a receptive field diameter the size of the response will change as shown in Figure 10.3a. When the intensity of the light is increased, the response is also increased. Clearly, position can be traded for intensity and the target cell does not "know" what the position or intensity of the spot may be. But now consider two overlapping receptive fields A and B as illustrated in Figure 10.3b (Leibovic, 1969, 1981). If we plot the simultaneous responses of the target cell of A versus those of B we get a series of curves such as in Figure 10.3c or d. In Figure 10.3c the position of the spot is fixed on each curve, but its intensity is variable. In Figure 10.3d the intensity is fixed on any one curve but the position is variable. We now see that the position and intensity of the spot are both specified by the simultaneous responses of A and B. There is still an ambiguity at the intersections of the receptive fields A and B in our figure. But this can be removed by adding a third receptive field C as shown.

The example is instructive. The overlapping region of the receptive fields diverges to two or three target cells. The convergence of each receptive field onto its target cell has destroyed the information on position and intensity. But this can be recovered through divergence, provided we read the simultaneous responses of the target cells.

The above case is concerned with the position and intensity of a spot on the retina. But it is more general and applies equally to computations in networks in different layers in the nervous system. Consider, for example, a receptive field which is tuned to direction of motion. Suppose the field is oval in shape and the target cell responds most strongly to a stimulus moving along the major axis, the preferred direction, and most weakly to a stimulus moving along the minor axis. Motion in one direction elicits a positive response, or excitation, while motion in the opposite direction elicits a negative response or inhibition. Let us also suppose that the response amplitude of the target cell is monotonic with the speed of motion. Let the retina be covered with such overlapping receptive fields in different orientations. Let the preferred direction of receptive field i ($i = 1,2, \ldots ,N$) be denoted by the unit vector $\mathbf{c}_i$ and the direction of stimulus motion by the unit vector $\mathbf{m}$, and suppose the target cell response is given by

$$r_i(\mathbf{m}) = r_{0i} + k_i\mathbf{c}_i \cdot \mathbf{m} \tag{7}$$

where r_{0i} is a maintained level of activity of cell i, k_i is a factor depending on stimulus speed, and $\mathbf{c}_i \cdot \mathbf{m}$ is the scalar product.

This case is formally equivalent to the one shown in Figure 10.3 and, as there, it needs only two or three overlapping receptive fields for direction and speed to be encoded in the simultaneous activity of the target cells.

For a large movement one may construct the following scheme with overlapping receptive fields covering the retina.

Form the weighting factors

$$w_i = r_i - r_{oi} = k_i\mathbf{c}_i \cdot \mathbf{m}$$

Then multiply each $\mathbf{c}_i$ by w_i and form the sum

$$\mathbf{P}(\mathbf{m}) = \sum_{i=1}^{N} k_i\mathbf{c}_i \cdot \mathbf{m}\, \mathbf{c}_i$$

$$= \sum_{i=1}^{N} w_i\mathbf{c}_i \tag{8}$$

$\mathbf{P}(\mathbf{m})$ is a vector which uniquely represents the speed and direction of a rectilinear movement over the retina.

We have adapted the above model from Georgopoulos et al. (1986) who found that it accurately predicts movement based on a knowledge of the firing of motorneurons.

The receptive fields we have considered were tuned to spots of light and rectilinear motion. We now consider a more general case which sheds additional light on the problem of filtering and preserving information (Leibovic, 1988). Our model illustrated in Figure 10.4a is now a "retina" of $n \times n$ pixels or picture elements and a superimposed set of $m \times m$ "receptive fields," each connected through some network to a target pixel at a succeeding level. The strength of activity of each pixel can be represented by a gray level. Let these gray levels be enumerated as x_{ij} according to their row and column position on the retina. Let the target cell responses be y_{ij}, enumerated according to the top left hand corner pixels of their receptive fields. We assume that the responses are given by

$$y_{ij} = \sum_{r=i}^{i+m-1} \sum_{s=j}^{j+m-1} K_{ij}^{rs} x_{rs} \tag{9}$$

where $i,j = 1,2,\ldots,[n - (m - 1)]$ and $r,s \leq [n - (m - 1)]$ and K_{ij}^{rs} are constants. For example, for $m = 2$, $n = 4$, and all coefficients $K = 1$, the Eq. (9) would read

$$\begin{array}{rcccccccccccc}
\text{Pixel\#:} & & 11 & 12 & 13 & 14 & 21 & 22 & 23 & 24 & 31 & 32 & 33 \\
y_{11} & = & x_{11} & +\,x_{12} & +\,0 & +\,0 & +\,x_{21} & +\,x_{22} & +\,0 & +\,0 & +\,0 & +\,0 & +\,0 \\
y_{12} & = & 0 & +\,x_{12} & +\,x_{13} & +\,0 & +\,0 & +\,x_{22} & +\,x_{23} & +\,0 & +\,0 & +\,0 & +\,0 \\
\cdots & & & & & & & & & & & & \\
y_{21} & = & 0 & +\,0 & +\,0 & +\,0 & +\,x_{21} & +\,x_{22} & +\,0 & +\,0 & +\,x_{31} & +\,x_{32} & +\,0 \\
\cdots & & & & & & & & & & & & \\
y_{33} & = & 0 & +\,0 & +\,0 & +\,0 & +\,0 & +\,0 & +\,0 & +\,0 & +\,0 & +\,0 & +\,x_{33}
\end{array}$$

In this representation we have written down the pixels according to their position on the retina at the top of the set of equations, and they form the columns of the coefficient matrix of (9). The condition on i,j ensures that the receptive fields do

---

▶

FIGURE 10.3. (a) Illustration showing how response amplitude of an idealized ON center target cell changes when a spot stimulus is moved across a diameter of its receptive field. The horizontal axis is the distance along the diameter of the field centered at zero. The vertical axis is the response amplitude. (b) Schematic representation of three overlapping, circular receptive fields A, B, C (top) and the target cell responses of A and B to spot stimuli of intensities I_1, I_2, I_3 in different positions along the receptive field diameter. (c) Illustrative plot of the simultaneous responses of A and B. Each curve is for a fixed position and varying intensity of the stimulating spot. (d) Same as (c) for a fixed intensity and varying position of the stimulating spot on each curve. The numbered points in (c) when connected by a smooth curve illustrate how (d) could be derived from (c). (b, c, and d From Leibovic 1981. © Hemisphere Publishing Co., by permission.)

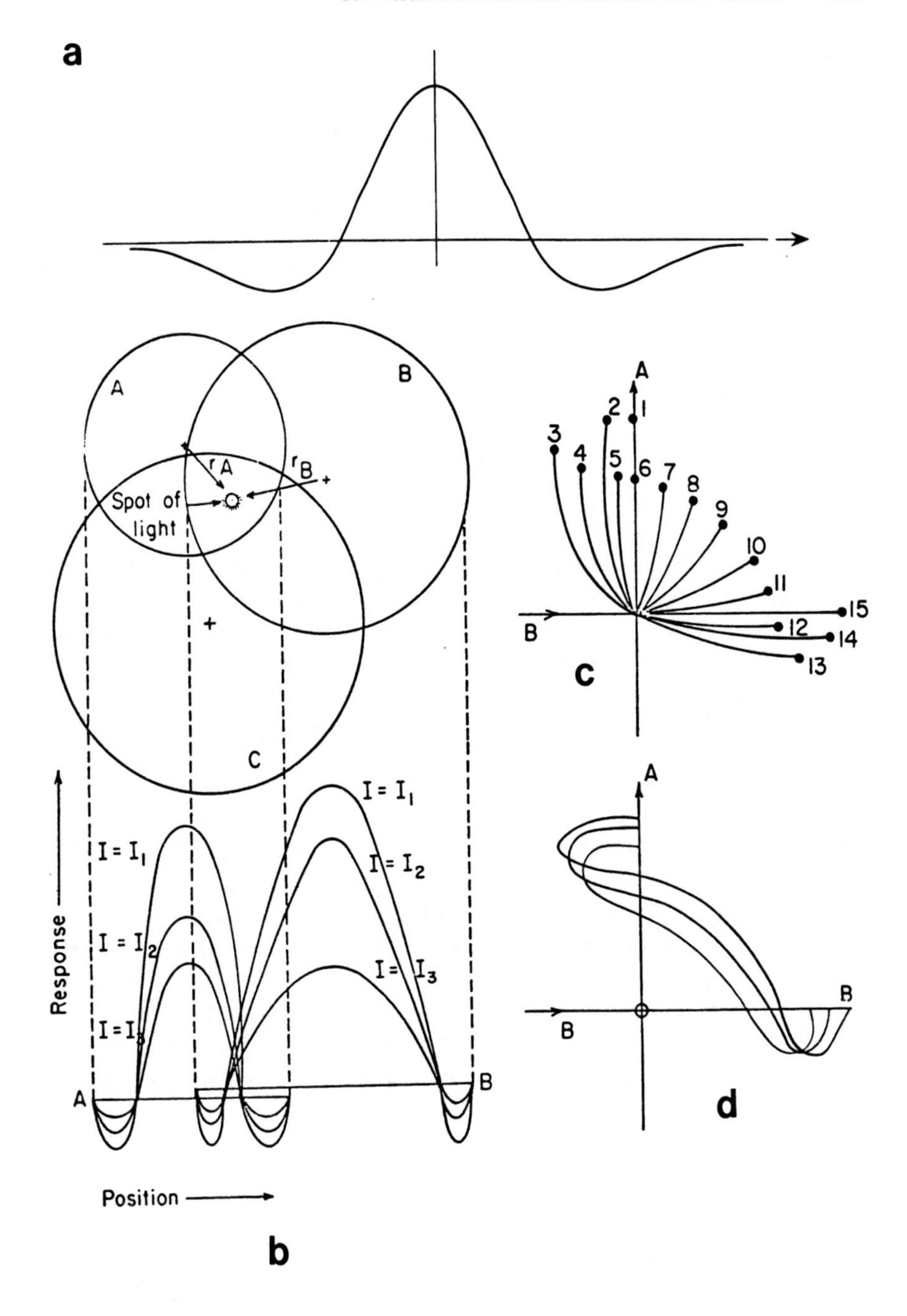
a
A
B
C
r A
r B
Spot of light
Response
Position
I = I1
I = I2
I = I3
b
c
d
1
2
3
4
5
6
7
8
9
10
11
12
13
14
15

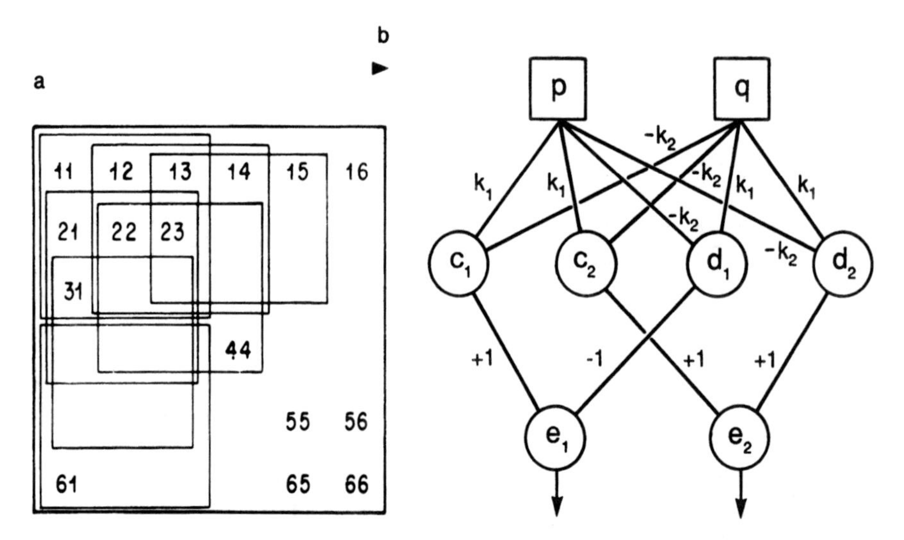

FIGURE 10.4. (a) The large square is a "retina" containing 6 × 6 pixels numbered 11, 12, ... 16, 21, ... 26, ... 61 ... 66. The small squares are 3 × 3 "receptive fields." (From Leibovic, 1988. © Kluwer Academic Publishers, by permission.) (b) Two elementary subnetworks with convergence and divergence. The strength of the connections or weights between p and c_1, p and c_2, q and d_1, q and d_2 are equal to k_1; the weights between p and d_1, p and d_2, q and c_1, q and c_2 are equal to $-k_2$. The weights between c_1 and e_1, c_2 and e_2, d_2 and e_2 are +1 and between d_1 and e_1 the weight is −1.

not spill over the boundary of the retina, and the condition on r,s ensures that the number of unknowns x_{rs} is equal to $[n - (m - 1)]^2$, the same as the number of equations. The latter condition is equivalent to setting $[2n - (m - 1)](m - 1)$ of the boundary pixels or their coefficients equal to zero, which is not a serious limitation if $m \ll n$.

The reason for choosing the form given by (9) for our set of equations is that the solution becomes easily apparent: In the determinant of the coefficients K_{ij}^{rs} all the elements below the principal diagonal are zero. Accordingly, the value of the determinant is

$$|K_{ij}^{rs}| = \prod_{i=1}^{n-(m-1)} K_{ii}^{ii} \tag{10}$$

i.e., it is the product of the elements in the principal diagonal. This is nonzero provided none of the K_{ii}^{ii} is zero. Therefore, there exists a unique solution of (9). In other words, the set of target cell responses implicitly and precisely carries the gray level values of the pixels on the retina.

It will be noted that in the above example, to have a unique solution, the number of receptive fields overlapping each pixel is m^2 —except near the boundary of the retina, where it is less. m^2 is the maximum possible number of distinct $m \times m$ fields overlapping a pixel. We summarize these results as follows:

A1. The maximum number of distinct $m \times m$ receptive fields overlapping a pixel is m^2.

A2. The maximum number of distinct receptive fields of size $m \times m$ on a retina of size $n \times n$ is $[n - (m - 1)]^2$, $m < n$. The term $(m - 1)$ is due to the stipulation that the receptive fields must not spill over the boundary. When $m \ll n$ as may be expected in practice, the maximum number of receptive fields is n^2 approximately.

A3. A unique solution exists for the pixel gray levels in an $n \times n$ retina containing $m \times m$ receptive fields with maximum overlap when the contributions of $(m - 1)[2n - (m - 1)]$ of the boundary pixels are negligible and set equal to zero.

The consequences of these results are as follows:

B1. We can have receptive fields of any size without losing information, provided there is sufficient overlapping of receptive fields. The larger (smaller) they are, the more (less) they will need to overlap.

B2. There is not necessarily a saving of target cells, but we have acquired computing power as compared to pointwise transmission. Each target cell in this model computes a weighted sum over its input, which may yield contrast, orientation, etc. The price for this is the greater number of lines in the network as represented by K_{ij}^{TS} in (9).

As regards B1 it disproves the frequently made statement that acuity or precision require small receptive fields. As regards B2 it shows that a target cell can carry information tuned to a stimulus class as well as information on position and intensity. For example, the target cell e_1 in Figure 10.4b will respond positively when the output $p > q$ and negatively when $p < q$ and it will not respond when $p = q$. Thus it distinguishes between two-sided contrast, on either the left or right. The target cell e_2, for $k_1 > k_2$, will respond positively when $p + q > 0$ and negatively when $p + q < 0$. Thus it registers a weighted mean level of activity above or below a zero baseline. At the same time, given the pair of responses e_1 and e_2, the values of p and q are uniquely determined.

In the model of Figure 10.4a and Eq. (9) we have not explicitly mentioned positional information. This is specified for each retinal pixel by the subset of target cells to which it diverges. Those pixels which diverge to different, not necessarily disjoint, sets of target cells are represented uniquely by these sets. In addition, in the nervous system we know that neighborhoods of cells at one level project to neighborhoods of cells at another level. Thus, for example, there is a topological map of the retina in V1. Positional information in the retina is mirrored in the positional information of V1.

Some Properties of Receptive Fields

Symmetric and Antisymmetric Receptive Fields

The response r of a target cell is often represented as a weighted sum as in (8) and (9) or as an integral, such as

$$\text{in two dimensions:} \quad r = \int_R K(x,y) f(x,y)\, dxdy$$
$$\text{or in one dimension:} \quad r = \int_R K(x) f(x)\, dx \tag{11}$$

where K is the kernel or transformation or computation applied by the receptive field network to f which represents the input on R. One form for K is the difference of two Gaussian functions corresponding to excitation and inhibition,

$$K(x) = Ae^{-ax^2} - Be^{-bx^2} \tag{12}$$

Experimentally determined response functions and shapes of receptive fields are often quite irregular and not like the idealized shapes and functions of Figure 10.3 or Eq. (12). The question therefore arises what, if any, relevance to reality the theoretical constructs may have. To this end consider two general types of theoretical receptive field functions: one symmetric and the other antisymmetric. Consider the symmetric field first.

Let $s(x) = s(-x)$ be an arbitrary normalized symmetric function with $\int s(x)\, dx = 1$; and let $t(x) = -t(-x) = A(x) - A(0)$ be an arbitrary antisymmetric function. Then

$$\int s(x)A(x)\, dx = A(0) \tag{13a}$$

If, instead of $t(x)$ we have a constant $S(0)$, then

$$\int s(x)S(0)\, dx = S(0) \tag{13b}$$

The relevance of (13) to our problem is that $s(x)$ can represent a symmetric receptive field function such as (12). $t(x)$, defined for $x \pm \Delta x$, or $S(0)$ can be piecewise linear approximations of an image projected onto a retina or of an activity pattern in a layer of cells in the nervous system. Then, according to (13) the target cell samples the activity at $x = 0$. A set of such target cells can sample the activity over a suitable set of points on the retina or in the layer of cells. Clearly, the accuracy of the individual sample depends on how good the linear approximation is and this depends on the size of the receptive field: A large receptive field encompasses a larger part of the image for which $t(x)$ or $S(0)$ is not an accurate approximation.

We next consider an antisymmetric receptive field. Let $a(x) = -a(-x)$ be an arbitrary, antisymmetric function so that $\int a(x)\, dx = 0$. If $r(x) = r(-x) = S(x) - S(0)$ is an arbitrary symmetric function then

$$\int a(x)S(x)\, dx = 0 \tag{14a}$$

while for an arbitrary antisymmetric function $t(x)$, as in (13),

$$\int a(x)t(x)\, dx = \int a(x)A(x)\, dx \neq 0 \tag{14b}$$

In other words, the receptive field kernel represented by $a(x)$ gives a response to antisymmetric inputs such as contrast or change, but not to unchanging, constant inputs.

Now let us return to an arbitrary function such as an experimentally determined receptive field response profile. It is well known that an arbitrary function $f(x)$ can be decomposed into the sum of a symmetric and an antisymmetric function, i.e., $f(x) = s(x) + a(x)$ where $s(x)$ and $a(x)$ are symmetric and antisymmetric as above. From our argument it therefore follows that an arbitrary receptive field kernel applied to the input will yield a target cell response which implicitly contains information on the magnitude of the input at the point of symmetry of $s(x)$ as in (13), as well as information on the weighted integral of the changes of the input, as in (14b), within the receptive field. This answers the question on the relevance of theoretical constructs to real receptive field kernels. In addition to the implicit information carried by the target cell there is the capability, in principle, for decomposing the kernels into symmetric and antisymmetric parts and displaying explicitly the magnitude and variation of the input. We must remember, however, that our analysis applies only to linear systems such as (9) or (11). On the other hand, due to its generality, our result is valid for any kernel or network extracting any input parameters.

Receptive Field Size

We have seen that receptive field size per se does not limit resolution, provided there is sufficient overlapping of receptive fields on each element or pixel. Thus, in the model of Eq. (9), each pixel must be covered by four receptive fields of size 2×2 or by nine of size 3×3 for maximum resolution. The larger the receptive fields the more are needed to overlap each pixel. In addition, large receptive fields have more connecting links between the input pixels and their target pixel. It follows that large receptive fields are more costly than small ones for achieving maximum resolution.

For this reason one would expect to see small receptive fields involved in computing local properties. On the other hand, large scale properties can better be computed by large receptive fields since in this case resolution is not a primary consideration and little if any overlap is required. One 10×10 field with one target cell covers as much as 25 disjoint 2×2 fields with 25 target cells.

Therefore, the presence of large or small receptive fields in different computational contexts is a matter of cost, or optimality, not of accuracy or resolution.

It has been found that multiscale computations make for efficient procedures in machine vision problems (for a review see, e.g., Ullman, 1986). From what we have said about large and small receptive fields, this is hardly surprising (see also Chapter 11).

Overlapping Receptive Fields and Reliability

Computer users often experience frustrations when their computers do not perform due to some imprecision, however slight, in their program instructions. A human being, on the other hand, can perform more or less satisfactorily even within a vaguely defined environment. A visual scene, if it appears puzzling or contradictory is given a second look.

Suppose there is some malfunction in the network subserving the receptive field of a target cell. We can divide the receptive field into subsets responding to the parameters to which it is tuned. For example, each small neighborhood within the receptive field of Figure 10.3 responds to a spot of light; and in the case of a field tuned to the speed and direction of motion, small subsets of the field also respond to these parameters. The target cell response represents a summation over all such neighborhoods. An error in one or two neighborhoods will be like a statistical deviation from a mean. Moreover, neighborhoods are in general overlapped by more than one receptive field. They are sampled, as it were, more than once, and this sampling is not only by a stationary set of receptive fields. As we saw in Chapter 6, our eyes execute small, involuntary movements even as we fixate a scene, and so the pattern on the retina is projected successively onto shifted sets of overlapping receptive fields. There is therefore a great potential for reliability built into the system. This is not achieved by a redundant duplication of units, but through convergence and divergence: The networks converging onto target cells are involved in local and global computations. The diverging connections, through overlap, make different computations possible on the same input cells and can at the same time preserve the fine grained information lost through convergence.

The degree of overlap depends on function: transmission of detail requires maximal overlap as in A1, while computation of extensive variables such as velocity may require less. However, maximal overlap may not always be convenient in computing all the variables which may be required. For example, in the "retina" of Figure 10.4a and Eq. (9) we may want to have a receptive field covering the whole "retina" to monitor directly the average (or summed) light intensity. Such a receptive field could be part of a system for adaptation, shifting the operating window of the pixels over a larger range as discussed in Chapters 2 and 7. But at the same time such a receptive field could also play a role in error detection. Here we shall indicate only briefly how this might be done.

We can write Eq. (9) as $y = Ax$, with $|A| \neq 0$, where y and x are column vectors whose components are y_{ij} and x_{rs}, respectively, A is the matrix whose elements are K_{ij}^{rs}, and $|A|$ is the determinant of A. Let $y_0 = kx$ be an additional equation where y_0 is a scalar, k a row vector with all elements equal to 1, and x the same column vector as before. Then y_0 is the target cell output of our new receptive field, summing all the pixel values. But k can be expressed as a linear combination, which we denote as (λ), of the rows of A and then y_0, being the target cell output, is the same linear combination of the y_{ij} in y. Now, suppose there is an error in the network so that either A or the output y is affected or y_0 is wrong. Then if we apply the linear combination (λ) to y it will not, in general, equal y_0, as it should. So the presence of an error is detected. The price for this is one additional network which applies the operation (λ) to y.

Of course, the existence of a mathematical procedure does not mean that it is applied in a biological system. In our case it demonstrates what information is implicitly available when certain kinds of receptive field are present. But considering that we know so little about actual "computations" performed by the brain networks, this is important information. It helps suggest to the experimenter not only which possible computations to look for but also saves his efforts in not searching for those that cannot be performed.

Summary of Consequences of Convergence and Divergence

Since target cells respond to several stimulus parameters and receptive fields are broadly tuned, single-cell responses cannot be either highly specific or selective. There are no grandmother cells which recognize granny: For even the cells which have been reported to respond to faces and hands respond to a variety of such faces and hands and maybe to other stimuli. Our experience with computers suggests that rigid requirements for high precision in the input would make it more difficult for an organism to function in a changing environment subject to statistical fluctuations. From this point of view multivariable broadly tuned units seem preferable if they can be recruited into appropriate patterns of activity in correspondence with the external world. This is what must be the case in the nervous system.

Our visual perception is concerned with color, movement, and form. The computations of their characteristics demand a convergence of activity onto target cells: The perception of color arises from the combined responses of photoreceptors tuned to wide wavelength bands, perception of movement from the activation of a network of cells tuned to a range of speeds and directions and so on. But convergence degrades resolution. It may seem counterintuitive that resolution can actually be restored through divergence.

As a consequence of convergence and divergence, information must be represented in the activities of functional groups of cells. To be coherent, such functional groups must arise from overlapping receptive fields. Computations are carried out in parallel from overlapping fields and are distributed to sets of target cells through divergence. This design has a degree of reliability built into it, and it can be achieved without unnecessary duplication. For receptive fields, even though they overlap, are not identical. The amount of overlap depends on the number of computations to be performed and on the resolution required in the area of overlap. The size of the receptive field varies depending on whether it is tuned to local or global properties of the input. It seems an altogether admirable design for information processing, radically different from present day man-made systems.

Conclusion

In this chapter we have discussed some features of biological design and their relation to function in information processing. We have considered three examples: diameter and spacing of photoreceptors, their outer segment lengths, and the convergence and divergence of fibers in the nervous system. We have seen that the first two examples can be viewed as design optimized for function. The third example, which may present something of a puzzle at first sight, turns out to have inherent possibilities for all kinds of parallel computation with high fidelity and reliability without undue redundancy. Here is a case where studies of a model show how structure implies certain modes in which the system can function.

We know little of the explicit information which the nervous system extracts and uses at different stages. But analyzing what information is there implicitly, we may be able to investigate experimentally whether and how it is used and whether it is made explicit at some point; and we can also see what kinds of information cannot be there.

There are many more examples of the relations between structure and function in biological information processing and, no doubt, we shall see further important advances in our understanding of this field in the future.

Acknowledgments. In writing this chapter I have benefited from discussions with Dr. Richard P. Wildes and my students. I am also grateful for earlier exchanges with Professors Roberto Moreno-Diaz and J. Mira-Mira.

References

Barlow HB (1972). Single units and sensation: A neuron doctrine for perceptual psychology? Perception **1**:371–394.

Baylor DA, Lamb TD, Yau K-W (1979). Responses of retinal rods to single photons. J. Physiol. **288**:613–634.

Baylor DA, Matthews G, Yau K-W (1980). Two components of electrical dark noise in toad retinal rod outer segments. J. Physiol. **309**:591–621.

Bodoia RD, Detwiler PB (1984). Patch-clamp recordings of the light sensitive dark noise in retinal rods from the lizard and frog. J. Physiol. **367**:183–216.

Born M, Wolf E (1987). *Principles of Optics*. Pergamon Press, Oxford, New York.

Bruce C, Desimone R, Gross CG (1981). Visual properties of neurons in a polysensory area in superior temporal sulcus of the macaque. J. Neurophysiol. **46**(2):369–384.

Campbell FW (1968). The human eye as an optical filter. Proc. IEEE **56**(6):1009–1014.

Eccles JC (1969). The dynamic loop hypothesis of movement control. In *Information Processing in the Nervous System* (KN Leibovic, ed.), pp. 245–269. Springer-Verlag, New York.

Fain GL (1976). Sensitivity of toad rods: Dependence on wavelength and background illumination. J. Physiol. **261**:71–101.

Georgopoulos AP, Schwartz AB, Kettner RE (1986). Neuronal population coding of movement direction. Science **233**:1416–1419.

Gross CG, Rocha-Miranda CE, Bender DB (1972). Visual properties of neurons in inferotemporal cortex of the macaque. J. Neurophysiol. **35**(1):96–111.

Harosi FI (1975). Absorption spectra and linear dichroism of some amphibian photoreceptors. J. Gen. Physiol. **66**:357–382.

Hughes A (1987). The schematic eye comes of age. In *Visual Neuroscience* (JD Pettigrew, K Sanderson, WR Levick, eds.), pp. 60–89. Cambridge University Press, Cambridge.

Kendrick KM, Baldwin BA (1987). Cells in temporal cortex of conscious sheep can respond preferentially to the sight of faces. Science **236**:448–450.

Leibovic KN (1969). Some problems of information processing and models of the visual pathway. J. Theoret. Biol. **22**:62–79.

Leibovic KN (1972). *Nervous System Theory.* Academic Press, New York.

Leibovic KN (1981). Principles of brain function: Information processing in convergent and divergent pathways. *4th European Meeting on Cybernetics and Systems Research* (F Pichler, R Trappl, eds.), pp. 91–99. Hemisphere, Washington, D.C.

Leibovic KN (1988). Parallel processing in nervous systems with converging and diverging transmission. In *Biomathematics and Related Computational Problems* (LM Ricciardi, ed.), pp. 65–72. Kluwer Academic Publishers, Dordrecht, The Netherlands.

Linksz A (1950). *Physiology of the Eye.* Vol. 1: *Optics.* Grune & Stratton, New York.

Matthews G (1986). Comparison of the light sensitive and cGMP sensitive conductances of rod photoreceptors: Noise characteristics. J. Neurosci. **6**(9):2521–2526.

Miller WH, Snyder AW (1973). Optical function of human peripheral cones. Vision Res. **13**:2185–2194.

Orban GA (1984). *Neuronal Operations in the Visual Cortex.* Springer-Verlag, Berlin.

Polyak S (1941). *The Retina.* Chicago University Press, Chicago.

Pugh EN (1988). Vision: Physics and retinal physiology. In *Stevens' Handbook of Experimental Psychology,* 2nd ed., Vol. 1. Wiley, New York.

Snyder AW (1979). Physics of vision in compound eyes. In *Handbook of Sensory Physiology,* Vol. VII/6A. Springer,Verlag, Berlin.

Ullman S (1986). Artificial intelligence and the brain. Annu. Rev. Neurosci. **9**:1–26.

Walls GL (1963). *The Vertebrate Eye and Its Adaptive Radiation.* Hafner, New York.

11
Computational Vision with Reference to Binocular Stereo Vision

RICHARD P. WILDES

Introduction

Biological and machine vision systems are clearly different at certain levels of analysis (e.g., neuronal vs. silicon-based circuitry). However, it is possible to develop an understanding of vision at a level that is independent of hardware-related issues. The key idea behind the computational approach is to view the study of vision as an inquiry into the type of information processing that is necessary to make useful inferences about properties of a three-dimensional world from corresponding visual images. This point of view allows one initially to abstract away from the algorithmic and implementational details of how any particular vision task is performed by any particular vision system. Instead, initial effort is placed into establishing formal relationships that serve to define exactly what problems are being solved and into specifying why those particular problems are being solved. It is believed that the computational paradigm can serve to define a set of precise yet fundamental problems for the vision sciences. In this way, the computational approach can bring together vision research from such diverse disciplines as computer science, neuroscience, and psychology.

In this chapter an overview of the computational approach to studying vision is provided. The developments unfold in four sections. In the first section, the basic tenets of computational vision are discussed. In the second section, these ideas are illustrated by taking a look at a particular area of computational inquiry: binocular stereo vision. This area has been chosen as it has been the subject of extensive computational investigation. The third section provides a brief look at current trends in computational vision. Finally, a closing summarizes the main points of the exposition.

The Computational Paradigm

This section explores the underlying principles of the computational approach to studying vision. The section begins by explaining how the computational paradigm derives from the general area of artificial intelligence. This discussion leads naturally into the second part of the section where the particulars of the computational paradigm are described. A brief recapitulation serves as a finish.

Relation to Artificial Intelligence

Computational vision derives its foundations and motivations from the field of artificial intelligence (AI). Therefore, it is useful to begin by exploring the two main goals of artificial intelligence: First, AI strives to identify and understand the principles of intelligence. Second, AI seeks to make more intelligent and hence more useful computers (Winston, 1984). Interestingly, while it is possible to state the goals of AI in this fairly straightforward fashion, it is rather difficult to define AI in a comparatively precise and terse manner. This is due to the seemingly heterogeneous set of behaviors that is commonly considered to be indicative of intelligence. (Consider, e.g., the task of planning a path of motion through a cluttered environment and its associated issues of collision avoidance and geometric reasoning as opposed to the task of understanding a sentence of a natural language and its associated issues of parsing and word meaning.) In fact, reflecting on the first stated goal of AI shows that one of the goals of the field is to precisely define what is meant by intelligence. In the next few paragraphs, these ideas are explored in some detail.

Consider the notion that AI is concerned with understanding the principles of intelligence. A basic tenet of research in the field is that intelligence can be studied independently of any particular system that might embody intelligence. This point of view arises naturally from the contemporary view of information processing where it is common to separate out an information processing task from a system that accomplishes that task (Weinberg, 1975). This idea is simply illustrated by noting that several different computer programs can be written in several different computer languages and run on several different computers, yet still produce essentially the same results (e.g., process payroll accounts). When this insight is applied to the study of intelligence, a science can evolve that is neither in favor of nor against explanations that bear resemblance to known facts about biological systems. Instead, the goal is to study intelligence in its own right, an area of inquiry that is subject to its own structures and rules.

A second set of issues revolves around the notion of building intelligent computers. Clearly, success in this realm will have great practical applications. However, the synthesis of intelligent systems also will shed significant light on more fundamental issues of intelligence. To be embodied in a working computer program, a theory must be precisely specified. This type of precision can play an important role in exposing hidden assumptions on the part of a scientist. Subsequent testing of a program can help demonstrate which ideas will and will not work in principle. Further, the design and testing of AI programs can help in developing an understanding of the sheer complexity of many information processing tasks. Indeed, one of the major lessons that has been learned so far in the history of AI is that many seemingly simple tasks require information processing at severe levels of complexity.

Another important point is that approaches to studying complex information processing that are limited to concerns of mechanism or behavioral observation can be inherently limited in the level of understanding that they ultimately achieve. In theory, one could even build a machine that duplicates a system's observed mechanisms and behaviors without understanding the principles behind the mimicked system. This is due to the inherent difficulties of extrapolating from primitive components of a system to the task that is being performed. An example that is often used to illustrate these matters is air flight (e.g., Hildreth and Hollerbach, 1987). Suppose a scientist wishes to understand the principles of air flight. A behaviorally based approach to this problem might begin by searching for an existing example of flight to study, say birds. Following sufficient experiment with this example, the scientist might find that a bird deprived of its feathers does not fly. Further, the scientist might conclude that the understanding of flight must somehow lie with feathers. Similarly, a mechanism-based approach might succeed through great care in dissection and duplication to build a mechanical bird that is capable of flight. This scientist might now claim to understand air flight. Unfortunately, it is not at all clear how either of these lines of inquiry will lead to the scientist's discovering the laws of aerodynamics and the formalization of concepts such as lift. These latter concepts go a long way toward an understanding of flight, while feathers and the mechanical model are simply tied to a particular system's way of manifesting these principles.

In summary, artificial intelligence is founded on the principle that to develop a complete description of a complex information processing problem it is necessary to consider more than mechanism and behavior. It also is necessary to inquire directly into the nature of the task that is under consideration. It is from these foundational ideas that the area of computational vision is derived.

The Computational Approach to Vision

The computational approach to vision follows in the path of artificial intelligence in its emphasis on capturing the underlying information processing structure of problems in its domain. The framework for computational vision was elegantly specified by David Marr (Marr, 1976, 1982; Marr and Poggio, 1977). Within the computational framework, an understanding of an information processing task is viewed as complete only when it can be specified at each of three distinct levels of explanation. One level of explanation is concerned with making a precise statement of what problem an information processing system is solving as well as why it is solving that particular problem; this level is called the computational theory. This level must also make a statement about how the physical properties of the world constrain the possible solutions to the problem. A second level of explanation is concerned with precisely stating a procedure that is capable of producing the solution to the problem that is isolated by the computational theory; this level is called the algorithm. The final level deals with a physical instantiation of the algorithm; this level is referred to as the implementation. The next few paragraphs expand on these points.

At the level of the computational theory, one essentially seeks to specify a formal mapping from some expected input information to some desired output information. For example, in the case of binocular stereo vision the mapping might be from a pair of two-dimensional images (retinal or camera) to useful assertions about the three-dimensional structure of the imaged scene. In attempting to specify the computational mapping for almost any problem in vision one faces the situation that the anticipated input information is inadequate to exactly specify the desired output information. Briefly, the problem is underconstrained. The solution is to understand how general properties of the physical world can provide additional constraint so that the solution is well defined. For the sake of reliability, constraints should reflect robust and generally true properties of the world or properties that can be independently verified. The notion of constraint can also be illustrated via stereo vision: Notice that if one has no knowledge of binocular image formation, then it is not clear how points in one image correspond to points in the other image. (This is called the correspondence problem.) However, in view of physical principles the problem becomes tractable. One useful principle is that an object can occupy only one point in space at any one time. This can serve to greatly simplify the correspondence problem as it motivates a constraining rule: A point in one image can correspond to no more than one point in the other image. This rule helps limit the possible correspondences that are established between the images in a binocular view.

Two points from the foregoing discussion are worth emphasizing: First, at the level of the computational theory, the emphasis is on developing an analysis that exposes the underlying structure of the information processing task at hand independently of any particular system for solving the task. Second, an integral part of any computational theory is an explicit statement of how the physical and mathematical structure of the problem domain constrains the computation so that there is an "in principle" solution to the problem of concern. In this light, the fact that the world has a structure is in some sense a precondition for perception (c.f. Bennett et al., 1989; Bobick and Richards, 1986; Marr, 1970). If the world did not have an underlying structure, then it is not at all clear what it would mean to perceive.

At the level of the algorithm the concern is with stating a step by step procedure that yields the solution that has been specified by the computational theory. At this level it also is necessary to choose explicit representations for the information that is input and output by the computation. Typically, it is possible to devise more than one set of procedures and representations for carrying out the necessary operations. In fact, it often is possible to specify both parallel and serial approaches to the computation of interest. This is important as it emphasizes the fact that parallelism vs. serialism is a concern at the level of the algorithm not at the level of the computational theory.

The level of implementation is concerned with the physical instantiation of the algorithm. In general, more than one implementation is possible for a given algorithm. Another concern at this level is that of testing. The demonstration that a computational theory can perform robustly in the face of real world image data is of great importance. This testing can often expose hidden assumptions on the part of even the most conscientious investigator.

It is important to note the manner in which the different levels of analysis relate. Clearly, the computational theory will go a long way in specifying the nature of the algorithm and implementation. Also, the types of mechanisms that are available can greatly influence the types of algorithms that are considered. However, the computational theory, founded in the notion of separating task from mechanism, is not directly influenced by algorithm or implementation. These issues can be illuminated by Marr's classic example involving Fourier analysis (Marr, 1976). For this case, the theory of the Fourier transform (a computational theory) is well developed and exists independently of any particular method for how it might be executed. At the level of algorithm there are many ways to carry out the needed operations. For example, there is the serial Fast Fourier Transform (Cooley and Tukey, 1965) as well as parallel optically based methods. Physical mechanisms are also numerous. However, the type of mechanism that is available can greatly influence the algorithm that is selected. If one has digital memory and adders, then one would be wise to opt for the Fast Fourier Transform. If one has a laser and holographic plates, then the optical methods are better. This discussion is not meant to imply that known facts about how a particular system operates are incapable of ruling against the application of a theory to that specific system. As an example, the fact that humans have only two eyes rules against the relevance of trinocular stereo (e.g., Hansen et al., 1988; Steart and Dyer, 1988) for human vision. However, this evidence against trinocular stereo in humans does not destroy its validity as an analysis of an information processing problem, namely stereo with three images.

When applying the computational approach to problems in visual information processing, it is probably folly to think of it as a rigid prescription for how to do research. That is, one rarely begins by developing a complete computational theory, followed by detailed specification of an algorithm, followed by the design and testing of a physical mechanism. Instead, research often proceeds along several lines at once or along one line for a while then over to another then to another. Nevertheless, the computational method does help build a crisp and rigorous approach to research if the following questions are kept in mind: First, at any time what level of explanation is being developed? Second, what types of constraints are operative at any one level of explanation? Third, when a research project is complete is it possible to point to explicit explanations of the given problem at each of the three levels: theory, algorithm, and implementation?

In addition to the notion of levels of analysis, another important principle of the computational approach is that computations proceed through several stages. For vision, the output can be taken as symbolic descriptions of the surrounding world, while the inputs are primitive measurements of light impinging on an optical transducer (see Fig. 11.1). Once this is recognized it should be clear that it would be unreasonable to expect that vision can be accomplished in a single step transformation. Instead, vision is viewed as being mediated by a set of relatively discrete functional modules (e.g., edge detection, stereo, motion, shape from

A

B

217	215	213	212	209	179	123	187	209	216	218	219	217
216	210	207	210	201	163	117	181	210	212	215	214	216
213	208	212	214	203	161	181	174	189	211	212	217	219
214	218	216	210	207	179	137	122	171	210	213	207	209
213	212	215	208	202	181	168	125	174	191	214	210	213
217	218	214	209	199	182	157	122	120	167	215	211	209
216	214	210	208	194	172	131	119	117	171	204	207	212
215	212	214	206	191	162	141	123	121	165	198	206	209
214	216	210	201	186	164	139	121	124	167	201	207	208
212	214	209	189	171	157	141	122	127	172	206	212	214
216	215	198	175	143	148	137	127	132	181	211	219	217
209	206	188	159	133	141	139	133	147	194	207	218	214
211	207	181	151	130	137	142	131	179	201	210	215	216
210	199	172	133	129	132	121	127	163	203	211	218	219
213	196	165	125	135	128	120	149	181	204	207	210	212
212	194	149	127	137	121	119	152	189	200	203	206	205
209	201	151	129	131	127	133	169	191	207	209	204	206

FIGURE 11.1. The input to vision can be taken as a set of measurements of light intensity that impinge on a photosensitive transducer. From this beginning a visual system ultimately constructs a rich symbolic description of the surrounding world. The left panel shows a gray level image that has been quantized to 256 levels. The numeric values for the portion of the image that is in the box are given in the right panel. (From Hildreth and Hollerbach, 1987.)

shading) that incrementally transforms measures of light intensity to rich symbolic descriptions of the world. Interestingly, one of the important general problems for computational vision is to precisely define what the functional modules should be as well as what intermediate steps are useful for any visual system.

The computational approach to studying vision clearly has intellectual antecedents. For example, Helmholtz (1925) clearly understood that vision has a physical and mathematical basis and Gibson (1950, 1979) was very explicit about the important role that world structure plays in perception. The key novel point that has been brought out by studies in computational vision is the absolute importance of the separation of an information processing task from a system for performing the task. The earliest examples of this approach in vision can be traced to Land and McCann's (1971) study of lightness and Horn's (1975) study of shape from shading. Marr's subsequent contribution was not only in developing several specific computational analyses, but especially in formally specifying the general research program in his levels of analysis and in insisting on the relevance of this approach to the study of biological systems (Marr 1976, 1982; Marr and Poggio, 1977).

One set of matters remains for this general introduction to computational vision: How does computational vision relate to other areas of vision science? In this regard, it is important to emphasize that computational vision does not substitute for the work that goes on in allied fields such as psychophysics, neuroscience, and applied machine vision. Instead, the role of computational vision can be seen as that of precisely defining a set of problems and "in principle" solutions for visual information processing. This can serve fields that are concerned with understanding biological vision systems in at least two ways: First, prediction can be made about what problems a visual system must functionally solve. Second, predictions can be made about the mechanisms that are of use in solving a given visual task. Applied machine vision can benefit from computational studies by garnering robust and theoretically motivated solutions to the problems that it desires to solve. At a more general level, computational vision already has made clear a very important fact for all of vision science. This fact is that vision is hard. The computations that are necessarily involved in performing even elementary visual tasks can be quite complex. This serves as an important message to the entire field of vision: Poorly motivated and simplistic solutions will not play an interesting part in a mature science of vision. Similarly, computational vision can gain from the results of other areas of visual science: Studies from biological systems can suggest problems that need to be understood at a more theoretical level as well as hint at possible solutions. Also, attempts to exploit the results of computational analyses in applied machine vision can provide severe real world tests of the robustness of the proposed solutions. In the best scheme, research in computational vision can serve to bring together research and researchers from across the varied approaches to studying vision.

Recapitulation

This section has provided a general overview of computational vision. The section began by describing how the basic tenets of computational vision are rooted in the field of artificial intelligence. In this regard, the most basic concern is that of developing an analysis of a complex information processing task independently of how any particular system deals with the task. This approach fosters results that explain how the task can be achieved in principle; these results are applicable to the study of any particular system aimed at carrying out the task. The particulars of computational vision discussed three levels of explanation that must be met for the complete understanding of a vision task: the computational theory that precisely states what is being done and why, the algorithm that gives a procedure for recovering the solution indicated by the theory, and the implementation that specifies a physical embodiment of the algorithm.

Computational Stereo Vision

This section provides a look at the computational investigation of binocular stereo vision. The motivations for this section are two-fold. First, the discussion familiarizes the reader with some of the important results that have been derived in computational stereo vision. Second, consideration of the computational investigation of a particular problem can illustrate how the general approach works in application. This section is not intended as a general review of computational stereo vision. Instead, it concentrates on a number of specific examples that are well suited to making key points about the problem domain. Other valuable discussions of stereo vision (including discussions of biological systems) can be found in Barnard and Fischler (1982), Binford (1984), Foley (1980), Gulick and Lawson (1976), Julesz (1971), Nishihara and Poggio (1984), Ogle (1950), and Poggio and Poggio (1984). Also, interesting collections and discussions of computational investigations of other visual processes can be found in Ballard and Brown (1982), Fischler and Firschein (1987), Horn (1985), Marr (1982), Richards (1988), Richards and Ullman (1987), and Ullman and Richards (1984).

Binocular Stereo Vision

The physical situation that leads to binocular stereo vision can be described as follows: An arrangement of surfaces in the three-dimensional world projects differentially onto a pair of two-dimensional imaging surfaces. To understand stereo vision would be to understand how the corresponding inverse mapping can take place. That is, given a pair of two-dimensional projections of a three-dimensional world, how is it possible to exploit the geometry of the situation to recover useful properties of the geometry of that world? At the current state of our understanding of stereo vision it is convenient to break the problem into three relatively independent parts: feature selection, correspondence, and disparity interpretation. Feature selection is concerned with determining what makes for good primitive elements as one seeks to compare the pair of images in a binocular projection. Correspondence deals with matching those elements in the two views that are projections of the same element in the three-dimensional world. Disparity interpretation is concerned with how the disparity in configurations

FIGURE 11.2. A random-dot stereogram. In the absence of any notion of constraints, any point in one image could be taken as corresponding to any point in the other image.

between corresponding elements can be mapped into useful descriptors of three-dimensional scene geometry.

Early attempts to model stereo vision as well as to build artificial stereo systems met with limited success. In retrospect, it can be seen that these studies were hampered by their overly simple approaches to the three basic parts of stereo vision. Essentially, these methods proceeded in the following fashion: First, extract a small patch in one image. Second, look for the most similar intensity patch in the other image and assign it as the match to the first patch. Third, measure the disparity between corresponding patches and use it to recover depth via triangulation (assuming that the relative viewing positions are known). This approach is known as gray-level correlation (see e.g., Bertram, 1969; Levine et al., 1973; Sperling, 1970; Sutro and Lerman, 1973).

It is possible to isolate two fundamental reasons why this type of approach is bound to failure. First, corresponding regions between two views are differentially distorted due to both photometric and geometric effects. For example, highlights depend critically on the position of a view and perspective will deform images according to center of projection. Therefore, it is not even theoretically correct to attempt exact matches of image brightness. A second type of difficulty is due to the inherent ambiguity that is involved in seeking stereo correspondence. This set of issues can be illustrated with the aid of Figures 11.2 and 11.3. Figure 11.2 is an example of a random-dot stereogram (Julesz, 1971). The left and right panels of this figure are the same except that a central region in one has been shifted relative to the other. When fused by a human observer, the disparity between the central regions yields the vivid perception of depth as a central patch appears to float in front of the background. There are two interesting points to note here: First, it is possible to establish the necessary correspondence for stereo in the absence of any monocular cues. Second, in the absence of any matching criteria other than "patch with most similar brightness" any black patch in one image could match any black patch in the other; the same type of ambiguity arises as one seeks to match white patches. Another illustration of the

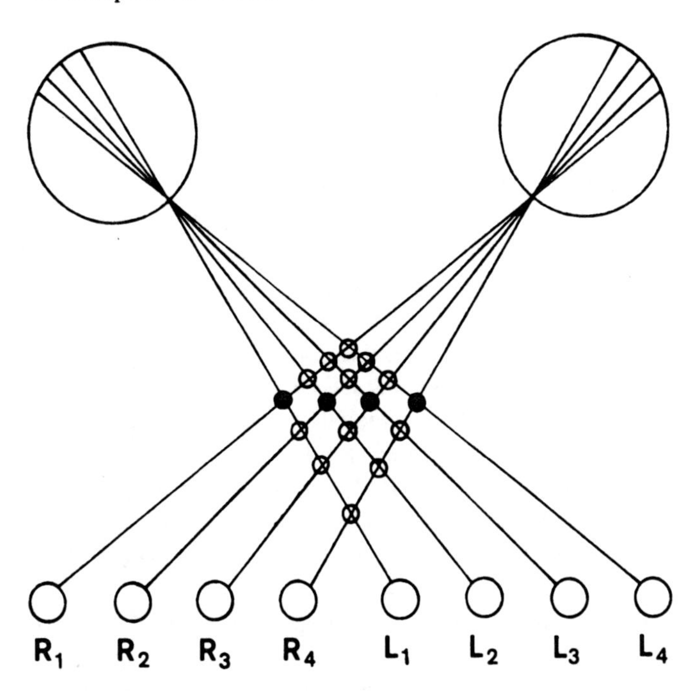

FIGURE 11.3. Various configurations of points in space can project to the same set of binocular projections. For this example, different combinations of the 16 circles in the center of the figure can give rise to the depicted projections. However, a human observer selects a single solution (indicated by the filled circles). How can stereo vision overcome the ambiguity and provide a coherent solution? (From Grimson, 1981.)

ambiguity in stereo is given in Figure 11.3. In this figure it is shown that different sets of points in the world can project to the same set of binocular projections. Nevertheless, a human observer is capable of deriving a single coherent interpretation in the face of this ambiguity. Clearly, the stereo vision problem as just described is underconstrained. In the next few paragraphs the three parts to stereo vision are examined from a computational perspective that emphasizes the use of constraints in posing problems and solutions for visual information processing. As there has been considerable interest in viewing biological systems in the light of these results (and vice versa), comments also will be made regarding some of the relevant investigations of biological stereo vision.

Feature Selection

Feature selection is concerned with isolating elements in the images of a binocular view that can serve as reliable primitives for matching between the images. As already noted, the effects of photometric and geometric distortions cast doubt on using raw image intensity as the features of choice. At another extreme, one could imagine postponing feature selection until objects had been recognized separately in each view on the basis of monocular information. Objects or object parts then could serve as features for stereo matching. However, the fact that it is possible to perceive stereoscopic depth in the absence of monocular cues (e.g., the random-dot stereogram of Fig. 11.3) shows that stereo can proceed without first recognizing objects.

In essence, these observations about primitives for stereo matching lead to two main conclusions: First, the primitives should be tied less to the details of imaging and more to the geometry of objects in the world. Second, recovery of the primitives should not rely on elaborate monocular processes. Many recent computational analyses of stereo vision have acknowledged these points. These approaches seek to uncover matching primitives that abstract away from image structure to world structure while remaining easily recoverable from input image intensity.

One tack on this problem has merged with studies of "edge detection." Edge detection is concerned with recovering tokens from image intensity that are indicative of interesting changes in the world, such as physical edges in surfaces or texture markings. Typically, these types of changes in the world are implicated by changes in the corresponding image intensity; in fact, a better term than "edge detection" might be "detection of significant intensity changes." Edge detection is an entire area of computational investigation in and of itself that is beyond the scope of this chapter. (See, e.g., Canny, 1986; Haralick, 1980; Hildreth, 1987; Marr and Hildreth, 1980; Mumford and Shah, 1985; Nalwa and Binford, 1986; Shanmugam et al., 1979; and Torre and Poggio, 1986 for useful discussions of and approaches to edge detection.) Nevertheless, it should be clear that edge detection as just defined should provide good features for stereo matching. Correspondingly, many existing stereo vision systems use the output of edge detectors or related approaches [e.g., the Moravec (1979) interest operator, which reports regions of high brightness variance in an image] as their matching primitives (e.g., Baker and Binford, 1981; Barnard and Thompson, 1980; Grimson 1981, 1985; Hoff and Ahuja, 1987; Marr and Poggio, 1979; Mayhew and Frisby, 1981; Medioni and Nevatia, 1985; Moravec, 1979; Ohta and Kanade, 1983; Poggio and Drumheller, 1986; Pollard et al., 1985; Pradzny, 1985). As an illustration of the type of information that can be delivered by an edge detector, the output of one well known edge detector (Marr and Hildreth, 1980) is illustrated in Figure 11.4.

A number of variants on this basic scheme also have been examined. Essentially, these extensions were motivated by a desire for richer feature sets than simplest edge detection can provide. The motivation for richer features is the likelihood that they can provide additional constraint on the stereo correspondence problem. For example, several approaches match not only on the basis of the edges but also on the basis of the features of edges, such as contrast and orientation (e.g., Arnold and Binford, 1980; Baker and Binford, 1981; Medioni and Nevatia, 1985). Another extension is to consider features that occur at several different levels of resolution (e.g., Grimson, 1981, 1985; Hoff and Ahuja, 1987; Marr and Poggio, 1979; Mayhew and Frisby, 1979; Moravec, 1979). At least two approaches develop the idea of first using edges as features for matching and then using the matched edges as "boundary conditions" for image intensity based

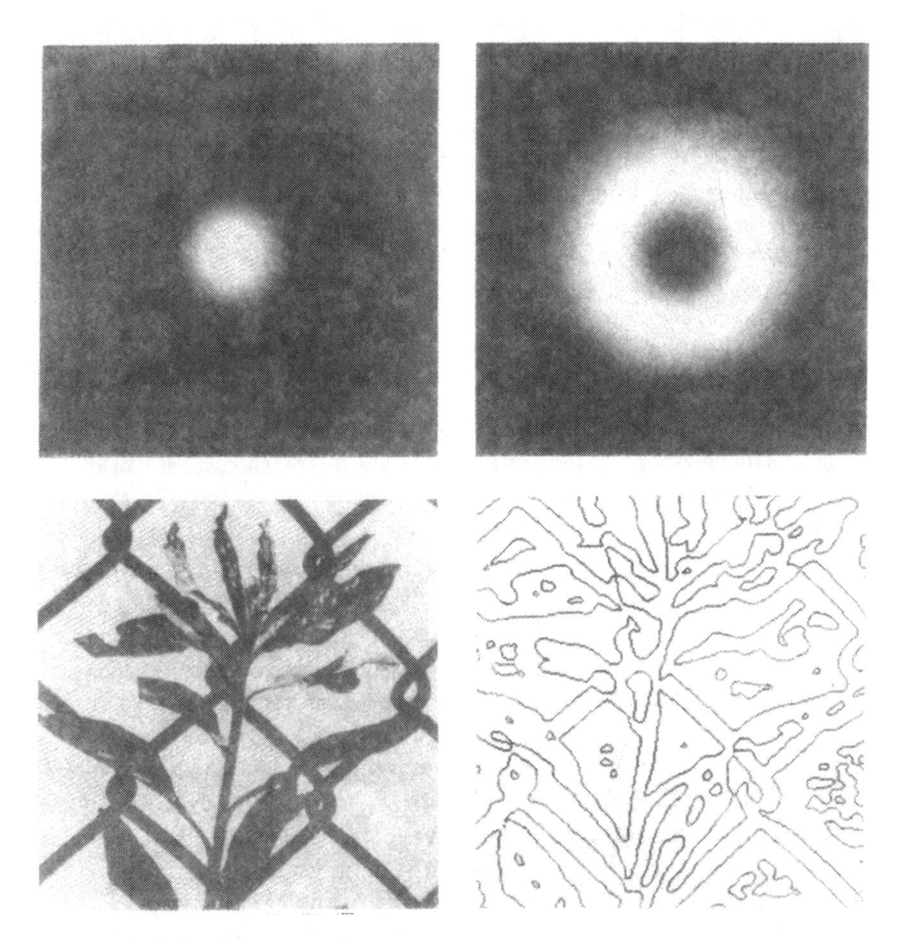

FIGURE 11.4. Many stereo vision algorithms match not on the basis of raw image intensity, but on some representation of significant change in intensity or "edges." The upper left and right panels show the profiles of the Marr-Hildreth edge operator in the spatial and frequency domains, respectively. The lower left panel shows a gray level image. The lower right panel displays the resulting "edges" from considering the zero-crossings of applying the operator to the image. (From Hildreth, 1980.)

matching (Baker and Binford, 1981; Henderson et al., 1979). It also is worth noting that not all approaches to finding robust features for stereo matching have turned to edge-like features. For example, Kass (1983) used features derived from multiple derivatives of image intensity and a stochastic model of image formation. The primitives provided by this approach tend to be tied more closely to image structure than are the edge-based approaches. Also, Nishihara (1984) developed an approach to establishing coarse stereo matching that essentially used the dual of an edge detection process (i.e., matching features were based in the image regions between edges). This approach was motivated not so much by theoretical considerations as by practical concerns in designing a high speed yet robust stereo system at the expense of accuracy.

Overall, the fact that there are a number of approaches to feature selection should not obscure the fact that at the level of the computational theory they share a common theme: Features for stereo matching should abstract away from image structure and toward structure in the world. Interestingly, at the levels of algorithm and implementation most of the various feature selection methods typically are instantiated in terms of relatively straightforward convolution or template matching operations.

Correspondence

The goal of stereo correspondence is to match those features in the binocular images that correspond to the same point in the three-dimensional world. The inherent ambiguities in this problem already have been illustrated with the aid of Figures 11.2 and 11.3. These ambiguities cannot be overcome solely with the use of good feature selection. However, recourse can be found by constraining possible stereo matches on the basis of physical properties of the world. A number of constraints have been proposed for addressing these issues. Perhaps the most widely employed constraints on stereo correspondence are the uniqueness constraint, the epipolar constraint, and the continuity constraint. The next few paragraphs examine each of these constraints in turn.

The uniqueness constraint is derived from the observation that an object in space can be at only one place at one time. This implies that each feature in one image should be matched to only one feature in the other image. In other words, a stereo match should be unique. Uniqueness was introduced earlier in this chapter to illustrate the general notion of physically motivated constraints for visual computation. Clearly, one mark of a good stereo matcher is that it return as few ambiguous correspondences as possible. The best case then can be considered a unique match. In this sense, many solutions to the correspondence problem have been implicitly concerned with the uniqueness constraint. However, it is generally best to be explicit about the use of constraints since this increases the perspicuity of one's approach. Approaches to stereo correspondence that explicitly exploit the uniqueness constraint include Grimson (1981, 1985), Marr and Poggio (1976, 1979), Mayhew and Frisby (1981), and Pollard et al. (1985).

Due to the nature of stereo projection, features that are located along a particular line in one image all will match features along a particular line in the other image. These lines are called the epipolar lines. Geometrically, the epipolar lines can be defined as follows: Consider the plane defined by a point in space and the centers of projection of the two imaging systems that make up the stereoscopic view. The intersections of this plane with the imaging surfaces of the two views are the epipolar lines. As a specific example, if the eyes are focused on the horizon, then the epipoles are roughly horizontal and at the same height in the two images. However, if the eyes are focused at a nearby object, then these two lines have different orientations in the two views. Nevertheless, if the positions of the two eyes in their orbits are known, then it is possible to define these linear correspondences. This constraint can greatly reduce correspondence ambiguities: Given a feature in one image, the search for its match in the other image is limited to a linear span. It appears that the use of the epipolar constraint in stereo matching was first exploited by Keating et al. (1975). Since that initial study a number of approaches to the correspondence problem have employed the epipolar constraint, including Arnold and Binford (1980), Baker and Binford (1981), Grimson (1981, 1985), Henderson et al. (1979), Mayhew and Frisby (1981), Ohta and Kanade (1983), Panton (1978), and Pollard et al. (1985).

Now consider the continuity constraint. The underlying notion here is that over most of a three-dimensional surface's extent the distance from a reference point varies slowly. In other words, the distance at nearby points is usually similar; the surface is continuous. Computational models have exploited this continuity constraint in a number of ways. Some researchers propose a restriction on the gradient of disparity that would arise from a given correspondence; when framed in this manner the continuity constraint is often called a disparity gradient limit (e.g., Pollard et al., 1985; Medioni and Nevatia, 1985). Other efforts have enforced slow changes along the extent of linear features in an image; when framed in this manner the continuity constraint is often called figural continuity (e.g., Baker and Binford, 1981; Grimson, 1985; Mayhew and Frisby, 1981; Ohta and Kanade, 1983). More generally, many approaches to reducing the ambiguity of stereo correspondence have employed some form of a continuity constraint (e.g., Arnold and Binford, 1980; Barnard, 1987; Barnard and Thompson, 1980; Eastman and Waxman, 1987; Grimson, 1981; Hoff and Ahuja, 1987; Levine et al., 1973; Marr and Poggio, 1976, 1979; Mori et al., 1973; Panton, 1978; Pradzny, 1985).

Now that a number of physically based constraints have been motivated, an essential part of a computational analysis of stereo correspondence is in hand. At this point it becomes interesting to inquire into the nature of algorithms that have been used to embody the various constraints. Perhaps the best known of the stereo correspondence algorithms is the Marr-Poggio (1979) algorithm and its extensions by Grimson (1981). This algorithm makes use of edge-like features for matching, in particular, features that are derived from the Marr-Hildreth edge detector (Marr and Hildreth, 1980) illustrated in Figure 11.4. The algorithm makes use of the three constraints that just have been described: uniqueness, epipolar, and continuity. A further attribute of this algorithm (as well as many other correspondence algorithms) is that it exploits a coarse to fine matching strategy. In this case, correspondences are initially established on the basis of features that have been derived at a coarse spatial resolution. Subsequent matches at finer resolutions are constrained by the coarser results. Also, matching is constrained so that disparities of relatively large magnitude are tolerated at the coarser scales while only smaller magnitude disparities are tolerated at finer scales. This algorithm has been implemented and successfully tested on a diverse set of binocular data. One example of its performance is provided in Figure 11.5. An example from a closely related algorithm that exploits the continuity constraint in terms of figural continuity (Grimson, 1985) is provided in Figure 11.6.

FIGURE 11.5. Results of applying a computer algorithm for recovering stereo disparity to a random dot stereogram. The upper left and right panels show the stereogram. The disparity in the stereogram implies a central planar patch that is closer to the viewer than the surrounding areas. The lower panel displays the recovered disparity plotted as height above a reference plane. (From Grimson, 1981.)

As is typically the case, more than one algorithm can be devised for manifesting a given set of constraints and computational theory. As a case in point, essentially the same three constraints have led to a rather different class of correspondence algorithms. These algorithms are based on iterative relaxation schemes where an initial set of correspondences is iteratively refined via spatially local interactions between disparity units. Relaxation algorithms that employ some or all of the three noted constraints on correspondence include Barnard (1987), Barnard and Thompson (1980), Dev (1975), Julesz (1971), Marr and Poggio (1976), Mori et al. (1973), Nelson (1975), Pollard et al. (1985), and Pradzny (1985).

Computational studies of stereo correspondence provide an excellent illustration of the use of constraints. Without any notion of constraining assumptions the correspondence problem is poorly defined. However, in view of physically based constraints (e.g., uniqueness, epipolar, and continuity) the matter of establishing stereo matches becomes much less ambiguous. Significantly, these constraints have been used in designing algorithms that have been implemented and tested with encouraging results.

Disparity Interpretation

Suppose the correspondence problem has been solved and that the resulting disparity has been measured. How can this disparity information be interpreted in

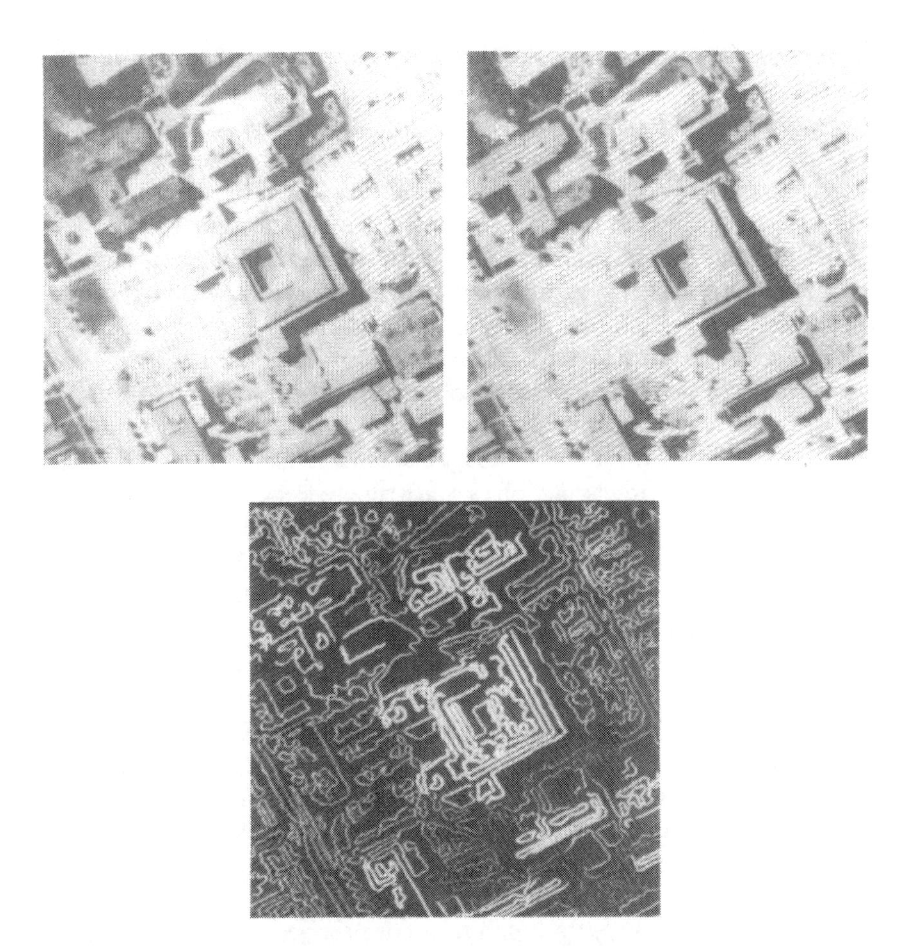

FIGURE 11.6. Results of applying a computer algorithm for recovering stereo disparity to a natural image stereogram. The upper left and right panels show a natural image stereo pair. The disparity in this example corresponds to the different heights of the buildings in the scene. The lower panel displays the recovered disparities plotted with increasing disparity shown as increasing in brightness. (From Grimson, 1985. © 1985 IEEE.)

terms of interesting and useful properties of the three-dimensional world? This is the question that must be answered by disparity interpretation. In the next few paragraphs computational approaches to this problem are described.

Perhaps the simplest and most straightforward approach to disparity interpretation is to perform a triangulation operation. Provided that the stereo viewing geometry is known, the triangulation can be used to map disparity measures into measures of the distance between the viewer and points in the three-dimensional world. However, a problem presents itself: The features that typically are used for stereo (e.g., edges) will provide disparity information at a sparse set of points. What sort of information should be returned between these data points? The most common solution to this problem is to exploit a continuity constraint. In particular, assume that distance varies smoothly between the available data points. The solution to dealing with sparse distance measures is then to interpolate or approximate the sparse measures with a smooth surface. The resulting output is of point by point distance. One form of instantiating the surface fitting idea is in terms of splines. Intuitively, this approach can be thought of as fitting a thin plate or membrane to the sparse data points and then allowing it to achieve equilibrium. The details of formulating tight theories as well as efficient algorithms and implementations for the application of splines to disparity interpretation have been the focus of considerable research (e.g., Blake and Zisserman, 1987; Boult, 1987; Terzopoulos, 1983, 1986). Another approach to interpolating or approximating sparse distance measures centers around fitting low order polynomials (essentially algebraic surface patches) to the available data (e.g., Eastman and Waxman, 1987; Hoff and Ahuja, 1987). As an example of the type of results that can be achieved by surface fitting, Figure 11.7 shows the results of fitting a spline surface to sparse distance information derived from a random-dot stereogram (Grimson, 1981).

In their simplest instantiations surface fitting approaches fail to make explicit the properties of surface orientation and discontinuity (in fact, discontinuities will tend to be smoothed over). Various attempts have been made to extend the surface fitting approaches to deal with these matters. One tack on recovering surface orientation has centered around numerically differentiating the recovered point by point distance representations (Brady et al., 1985; Medioni and Nevatia, 1984). Other researchers have sought to recover surface orientation in the spline-based approaches via a cascade of differencing operations that are in effect during the recovery process (Harris, 1987; Terzopoulos, 1986). Also, the recovered coefficients of the surface patch methods can often be interpreted in terms of surface orientation or curvatures (Eastman and Waxman, 1987; Hoff and Ahuja, 1987). The majority of approaches to recovering surface discontinuities in this paradigm fall into one of two classes. Several methods focus on studying the residuals of a surface fit in the vicinity of a discontinuity (e.g., Grimson and Pavlidis, 1985; Eastman and Waxman, 1987; Hoff and Ahuja, 1987; Lee and Pavlidis, 1987). In application these approaches first fit a surface and then look for regions where a discontinuity was likely smoothed over. Other approaches seek to understand how it is possible to let discontinuities form in a surface while it is being fit (e.g., Terzopoulos, 1986b; Marroquin, 1984; Koch et al., 1985; Blake and Zisserman, 1987).

There also is another important set of approaches to disparity interpretation. These approaches study how it is possible to recover surface discontinuities, orientations, curvatures, and more qualitative descriptors (in addition to distance) directly from stereo disparity, without an intervening representation of distance. Typically, the computational theories for this approach are based on detailed analyses of disparity geometry and its relation to the world's three-dimensional structure. Interestingly, while some of these analyses are rather technical, they often lend themselves to algorithms and implementations that are quite simple. This line of research has produced a number of intriguing results. As examples, surface orientation can be recovered from measures of orientational

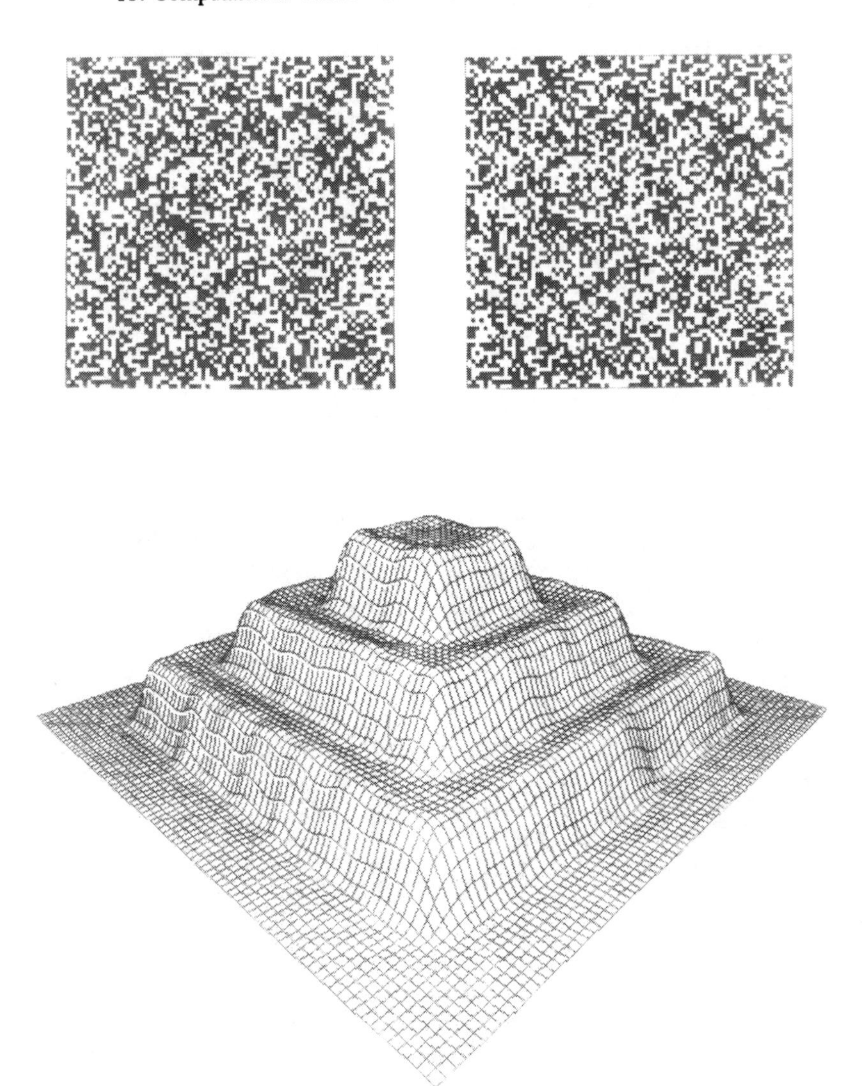

FIGURE 11.7. Interpolation of sparse depth measurements as an approach to the disparity interpretation problem. The upper left and right panels show a random dot stereogram. The lower panel shows the result of converting recovered disparity to distance and then interpolating with a smooth surface. (From Grimson, 1981.)

disparity, the difference in the orientation of linear features as projected to two views (Koenderink and van Doorn, 1976; Wildes, 1989b). It is possible to recover surface curvature from stereo disparity (Rodgers, 1987; Stevens and Brookes, 1987; Wildes, 1989c). Qualitative depth orderings can be derived directly from stereo disparity (Weinshall, 1988). As an explicit example of the type of results that can be obtained from these approaches, Figure 11.8 shows the

FIGURE 11.8. Recent approaches to interpreting stereo disparity concentrate on directly recovering interesting geometric scene properties directly from stereo disparity. This example illustrates the results of an algorithm for recovering the three-dimensional configuration of surface discontinuities from disparity. The top half of the figure displays a natural image stereo pair. The bottom left panel shows the corresponding features along which disparity information is made explicit. The bottom right panel shows the recovered regions of discontinuity. The recovered depth from the discontinuity to the viewer is displayed in terms of grey levels with black the furthest and white the closest.

results of exploiting an analysis of stereo disparity interpretation with the goal of recovering surface discontinuities from disparity (Wildes, 1989a). In essence the algorithm operates by comparing locally adjacent disparity measures for violations of a type of continuity constraint that assumes the three-dimensional surface is locally planar. Additional approaches to recovering surface discontinuities from stereo disparity are detailed in Smitly and Bajcsy (1984), Spoerri and Ullman (1987), Stevens and Brookes (1987), and Yeshurun and Schwartz (1989).

Finally, while many of these later approaches to disparity interpretation have considerable potential, as a whole they have not yet received adequate implementation and testing to be adequately assessed.

In attempting to take account of this range of attacks on disparity interpretation, it is useful to consider the basis of computational theories. In particular, recall that an important concern in a computational theory is the development of an understanding of why a particular computation is being performed. Presumably, disparity interpretation is being performed so its output can feed useful information to subsequent processes such as visually based navigation, manipulation, and object recognition (e.g., Gennery, 1980; Moravec, 1979). In this light, it is not at all clear that dense representations of distance are very interesting. Instead, direct information about surface discontinuities, orientations, and curvatures as well as more qualitative characterizations would be considerably more vital. Correspondingly, disparity interpretation methods that directly recover these more interesting parameters should be favored over approaches that concentrate on the recovery of dense distance data.

Relation to Biological Stereo Vision

Just as there has been a considerable amount of research devoted to understanding the computational aspects of binocular stereo vision, there has been a great amount of investigation into the processing of stereoscopic information in biological systems. Further, it sometimes has been the case that investigations along one of these lines has been closely related to or inspired by the other line of attack. The next few paragraphs provide a brief look at studies of biological stereo that relate to computational stereo vision. In turn, consideration will be given to matters of feature selection, correspondence, and disparity interpretation.

A number of studies of biological visual systems address the question of what features serve as the substrate for biological stereo matching. There is psychophysical evidence that humans exploit features that abstract away from image intensity. For example, Julesz (1971) presents evidence that image regions of different contrasts can be matched, provided the sign of contrast is preserved between left and right views. Other psychophysical research has addressed the issue of characterizing the operations from which features are derived. In this regard, there is evidence for and against the notion that stereo features are derived from spatial filters with circularly symmetric profiles (e.g., circularly symmetric center-surround operators). Mayhew and Frisby (1978, 1979) provide evidence from masking as well as stimuli with steep variation in depth in favor of the notion of circularly symmetric filters. In contrast, Wilson (1988) reports evidence from studies of human disparity thresholds in support of orientationally tuned filters. Additional evidence suggests that the features that are derived from the filtering may directly be related to the zero-crossings and extrema of the filtered output (Mayhew and Frisby, 1981). Further evidence is in support of the notion that features for human stereo are derived at a number of spatial scales (e.g., Felton et al., 1972; Frisby and Mayhew, 1977; Julesz and Miller, 1975; Kulikowski, 1978; Levinson and Blake, 1979; Mayhew and Frisby, 1981; Wilson, 1988). Taken together, this research can be interpreted as being in accord with ideas from computational stereo research that features should be derived at a number of spatial scales and be concerned with capturing some notion of significant intensity change. However, there is some psychophysical evidence that humans can derive a weak perception of depth from stimuli that lack any edge-like features (Bülthoff and Mallot, 1987). To a limited extent, research from neurophysiology can be brought to bear on the question of feature selection for stereo. Neural recordings in cortical areas 17 and 18 in cat (Barlow et al., 1967; Ferster, 1981; Fischer and Kreuger, 1979; Hubel and Wiesel, 1973; Ohzawa and Freeman, 1986a,b; Pettigrew et al., 1968; von der Heydt et al., 1978) and monkey (Hubel and Wiesel, 1970; Motter and Poggio, 1982; Poggio, 1984, 1985; Poggio and Fischer, 1977; Poggio and Talbot, 1981) have revealed cells that are sensitive to binocular disparity. Further, the Hubel and Wiesel (1977) complex/simple distinction can be used to characterize the receptive field types of the disparity selective cells (e.g., Ferster, 1981; Poggio, 1984). To the extent that simple cells can be taken to be indicative of edge or bar detectors, this evidence can be taken in support of the notion of edge-like features in biological stereo systems. In contrast, the complex cells could be taken in support of more complicated features for stereo. For example, the complex cells could be serving to provide features that are based on texture patches (e.g., Burr et al., 1981; Hammond, 1985; Poggio, 1984). In any case, most of the neurophysiological evidence is in support of the notion that the features for stereo are derived at a number of spatial scales. Finally, while experiments using random-dot stereograms provide both psychophysical (Julesz, 1981) and neurophysiological (Poggio, 1985) evidence that stereo can work in the absence of obvious monocular cues, other evidence shows that the perception of depth in humans is significantly stronger if monocular cues are present (Kulikowski, 1978; Richards, 1977). This observation from biology should be a cause for thought in the computational stereo community. A considerable amount of computational research has assiduously avoided exploiting monocular cues for stereo vision. This limitation has been partly inspired by the fact that biological systems can function in the absence of such cues. However, it may well be time to follow the other hint provided by biology: A rich natural world stimulus should be exploited for all its available structure, monocular or not.

It is currently rather difficult to directly address the issue of whether biological systems exploit the constraints on stereo correspondence that have been studied in computational vision. This is due to the lack of experiments aimed at addressing these questions in a direct fashion. Nevertheless, several observations can be made. There is psychophysical evidence in favor of the notion that human observers enforce some type of continuity constraint as they establish stereo correspondence. In particular, there appears to be an upper bound on the magnitude of the disparity gradient that can be tolerated by human observers (Burt and Julesz,

1980; Tyler, 1973, 1974, 1975). Interestingly, psychophysical results regarding the human disparity gradient limit have inspired the way in which some computational studies have approached the continuity constraint (e.g., Pollard et al., 1985). Other psychophysical evidence in favor of a continuity constraint in human vision can be found in the work of Mitchison and McKee (1985, 1987a,b). It also is possible to derive some evidence in support of an epipolar constraint in human stereo. First, there are reports that human eye movements typically operate to eliminate vertical disparity (Duwaer, 1981; Duaer and van der Brink, 1981). Second, there is evidence that in the absence of eye movements even moderate vertical disparities (e.g., 4′–7′) lead to difficulties in fusion (Nielsen and Poggio, 1984). These types of results indicate that human observers favor binocular image alignment that would allow for the easiest execution of epipolar-based matching.

There is evidence from biological stereo systems in support of multiresolution operations in correspondence. For example, there is psychophysical evidence that the outputs of coarse spatial filters for stereo features feed into similarly coarse disparity measurements, while finer feature filters feed into finer disparity measures (Felton et al., 1972; Kulikowski, 1978; Levinson and Blake, 1979). Evidence from neurophysiological studies of disparity selective cells in cat cortex supports a similar type of scheme (Ferster, 1981; Pettigrew et al., 1968). In monkey, it appears that disparity selective cortical cells are grouped into pools of cells that are narrowly tuned to zero (point of fixation), far (beyond fixation), and near (in front of fixation) disparity (Poggio, 1984; Poggio and Fischer, 1977; Poggio and Talbot, 1984). Significantly, Richards (1971) had hypothesized a similar organization in humans on the basis of stereoanomolies. These types of results are in support of some notion of multiresolution stereo processing in biological systems. In particular, this evidence can be marshalled in favor of a correspondence scheme that employs multiresolution spatially constrained matching. In this regard, it has been demonstrated that in humans correspondences at one spatial scale are not established independently of correspondences at other spatial scales (Frisby and Mayhew, 1977; Mayhew and Frisby, 1981).

Although it is clear that binocular disparity provides the basis for a sense of depth in biological systems, the details of how this is accomplished are far from clear. Psychophysical studies suggest that while there is a relation between horizontal disparity and perceived depth, the relationship is not straightforward. In particular, many studies show that when observers are presented with binocular horizontal disparity the resulting perception is not what would be predicted if the stereo system was performing a simple triangulation with all viewing parameters known in advance; a constant binocular disparity corresponds to neither a constant perceived depth nor a constant perceived distance ratio (Braddick, 1979; Foley, 1980; Leibovic et al., 1971; Mitchison and Westheimer, 1984). With regard to surface interpolation, Collett (1983) has provided data in support of the thesis that humans are capable of interpolating sparse stereo data; however, the exact form of the interpolated surface is unclear. Further, it is interesting to note that there is psychophysical (DeValois et al., 1975; Ninio,

1985; von der Heydt, 1979; von der Heydt et al., 1981) and neurophysiological (Blakemore et al., 1972; Hänny et al., 1980; von der Heydt et al., 1977) evidence in support of the biological computation of orientation disparity. Significantly, there are results supporting the use of orientational disparity in the recovery of three-dimensional surface orientation (Ninio, 1985). Also, there is psychophysical evidence that stereoscopic stimuli corresponding to discontinuities in depth are powerful stimuli for biological systems (e.g., Gillam et al., 1983, 1988; Mitcheson, 1987; Stevens and Brookes, 1987). Finally, it is possible to draw one general conclusion from the biological studies of disparity interpretation: Biological systems interpret disparity as more than simple point by point depth. This conclusion is in accord with the observations from computational analyses that emphasize the importance of recovering more than simple distance information.

On the whole, many results from biological vision support the thesis that the principles analyzed in computational stereo vision also are exploited in biological stereo systems. Clearly, the correspondence between all the lines of research is not exact. However, there is a strong sense that many investigations are converging toward similar questions and answers.

Recapitulation

This section has provided a look at computational stereo vision. The discussion began by stressing the inherent difficulties involved in binocular stereo vision. Following this, the three major computational steps to stereo vision (feature selection, correspondence, and disparity interpretation) were considered in turn. For each step computational solutions were presented as were the results of algorithms and implementations that have served to embody the theoretical analyses. When possible, these theoretical and empirical results have been compared and contrasted to known facts about how biological systems deal with binocular stereo.

Trends

With a basic understanding of computational vision, it becomes interesting to look at current trends in this area of investigation. In this section five specific directions are noted. The first is concerned with taking a novel look at the very nature of perception. The second example deals with new answers to the question: What types of output information should be expected from visual processes? A third trend focuses on developing an increasingly detailed understanding of how visual operations behave in the face of imperfect input. A fourth trend concerns new light that has been shed on detailed physical mechanism with the aid of computational vision. Finally, a more overall trend is mentioned.

One intriguing new line of research is Observer Theory (Bennett et al., 1989). This research program is concerned with providing a rigorous framework for what it means to make a perceptual inference in the face of uncertain or incomplete information. Many of the important concerns of computational vision can be embedded in this theory: The notion of constraining assumptions for visual information processing can be cast as necessary preconditions for perception. Also, issues of imperfect or noisy input to a vision system and its effects on computation can be captured in terms of uncertainty in inference. Interestingly, the researchers in this area show that their definition of an "observer" properly contains the class of Turing machines that serves as the theoretical basis for automated computation. Investigations along these lines have the potential for greatly influencing the way perception is conceptualized.

Another set of trends is concerned with examining the question of what constitutes useful and interesting output information at various stages of processing. In the past, there has been considerable concentration on the recovery of surface representations that are essentially dense maps of surface orientation or distance from an observer (see, e.g., Barrow and Tennenbaum, 1978; Marr, 1982). However, recently investigators have started to ask if this type of representation is necessary or useful for vision. If the goal of vision is to produce rich symbolic descriptions of a surrounding world, then it may be desirable to move more directly to more qualitative assertions about the world's structure. This is not to say that the analyses of the problems should be any less rigorous. Rather, the suggestion is that the emphasis of the analyses should be on recovering properties that are more directly applicable to providing useful information about the world. Some examples of this type of result were provided in the previous section of this chapter. There it was noted that recent results in disparity interpretation have concentrated on finding properties such as surface discontinuities (Wildes, 1989a) and simple depth orderings (Weinshall, 1988) in a direct fashion from stereo disparity. In motion vision there has been recent criticism of the general attempt to exactly recover optical flow (Poggio and Verri, 1987) as well as the development of methods that recover the motion of an observer through the environment directly from time varying image brightness (Horn and Weldon, 1988). Another interesting example can be found in topological approaches to visual object representation and recognition (e.g. Bowyer et al., 1989; Koenderink, 1987; Ponce and Kriegman, 1989; Rieger, 1987; Sripradisvarakul and Jain, 1989), where the topologically defined qualitative structure of an object serves as the basis for its representation and recognition. This rethinking of the types of results that can be expected from visual processes should provide the impetus to a considerable body of novel and interesting results in visual information processing.

Recent research in computational vision also is paying increased attention to analyzing the stability of various methods for recovering world structure from visual input. The desire here is to develop rigorous analyses of how visual recovery methods will behave in the face of the type of imperfect input that can be expected in real world situations. This type of analysis can serve to provide precise answers to questions such as: Where will a given approach fail? How will a given approach fail? And significantly, exactly why will a given approach fail? It also is often the case that these types of results can provide insight into how unstable situations can be diagnosed and avoided. There has always been some interest in developing these types of results in computational vision. However, it seems that just now this line of endeavor is coming to be recognized for its considerable importance. Some of the recent examples of stability analyses can be found in studies of stereo (Blostein and Huang, 1987; McVey and Lee, 1982; Mohan et al., 1989; Verri and Torre, 1986; Wildes, 1989) and motion (Barron et al., 1985; Brosh et al., 1989; Kearney et al., 1987; Koenderink and van Doorn, 1987; Verri et al., 1989; Weng et al., 1989).

Yet another intriguing trend can be discerned with regard to the physical instantiation of vision algorithms. This trend is being manifest by two (not entirely independent) research paths: First, there is serious interest in building and testing special purpose hardware for embodying vision algorithms. For example, circuitry has been built for distance sensing, surface interpolation, image smoothing, and motion sensing (Kanade, 1989; Luo et al., 1988; Sage and Lattes, 1987; Tanner and Mead, 1986). Second, there is serious interest in understanding the detailed nature of how vision algorithms can be embodied in biophysical mechanism. Here, proposals have been made and tested for a number of cases including gain enhancement, temporal differentiation, and motion selectivity (Koch et al., 1986; Koch and Poggio, 1988; Koch and Segev, 1989). Significantly, the biophysical models make specific predictions about operations not only at the level of the neuron, but at the level of membrane patches, individual synapses, dendrites, and spines. In general, investigations into the physical instantiation of vision algorithms are quite exciting as they demonstrate how computational analyses can play a role in understanding the fine grained details of how vision systems operate.

It is possible to note a more general trend in and around computational vision. This trend is that there is a growing exchange of ideas between the various visual sciences. This is certainly a positive development and one that should be encouraged to continue.

Summary

Computational vision inquires into the basic nature of visual information processing. It is particularly concerned with developing an understanding of vision at a level of analysis that is independent of how any particular visual system operates. However, the computational approach is not only concerned with abstract theories but also with the development and testing of algorithms and implementations that embody the theories. Further, the general principles that are uncovered in this research program can be used to aid in the study or design of specific visual systems. This chapter has developed these ideas in three sections. In the initial section, the fundamental tenets of computational vision were delineated. In the middle section, binocular stereo vision was used as an example to illustrate the style and concerns of computational vision. A final section served to discuss current trends in the paradigm. In conclusion, it can be noted that the overarching goal of computational vision is to provide a rigorous yet applicable basis for investigations into visual information processing. In this regard significant strides have been made. Perhaps it is needless to say that much more lies ahead.

Acknowledgments. The author thanks K.N. Leibovic, B. McIntosh, L. Parsons, and W. Richards for comments on this chapter.

References

Arnold RD, Binford TO (1980). Geometric constraints in stereo vision. *Soc. Photo-Illum. Engineer.* **238**:281–292.

Baker HH, Binford TO (1981). Depth from edge and intensity based stereo. *Proc. Int. Joint Conf. Artificial Intell.* 631–636.

Ballard DH, Brown CM (1982). *Computer Vision*. Prentice-Hall, Englewood Cliffs, NJ.

Barlow HB, Blakemore C, Pettigrew JD (1967). The neural mechanism of binocular depth discrimination. *J. Physiol.* **193**:327–342.

Barnard ST (1987). A stochastic approach to stereo vision. In *Readings in Computer Vision: Issues, Problems, Principles and Paradigms* (MA Fischler, O Firschein, eds. pp. 21–25.). Morgan Kaufman, Los Altos, CA.

Barnard ST, Fischler MA (1982). Computational stereo. *Comput. Surv.* **14**(4):553–572.

Barnard ST, Thompson WT (1980). Disparity analysis of images. *IEEE Transact. Pattern Anal. Machine Intell.* **2**:333–340.

Barron JL, Jepson AD, Tsosos JK (1985). The sensitivity of motion and structure computations. *Proc. Int. Joint Conf. Artificial Intell.* 700–705.

Barrow HG, Tenenbaum JM (1978). Recovering intrinsic scene characteristics from images. In *Computer Vision Systems* (AR Hanson, EM Riseman, eds. pp. 475–487.). Academic Press, New York.

Bennett BM, Hoffman DD, Prakash C (1989). *Observer Mechanics*. Academic Press, New York.

Bertram S (1969). The UNMACE and the automatic photomapper. *Photogram. Engineer.* **35**(6):569–576.

Binford TO (1984). Stereo vision: Complexity and constraints. In *Robotics Research I*. (JM Brady, R Paul, eds., pp. 475–487.) MIT Press, Cambridge, MA.

Blake A, Zisserman A (1987). *Visual Reconstruction*. MIT Press, Cambridge, MA.

Blakemore C, Fiorenteni A, Maffei L (1972). A second neural mechanism of binocular depth discrimination. *J. Physiol.* **226**:725–749.

Blostein SD, Huang TS (1987). Error analysis in stereo determination of 3-D point positions. *IEEE Transact. Pattern Anal. Machine Intell.* **9**(6):752–765.

Bobick A, Richards W (1986). *Classifying Objects from Visual Information*. Massachusetts Institute of Technology Artificial Intelligence Laboratory Memo 879, Cambridge, MA.

Boult TE (1987). *Reproducing Kernels for surface Interpolation*. Columbia University Artificial Intelligence Memo, New York.

Bowyer K, Eggert D, Stewman J, Stark L (1989). Developing the aspect graph representation for use in image understanding. Proc. Image Understanding Workshop, 831–849.

Braddick OJ (1979). Binocular single vision and perceptual processing. *Proc. R. Soc. London B* **201**:503–512.

Brady JM, Ponce J, Yuille A, Asada H (1985). *Describing Surfaces*. Massachusetts Institute of Technology Artificial Intelligence Laboratory Memo 822, Cambridge, MA.

Brosh M, Kamgar-Parsi B, Kamgar-Parsi B (1989). *Reliability Analysis of the Closed-Form Solution to the Image Flow Equations for 3D Structure and Motion (Planar Patch)*. University of Maryland Center for Automation Research Technical Report CAR-TR-431. College Park, MD.

Bülthoff HH, Mallott HP (1987). Interaction of different modules in depth perception. *Proc. Int. Conf. Comput. Vision* 295–306.

Burr DC, Morrone C, Maffei L (1981). Intra-cortical inhibition prevents simple cells from responding to textured visual patterns. *Exp. Brain Res.* **43**:455–458.

Burt P, Julesz B (1980). A disparity gradient limit for binocular fusion. *Science* **208**: 615–617.

Canny JF (1986). A computational approach to edge detection. *IEEE Transact. Pattern Anal. Machine Intell.* **8**:679–698.

Collett T (1983). Interpolation and extrapolation of surfaces in human vision. *Phil. Transact. R. Soc. London Ser. B.*

Cooley JM, Tukey JW (1965). An algorithm for the machine computation of complex Fourier series. *Math. Comp.* **19**:297–301.

Dev P (1975). Perception of depth surfaces in random-dot stereograms: A neural model. *Int. J. of Man-Machine Studies* **7**:511–528.

De Valois KK, von der Heydt R, Adorjani CS, De Valois RL (1975). A tilt aftereffect in depth. *Assoc. Res. Vis. Ophthalmol.* **15**:90.

Duwaer AL (1981). *Binocular Single Vision: Psychophysical Studies on Underlying Sensory and Motor Processes*. Delfta University Press, Rotterdam.

Duwaer AL, van der Brink G (1981). Diplopia thresholds and the initiation of vergence eye movements. *Vision Res.* **21**:1727–1737.

Eastman RD, Waxman AM (1987). Using disparity functionals for stereo correspondence and surface reconstruction. *Comp. Vision, Graphics Image Process.* **39**:73–101.

Felton B, Richards W, Smith A (1972). Disparity processing of spatial frequencies in man. *J. Physiol.* **225**:319–362.

Ferster D (1981). A comparison of binocular depth mechanism in areas 17 and 18 of the cat visual cortex. *J. Physiol.* **311**:623–655.

Fischer B, Kreuger J (1979). Disparity tuning and binocularity of single neurons in the cat visual cortex. *Exp. Brain Res.* **35**:1–8.

Fischer B, Poggio GF (1979). Disparity selectivity in cortical neurons. *Proc. R. Soc. London Ser. B* **204**:409–419.

Fischler M, Firschein O (eds.) (1987). *Readings in Computer Vision: Issues, Problems, Principles and Paradigms*. Morgan Kaufmann, Los Altos, CA.

Foley JM (1980). Binocular distance perception *Psychol. Rev.* **87**(5):411–434.

Foley JM, Richards WA (1972). Effects of voluntary eye movements and convergence on the binocular appreciation of depth. *Percept. Psychophys.* **11**:423–427.

Frisby JP, Mayhew JEW (1977). Global processes in stereopsis: Some comments on Ramachandran & Nelson (1976). *Perception* **6**:195–206.

Gennery DB (1980). Object detection and measurement using stereo vision. *Proc. DARPA IU Workshop* 161–167.

Gibson JJ (1950). *The Perception of the Visual World*. Houghton Mifflin, Boston.

Gibson JJ (1979). *The Ecological Approach to Visual Perception*. Houghton Mifflin, Boston.

Gillam B, Flagg T, Finlay D (1983). Evidence for disparity change as the primary stimulus processing. *Percept. Psychophys.* **36**(6):559–564.

Gillam B, Chambers D, Russo T (1988). Postfusional latency in stereoscopic perception and the primitives of stereopsis. *J. Exp. Psychol. Human Percept. Perform.* **14**(2): 163–175.

Grimson WEL (1981). *From Images to Surfaces*. MIT Press, Cambridge, MA.

Grimson WEL (1983). Surface consistency constraints in vision. *Comp. Vision, Graphics Image Process.* **24**(1):28–51.

Grimson WEL (1985). Computational experiments with a feature based stereo algorithm. *IEEE Trans. Pattern Anal. Machine Intell.* **7**:17–34.

Grimson WEL, Pavlidis T (1985). Discontinuity detection for visual surface reconstruction. *CVGIP* **30**:316–330.

Gulick WL, Lawson RB (1976). *Human Stereopsis: A Psychophysical Analysis*. Oxford University Press, New York.

Hammond P (1985). Visual cortical processing: Textural sensitivity and its implications for classical views. In *Models of the Visual Cortex* (D Rose, VG Dobson, eds.). Wiley, New York.

Hansen C, Ayache N, Lustman F (1988). Towards real-time trinocular stereo. *Proc. Int. Conf. Comp. Vision* 129–133.

Hänny P, von der Heydt R, Poggio G (1980). Binocular neuron responses to tilt in depth in the monkey visual cortex: Evidence for orientation disparity processing. *Exp. Brain Res.* **41**:A26.

Haralick RM (1980). Edge and region analysis for digital image data. *Comp. Graphics Image Process.* **12**:60–73.

Harris JG (1987). The coupled depth slope approach to surface reconstruction. *Proc. Int. Conf. Comp. Vision* 297–303.

Helmholtz H von (1925). *Treatise on Physiological Optics*. Dover, New York.

Henderson RL, Miller WJ, Grosch CB (1979). Automatic stereo reconstruction of man-made targets. *Soc. Photo Illum. Engineer.*

Hildreth E (1980). *Implementation of a Theory of Edge Detection*. Massachusetts Institute of Technology Artificial Intelligence Lab Technical Report 579, Cambridge, MA.

Hildreth EC (1987). Edge detection. In *Encyclopedia of Artificial Intelligence* (S. Shapiro, ed., pp. 257–267.). Wiley, New York.

Hildreth EC, Hollerbach JM (1987). Artificial intelligence: Computational approach to vision and motor control. *Handbook Physiol. Nervous Syst.* **V**:605–642.

Hoff W, Ahuja N (1987). Extracting surfaces from stereo images. *Proc. Int. Conf. Comp. Vision* 284–294.

Horn BKP (1975). Obtaining shape from shading information. In *The Psychology of Computer Vision* (P H Wilston, ed., pp. 115–155.) McGraw-Hill, New York.

Horn BKP (1985). *Robot Vision*. MIT Press, Cambridge, MA.

Horn BKP, Weldon EJ (1988). Direct methods for recovering motion. *Int. J. Comp. Vision* **2**:51–76.

Hubel DH, Wiesel TN (1970). Cells sensitive to binocular depth in area 18 of the macaque monkey cortex. *Nature* (London) **225**:41–42.

Hubel DH, Weisel TN (1973). A reexamination of stereoscopic mechanisms in the cat. *J. Physiol.* **232**:29–30.

Hubel DH, Wiesel TN (1977). Responses of cells and organization of visual cortex. *Proc. R. Soc. London B* **198**:1–59.

Julesz B (1971). *Foundations of Cyclopean Perception*. The University of Chicago Press, Chicago.

Julesz B, Miller J (1975). Independent spatial-frequency-tuned channels in binocular fusion and rivalry. *Perception* **4**:125–143.

Kanade T (1989) Invited paper presented at IEEE Workshop Int. 3D Scenes.

Kass M (1983). A computational framework for the visual correspondence problem. *Proc. Ad. Res. Projects Agency Image Understanding Workshop.*

Kearney JK, Thompson WB, Boley DL (1987). Optical flow estimation: An error analysis of gradient-based methods with local optimization. *IEEE Transact. Pattern Anal. Machine Intell.* **9**(2):229–244.

Keating TJ, Wolf PR, Scarpace FL (1975). An improved method of digital image correlation. *Photogram. Eng. Remote Sensing* **41**(8):993–1002.

Koch C, Poggio T (1988). Biophysics of computational systems: Neurons, synapses and membranes. In *New Insights into Synaptic Function* (GM Edelman, WE Gall, WM Cowan, eds.). Wiley, New York.

Koch C, Sagev S (eds.) (1989). *Methods in Neuronal Modeling*. MIT Press, Cambridge, MA.

Koch C, Marroquin J, Yuille A (1985). *Analog "Neural" Networks in Early Vision.* Massachusetts Institute of Technology Artifical Intelligence Lab Memo 751, Cambridge, MA.

Koch C, Poggio T, Torre V (1986). Computations in the vertibrate retina: Gain enhancement, differentiation and motion discrimination. *Trends Neurosci.* **9**:2004–2011.

Koenderink JJ (1987). An internal reprsentation for solid shape based on the topographical properties of the apparent contour. In *Image Understanding 1985–1986* (W Richards, S Ullman, eds., pp. 257–285.) Ablex, Norwood, NJ.

Koenderink JJ, van Doorn AJ (1976). Geometry of binocular vision and a model for stereopsis. *Biol. Cybernet.* **21**:29–35.

Koenderink JJ, van Doorn AJ (1987). Facts on optic flow. *Biol. Cybernet.* **56**:247–254.

Kulikowski JJ (1978). Limit of single vision in stereopsis depends on contour sharpness. *Nature* (London) **275**:126–127.

Land EH, McCann JJ (1971) Lightness and retinex theory. *J. Opt. Soc. Am.* **61**:1–11.

Lee D, Pavlidis T (1987). One-dimensional regularization with discontinuities. *Proc. ICCV*, 572–577.

Leibovic KN, Balslev E, Mathieson TA (1971). Binocular vision and pattern recognition. *Kybernetik* **8**:14–23.

Levine MD, O'Handley DA, Yagi GM (1973). Computer determination of depth maps. *Comp. Graphics Image Process.* **2**(2):131–150.

Levinson E, Blake R (1979). Stereopsis by harmonic analysis. *Vision Res.* **19**:73–78.

Luo J, Koch C, Mead C (1988) An experimental subthreshold analog (MOS 2D surface interpolation circuit. Proc. IEEE Neural Info. Proc. Sys.

Marr D (1970). A theory for cerebral neocortex. *Proc. R. Soc. London Ser. B* **176**:161–234.

Marr D (1976). Artificial intelligence—A personal view. *Artif. Intell.* **9**:37–48.

Marr D (1982). *Vision: A Computational Investigation into the Human Representation and Processing of Visual Information*. Freeman, San Francisco.
Marr D, Hildreth EC (1980). Theory of edge detection. *Proc. R. Soc. London Ser. B* **207**:187–217.
Marr D, Poggio T (1976). Cooperative computation of stereo disparity. *Science* **194**: 283–287.
Marr D, Poggio T (1979). A computational theory of human stereo vision. *Proc. R. Soc. London Ser. B* **204**:301–328.
Marr D, Poggio T (1977). From understanding computation to understanding neural circuitry. *Neurosci. Res. Prog. Bull.* **15**:470–488.
Marroquin JL (1984). *Surface Reconstruction Preserving Discontinuities*. Massachusetts Institute of Technology Artificial Intelligence Lab Memo 792, Cambridge, MA.
Mayhew JEW, Frisby JP (1978). Stereopsis masking in humans is not orientationally tuned. *Perception* **7**:431–436.
Mayhew JEW, Frisby JP (1979). Surfaces with steep variations in depth pose difficulties for orientationally tuned disparity filters. *Perception* **8**:691–698.
Mayhew JEW, Frisby JP (1981). Psychophysical and computational studies towards a theory of human stereopsis. *Artif. Intell.* **17**:349–385.
McVey ES, Lee JW (1982). Some accuracy and resolution aspects of computer vision distance measurements. *IEEE Transact. Pattern Anal. Machine Intell.* **4**(6):646–649.
Medioni GC, Nevatia R (1984). Description of three-dimensional surfaces using curvature properties. *Proc. DARPA IU Workshop* 219–229.
Medioni GG, Nevatia R (1985). Segment-based stereo matching. *Comp. Vision Graphics Image Process.* **31**:2–18.
Mitchison G (1987). *Planarity and Segmentation in Stereoscopic Matching*. Massachusetts Institute of Technology Center for Biological Information Processing Memo 23, Cambridge, MA.
Mitchison GJ, McKee SP (1985). Interpolation in stereoscopic matching. *Nature* (London) **315**:402–404.
Mitchison GJ, McKee SP (1987a). Interpolation and the detection of fine structure in stereoscopic matching. *Vision Res.* **27**:295–302.
Mitchison GJ, McKee SP (1987b). The resolution of ambiguous stereoscopic matches by interpolation. *Vision Res.* **27**:285–294.
Mitchison GJ, Westheimer G (1984). The perception of depth in simple figures. *Vision Res.* **24**(9):1063–1073.
Mohan R, Medioni G, Nevatia R (1989). Stereo error detection, correction and evaluation. *IEEE Transact. Pattern Anal. Machine Intell.* **11**(2):113–121.
Moravec H (1979). Visual mapping by a robot rover. *Proc. Int. Joint Conf. Artif. Intell.* 598–600.
Mori K, Kidode M, Asada H (1973). An interative prediction and correction method for automatic stereocomparison. *Comp. Graphics Image Process.* **2**:393–401.
Motter BC, Poggio GF (1982). Spatial invariance of receptive field location in the presence of eye movements of fixation for neurons in monkey striate cortex. *Soc. Neurosci. Abstr.* **8**:707.
Mumford D, Shah J (1985). Boundary detection by minimizing functionals. *Proc. Comp. Vision Pattern Recog.*
Nalwa VS, Binford TO (1986). On detecting edges. *IEEE Transact. Pattern Anal. Machine Intell.* **8**:699–714.

Nelson JI (1975). Globality and stereoscopic fusion in binocular vision. *J. Theoret. Biol.* **49**:1–88.

Nelson JI, Kato H, Bishop PO (1977). Discrimination of orientation and position disparity by binocularly activated neurons in cat striate cortex. *J. Neurophysiol.* **40**:260–283.

Nielsen KR, Poggio T (1984). Vertical image registration in stereopsis. *Vision Res.* **24**: 1133–1140.

Ninio J (1985). Orientational versus horizontal disparity in the stereoscopic appreciation of slant. *Perception* **14**:305–314.

Nishihara HK (1984). Practical real-time imaging stereo matcher. *Opt. Engineer.* **23**(5): 536–545.

Nishihara HK, Poggio T (1984). Stereo vision for robotics. In *Robotics Research I* (JM Brady, R Paul, eds., pp. 489–505.) MIT Press, Cambridge, MA.

Ogle, K. N. (1950) *Researches in Binocular Vision*. Saunders, Philadelphia.

Ohta Y, Kanade T (1983). *Stereo by Intra- and Inter-Scanline Search Using Dynamic Programming*. Carnegie-Mellon University Department of Computer Science Technical Report CMU-CS-83-162, Carnegie-Mellon University, Pittsburgh.

Ohzawa I, Freeman RD (1986a). The binocular organization of complex cells in the cat's visual cortex. *J. Neurophysiol.* **56**(1):243–259.

Ohzawa I, Freeman RD (1986b). The binocular organization of simple cells in the cat's visual cortex. *J. Neurophysiol.* **56**(1):221–242.

Panton DJ (1978). A flexible approach to digital stereo mapping. *Photogram. Engineer. Remote Sensing* **44**:1499.

Pettigrew JD, Nikara T, Bishop PO (1968). Binocular interaction on single units in cat striate cortex: Simultaneous stimulation by single moving slits with receptive fields in correspondence. *Exp. Brain Res.* **6**:391–410.

Poggio GF (1984). Processing of stereoscopic information in primate visual cortex. In *Dynamic Aspects of Neocortical Function* (WM Cowan, WE Gall, eds.). Wiley, New York.

Poggio GF (1985). Responses of neurons in visual cortex of the alert macaque to dynamic random-dot stereograms. *Vision Res.* **25**:397–406.

Poggio GF, Fischler B (1977). Binocular interaction and depth sensitivity of striate and prestriate cortical neurons of the behaving rhesus monkey. *J. Neurophysiol.* **40**:1392– 1405.

Poggio GF, Poggio T (1984). The analysis of stereopsis. *Annu. Rev. Neurosci.* **7**:379–412.

Poggio GF, Talbot WH (1981). Mechanisms of static and dynamic stereopsis in foveal cortex of the rhesus monkey. *J. Physiol.* **315**:469–492.

Poggio T, Drumheller M (1986). On parallel stereo. *Proc. IEEE Int. Conf. Robotocs Automat.*

Poggio T, Verri A (1987). Against quantitative optical flow. *Proc. Int. Conf. Comp. Vision*, pp. 171–180.

Poggio T, Torre V, Koch C (1985). Computational vision and regularization theory. *Nature* (London) **317**:314–319.

Pollard SB, Mayhew JEW, Frisby JP (1985). PMF: A stereo correspondence algorithm using the disparity gradient limit. *Perception* **14**:449–470.

Ponce J, Kriegman D (1989). On recognizing and positioning curved 3D objects from image contours. Proc. IEEE Workshop on Int. 3D Scenes, 61–67.

Prazdny K (1985). Detection of binocular disparities. *Biol. Cybernet.* **52**:93–99.

Richards WA (1971). Anomolous stereoscopic depth perception. *J. Opt. Soc. Am.* **61**: 410–414.

Richards WA (1977). Stereopsis with and without monocular cues. *Vision Res.* **17**: 967–969.

Richards WA (ed.) (1988). *Natural Computation*. MIT Press Cambridge, MA.

Richards WA, Ullman S (1987). *Image Understanding 1985–86*. Ablex, Norwood, NJ.

Rieger JH (1987). On the classification of views of piecewise smooth objects. Image and Vision Computing **5**:91–97.

Rogers BJ (1987). Paper presented at the International Conference on Computer Vision.

Sage JP, Lattes AL (1987). A high-speed 2D CCD Gaussian Image convolver. Solid State Research, Quarterly Research Report, MIT Lincoln Lab, 49–52.

Shanmugan KS, Dickey FM, Green JA (1979). An optimal frequency domain filter for edge detection in digitall pictures. *IEEE Transact. Pattern Anal. Machine Intell.* **1**:37–49.

Smitley T, Bajcsy R (1984). Stereo processing of aeriel urban images. *Proc. Int. Conf. Pattern Recog.* 405–409.

Sperling G (1970). Binocular vision: A physical and a neural theory. *Am. J. Psychol.* **83**:461–534.

Spoerri A, Ullman S (1987). Early detection of motion boundaries. *Proc. ICCV* 209–218.

Sripradisvarakul T, Jain R (1989). Generating aspect graphs for curved objects. Proc. IEEE Workshop Int. 3D Scenes, 109–115.

Stevens KA, Brookes A (1987). Depth reconstruction in stereopsis. *Proc. Int. Conf. Comp. Vision* 549–603.

Stewart CV, Dyer CR (1988). The trinocular general support algorithm: A three-camera stereo algorithm for overcoming binocular matching errors. *Proc. Int. Conf. Comp. Vision* 134–138.

Sutro LL, Lerman JB (1973). *Robot Vision*. Internal Report R-635, Charles Stark Draper Lab, Cambridge, MA.

Tanner JE, Mead C (1986). An integrated optical motion sensor. VLSI Sig. Proc. II, 59–76.

Terzopoulos D (1983). Multilevel computational processes for visual surface reconstruction. *Comp. Vision, Graphics Image Process.* **24**:52–96.

Terzopoulos D (1986a). Integrating visual information from multiple sources. In *From Pixels to Predicates*. (A Pentland, ed., pp. 111–142.). Ablex, Norwood, NJ.

Terzopoulos D (1986b). Regularization of inverse problems involving discontinuities. *IEEE Transact. Pattern Anal. Machine Intell.* **8**(4):413–424.

Torre V, Poggio T (1986). On edge detection. *IEEE Transact. Pattern Anal. Machine Intell.* **8**:147–163.

Tyler CW (1973). Stereoscopic vision: Cortical limitations and a disparity scaling effect. *Science* **181**:276–278.

Tyler CW (1974). Depth perception in disparity gratings. *Nature* (London) **251**:140–142.

Tyler CW (1975). Spatial limitations on human stereoscopic vision. *Soc. Photo-Illum. Engineer.* **120**:36–42.

Ullman S, Richards W (1984). *Image Understanding 1984*. Ablex, Norwood, NJ.

Verri A, Torre V (1986). Absolute depth estimate in stereopsis. *J. Opt. Soc. Am. A* **3**(3):297–299.

Verri A, Girosi F, Torre V (1989). Mathematical properties of the two-dimensional motion field: From singular points to motion parameters. *JOSA A* **6**(5):698–612.

von der Heydt R (1979). *Stereoskopische Wahrnehmung der Orientierungsdisparation*. Doctoral dissertation, Eidgenossische Technische Hochschule, Zurich, Switzerland.

von der Heydt R, Adorjani C, Hänny P (1977). Neuronal mechanisms of stereopsis: Sensitivity to orientational disparity. *Experientia* **33**:786.

von der Heydt R, Adorjani C, Hänny P, Baumgartner G (1978). Disparity sensitivity and receptive field incongruity of units in the cat striate cortex. *Exp. Brain Res.* **31**:423–545.

von der Heydt R, Hänny P, Dursteller MR (1981). The role of orientation disparity in stereoscopic perception and the development of binocular correspondence. In *Advances in Physiological Science*, Vol. 16. (E Grastyan, P Molnar, eds). Pergamon, Oxford.

Weinberg G (1975). *An Introduction to General Systems Thinking*. Wiley, New York.

Weinshall D (1988). Application of qualitative depth and shape from stereo. *Proc. Int. Conf. Comp. Vision* 144–148.

Weng J, Huang TS, Ahuja N (1989). Motion and structure from two perspective views: Algorithms, error analysis and error estimation. *IEEE Transact. Pattern Anal. Machine Intell.* **11**(5):451–476.

Wildes RP (1989a). *On Interpreting Stereo Disparity.* Massachusetts Institute of Technology Artificial Intelligence Lab Technical Report 1112, Cambridge, MA.

Wildes RP (1989b). Surface orientation from binocular stereo orientation disparity. *Proc. SPIE Intell. Robotics Comp. Vision VIII*, pp. 309–317.

Wildes RP (1989c). Three-dimensional surface curvature from binocular stereo curvature disparity. *Opt. Soc. Am. Tech. Digest* **18**:62.

Wilson HR (1988). Spatial masking in the third dimension. *Opt. Soc. Am. Tech. Digest* **11**:155.

Winston PH (1984). *Artificial Intelligence*, 2nd ed. Addison-Wesley, Reading, MA.

Yeshurun Y, Schwartz EL (1989). Cepstral filtering on a columnar image architecture: A fast algorithm for binocular stereo segmentation. *IEEE Transact. Pattern Anal. Machine Intell.* **11**(7):759–767.

12
Computer Vision Analysis of Boundary Images

DEBORAH WALTERS

This chapter gives a brief introduction to the field of computer vision by describing different approaches to computer vision research, and then illustrates several of the approaches by describing some current research concerned with the computer analysis of boundary images.

Introduction to Computer Vision

Computer vision is the process of using computers to extract from images useful information about the physical world including meaningful descriptions of physical objects (Ballard and Brown, 1982; Horn, 1986). For example, if an image sensor such as a digitizing video camera captured an image of a physical scene, and the digital image was input to a computer vision system, the desired output would be a description of the physical scene in terms that would be useful for the particular task at hand. Computer vision has many applications including robotics, industrial automation, document processing, remote sensing, navigation, microscopy, medical image analysis, and the development of visual prostheses for the blind (Fischler and Firschein, 1987).

Basic Definitions

Visual information initially consists of light energy from the environment. But before the information can be used, it must be transformed into a form that is accessible to the visual processor. In human vision this task is performed by the rods and cones of the retina. The light energy that strikes the photoreceptors is converted into the electrochemical activity that is the substrate of computations in the brain. The same function can be performed for a computer by a video camera, or some other photosensitive device that converts the photon energy into electrical signals that can then be sampled and expressed as digital values in a computer. We can think of this process as being analogous to the process of using a camera to generate a photographic image that captures visual information about the environment. In a computer the sampling is generally done along a rectangular grid to

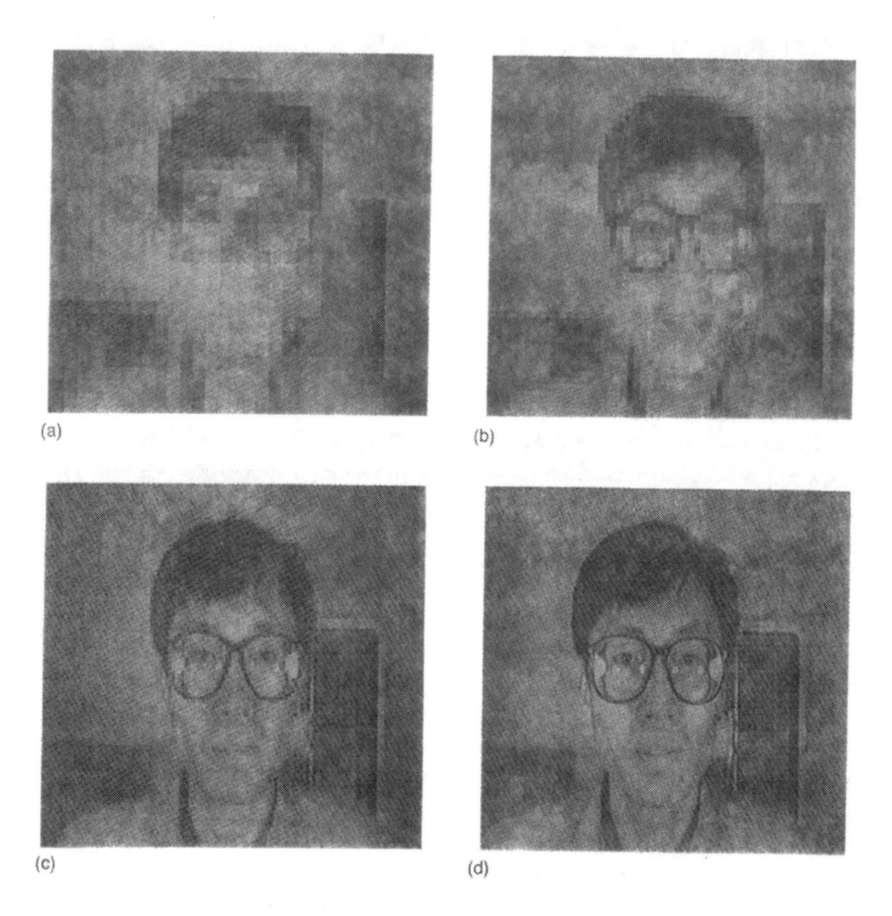

FIGURE 12.1. These images illustrate the discrete spatial sampling of images used in computer vision: (a) 32 × 32 pixels; (b) 64 × 64 pixels; (c) 256 × 256 pixels; and (d) 512 × 512 pixels. (Each image has the possibility of 64 different intensity levels.)

produce a two-dimensional array of values that corresponds to light energy. Each element of the array is referred to as a "pixel," and the value of a pixel is referred to as its "intensity" or "gray-level." If the photosensitive device used was selectively sensitive to wavelength, then the value can be referred to as the pixel's "color." Figure 12.1 shows an image that was sampled using grids of different sizes. The smallest grid size is 512 × 512 pixels, and creates an image in which it is hard to distinguish the individual pixels, while the largest grid size of 32 × 32 pixels clearly shows the individual pixels. Figure 12.2 illustrates the concept of pixel gray levels by showing an image sampled at 512 × 512 pixel resolution, but with the number of levels of gray-scale sampling varied from 2 (which produces the "binary" image) to 16.

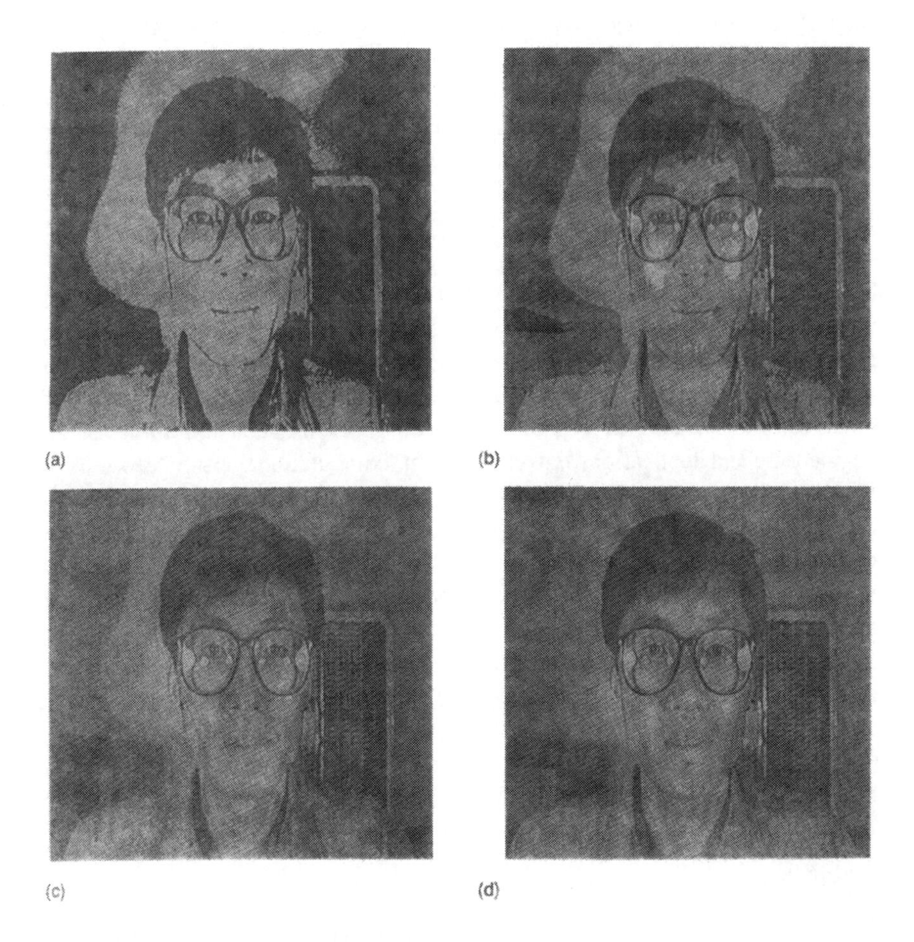

FIGURE 12.2. These images illustrate the discrete intensity sampling of images used in computer vision: (a) 2 intensity levels; (b) 4 intensity levels; (c) 8 intensity levels; and (d) 16 intensity levels. (Each image has 512 × 512 pixels.) Compare these images to Figure 12.1d which has 64 intensity levels.

In computer vision systems it is common to use the rectangular sampling grid as illustrated in Figures 12.1 and 12.2. But, as preceding chapters have shown, the mammalian visual system does not uniformly sample the visual field: sampling is densest in the fovea and becomes less dense in the periphery. So why do computer vision systems use a different sampling scheme? One reason could be that the nonhomogeneous sampling of the mammalian system is due to some idiosyncrasies of the physiological wetware. If that were the case, then it may not make sense for a computer vision system to incorporate such sampling. But another reason is again related to our lack of conscious awareness of the initial stages of our visual processing; we perceive the visual array to be uniformly sampled as we think we see all aspects of a scene at a single resolution. This percept was observed by computer scientists, and led to the use of uniformly sampled rectangular arrays. However, recently the use of the biological style of inhomogeneous sampling has been studied and it was found that there are computational advantages in using such sampling (Browse and Rodriques, 1987). Thus it may be possible to improve the performance of computer vision systems by using the biologically inspired sampling grid.

Terminology

There are various terms used to refer to the field of computer vision: machine vision, computational vision, image understanding, robot vision, image analysis, and scene analysis. Each of these terms has a different historical perspective and some retain a difference in emphasis. For example, the term "machine vision" is most commonly used in engineering disciplines and thus has more of an engineering and applications flavor. The term "computational vision" arose from interdisciplinary research by computer scientists, visual psychophysicists, physicists, and neuroscientists. There are two goals of computational vision: one concerns the creation of computer systems that can "see," and the other concerns understanding biological vision. The unifying principle of computational vision is the concept that it is possible to understand vision independently of whether it is implemented in computer hardware or in biological wetware. More specifically, the goal of computational vision is to express the process of vision in terms of computations. This sense of computation is not limited to the numerical computations performed on a calculator, but includes all the more abstract computations that can be performed by an abstract algorithmic processing system.

Related Fields

There are several fields to which computer vision is closely related: image processing, which involves image to image transformations; computer graphics, which involves description to image transformations (which is the inverse of computer vision's image to description transformations); and pattern recognition, which involves pattern to class transformations. Computer vision is a subfield of artificial intelligence (AI) and the process of retracting information from images requires the same types of knowledge acquisition and cognitive reasoning as other AI subfields. However, in addition vision requires significant perceptual preprocessing of the visual input before the cognitive analysis.

Levels of Computer Vision Processing

Computer vision processing is generally divided into two levels: early vision and scene analysis. Early vision, otherwise known as low level vision, involves the first stages of processing required for a visual task. One aspect of this first stage is feature analysis, whereby information about color, motion, shape, texture, stereo depth, and intensity edges is extracted. Another aspect of early vision is image segmentation, whereby the feature information is used to segment the image into regions that have a high probability of having arisen from a single physical cause. For example, suppose a scene consisted of a single orange resting on an infinitely large flat white surface illuminated by a diffuse light source. An image of this scene could be segmented based on color information alone to form two regions, one corresponding to the orange and the other corresponding to the flat surface.

The second level of processing, scene analysis, involves taking the featural descriptors generated by early vision and constructing higher level descriptions of the scene. Some components of this task are shape analysis, object recognition, and object localization. This level is also referred to as high level vision, and involves more knowledge-based processing than early vision. In the example of the image of an orange, the scene analysis level of processing would involve recognizing that the circular orange-colored region was an image of an orange. This recognition must be based on the system having knowledge about the nature of oranges, and the ability to make inferences based on the visual information.

The division between early vision and scene analysis is not firm. Many computer vision systems have partial information generated by the scene analysis processing feedback to the early vision processing stages. The information fed back is used in refining the initial descriptions to make them more useful for the scene analysis processing. Several iterations through this feedback and feedforward process may be required to generate the final scene descriptions. Another sense in which the division is not firm is that some researchers refer to three levels of vision: low level, intermediate, and high level. Again the exact boundaries between these three levels are not distinct.

Why Vision Is Difficult

The goal of creating computer vision systems that can "see" was initially thought to be rather easy. The argument was made that computers are very powerful. For example, even though the solution of simultaneous differential equations is difficult for humans, computers can readily solve them. So if we take a task that is trivially easy for humans, such as vision, it should be even easier to implement it on a computer. Yet when computer scientists and engineers first attempted to give computers a visual sense, they failed completely. The problem was that there is a fallacy in the simple argument used above. While humans are conscious of most of the stages of processing involved in solving simultaneous differential equations and thus can realize the complexity involved, most of the processing involved in visual perception remains subconscious. So while the process of encountering, say, a yellow Volkswagen in the environment and using our visual sense to determine that there is a yellow Volkswagen currently present in our environment may feel like a simple process, it actually involves many interrelated levels of computational processing. Uncovering just what those levels of processing are and expressing them as algorithms is one goal of computer vision.

Another reason that vision is difficult is because it is an underconstrained problem. For example, an image is a two-dimensional projection of a three-

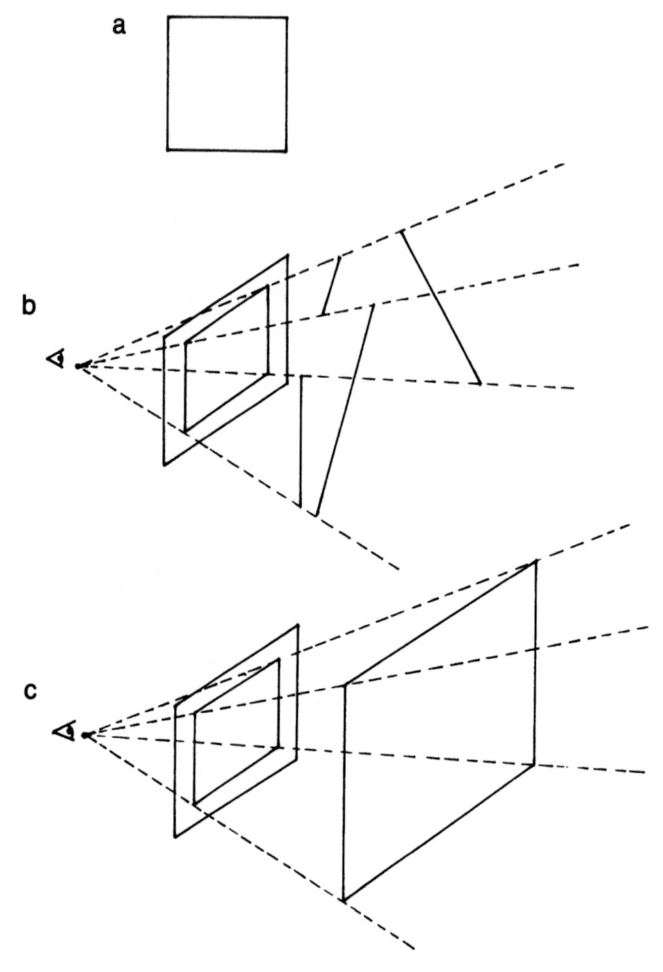

FIGURE 12.3. (a) An ambiguous image. (b) One possible scene that would project to form the image in (a). (c) Another possible scene that would project to form the image in (a).

dimensional scene, but there can be infinitely many three-dimensional scenes that project onto any two-dimensional image. Thus, given just the single image, it is impossible to determine which of the possible scenes is depicted in the image. For example, the image in Figure 12.3a appears to depict a rectangle, but the actual scene from which this image arose is seen in Figure 12.3b: it consists of four thin wires that do not touch. But the image in Figure 12.3a could equally well have been the projection of a scene containing a rectangle (Fig. 12.3c). This simple example illustrates that it is possible to have two scenes that both project to the same image. But notice that the image in Figure 12.3a does not appear to be ambiguous; we do not perceive all possible scenes. Thus humans are either using some additional high-level information about the world to interpret images unambiguously (such as knowledge about rectangles, for example), or they are using some general constraints to rule out multiple interpretations. There is psychophysical evidence that humans use both strategies, but the surprising result is that the high level knowledge appears to provide less information for disambiguating scenes than the lower level general constraints. This suggests the idea that there must be some additional general constraints that humans use in perceiving images. When these constraints are based on the general physical properties of the world and of image projection, they can be of great use in creating computer vision systems. Unless such constraints are explicitly understood, creating a computer vision system will be very difficult.

Determining Constraints for Vision Theory

One of the primary tasks in computer vision is to find a set of constraints that would allow a computer to unambiguously interpret images. The constraints can be either features in the image that can be used to make inferences about the scene, or regularities of nature that can be exploited. A general theory of vision can be developed from appropriate constraints. There are four main techniques for determining such constraints: the engineering approach, the statistical approach, the biological approach, and the physical approach.

Engineering Approach to Determining Constraints

The engineering approach relies on the intuitions, introspections, and prior knowledge of the system designer as to what the important image features should be, and how such features should be interpreted. This approach was used in much of the early work in computer vision, and continues to be used in many machine vision systems. Although this approach has been successful in some applications, its lack of a theoretical basis makes it less desirable than some of the other approaches.

Statistical Approach to Determining Constraints

Statistical analysis provides the second approach to determining image constraints. The basic idea behind this approach is that it is possible to design a system that can "learn" what the constraints are simply by observing the world through sensory input. This is the approach used in statistical pattern recognition, and, more recently, in artificial neural networks. In the statistical approach the relevant aspects of the visual environment are sampled, and the sample statistics used to find image features that can be used to provide the necessary constraints on image interpretation. This approach can be successful when the input stimuli can be recognized from descriptions that are directly constructed from the image. However, in most vision problems the direct descriptions are further processed into higher level descriptions, which are then further processed into even higher level descriptions, etc., and only after several such intervening levels of successively more derived descriptions does recognition occur. It is not yet clear how such complex structure could emerge in an artificial system that must "learn" its structure. One solution would be to design such structure into the network, but this then begs the question of what image features are important, as at least the basic level features would have to be predetermined. So although the statistical approaches may produce interesting results for early vision, they are not well suited for scene analysis.

Biological Approach to Determining Constraints

The third approach to finding useful image constraints involves studying biological vision systems with the aim of uncovering the constraints that they use in image interpretation. In some cases the biological solution may be constrained by the neurophysiological implementation mechanisms, and in such cases the biological approach may fail. However, in other cases the biological solution may be constrained by the general problem of vision and may provide useful insights for computer vision.

In one sense all computer vision uses the biological approach, as vision is defined in terms of the human visual sense. But as mentioned previously, most stages of visual processing are not open to conscious introspection so in order to systematically use the biological approach, other means of determining how the mammalian visual system functions must be used. This is possible using techniques from the disciplines of visual perception, psychophysics, neurophysiology, and neuroanatomy. Over the past 10 years there has been an increasing degree of interactions between those interested in biological vision and those primarily interested in machine vision, and the contributions from each side enhance the research of the other.

Physical Approach to Determining Constraints

In the physical approach the basic idea is to determine properties of the physical world that can be used to constrain image interpretation. This has been a very successful approach and has led to many useful constraints. For example, David Marr frequently used continuity as a constraint (Marr, 1982). The continuity constraint makes use of the fact that the physical world is basically continuous, thus neighboring image points have a high probability of having arisen from the same physical entity. This constraint has been used in a simple cooperative algorithm to compute the relative depth of different image regions. Another example of a frequently used constraint is the assumption of a general viewpoint. This assumption is based on the fact that for a given scene, for small changes in the viewpoint, there will generally be only small changes in the projected image and thus many properties of the projected image will remain constant over most viewpoints (Lowe, 1984). So if it is possible to determine which image properties are invariant over viewpoint, except at a limited number of viewpoints, then it may be possible to make useful inferences about the scene based on these properties. For example, if several points in an image are collinear, then there is a high probability that they depict collinear points in three space. This will not always be true: if an arc of a circle exists in three space, then there will be a limited number of viewpoints such that the points on the arc project onto a straight line in an image, but for the vast majority of possible viewpoints the collinearity assumption will be true. Vision constraints often have the property of being true only in general, yet they can be important in providing inferences useful for the correct interpretation of an image.

In computational vision the major emphasis is placed on the physical approach to determining image constraints, as such constraints should be relevant to both biological and machine vision. In addition the biological approach is frequently used, especially as a means of providing inspiration for the physical approach. Thus, just the fact that a certain image feature appears to be important for human perception does not by itself help to develop a general theory of vision. But the fact that such a feature is used by humans can motivate an analysis of the underlying physical constraints relevant to such a feature, and thus may lead to the understanding necessary for a general vision theory.

Algorithms and Implementations

Once a theory of vision has been established it can be tested by developing algorithms based on the theory, and then analyzing the algorithms or implementing the algorithms and testing the results. For a given theory there will be many different types of possible algorithms that embody the theory, and for a given algorithm there will be many different methods of implementation.

There are two basic ideologies in cognitive science concerning how to study the computations involved in human cognition and perception. One is the symbolic belief that argues that the correct level to study cognition is the level of the mind, as that is the level at which humans perform symbolic processing that is the essence of cognition (Pylyshyn, 1980). In the symbolic ideology digital computers can provide a useful model of cognitive computations, and the differences between brains and computers can be ignored as an irrelevant detail of the hardware implementation. The opposing ideology argues that it is useful to study the type of massively parallel computations that could be performed in the brain, as they may provide not only new models of implementation, but may offer new insights into higher cognitive processing. The basic idea is that connectionist formalisms may provide a new language in which to express the processes of cognition. Just as the invention of calculus enabled scientists to think differently about physical processes, a less serial formal language may provide insights into cognitive processes. Symbolic processing as we now understand it may be just one limited form of all possible types of computation.

Connectionist or artificial neural networks have been inspired by the anatomy and physiology of the brain, with the basic motivation of trying to understand the types of computations the brain could perform. This is necessary because most knowledge about computation comes from the study of either theoretical symbol-processing machines (such as automata), or through the study of the digital computer. But while a digital computer consists of a small number of processing elements that can perform millions of individual operations per second, a brain consists of very large numbers of processing elements (neurons) that are highly interconnected and can perform only a few operations in a second. Thus the model of computation as provided by studying computers is not well suited to study the type of massively parallel computations being performed by the brain. Artificial neural networks are an attempt to study the highly parallel neural computations even though they are only very loosely based on known neurophysiology. For example, neurons are often modeled as simple threshold units, which can interact with each other only through simple weighted connections. Real neurons are obviously much more complex than this, and may be individually performing computations as complex as those currently modeled by an entire network of artificial units. However, the hope is that by understanding the complex interactions that arise in even a simple artificial network of simple elements, our general understanding of massively parallel computation will be improved.

The following example will help to distinguish between the symbolic and the neural network or connectionist style of computation. Consider the problem where there are three possible colors for a given region of a map, and the goal is to assign colors to the regions such that no adjacent regions have the same color. Although, in general, this is impossible, it can be an easy problem in particular instances. In these cases, using symbolic computations, it is possible to write out an algorithm for this problem using the constraints that adjacent regions cannot be the same color and that a single region can be only one color. The only possible problem will be that it might take the algorithm several iterations to find a solution to the problem. In the connectionist approach the constraints would not be expressed procedurally as in the symbolic case, but would be implicitly expressed in the connections of the network. For example, the connectionist network for coloring the map shown in Figure 12.4a is seen in Figure 12.4b. For each region there are one to three simple processing units, each representing one of the possible colors for that region. Each unit receives excitatory and inhibitory inputs from neighboring units, sums those inputs, and produces an output based on that sum. There is a weight associated with each connection between units, and the task constraints are implicitly expressed in the values of those weights. The constraint that each region can have only one color is expressed by having inhibitory connections between the three units within each region. The constraint that adjacent regions have different colors can be expressed as excitatory connections between units representing different colors in adjacent regions. If, for example, all the excitatory weights had a value of one and all the inhibitory weights had a value of negative two, then given some random initial values of the units, the network would eventually converge on a correct solution where only one unit within each region would produce a positive output (Feldman and Ballard, 1982).

The map-coloring network illustrates how constraints can be implicitly expressed in terms of connections, rather than being explicitly expressed as procedures in a symbolic computation. For some problems the symbolic expression may be more natural, while for other problems there may be an advantage to using the connectionist expression.

Another important aspect of connectionist networks is their ability to "learn." In the map coloring network, the connection weights were arbitrarily set by the

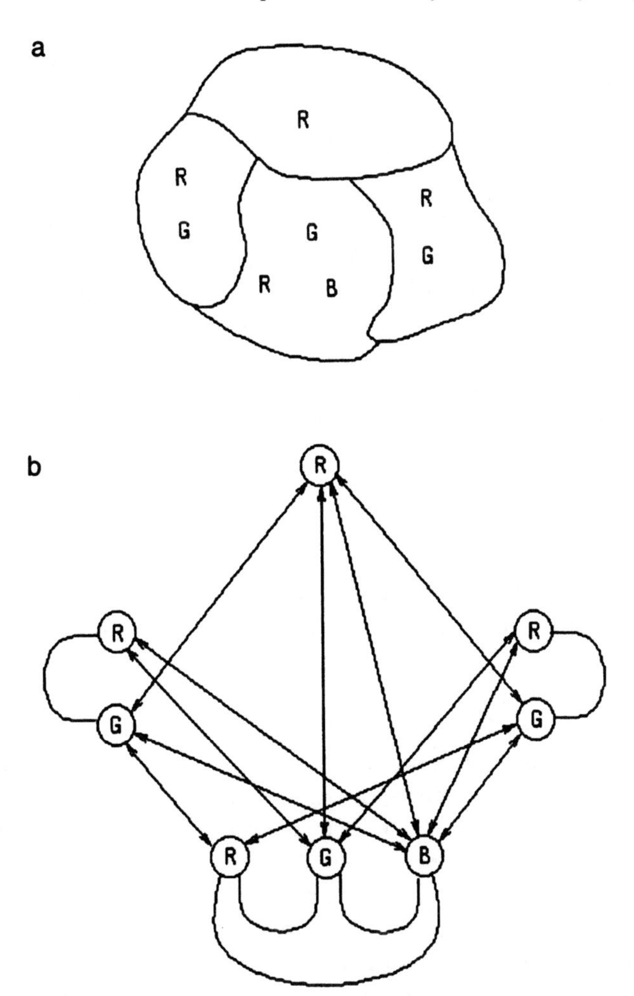

FIGURE 12.4. (a) An example map to be colored with the added constraints on the possible colors of each region indicated by the letters "R," "G," and "B," indicating the colors red, green, and blue, respectively. (b) A connectionist network for solving the map coloring problem in (a). (See text for a detailed explanation.)

network designer. It is only when a network has the ability to modify its connections that it can exhibit "learning," and it is such networks which are used to find image features that can provide the necessary constraints on image interpretation. Although this approach has attracted considerable attention in recent years, it is not clear that it will lead to new insights. One problem is that biological vision systems contain considerable structure, such as several retinotopic mappings of various image parameters with many feedback and feedforward pathways through the system. It is not clear how such complex structure could emerge in a homogeneous artificial system that must "learn" its structure. One solution would be to design such structure into the network, but this as we saw earlier begs the question of which image features are important, as at least the basic level features would have to be predetermined. One possible solution is based on the distinction between evolutionary "learning" and the "learning" that occurs within the lifetime of an individual. Many neural networks are attempting to produce both types of learning in a single system. A more realistic approach might be to hardwire into a network the inherited structure, and then to allow the network to learn based on the image representation implicit in the inherited structure. Several visual neural networks have been based on this principle and are used to study the development of organization in the visual system (von der Marlsburg, 1973; Linsker, 1986; Yuille et al., 1989), to study the early levels of visual perception without learning (Walters, 1987), and to learn to recognize objects based on the inherited features (Walters, 1988).

Interpretation of Line Art

In order to illustrate these ideas about computer vision, a specific problem will be discussed in detail, with examples of the physical and biological approaches to determining useful constraints and examples of connectionist networks as implementation devices.

When humans view images or scenes, there are many aspects of the visual stimulation that convey information useful for interpreting the scene. Among these are color, motion, stereo depth, texture, shading, and edges. But are all of these properties necessary to interpret an image? Obviously not, as humans can readily interpret line art such as cartoons, technical drawings, and maps even when they contain only edge information. Much of the early work in computer vision concentrated on interpreting line drawings as it was felt that such images would be easier to interpret than natural images, yet would yield results that would also be applicable to natural images. In most cases the domain was further restricted to interpretation of "blocks world" images: images depicting scenes created from blocks of several simple shapes. Using such restricted domains several interesting computational techniques were developed [e.g., Waltz' constraint satisfaction algorithm (Waltz, 1975)]. However, when attempts were made to extend the domains to include images containing curves it was found that the techniques could not be extended to deal with the extended domains. This led to a general lack of interest in the study of algorithms dealing with line art as a means of uncovering basic algorithms that can be extended to the analysis of natural images.

Recently a new interest in line art algorithms has developed because of the converse problem: it has been found that the computer vision algorithms that deal well with natural images are not adequate for analyzing line art. For example, current algorithms for natural images assume that no single feature class (edges, texture, color, motion, etc.) can give adequate information for object recognition, and that only through combining information from several feature classes can the problem be solved (Edelman and Poggio, 1989). But in line art there is only a single source of information—line edges. Therefore line art interpretation may turn out to be more difficult than natural image interpretation. As line art images are a commonly used means of communication (e.g., technical drawings, contour maps, cartoons) it would be useful for computers to be able to interpret such images.

Two Specific Line Art Problems

In computer vision research one of the first tasks to be dealt with is the specification of the particular aspect of vision to be studied. This can even be one of the most difficult tasks, as it requires the process of vision to be analyzed in such a manner that it is possible to specify particular components of the processing.

In this chapter two problems in line art perception will be discussed. The first problem concerns how the most perceptually significant components of an image can be selected, and the second concerns how the components of an image can be grouped into sets that have a high probability of corresponding to a single object or object part. These two problems will be explained in the following sections.

Selection of Perceptually Significant Image Components

Images contain too much information for humans or machines to process all of it in detail. The human visual system solves this problem by performing an initial, cursory analysis of the entire image that allows it to pick out automatically what is important in the image (Treisman, 1985), and then to selectively process that information in preference to the rest of the information. That is, a rapid, parallel analysis of the entire image indicates which image components are likely to contain the most useful information, and then the following stages of analysis, which require more focused, serial processing can concentrate primarily on the preselected regions. This can also be a useful approach for a computer vision system, and in fact the parallel computation of intrinsic images can be viewed as an example of the first stage (Barrow and Tennenbaum, 1981).

The first problem to be addressed here in line art understanding will be how to determine which aspects of an image are the most perceptually significant. For example, when subjects view the contour image in Figure 12.5a, they indicate that the closed contour is more significant than the other contours in the image. The first goal of the computer vision algorithms to be discussed here is to take any line art image, and to create a new image where the most perceptually significant image components have been identified. In fact, the algorithms described here will create from an image a perceptual significance hierarchy: image contours at the top of the hierarchy will be the most perceptually significant, those at the bottom of the hierarchy will be the least perceptually significant, and those at the intermediate levels will have a corresponding intermediate level of significance.

Grouping Image Components into Perceptually Meaningful Groups

The second problem of line art analysis concerns grouping the image components into sets that are likely to correspond to single objects or object parts. For

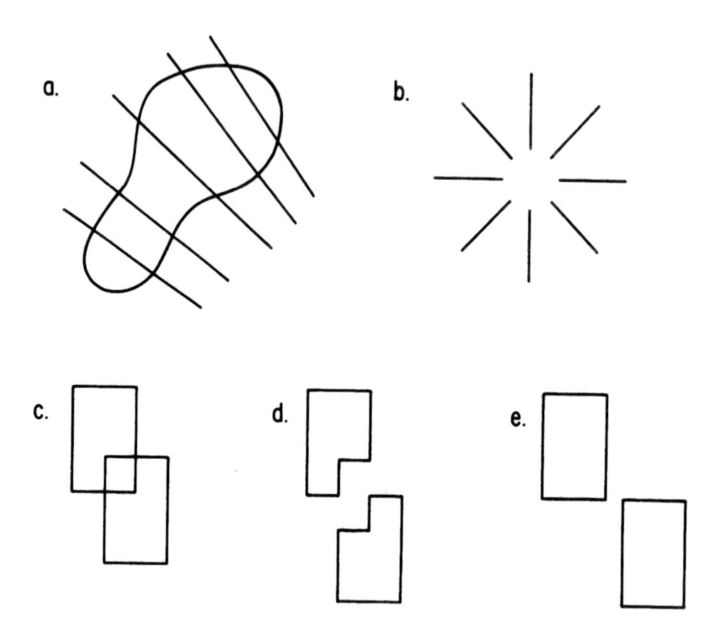

FIGURE 12.5. (a) Contours used to illustrate the concept of perceptual significance (see text for a detailed explanation). (b) An illusory contour example. (c) An original image. (d) One possible grouping of (c). (e) Another possible grouping of (c).

example, subjects indicate that the image in Figure 12.5c consists of two rectangles. Thus they are grouping the object components (lines) into two sets, each corresponding to a single rectangle (see Fig. 12.5e). But this grouping is not the only mathematically possible grouping of the lines: Figure 12.5d shows another possible grouping that for some reason does not appear as perceptually meaningful as the rectangle grouping. The second goal of the computer vision research to be presented in this chapter is to study the grouping of object components into sets that are perceptually meaningful.

Selection of Image Features

Determining which aspects or features of an image contain the most perceptually significant information, and thus would be useful for determining which image components are perceptually significant and for determining how the image components should be grouped into meaningful sets, is difficult because there are nearly an infinite number of potential features. Physical variables are not necessarily directly relevant to perception as perceptual features may require some transformations of the physical variables. For example, the perceived color of a region depends not only on the wavelength and intensity of the light reflected from it, but also on the relative intensity and relative wavelength between it and neighboring regions. So how can perceptually relevant features be found?

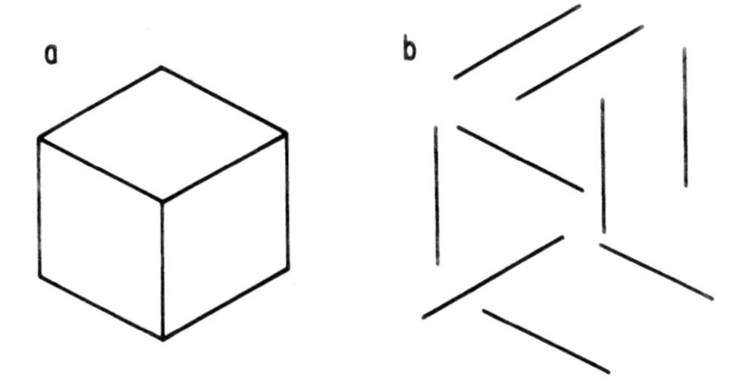

FIGURE 12.6. Example of patterns used in perceived contrast experiments. Both are formed of lines of identical number, length, and orientation. (From Walters DKW [1987], with permission.)

The question is further complicated as one set of features may be ideal for one task, but useless for another. Many algorithms are based on the color, shading, texture, and shape information. Line drawings are simpler in that they do not necessarily contain all of that information, yet even for simple line drawings, it is not obvious which features should be used.

As the goal is to find perceptually significant components of an image, the biological approach can be used to determine which image features are perceptually significant to humans. It is not possible to just introspect about possible features, as the relevant preattentive stages of human visual processing are not available for conscious introspection (Julesz and Schumer, 1981). It is possible to use psychophysical experiments to explore preattentive vision and to discover image features used by humans. The hope is that the features used by humans will then be useful for the computer vision algorithms. This will be true if the features used by humans are constrained by the general problem of vision, rather than by the particular neurophysiological implementation.

Perceived Contrast Psychophysical Experiments

The results of perceived contrast experiments can be used to suggest one set of image features that appears to be of perceptual significance to humans (Walters and Weisstein, 1982a). The patterns in Figure 12.6 can be used to explain the perceived contrast of lines phenomenon. When viewed at low contrast and for brief exposures on a video monitor the lines in the cube (Figure 12.6a) appear to have slightly higher contrast than the lines in Figure 12.6b. (This difference in perceived contrast will not be evident in the high contrast, nonluminous patterns in Fig. 12.6). If these differences in perceived contrast are correlated with the presence of particular image features, it would suggest that stimuli with those features are preferentially processed, which in turn suggests that such stimuli may be of greater perceptual significance than stimuli lacking such features. Psychophysical

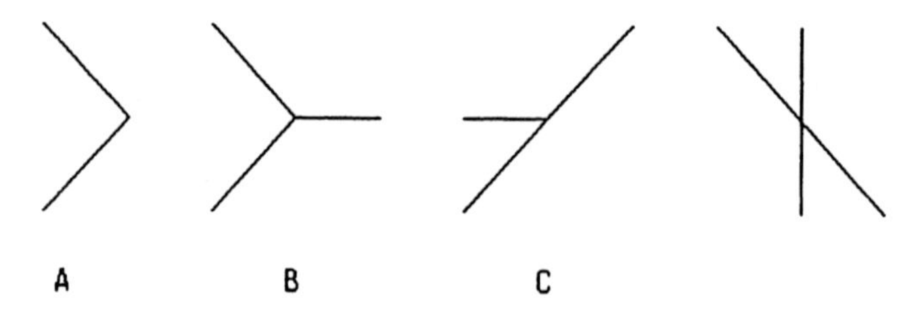

FIGURE 12.7. Various end connections with differing perceived contrasts when viewed as low contrast luminous lines.

experiments have been performed to isolate such features (Walters and Weisstein, 1982b; Walters, 1984, 1985). It is not possible to perform such experiments just by asking subjects which contours appear to have higher contrast. Many controls are required, and standard psychophysical procedures must be used. In the perceived contrast experiments, the results of which will be used here, the method of constant stimuli was used in a two-alternative forced choice paradigm. In addition, in any experiment in which the effect of a given property on perceived contrast is studied, all other variables must be held constant.

By looking at many pairs of patterns designed to differ in terms of various global and local properties, it was found that the difference in perceived contrast did not correlate with any of the global features studied. For example, closure, area, global connectivity, perceived three dimensionality, and objectness did not correlate with perceived contrast. Local features were also explored, and the presence of angles and the number of free line ends were ruled out. The only two features (apart from the obvious feature of physical contrast) that did correlate with perceived contrast were line length and the local connections between the ends of the line segments.

For lines that subtended less than 1° of visual arc, perceived contrast was a positive function of line length. For longer lines, there was no correlation between perceived contrast and length. The other local property is the way in which line ends are connected, and the experiments showed that there is actually a hierarchy of end connections, as seen in Figure 12.7. Two lines connected end-to-end (type A connection) have higher contrast than three lines connected end-to-end (type B connection), which have higher contrast than lines which connect end-to-middle (type C connection), which have higher contrast than the lines that intersect.

Computational Analysis

The results of perceived contrast experiments suggest that the length of short lines and the different types of line end connections are perceptually significant features for humans. But such empirical results do not provide a theory for perceptual significance; there is no evidence as to why such features are important for perception. If the goal of the computer vision research is to understand the process of vision, then behavioral results such as those just described do not in themselves provide a theory. Yet, such experimental evidence often drives the theoretical analysis by providing insights into human vision, which can then be explored using mathematical or physical approaches to formulate the theory. As an example, this chapter will show how a computational analysis of contour perception can be used to construct an underlying visual theory. Two components of the computational analysis will be illustrated: the determination of the completeness of the feature set and the theoretical basis for the perceptual significance of the feature set.

Completeness of the Feature Set

Any time that a set of features is proposed it must be determined if the set is complete. In this case it is necessary to know if all possible end connections can be classified as one of the four connection types: type A, type B, type C, or not a connection. The following analysis will demonstrate the completeness of the perceptually significant end-connection features.

The first step in the computational analysis of end connection features is to develop a definition of a line. As is often the case, a simple term commonly used in our language can be difficult to define precisely. In terms of Euclidean geometry there is a standard definition of a line. But in both human and machine vision, an image does not have the continuous nature assumed in the Euclidean definitions: the image is a discrete sampling of the radiant energy stimulus. Thus computer scientists often define images as two-dimensional arrays of pixels, as mentioned earlier. Now the problem is how to define a line or contour in the discrete image array. Rosenfeld has provided the standard definition of a contour (which he calls an "arc"): a connected set of pixels such that exactly two pixels are connected to just one other pixel in the set, and the rest of the pixels are connected to exactly two pixels (Rosenfeld, 1979). But this definition is problematical for the goal of creating perceptually significant line art analysis algorithms. The patterns in Figure 12.8 illustrate the problem. In part a of the figure are two patterns that when discretely sampled, would each satisfy the definition of arcs. But when the same two patterns are superimposed, as in part b of the figure, neither now satisfies the definition, yet human subjects still describe the figure as consisting of two arcs. The problem can be pinpointed to the concept of connectivity used in the definition. It is an image-based connectivity: two pixels are said to be connected if they have in common any portion of their boundaries. If instead a more perceptually consistent concept of connectivity were used, the problem would disappear. Such a concept is developed below.

The computational approach can now be assisted by the biological approach. A new definition of connectivity is suggested by considering the manner in which images appear to be represented in the visual cortex. Hubel and Wiesel found that there are a number of orientation-selective edge neurons for each spatial location, and each neuron is sensitive to edges or lines within a narrow range of orientations (Hubel and Wiesel, 1968). Thus the images are not only sampled in terms of spatial

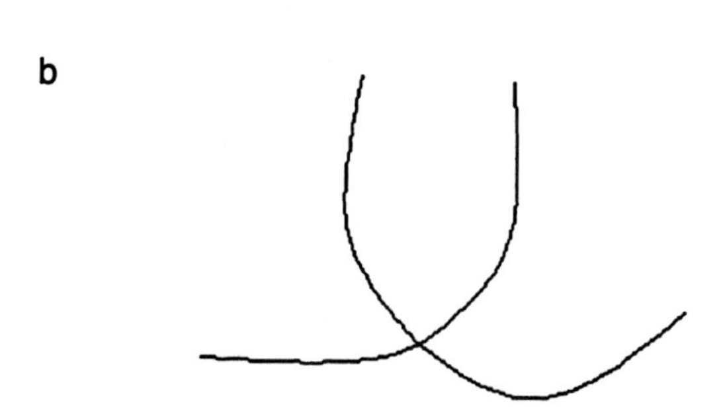

FIGURE 12.8. (a) Two discretely sampled sets of pixels that satisfy the Rosenfeld definition of arcs. (b) Translation of one set of pixels in (a) results in no sets satisfying the Rosenfeld definition.

position, but are also sampled in terms of the local orientation of image contours and edges (as well as along other featural dimensions). In computer vision terms, this corresponds to considering the image as a three-dimensional array of pixels, where two of the dimensions correspond to the sampled spatial dimensions of the image, and the third dimension corresponds to the local sampled orientation. Such a three-dimensional image array has been referred to as the rho space image (Walters, 1987). Figure 12.9 graphically shows the basic form of rho space. Actually, the rho space is topologically a torus, as the top and bottom planes of the space are assumed to be connected. Note that only contours are represented in rho space, thus each image point must be classed as either a contour point or a noncontour point, and then the local orientation of each contour point must be made explicit before it can be expressed in the rho space representation.

As an example, consider how the two intersecting contours of Figure 12.8b would be represented in rho space. As each point is represented in terms of both spatial position and orientation, it can be seen that the points belonging to each

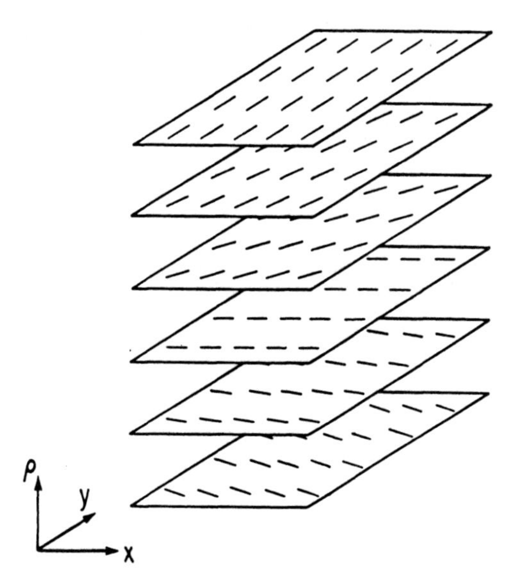

FIGURE 12.9. A graphical illustration of a rho space image. The x and y axes represent the discretely sampled spatial dimensions, and the rho axis represents the discretely sampled orientation dimension. Thus each plane in rho space represents an image consisting of only those pixels in the original image whose local orientations lie within a specified range of orientations. In this example only six possible ranges of orientations are used. (From Walters DKW [1987], with permission.)

contour would form an arc through the rho space, but that the points from one of the contours would not anywhere touch the points from the other contour in rho space. The point at the intersection of the two contours would be represented twice in rho space; once with the local orientation of the one contour, and again with the local orientation of the other contour. It is also interesting to imagine what the rho space representation of a circle would be. It is a double helix in the rho space.

Using rho space a different definition of connectivity has been constructed that is in closer agreement with perceptual connectivity. The new definition of "connected" is orientation dependent. Line pixels can be connected only to other line pixels that lie within a certain x,y distance, and a certain orientation distance. The greater the x,y distance, the greater the possible orientation distance that can yield a connection. This connectivity is defined in rho space by defining a three-dimensional neighborhood about each pixel, and indicating which pixels in the neighborhood are considered to be connected to that pixel.

Using the rho space definition for connectivity in Rosenfeld's definition of a line results in a solution to the problem of superimposed lines. Thus the use of the biological approach in conjunction with the computational approach has proven useful here. The previous definitions of discrete lines had the disadvantage of not agreeing with human perception. It might be argued that this is not necessarily a

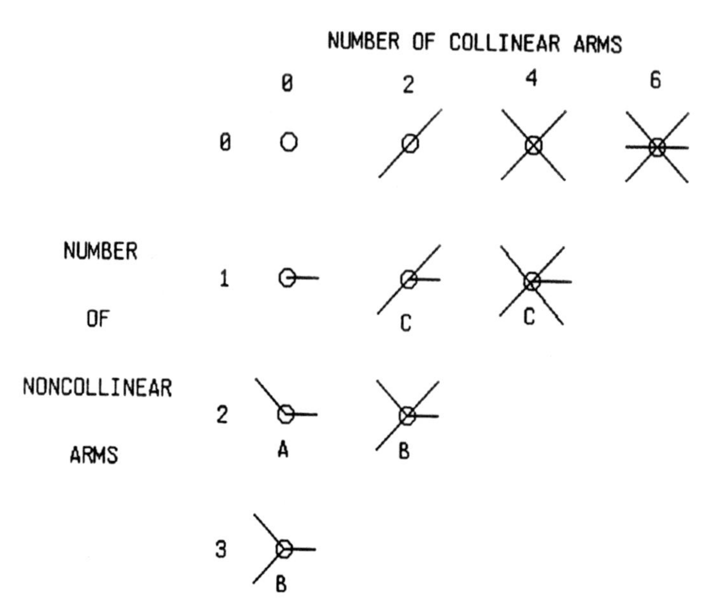

FIGURE 12.10. Graphical representation of all possible end connections with just three possible line orientations. The number of collinear arms corresponds to the number of "m" labels at the junction, and the number of noncollinear arms correspond to the number of "e" labels. For each pattern, only the connection at the large dot is being described.

disadvantage as a machine vision system need not exhibit the same behavior as the human visual system. But if the goal of creating computer-based line art interpretation systems is to enable humans and machines to communicate via line art, then what is perceived as a line by a human should also be perceived as a line by the machine.

Having a definition of a line, it is possible to define a line end point, "e," as those pixels on a line that have connections in only one-half of their rho space neighborhood. A line middle pixel, "m," is defined as a line pixel that is connected on both sides of its neighborhood.

Connections can be defined in terms of these "e" and "m" pixels. For example, for a type A connection located at point $(x1,y1)$, examining the $x1,y1$ position in each plane of rho space would yield exactly two "e" pixels, and no others. Figure 12.10 shows all the possible connections for straight lines of just three possible orientations, in terms of the number of "e" and "m" pixels at the center of the patterns (as indicated by the dots in the figure). All of the patterns in the top row are not connections. Three of the connections were used in the psychophysical experiments, and there are two additional connections needed to cover the space. The new connections are hypothesized to belong to the classes as labeled. All patterns in the second row are classed as type C connections. The pattern with just two "e" pixels is a type A connection and the rest are type B connections.

There are only a limited number of orientations in Figure 12.10. For more orientations, the set would be extended, and the labels can also be extended. In particular, consider the case in which an infinite number of orientations are present. Then none of the infinite number of patterns in the top row is a connection. Everything out to infinity in the second row is a type C connection. The type A connection remains unchanged, and everything else out to infinity in the other rows is a type B connection. This demonstrates that it is possible to label all possible patterns as belonging to one of the four feature classes. Thus the set of end connection features is geometrically complete. This is an important result as it means that any theory of visual perception based on the end connection types will be complete as all possible junctions of image contours can be classified.

Computational Significance of the Connection Features

The second issue to be addressed by the computational analysis concerns why the connection features are of significance. What information about the scene is being conveyed by the connection features? Answers to these questions that are independent of algorithms and implementations can form the basis of a theory of perceptual significance for line art perception.

Uses of Image Features

Various theories concerning the use of features have been proposed in computer vision. One conceptually simple use of features is to represent objects in terms of a list of features (Feldman, 1985). A model of an object can be expressed in terms of features and the relations between them, and then portions of an image can be compared to the model to see if the object is present. This is similar to the way line drawing junctions were first used by Roberts (Roberts, 1965). But this type of use involves domain-specific knowledge: to recognize an object, a description (a model) of the object must be present in the system. This is a major drawback of this use as it is not easily extendible to deal with arbitrary images, where it is not known in advance what objects will be present in the scene.

Guzman (1969), Kanade (1981), Draper (1981), and Lee et al. (1985) have used very similar line drawing features in their boundary image interpretation algorithms. But the features are used in constraint satisfaction systems similar in essence to the constraint satisfaction described earlier in the map coloring problem.

A third use of features is suggested by the psychophysical experiments. Lines appear to be selectively processed based on the presence of a few basic features. Why would this be useful for a visual system? Well, as previously mentioned, a major problem for a visual system is that there is too much information in a visual image to process all of it in detail. One solution is to have some automatic preprocessing system that determines which lines or areas contain the most important information, and then to concentrate the serial processing on those areas, while ignoring other potentially less fruitful areas. With the theories described here those lines that have a high probability of being part of object contours, rather than just part of texture or noise, can be automatically selected from an image. If the next stage of processing has to be selective, it can "attend" only to the enhanced lines and thus not waste resources processing potentially spurious contours.

Perceptual Significance of Selective Enhancement

A perceptual significance hierarchy theory can be developed from simple assumptions about the physical properties of objects in scenes and about the physics of the generation of images from scenes. Thus this component of the analysis is an example of the physical approach discussed earlier in this chapter. The assumptions are relevant to the generation of natural images from scenes. But it will be further assumed that artists generally do not violate the assumptions in creating line art images. From the assumptions two sets of inferences are made: one relates to the relative perceptual significance of different image components containing certain connection features, and the other relates to the grouping of image components into sets likely to have arisen from a single object or object part, and thus likely to form a single perceptual group or subgroup.

Assumptions

There are two types of assumptions: those concerning the objects in a scene, and those concerning the projection physics involved in generating an image from the scene.

Assumption: Objects occupy compact, closed regions of space.
Assumption: Different objects generally have different reflective properties.
Assumption: Object self-occlusions are rare.
Assumption: Viewing position is almost always representative.

This means we assume we generally are looking at an object along a viewing direction that is not one of the few viewing directions that results in the accidental alignment of object boundaries or wires in a scene (Binford, 1981; Cowie, 1982).

Assumption: Object position is almost always representative.

This means we assume objects or wires in a scene are generally not accidentally aligned. The previous assumption concerns looking at objects in such a way as to make them appear to be accidentally aligned. This assumption concerns cases where the objects are in some form of accidental alignment with each other, independent of the viewing position.

The assumptions about nonaccidental alignment do not mean that images with accidental alignment cannot be enhanced or segmented using this algorithm. It just means that the most general interpretation for a basic feature will be utilized. Thus in the majority of cases, the correct interpretation will arise, while a few cases may exist where algorithms based on the assumptions give an incorrect interpretation.

Inferences Concerning Perpetual Significance

From the assumptions the following inferences about the perceptual significance of image components can be made.

Inference: Image contours terminating in type A connections have a higher probability of having arisen from the outer, occluding contours of objects than have any other image contours.

Inference: Image contours terminating in either type A or type B connections have a higher probability of having arisen from object contours than have other image contours.

Inference: Image contours terminating in type C connections have a high probability of corresponding to an occluded scene contour.

Inference: Image contours with no end connections have a higher probability of having arisen from texture or pattern image contours, or of corresponding to wire-like objects than have other image contours.

These assumptions and inferences form the theoretical basis of the perceptual significance hierarchy theory. In particular, as objects can generally be recognized from just their silhouettes, which contain only the outer contour information, it is likely that the most perceptually significant contours in an image would correspond to such outer contours. Based on these inferences and the definition of the perceptual significance hierarchy, it can be seen that the contours at the top of the hierarchy have the highest probability of having arisen from the outer contours of objects, and thus should have the highest probability of being perceptually significant. As object contours will be more significant than nonobject contours, and as there is a high probability that image contours in the upper levels of the perceptual significance hierarchy correspond to object contours, the relative perceptual significance of the upper levels of the perceptual significance hierarchy appear to capture one aspect of perceptual significance. As occluded object contours are generally the least significant of the object contours, the relative perceptual relevance of contours terminating in the type C connections is seen. Similarly it can be argued that image contours having arisen from texture or noise are not as significant as other image contours, and the fact that they appear only in the bottom levels of the perceptual significance hierarchy thus appears to agree with human perception.

The assumptions and inferences, along with the definition of the perceptual significance hierarchy in terms of the presence of end connections, form a theory of the relative perceptual significance of contours in line art.

Inferences Concerning Perceptual Grouping

From the same assumptions another set of inferences is possible that is relevant to a theory of perceptual grouping of line art contours.

Inference: Contours terminating at a type A connection have the highest probability of having arisen from the same object.
Inference: Contours terminating at a type B connection have the second highest probability of having arisen from the same object.
Inference: Contours terminating at a connection do not necessarily belong to the same object as contours that pass through a connection.
Inference: The fact that two contours cross does not provide information either for or against their having arisen from the same object or object part.

The assumptions and inferences, along with the definition of the perceptual significance hierarchy in terms of the presence of end connections, form a theory of the perceptual grouping of contours in line art.

Computer Vision Algorithms Based on the Perceptual Significance and Perceptual Grouping Theories

The psychophysical experiments provide evidence that the length of lines and the connections between the ends of lines are basic features for human vision. The computational analysis shows that the set of end connection features is complete, and provides a theory for why such features would be useful to a vision system. This section describes some simple computer vision algorithms based on the features that can produce the perceptual significance hierarchy, and can group image contours into sets that have a high probability of corresponding to objects or object parts.

Detection of Connection Features

The algorithms will assume that the input image has been preprocessed and is in the form of the rho space representation. In addition it will be assumed that each contour point in rho space is labeled as being either a contour end point or a contour middle point. For each spatial position in rho space, the following algorithm can be implemented to detect the various end connections.

1. All spatial positions containing exactly two "e" labels have a type A connection.
2. All spatial positions containing two "e" labels, and at least one additional "e" or "m" label, have a type B connection.
3. All spatial positions containing exactly one "e" label and one or more "m" labels have a type C connection.

Creation of Perceptual Significance Hierarchy

It will be assumed that the perceptual significance of a line is a function of the type of connection at each end of the line. Three base levels of significance are specified: "*a*," "*b*," and "*c*," which are in decreasing order of significance. Each line will have two significance levels assigned to it based on the connection type at either end, and the final significance of the line is the algebraic sum of the two levels. The sum is defined such that the final perceptual significance hierarchy has 10 levels, which are, in decreasing order of significance, $a+a$, $a+b$, $a+c$, a, $b+b$, $b+c$, b, $c+c$, c, and no assigned significance. Then the following rules encapsulate an algorithm for creating a perceptual significance hierarchy.

1. Lines with type A connections are assigned significance a.
2. Lines with type B connections are assigned significance b.
3. Lines with type C connections are assigned significance c.
4. Lines with no 'e' points are assigned significance $a+a$.

As each contour has exactly two ends (or no ends), and as its significance is a function of the classification of both end points, there are 10 possible assigned significances.

The perceptual significance hierarchy is created by placing all contours with significance of $a+a$ in the top layer of a three-dimensional array, placing all contours with significance greater than or equal to $a+b$ in the second layer, etc., with all contours with significance greater than or equal to no assigned significance in the bottom layer. Thus there are 10 levels in the perceptual significance hierarchy.

Grouping Algorithm

The following algorithm has the effect of grouping image contours into sets that are likely to correspond to a single object or object part. A similar algorithm is used to segment the image contours into sets that have a high probability of having arisen from a single object surface or face. The basic idea of these algorithms is to assign a numeric label to each contour such that all contours that have a high probability of having arisen from the same object or object part have the same numeric label.

For each level i of the perceptual significance hierarchy, starting with the top layer:

Assign a unique numeric label to each connection of level i and to associated lines.

Assign labels for other contours recursively in parallel until no further contours appear, as follows:

At type A connections all contours take the lowest of the associated labels, and make like labels equivalent to the new label.

Unlabeled lines at type B connections take label(s) of labeled line(s) at the connection.

Labeled lines at type B connections remain unchanged.

Example results of both the grouping and the segmentation algorithms are presented below.

Connectionist Implementation of the Computer Vision Algorithms

The algorithms can be implemented in parallel using a connectionist architecture. Imagine a simple processor at each pixel in rho space, such that each processor receives input only from a small neighborhood of rho space pixels. Each processor can compute whether certain simple spatial relations hold between contours in its neighborhood, and, if so, can send the appropriate information out along to the appropriate pixels. Such a theoretical machine is called a rho space machine, and can be used to implement the computer vision algorithms. The exact details of the connectionist implementation can be found elsewhere (Walters, 1990).

An image can be transformed into the rho space representation by convolving the image with a separate oriented edge operator for each orientation plane. It is possible to construct any number of separate orientation planes for this representation, and for the examples presented in this chapter either 8 or 16 orientation planes were used. However, the transformation cannot consist simply of the convolutions, because convolution operators respond not only to image contours, but also to image points that lie close to contours, etc. Thus significant preprocessing of the convolution results is required to produce the necessary form for the algorithms. This processing also occurs in the rho space machine. The input to the preprocessing is the two-dimensional contour image, and the output is the three-dimensional rho space representation of the image where each contour point contains information about spatial position and local orientation. It may seem surprising that such preprocessing is required, especially as a visual display of the input to the preprocessing looks nearly identical to a visual display of the output, when that visual display is a projection of the rho space representation onto the spatial image plane as is done in the examples presented in this chapter. But the preprocessing is required to make explicit to the computer what is immediately explicit to humans who view such images. This is the underlying goal of many computer vision algorithms, and again it is the fact that humans are not consciously aware of the related types of preprocessing that are necessary for visual perception that can cause a person who has not had direct experience with computer vision to not understand the complexity of such preprocessing. In particular in this preprocessing, the small gaps in the contours are filled, the responses of the convolution operators are disambiguated, the pixels are grouped into contours, intersecting and connecting contours are disambiguated, and, most importantly, the local orientation of each point on a line is made explicit.

Algorithm Results

Figure 12.11 shows the output of the perceptual significance hierarchy algorithm for four of the psychophysical patterns. The most significant contours appear at the top of the figure. In each subsequent row the next most significant contours appear, until at the bottom of the hierarchy all of the lines that were present in the patterns appear.

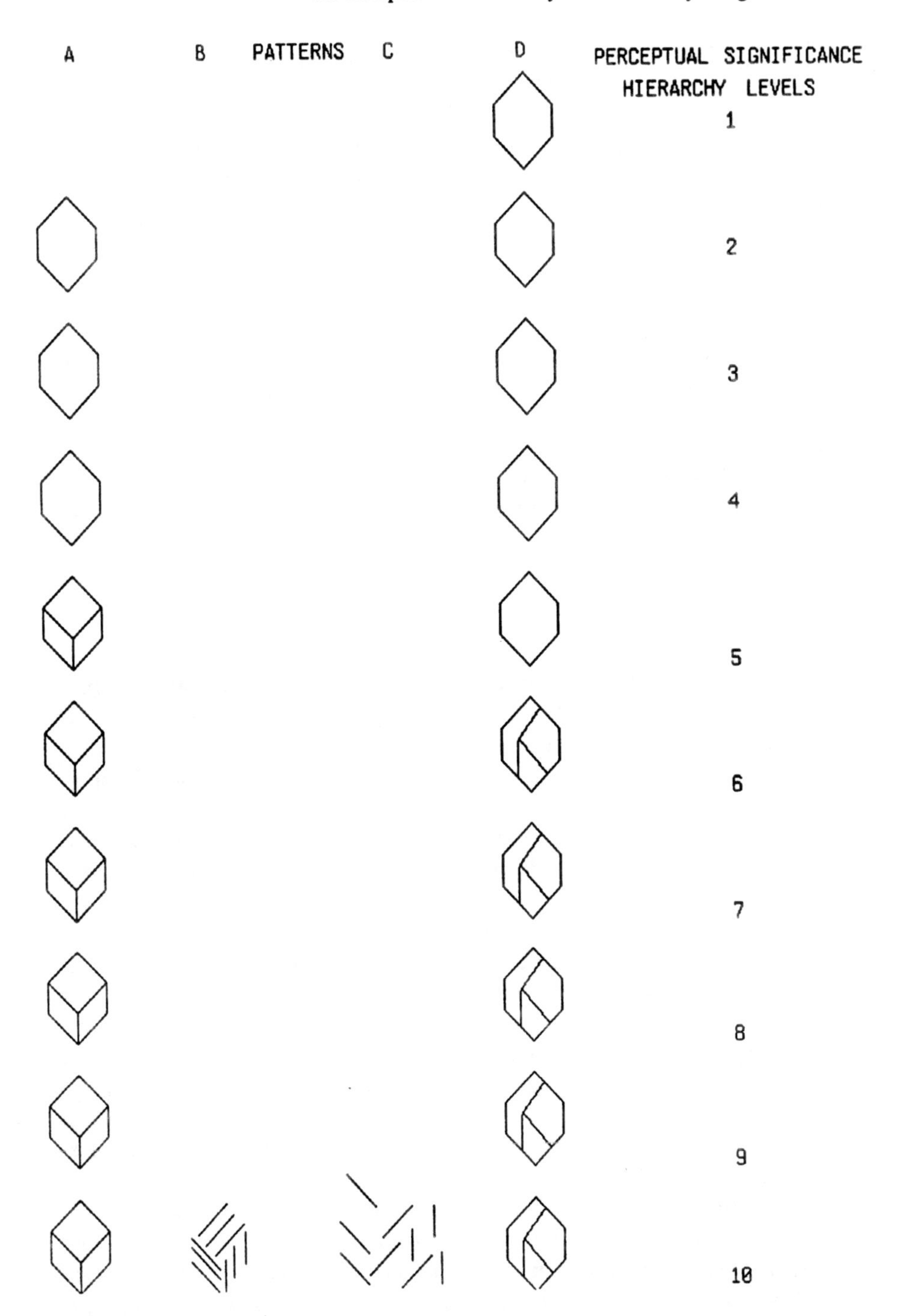

FIGURE 12.11. Perceptual significance hierarchy representation of four patterns used in the perceived contrast experiments. At the top are the contours present at the top of the hierarchy, at the bottom are those present at the bottom, and the intermediate levels are seen in the middle.

The algorithm results agree with the experimental results for all of the patterns used in the psychophysical experiments. The object contours are judged by the algorithm to be more significant than lines that appear as texture or noise. The outer contours of objects are enhanced relative to the inner contours. This pattern of results generally holds, and Figure 12.12 shows the perceptual significance hierarchy for a simple two-dimensional scene.

Segmentation Examples

Figure 12.13 shows some examples of applying the segmentation algorithm, which is a variation of the grouping algorithm, to 2-D, origami, and 3-D objects. The drawings are correctly segmented in all three cases, as indicated by the different line styles for the different objects. Again the outer contours of the objects are judged to be more significant than the inner contours, as are objects in the foreground relative to occluded objects. Note again that the one set of rules can deal with the three separate domains.

Grouping Performance on an Arbitrary Line Drawing

To demonstrate the ability of the perceptual significance hierarchy and grouping algorithms to deal with an arbitrary line drawing, a cartoon from the New Yorker was processed in accord with the algorithm. Figure 12.14a shows the original cartoon. If only the most significant groups of lines are displayed, those from the top four levels of the perceptual significance hierarchy, then object recognition is possible (Fig. 12.14b). If just the least significant lines are displayed, object perception is not possible—which is weak evidence that the algorithm picks out the most perceptually salient lines.

The grouping at the high level stage is depicted by the different line styles in Figure 12.14b. Sixteen of the 23 separate objects or object parts are represented at this stage. (Due to reproduction limitations, only four line styles are used in the figure, however, each instance of line style indicates a separate set of lines.) Again the algorithm is effective in reducing the complexity of the drawing in terms of the number of lines, without diminishing the grouping capabilities.

Figure 12.14c shows the final grouping of the cartoon. The sets all correspond to object or object parts that are readily named by humans, i.e., "crown," "robe," "cuff," "sleeve," "foot," etc. There are no groups that would have to be described as "the upper right hand portion of object x," which again suggests that the grouping has perceptual significance. The algorithm could be used as a powerful preprocessor for a scene analysis system, as it accomplishes a lot, given just a handful of simple rules. Later stages of analysis could use the grouped contours as input to an object recognition system (Pentland, 1985, Biederman, 1985).

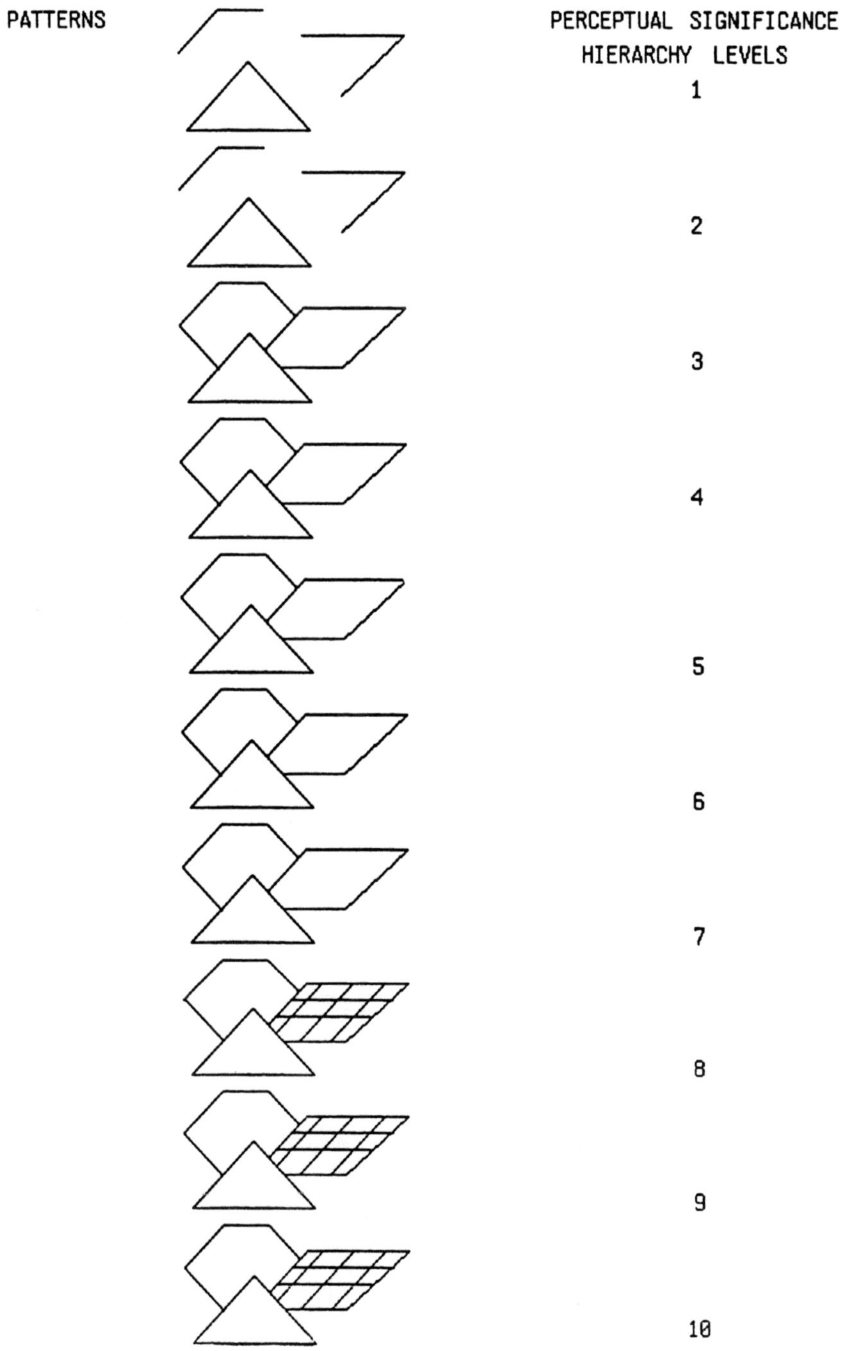

PATTERNS

PERCEPTUAL SIGNIFICANCE HIERARCHY LEVELS

1

2

3

4

5

6

7

8

9

10

FIGURE 12.12. Perceptual significance hierarchy representation for a simple scene.

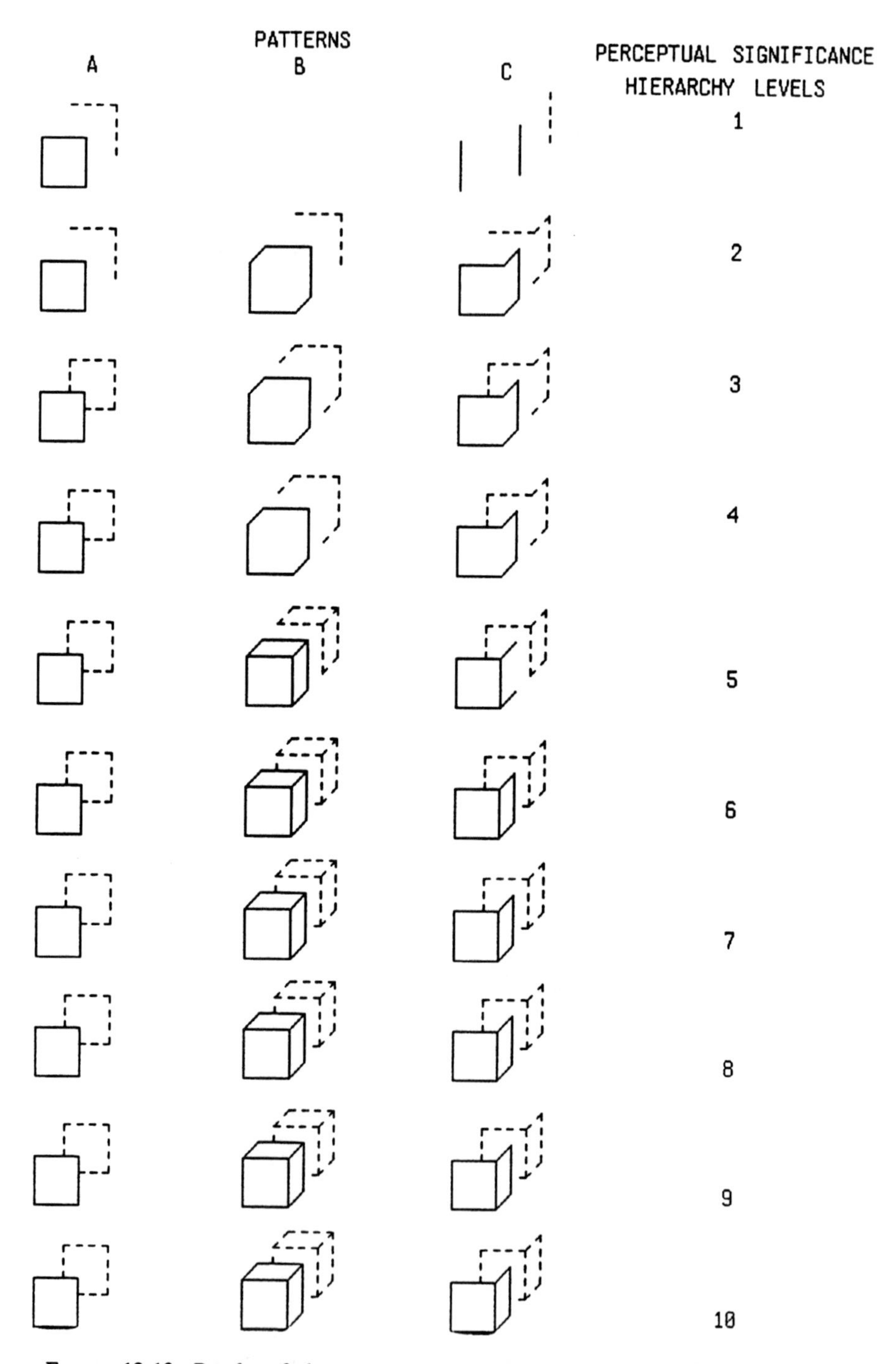

FIGURE 12.13. Results of the segmentation algorithm, expressed as perceptual significance hierarchies. The different line styles represent different segments as determined by the algorithm.

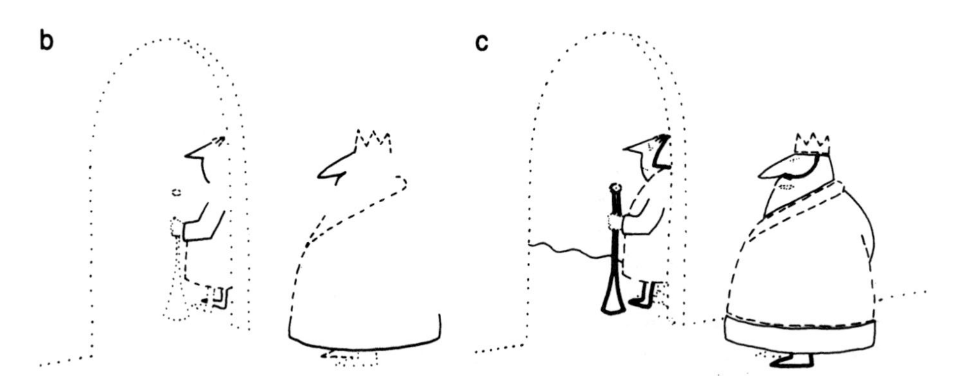

FIGURE 12.14. Results of the grouping algorithm. (a) Original Barsotti cartoon reprinted with permission of the New Yorker. (b) Groupings based on top levels of the perceptual significance hierarchy. (c) Groupings based on all levels. (From Walters DKW [1987], with permission.)

Conclusion

A theory of line drawing analysis has been presented that can be used to create computer vision algorithms to determine the relative perceptual significance of image contours, and to group the contours into sets likely to correspond to objects or object parts. The algorithms can be implemented in a connectionist architecture. The results of the implementations show that grouping can occur without any knowledge about the objects being depicted in an image. Thus no domain-dependent knowledge is required, and only bottom-up, data-driven processing is required. Object models are obviously necessary for some stages of object recognition, but the algorithms described here demonstrate that some steps that were previously thought to require model matching do not. Thus the computational analysis of image features that appear to be used in the human visual system can improve the performance of computer vision systems.

The contour analysis has been used to illustrate several of the approaches to computer vision research and to show how the different approaches can be used together. Through the psychophysics of the biological approach, a set of image features was uncovered that appear to be used in human visual perception. The computational analysis of the physical approach was used to create a theory of contour perception that shows why such features are useful. Then neurophysiological results of the biological approach were used to suggest a data representation for the connectionist implementation. This mixture of different approaches within a single research project is becoming more common in vision research, as the field becomes more interdisciplinary and researchers continue to realize the particular benefits of the different approaches and the synergistic relationships they have with each other.

References

Ballard DH, Brown CP (1982). *Computer Vision*. Prentice-Hall, Englewood Cliffs, NJ.

Barrow HG, Tennenbaum (1981). Interpreting line drawings as three-dimensional surfaces. J. Artif. Intell. **17**:75–116.

Biederman I (1985). Human image understanding: Recent research and a theory. J. Comp. Vision, Graphics Image Process. **32**:29–73.

Binford TO (1981). Inferring surfaces from images. J. Artif. Intell. **17**:205–244.

Browse RA, Rodrigues MG (1987). Propagation of interpretations based on graded resolution input. *Proceedings of First International Conference on Computer Vision*, 405–410. IEEE Computer Society Press.

Cowie R (1982). Modeling people's interpretation of line drawings. D.Phil. Thesis, University of Sussex.

Draper SW (1981). The use of gradient and dual space in line-drawing interpretation. J. Artif. Intell. **17**:461–508.

Edelman S, Poggio T (1989). Integrating visual cues for object segmentation and recognition. Optic News, 8–16.

Feldman JA (1985). Four frames suffice: A provisional model of vision and space. Behav. Brain Sci. **8**:265–289.

Feldman JA, Ballard DH (1982). Connectionist models and their properties. Cog. Sci. **6**:205–254.

Fischler M, Firschein O (1987). *Readings in Computer Vision: Issues, Problems, Principles and Paradigms*. Morgan Kaufmann, Los Altos, CA.

Guzman A (1969). Decomposition of a visual scene into three-dimensional bodies. In *Automatic Interpretation and Classification of Images* (A Greselli, ed.). Academic Press, New York.

Horn B (1986). *Robot Vision*. McGraw-Hill, New York.

Hubel DH, Wiesel TN (1968). Receptive fields and functional architecture of monkey striate cortex. J. Physiol. **195**:215–243.

Julesz B, Schumer RA (1981). Early visual perception. Annu. Rev. Psychol. **32**:575–627.

Kanade T (1981). Recovery of the three-dimensional shape of an object from a single view. Artif. Intell. **17**:409–446.

Lee SH, Haralick RM, Zhang MC (1985). Understanding objects with curved surfaces from a single perspective view of boundaries. J. Artif. Intell. **26**:145–169.

Linsker R (1986). From basic network principles to neural architecture: Emergence of orientation-selective cells. Proc. Natl. Acad. Sci. U.S.A. **83**:8390–8394.

Lowe DG (1984). Perceptual Organization and Visual Recognition. Ph.D. Thesis, Stanford University.

Marr D (1982). *Vision*. Freeman, San Francisco.

Pentland AP (1985). Perceptual organization and the representation of natural form. SRI Technical Note **357**.

Pylyshyn Z (1980). Computation and cognition: Issues in the foundations of cognitive science. Behav. Brain Sci. **3**(1):111–169.

Roberts LG (1965). Machine perception of three-dimensional objects. In *Optical and Electro-optical Information Processing* (JP Tippett et al., eds.). MIT Press, Cambridge, MA.

Rosenfeld A (1979). *Picture Languages*. Academic Press, New York.

Treisman AM (1986). Preattentive processing in vision. In Human and Machine Vision Vol 2:313–334. A. Rosenfeld (Ed). Academic Press, Orlando, FL.

von der Marlsburg (1973). Self-organization of orientation sensitive cells in the striata cortex. Kybernetic **14**:85–100.

Walters DKW (1984). Local connections in line drawing perception: a computational model. Invest. Ophthalmol. Visual Sci. **25**:200.

Walters DKW (1985). The use of natural constraints in image segmentation. In *Applications of Artificial Intelligence* (J.F. Gilmore, ed.), Vol. 11. SPIE Proc. **548**:27–34.

Walters DKW (1987). Selection and use of image primitives for general-purpose computer vision algorithms. Comp. Vision, Graphics Image Proc. **37**:261–298.

Walters DKW (1988). Opponent mechanisms in vision. Neural Networks **1**(Suppl. 1): 529.

Walters DKW (1990). Parallel computations in rho-space. In preparation.

Walters DKW, Weisstein N (1982a) Perceived brightness is influenced by structure of line drawings. Invest. Ophthalmol. Visual Sc. **22**:124.

Walters DKW, Weisstein N (1982b) Perceived brightness is a function of line length and perceived connectivity. Bull. Psychon. Soc. 130.

Waltz DI (1975) Understanding line drawings of scenes with shadows. In *The Psychology of Computer Vision*, PH Winston, ed. McGraw-Hill, New York.

Yuille AL, Kammen DM, Cohen DS (1989). Quadrature and the development of orientation selective cortical cells by Hebb rules. Biol. Cybern. **61**(3):183–194.

13

A Representation for Qualitative 3-D Object Recognition Integrating Object-Centered and Viewer-Centered Models

SVEN J. DICKINSON, ALEX P. PENTLAND, AND AZRIEL ROSENFELD

Introduction

In the context of computer vision, the recognition of three-dimensional objects typically consists of image capture, feature extraction, and object model matching. During the image capture phase, a camera senses the brightness at regularly spaced points, or pixels, in the image. The brightness at these points is quantized into discrete values; the two-dimensional array of quantized values forms a digital image, the input to the computer vision system. During the feature extraction phase, various algorithms are applied to the digital image to extract salient features such as lines, curves, or regions. The set of these features, represented by a data structure, is then compared to the database of object model data structures in an attempt to identify the object. Clearly, the type of features that need to be extracted from the image depends on the representation of objects in the database.

In most cases, the features extracted from the image are considerably less complex than the object models with which they are compared; as a result, many models may contain a particular type of feature. The comparison of sets of image features to object models actually consists of two phases. In the bottom-up, or indexing, phase each image feature serves as an index into the object model database to select candidate object models containing that feature. In the top-down, or verification, phase the candidate model(s) are used to verify feature interpretations and to constrain further feature extraction.

In three-dimensional (3-D) object recognition by computer, two important issues pertain to the representation of objects. The first issue is the choice between object-centered and viewer-centered representations. Object-centered representations model objects as constructions of 3-D primitives, such as planar faces or generalized cylinders. Recognition consists of matching image features to the predicted projections of specific 3-D model features, a process requiring the determination of the object's position and orientation with respect to the camera. Viewer-centered representations model objects as a set of 2-D characteristic views, or aspects. Recognition consists of matching image features against the set of aspects; the view most closely resembling the features in the image defines the object and its orientation. The major advantage of viewer-centered recognition is that it reduces the 3-D recognition problem to a 2-D recognition problem. However, with each model object having potentially many aspects, matching becomes less efficient than with object-centered models.

The second issue in 3-D object recognition concerns the amount of detail inherent in object models. If a quantitative representation specifies the exact dimensions and shape of an object, then simple model-based verification procedures can be employed to confirm or reject object hypotheses. However, exact representations of objects result in complex object models. If a qualitative representation captures only the gross shape characteristics of an object, then model-based verification will fail to predict the exact location of the object's features. Instead, the bottom-up stage of recognition must be extended to extract higher order features with which to index into the database. The advantage of qualitative models is that they are less complex than quantitative models, and are invariant to minor changes in shape.

This chapter proposes a modeling paradigm for 3-D object recognition from two-dimensional (2-D) images integrating object-centered and viewer-centered models. Object models are constructions of 3-D volumetric primitives, offering an efficient indexing mechanism for large object databases. The 3-D primitives, in turn, are mapped into a set of viewer-centered aspects. During recognition, image contours are matched to the contours comprising an aspect, defining the primitive type and constraining its orientation. The 3-D primitive is then used to index into the object database. When objects are composed of multiple primitives, occluded primitives from a given viewpoint project to occluded, and hence incomplete, aspects in the image. To accommodate the matching of occluded aspects, we have developed a hierarchical aspect representation based on the visible faces of the primitives. At the top of the hierarchy lie connected face structures, while at the bottom lie contour features of the component faces. Thus, aspect inferences can be made from incomplete projections.

With a large database of objects, we might expect there to be a great variety of primitive shapes and sizes. Representing every single primitive with a different set of aspects would make matching image contours to primitive aspects intractable. To minimize the size of the aspect set, we constrain the aspects to be invariant to minor changes in primitive shape, forcing the primitives to be qualitative in nature. The size of the resulting aspect set depends only on the size of the set of primitives, not on the number of object models or on object model complexity. The primitives that we have selected are based on Biederman's geons (Biederman, 1985), offering a rich vocabulary with which to construct objects.

A qualitative recognition paradigm has advantages over and above that of restricting the number of aspects we require to represent our primitives. Such a system would be of great value in many robotic vision applications requiring object identification. For example, the sorting of distinct objects by a robot may require only that the objects be classified; position and pose determination may be unnecessary. Or, for an autonomous vehicle, quickly identifying the objects in the field of view may guide the vision system to select a course of action. For example, noticing a tree on the side of the road may not alarm the system, while noticing another vehicle may invoke modules to estimate its velocity. Finally, a qualitative object recognition engine could provide a coarse front end to a more quantitative recognition engine; identification of the object's generic class could be used to invoke specific modules to distinguish among instances of subclasses. A coarse-to-fine approach to object recognition incurs the cost of extracting finer detail only when necessary.

In this chapter, we present only our object modeling paradigm; techniques for primitive extraction and model matching will not be presented. The second section discusses some of the issues in selecting an object modeling scheme and gives the motivation for our choice. The third section presents the object-centered component of our representation, while the fourth section presents the viewer-centered component. In the fifth section, we tie together the two representations with some probabilistic results based on an extensive analysis of the primitives over the viewing sphere; the sixth section evaluates the integrated representation. In the seventh and eighth sections we discuss related work and draw conclusions about this approach.

Object Modeling for 3-D Recognition

Many object modeling techniques have been applied to the task of 3-D object recognition (e.g., Requicha, 1980; Srihari, 1981; Binford, 1982; Besl and Jain, 1985; Chin and Dyer, 1986). In any object modeling scheme, an object is composed of one or more features or primitives; examples include lines, vertices, surface patches, generalized cylinders, and superquadrics. Models may be object-centered constructions of 3-D primitives, or viewer-centered constructions of 2-D primitives. In the former, the model is independent of viewpoint, while in the latter, each distinct view of the model generates a unique representation. When selecting a modeling technique for 3-D object recognition, a number of trade-offs must be considered. For example, complex primitives such as generalized cylinders are more difficult to extract from an image than simple primitives such as lines and curves. However, it is more efficient to search a database of objects, each composed of a relatively small number of complex primitives, than a database of objects, each composed of a relatively large number of simple primitives. Figure 13.1 illustrates the trade-offs in selecting a representation scheme.

Many approaches to 3-D object recognition (e.g., Lowe, 1985; Huttenlocher and Ullman, 1987; Thompson and Mundy, 1987; Lamdan et al., 1988) limit the bottom-up feature extraction process to 2-D primitives such as line segments, corners, zeroes of curvature, and 2-D perceptual structures. These features are appealing due to their viewpoint invariance; however, they suffer the shortcoming of requiring complex models. Since a model in these representations typically consists of a large number of very similar primitives, searching a large model database becomes inefficient. As a result of this limitation, these recognition systems have only been successfully applied to object databases containing one or two objects. In addition, the simplicity and 2-D nature of the indexing primitives

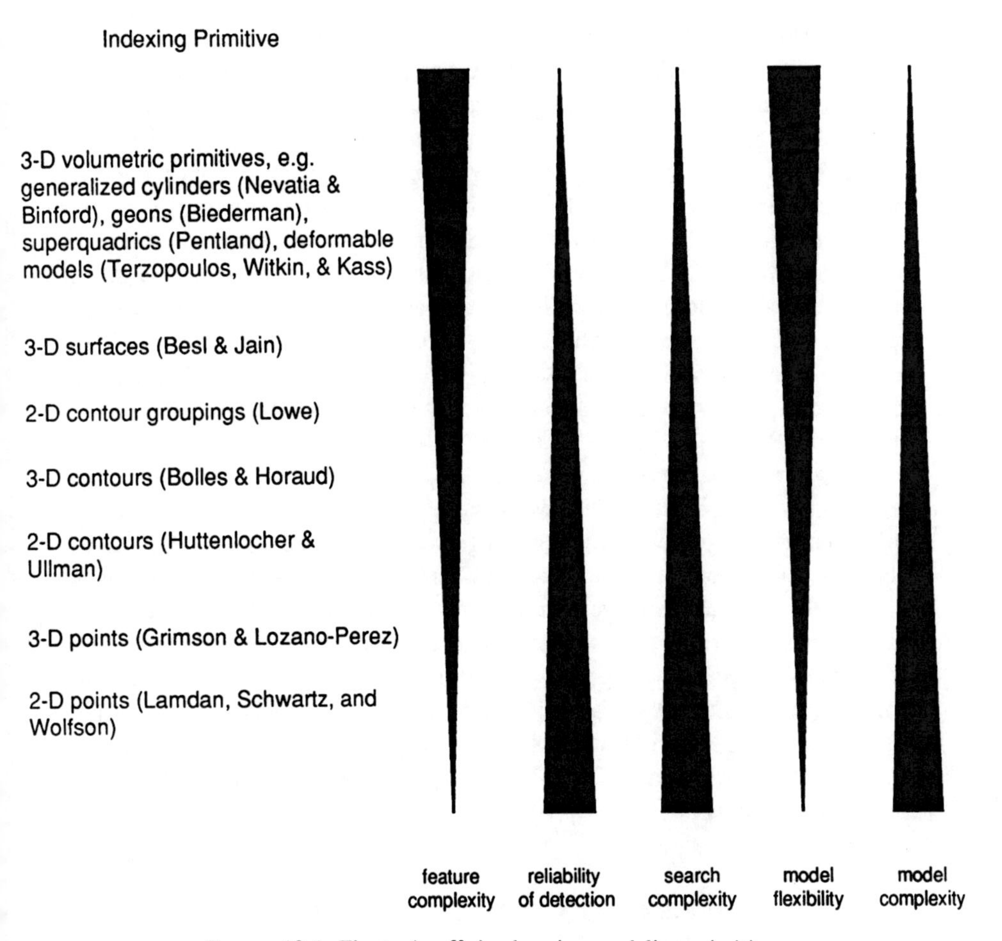

FIGURE 13.1. The trade-offs in choosing modeling primitives.

require a complex verification procedure involving the determination of the object's pose with respect to the image. Indexing with weak hypotheses shifts the burden of recognition to verification, resulting in a top-down system. Our approach is to extend the bottom-up process to the point of inferring 3-D volumetric primitives with which we index into the database. Since our primitives capture more information than simple point or line primitives, they provide more discriminating indices to less complex models. These higher order indices provide a foundation for a more bottom-up unexpected-object recognition system (Rosenfeld, 1986).

A 3-D recognition system based on the bottom-up extraction of 3-D volumetric indexing primitives raises the obvious question: How do we extract the primitives from the image? To meet this requirement, we employ a viewer-centered aspect representation to model an object's primitives. This differs from traditional aspect-based recognition systems where the entire object is modeled as a set of aspects (e.g., Chakravarty and Freeman, 1982; Ikeuchi and Kanade, 1988). The advantage of aspect representations is that they reduce the 3-D matching problem to a 2-D matching problem; each 2-D aspect defines the object's identity and 3-D pose. The disadvantage of aspect representations is that they incur a high cost of matching as each 3-D object must be represented by many different views. Moreover, as the complexity of an object increases, so do the number of views required to represent it. To restrict the number of aspects representing the model primitives, we model primitive classes rather than primitive instances. Thus, for each primitive, we define a set of aspects that is invariant to minor changes in primitive shape. More specifically, we define our aspects to be invariant to changes in line length, curvature, and angle. Consequently, a primitive may undergo changes in surface dimension and surface curvature without introducing new aspects. This constrains our primitives to possess a qualitative nature, capturing only the gross shape characteristics of the object.

The integration of object-centered and viewer-centered models combines the advantages of each scheme while avoiding their disadvantages. The extraction of 3-D volumetric primitives provides a highly selective index into a database of compact object representations. Extraction of the primitives from the image is performed by matching image contours to the contours comprising a set of 2-D aspects; pose determination is inherent in the 2-D matching process. To restrict the explosion of primitive aspects, we represent each primitive class, rather than each primitive instance, with a set of aspects. Unlike traditional aspect-based recognition systems, the resulting number of aspects is constant and independent of the size of the model database. In the following sections, we discuss the representation in more detail.

The Object-Centered Modeling Component

The goal of the object-centered modeling component is to define a set of three-dimensional volumetric primitives that, when assembled together, comprise a large set of concrete objects in the world. The constraints on these primitives are two-fold: they must be rich enough to describe real objects, yet simple enough to be reliably extracted from a contour image. The primitives, in turn, will be mapped into a set of viewer-centered aspects. Any selection of modeling primitives would support our approach; however, we seek a set of primitives whose aspect set will remain stable under minor changes in primitive shape. For example, the aspects should be invariant to changes in the primitive's scale, dimensions, and curvature (if the primitive is curved). Otherwise, the resulting large aspect set introduces the same matching inefficiency as traditional aspect-based recognition systems. To meet these requirements, we have chosen an object representation based on Biederman's Recognition by Components (RBC) theory (Biederman, 1985). RBC suggests that from nonaccidental relations in the image, a set of contrastive dichotomous (e.g., straight vs. curved axis) and trichotomous (e.g., constant vs. tapering vs. expanding/contracting cross-sectional sweep) 3-D primitive properties can be determined. The values of these properties give rise to a set of 36 shapes, called geons. Biederman claims that these geons constitute a rich set of primitive volumetric components that, when assembled together, can be used to model real world objects for the purpose of fast object recognition.

Many 3-D object recognition systems employ 3-D volumetric primitives to construct objects. Biederman's geons are a restricted class of generalized cylinders (e.g., Binford, 1971; Agin and Binford, 1976; Nevatia and Binford, 1977; Brooks, 1983) whose cross-section, axis, and sweep properties are arbitrary functions. Superquadrics (Gardiner, 1965) provide a volumetric representation requiring fewer parameters than generalized cylinders. Pentland (1986) first applied superquadrics to primitive modeling for object recognition, while Pentland (1987a) and Solina (1987) have achieved considerable success in deriving superquadric primitives from range data. Terzopoulos et al. (1987) propose symmetry-seeking deformable 3-D shape models, which they have successfully applied to the recovery of 3-D shape and nonrigid motion from natural imagery (Terzopoulos et al., 1988). Although generalized cylinders, superquadrics, and active models provide a rich language for describing parts, their extraction from the image is computationally complex. Biederman's geons, requiring only a few parameters, are an appropriate selection for the purposes of qualitative object modeling.

The Primitives

A generalized cylinder is defined by a cross-section function, an axis function, and a sweeping function; its shape results from sweeping the cross-section along the axis. Biederman (1985) mapped these three continuous functions to dichotomous and trichotomous properties. His 36 geons result from the Cartesian product of the possible values of these properties, which are defined as follows:

1. **Cross-Section Shape:** The cross-section shape can be either straight edged or curved edged.
2. **Cross-Section Symmetry:** The cross-section shape can be either rotationally symmetric, reflectively symmetric, or asymmetric.
3. **Axis Shape:** The axis can be either straight or curved.
4. **Cross-Section Sweep:** The cross-section can either remain constant, increase in size, or increase and then decrease in size as it is swept along the axis.

As a basis for initial investigation, we have defined a set of 10 primitives representing a restricted subset of Biederman's geons:

1. rectangular cross-section, straight axis, and constant cross-section size
2. rectangular cross-section, straight axis, and linearly increasing cross-section size not starting from a point
3. rectangular cross-section, straight axis, and linearly increasing cross-section size starting from a point
4. rectangular cross-section, curved axis, and constant cross-section size
5. elliptical cross-section, straight axis, and constant cross-section size

6. elliptical cross-section, straight axis, and linearly increasing cross-section size not starting from a point
7. elliptical cross-section, straight axis, and linearly increasing cross-section size starting from a point
8. elliptical cross-section, straight axis, and ellipsoidally increasing then decreasing cross-section size, neither starting nor ending with a point
9. elliptical cross-section, straight axis, and ellipsoidally increasing then decreasing cross-section size, starting and ending with a point
10. elliptical cross-section, curved axis, and constant cross-section size

Our three-property characterization resembles Biederman's four-property taxonomy; however, we have imposed additional restrictions in an effort to reduce the number of aspects and simplify the investigation. Nevertheless, the above primitive set forms a basis from which we can model a significant number of objects. The 10 primitives have been modeled using Pentland's SuperSketch 3-D modeling tool (Pentland, 1987b), and are illustrated in Figure 13.2. More primitives can easily be added to enrich the vocabulary.

Primitive Attachment

Having defined a set of modeling primitives, we must decide how to connect them to construct objects. We adopt a convention based on a labeling of each primitive's attachment surfaces. For example, the truncated cone primitive (primitive 6) has three attachment surfaces: the small end, the large end, and the side. Similarly, the curved block primitive (primitive 4) has six attachment surfaces: the concave side, the convex side, the two planar sides, and the two planar ends. The attachment surface labels for the 10 primitives can be found in Dickinson et al. (1989). We restrict any junction of two primitives to involve exactly one attachment surface from each primitive. Figure 13.3 presents an example object and its representation.

Both the primitive description and the interconnection description have been oversimplified to demonstrate the approach. Many enhancements are possible that would provide a much richer vocabulary for describing objects. For example, although not viewpoint invariant, additional properties such as cross-section extent, axis extent, and axis curvature provide important cues for recognition. Although these properties are quantitative, we could treat them as symbolic, based on a qualitative partitioning of the property range. For example, an axis might be "slightly curved" or "strongly curved" depending on its average curvature value.

In addition to specifying the two surfaces participating in the junction of two primitives, we could specify the position of the join on each surface. For example, a primitive attached to a rectangular planar surface may be attached near the middle, the sides, the corners, or the ends of the surface. A primitive attached to

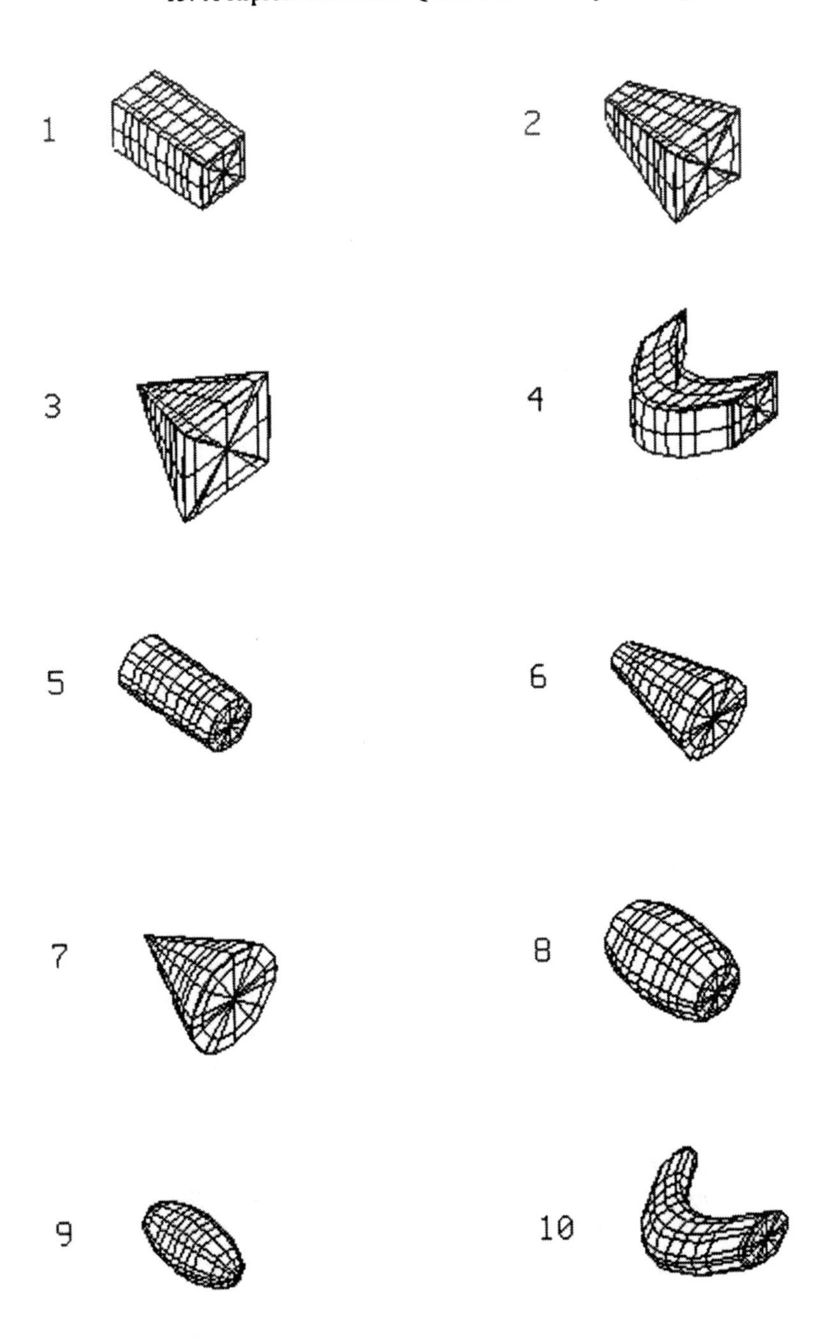

FIGURE 13.2. The 10 object modeling primitives.

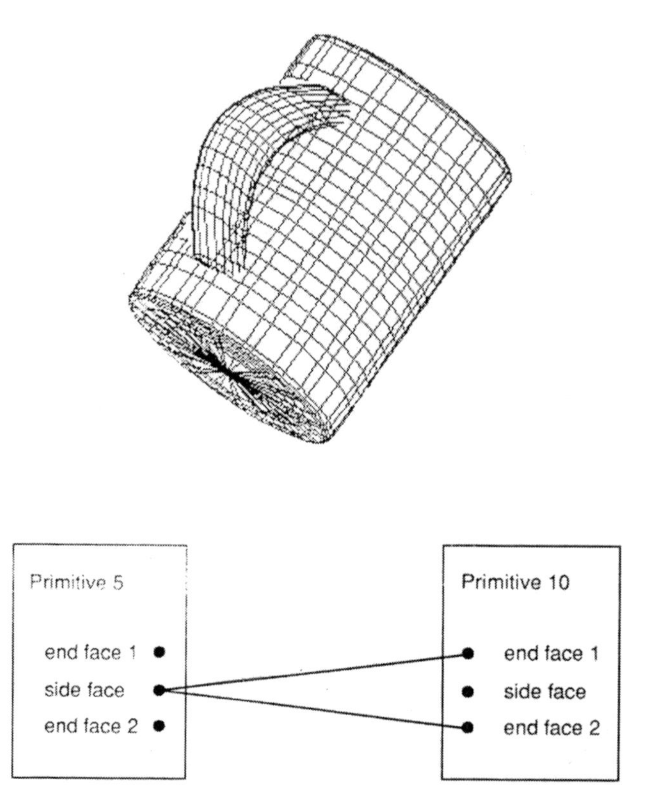

FIGURE 13.3. Example object and its representation.

an ellipsoidal primitive may be attached near the middle, the sides, or the ends of the ellipsoid. Again, we seek a qualitative localization of interconnection; we want to avoid an exact quantitative specification. Another enhancement to our vocabulary would be to describe the relative sizes of the primitives and the angles at which they join. For example, relative size measures such as "much larger than," "slightly larger than," or "roughly equal," and join angles such as "acute" or "perpendicular," are additional interprimitive relations that enhance the description of the object. The resolution of these descriptors would depend on the similarity of the objects in the database; perhaps having both coarse and fine descriptors would maximize matching efficiency.

Viewer-Centered Modeling Component

For each of the 10 primitive classes, we define a set of 2-D characteristic views, or aspects. Each aspect represents a set of topologically equivalent views of the primitive. To extract instances of the primitives from the image, we match image

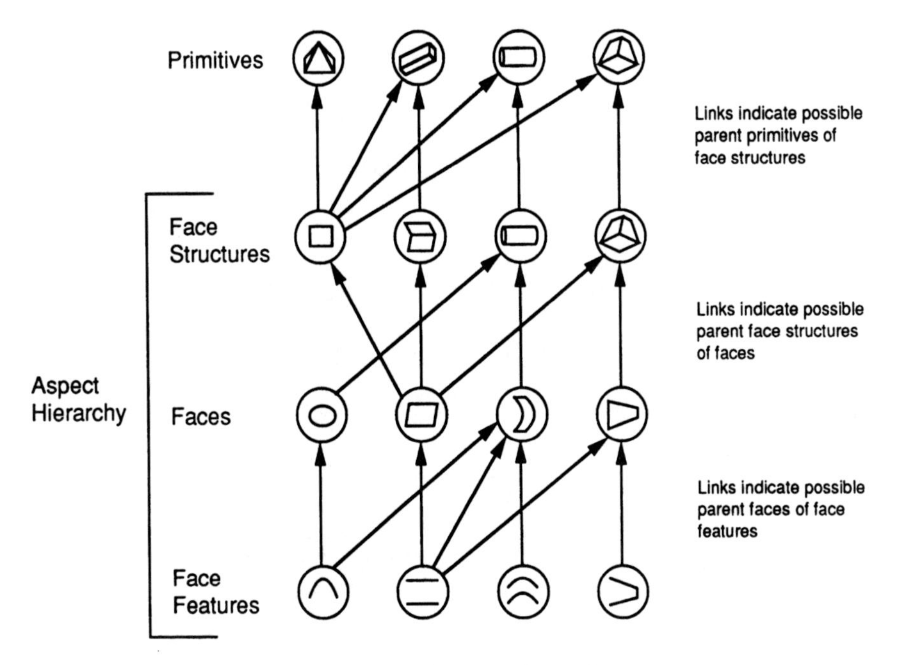

FIGURE 13.4. The aspect hierarchy.

contours against the set of aspects; a match not only identifies the primitive but qualitatively specifies its orientation. Unfortunately, if a primitive is occluded from a given 3-D viewpoint, its projected aspect in the image will also be occluded. In addition, the intersection of two primitives may alter the projected aspect of either primitive. To accommodate the matching of partial aspects to the set of aspects, we introduce a representation called the aspect hierarchy.

The aspect hierarchy consists of three levels, based on the faces appearing in the aspect set. A face is defined to be a closed cycle of image contours, e.g., a polygon, containing no other cycles. At the top level of the aspect hierarchy, we have the set of aspects, which we call face structures. Ideally, we would like to match image contours directly to the face structures. However, due to occlusion, it is unlikely that complete face structures will be visible. Some component faces of a face structure may be completely occluded, others partially occluded. The set of component faces of all face structures represents the middle level of the aspect hierarchy. Hence, we reduce face structure extraction to face extraction. However, we again run into the problem of occlusion, resulting in faces appearing in the image that are not included in the face level. Nevertheless, there may be subsets of face contours that survive occlusion and offer a mechanism for matching. Thus, at the bottom level of the aspect hierarchy, we have the face features that comprise the set of faces. Figure 13.4 illustrates a portion of the aspect hierarchy, while the following subsections describe the levels of the aspect hierarchy in more detail.

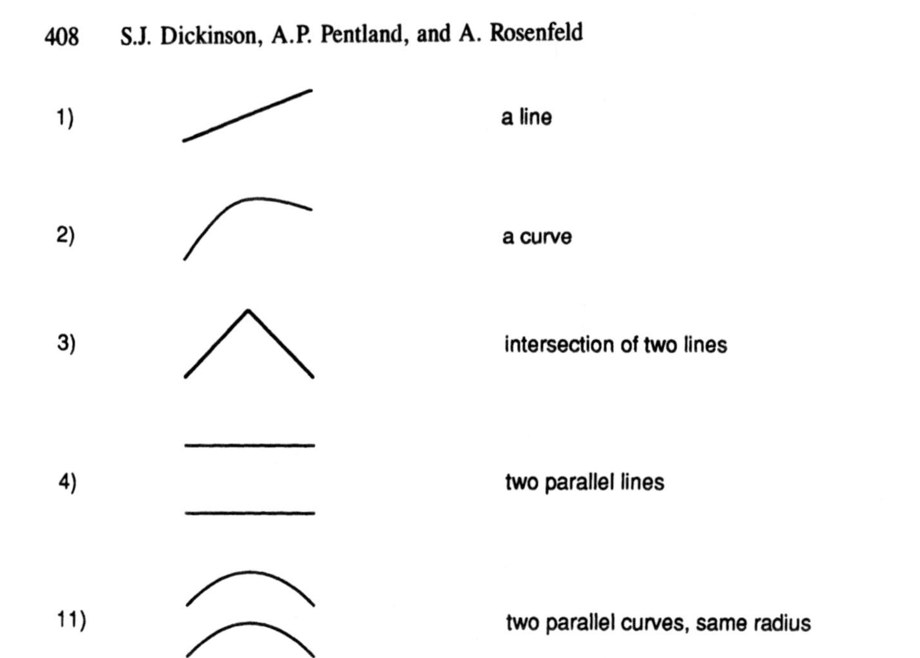

FIGURE 13.5. Examples of face features.

Face Features

The face features represent all subsets of lines and curves comprising the faces. Figure 13.5 illustrates a few of the 31 face features based on our 10 primitives; the complete set of face features can be found in Dickinson et al. (1989). The relations between face feature components (lines and curves) represent nonaccidental properties of lines including parallelism, symmetry, and cotermination. These relations, described by Biederman (1985) as a basis for the extraction of geons, are a subset of the nonaccidental properties suggested by Lowe (1985) and Witkin and Tenenbaum (1983). The important characteristic of the face features is that they represent *qualitative* relationships among *qualitative* lines; exact lengths of lines, distances between lines, angles between lines, curvature, etc., are not represented.

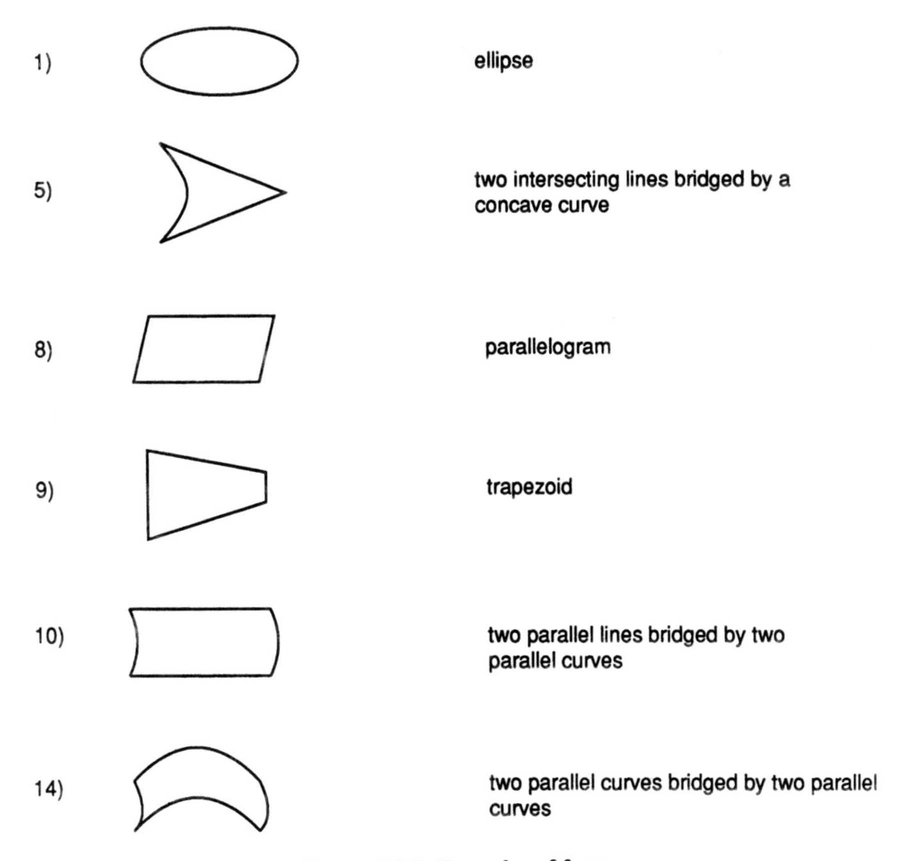

FIGURE 13.6. Examples of faces.

Faces

The faces represent the set of polygons appearing in the aspects. Figure 13.6 illustrates a few examples of our 16 faces; the complete set of faces can be found in Dickinson et al. (1989). As mentioned earlier, the faces form the backbone of the aspect hierarchy. Each differs in the number of constituent lines, the types of lines, or the nonaccidental relations between the lines. Since a face definition is invariant to changes in constituent line length and angle (provided the defining line relationships still hold), each face in Figure 13.6 represents only one of many possible instances defining the class.

Face Structures

The face structures represent connected sets of faces; each face in the structure shares a line with at least one other face in the structure. Figure 13.7 illustrates

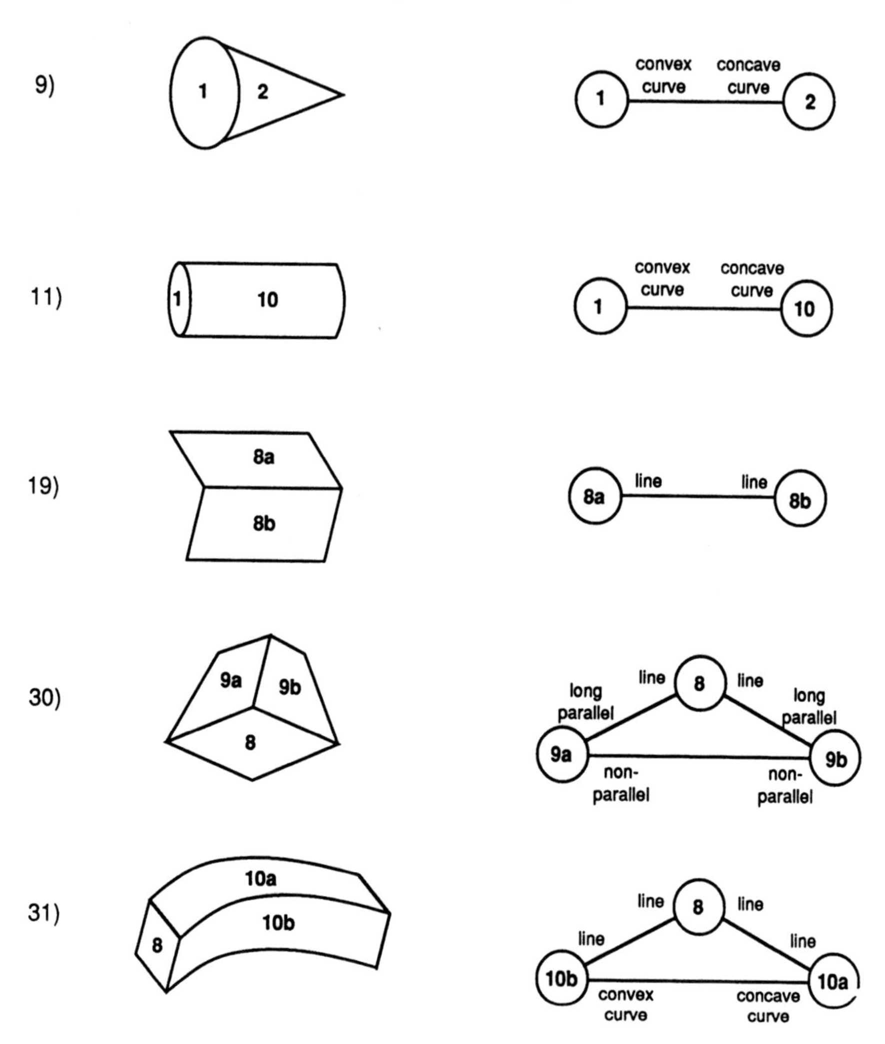

FIGURE 13.7. Examples of face structures.

a few examples of our 37 face structures; the complete set of face structures can be found in Dickinson et al. (1989). A face structure can be represented by a connected graph, with the nodes representing faces and the arcs representing the sharing of lines between faces; arc labels indicate which line is being shared.

Combining the Two Components

A given face feature may be common to a number of faces. Similarly, a given face may be a component of a number of face structures, while a given face structure may be the projection of a number of primitives. To capture these ambiguities, a matrix maps face features to faces, while another matrix maps faces to face struc-

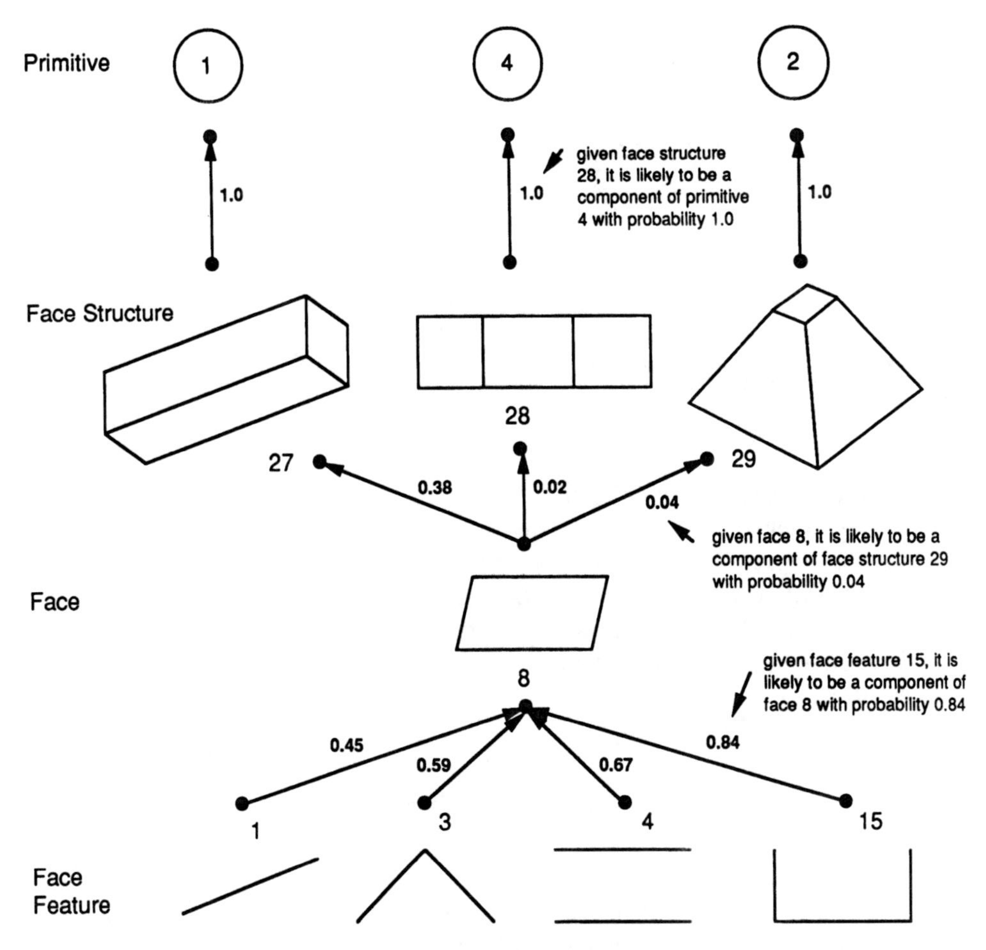

FIGURE 13.8. Combining the object-centered and viewer-centered models.

tures. To tie together the object-centered and viewer-centered representations, we define a third matrix mapping the top level of the aspect hierarchy, the face structure level, to the primitives.

In many cases, a feature (face feature, face, or face structure) could be a component of more than one parent feature at the next higher level; however, some parents might be more likely than others. The entries in the three matrices capture this likelihood. For example, consider the matrix mapping faces to face structures; the rows represent faces while the columns represent face structures. If a particular face can be a component of 10 different face structures, then those 10 column entries corresponding to the 10 face structures contain a value from 0 to 1.0, indicating the probability that the face is part of that particular face structure. Thus, the entries along each row sum to 1.0. Figure 13.8 presents a portion of the aspect hierarchy and related primitives along with the corresponding portions of the matrices.

TABLE 13.1. Superquadric definitions of the 10 primitives.

	Primitive									
Parameter Value	1	2	3	4	5	6	7	8	9	10
x-size	15	15	15	15	15	15	15	15	15	15
y-size	15	15	15	15	15	15	15	15	15	15
z-size	30	30	30	30	30	30	30	30	30	30
ϵ_1	.05	.05	.05	.05	1.0	1.0	1.0	1.0	1.0	1.0
ϵ_2	.05	.05	.05	.05	.05	.05	.05	.05	1.0	.05
z-axis bend	0.0	0.0	0.0	0.0	0.0	0.0	0.0	0.0	0.0	0.0
z-axis taper	0.0	0.5	1.0	0.0	0.0	0.5	1.0	0.0	0.0	0.0
z-axis pinch	0.0	0.0	0.0	0.0	0.0	0.0	0.0	0.5	0.0	0.0

To generate the probabilities in the tables mapping face features to faces, faces to face structures, and face structures to primitives, we first modeled our 3-D volumetric primitives using the SuperSketch modeling tool (Pentland, 1987b). SuperSketch models each primitive with a superquadric surface subject to deformation. The superquadric with length, width, and breadth a_1, a_2, and a_3 is described (adopting the notation cos $\eta = \mathbf{C}_\eta$, sin $\omega = \mathbf{S}_\omega$) by the following equation:

$$\mathbf{X}(\eta,\omega) = \begin{pmatrix} a_1\mathbf{C}_\eta^{\epsilon 1}\mathbf{C}_\omega^{\epsilon 2} \\ a_2\mathbf{C}_\eta^{\epsilon 1}\mathbf{S}_\omega^{\epsilon 2} \\ a_3\mathbf{S}_\eta^{\epsilon 1} \end{pmatrix}$$

where $\mathbf{X}(\eta,\omega)$ is a three-dimensional vector that sweeps out a surface parameterized in latitude η and longitude ω, with the surface's shape controlled by the parameters ε_1 and ε_2. The superquadric can be deformed by stretching, bending, twisting, or tapering. The SuperSketch superquadric definitions for the 10 primitives are given in Table 13.1.

The next step in generating the probability tables involves rotating each superquadric primitive about its internal x, y, and z axes in 10° intervals. The resulting quantization of the viewing sphere gives rise to 648 different views per superquadric primitive. However, we can exploit symmetries of the primitives to significantly reduce the number of views (688 views for all primitives). For each view, we orthographically project the superquadric primitive into the image plane. The final step involves a manual analysis of the images, noting each feature (face feature, face, and face structure) and its parent. The resulting frequency distribution gives rise to the three probability matrices given in Dickinson et al. (1989).

It should be emphasized that the results offer only a crude approximation to the true probabilities. A more thorough analysis would vary the dimensions, curvature, expansion rate, etc. of the primitives at a finer resolution on the viewing sphere. The resulting explosion of views would require an automated tool to perform the analysis and generate the probabilities. However, we believe that the probabilities will not change significantly in a more thorough analysis.

The three matrices can be used to guide the process of extracting the primitives from image contours. For example, if a face extracted from the image fails to match one of the 16 face types due to occlusion, the matrix mapping face features to faces can predict the most likely face type given a face feature belonging to the face. Once a face has been identified, the matrix mapping faces to face structures can predict the most likely face structure containing that face. Finally, given a face structure in the image, the matrix mapping face structures to primitives can predict the most likely primitive to which the face structure belongs. At each level, the matrices provide a heuristic to guide the search through the interpretations. The details of the processes used in both primitive extraction and model matching will not be presented in this chapter.

Evaluating the Aspect Hierarchy

An alternative approach to our hierarchical aspect representation would be to map the features at the lowest level (in our case, the face features) directly to the 3-D models (in our case, the 3-D primitives), an approach advocated by Lowe (1985), Huttenlocher and Ullman (1987), and Lamdan et al. (1988). In such a scenario, a given face feature index would return a set of candidate primitives containing that face feature. This approach has several drawbacks. First, the complexity of the primitives would increase to accommodate constituent face features. We would also face the problem that a weak hypothesis based on simple features requires a top-down verification step; the more qualitative the primitive, the more difficult the verification. Finally, simple indexing features do not provide strong orientation constraints on the primitives. For example, a pair of intersecting lines may lead to primitive hypotheses in many different orientations, whereas the face structure encompassing the lines constrains the orientation. In fact, we use the face structure label as a qualitative specification of a primitive's orientation.

In addition to the above heuristic arguments, we can make a quantitative case for the aspect hierarchy based on the three mapping matrices. By multiplying together matrices representing adjacent levels of the aspect hierarchy, we can generate new matrices mapping face features to face structures, faces to primitives, and face features to primitives. The results can best be seen in a set of histograms. To generate the histogram between two levels in the aspect hierarchy, we retain only the strongest probability arc emanating from each feature at the lower level. This indicates the degree of ambiguity in the mapping. For example, a node having two emanating arcs with values 0.50 and 0.50 is clearly inferior to a node having three emanating arcs with values 0.90, 0.05, and 0.05. Making an inference at the node with two emanating, equal probability arcs is a 50–50 guess, whereas there is a clear choice at the node with three emanating arcs. Clearly, having fewer emanating arcs is not as important as having a distinctly high probability arc. Once the strongest probability arc emanating from each node has been retained, we simply count the number of remaining arcs falling in each of 10 probability intervals. The resulting histograms (percentage of nodes whose emanating highest probability arc falls within given probability range) are presented in Figure 13.9.

Working backwards from the primitives, we can compare mappings from the face structures, faces, and face features to the primitives; the three histograms are illustrated in Figure 13.9c, e, and f, respectively. The face structure to primitive mapping is the strongest, with 90% of the face structure nodes having a high probability (0.80–1.0) arc. At the face structure level, we can compare mappings from the faces and face features; the two histograms are illustrated in Figure 13.9b and d, respectively. In this case, the mapping from the faces is much less ambiguous than the mapping from the face features. The final mapping from face features to faces is illustrated in Figure 13.9a.

The aspect hierarchy effectively prunes the mapping from face features to primitives by introducing intermediate constraints in the form of faces and face structures. The histograms suggest that for a typical set of 3-D modeling primitives, image regions, or faces, are the most appropriate features for recognition. Moreover, these faces should be grouped into the more complex face structures, providing a less ambiguous mapping to the primitives and further constraining their orientation. Only when a face's shape is altered due to primitive occlusion should we descend to the face feature level.

Related Work

Brooks' ACRONYM system (Brooks, 1983) exemplifies the object-centered approach to object recognition. In ACRONYM, objects are represented as constructions of generalized cylinders. Recognition of a particular model object consists of predicting the projected appearance in the image of the object's components; constraints on the 3-D parts of the model are mapped to constraints on the 2-D parts of the projection. The image contours are then examined, subject to these constraints, and matched contours are used to further constrain the size and orientation of the 3-D parts. The top-down nature of ACRONYM makes it unsuitable for unexpected-object recognition; ACRONYM can only confirm or deny the existence in the image of a user-specified object. In addition, the quantitative nature of ACRONYM's constraints requires the overhead of a complex constraint manipulation system. ACRONYM is appropriate for recognizing the subclasses of a particular airplane, while our system cannot; however, in distinguishing an airplane from, say, a horse, we avoid detailed quantitative constraints.

In contrast to ACRONYM's top-down approach, Lowe's SCERPO system (Lowe, 1985) takes a more bottom-up approach to object-centered recognition. In SCERPO, objects are represented as polyhedra, or constructions of 3-D faces. Image contours are first grouped according to perceptual organization

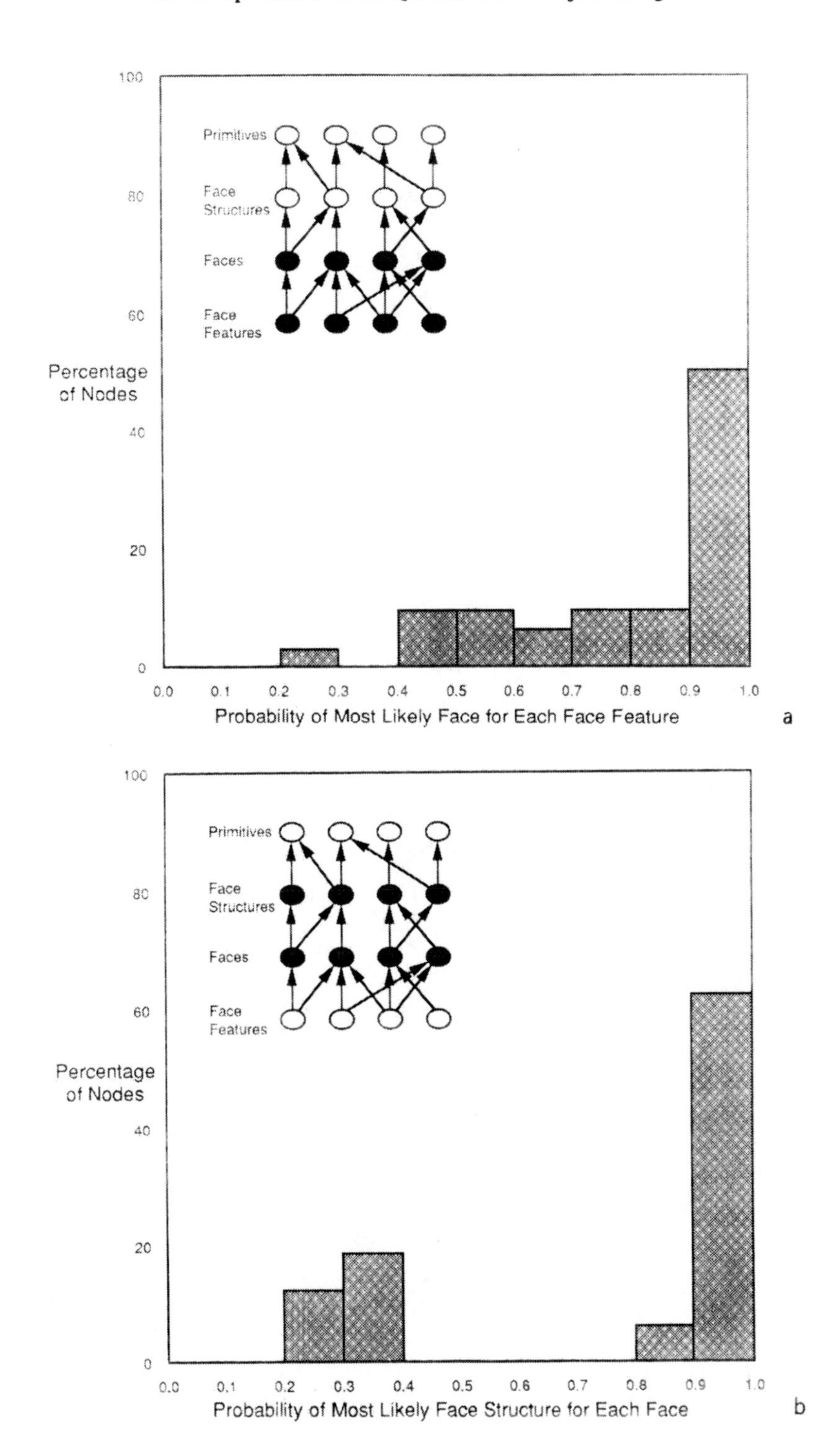

FIGURE 13.9. (a) Face feature to face mapping. (b) Fact to face structure mapping.

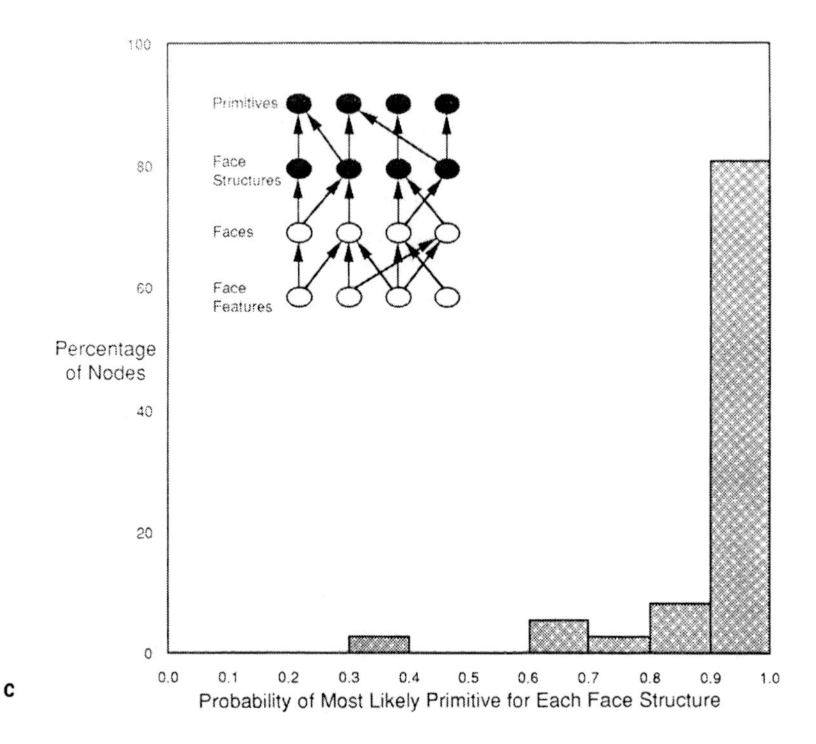

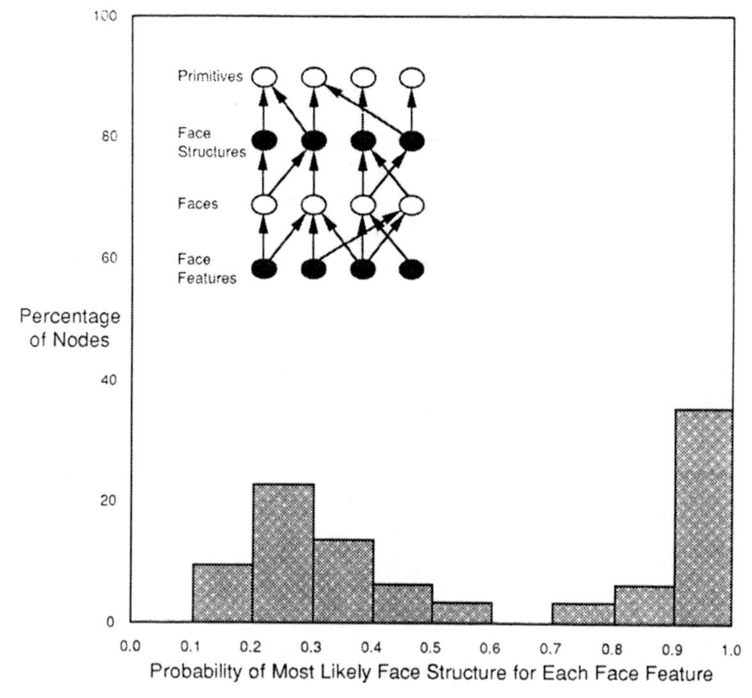

FIGURE 13.9. (c) Face structure to primitive mapping. (d) Face feature to face structure mapping.

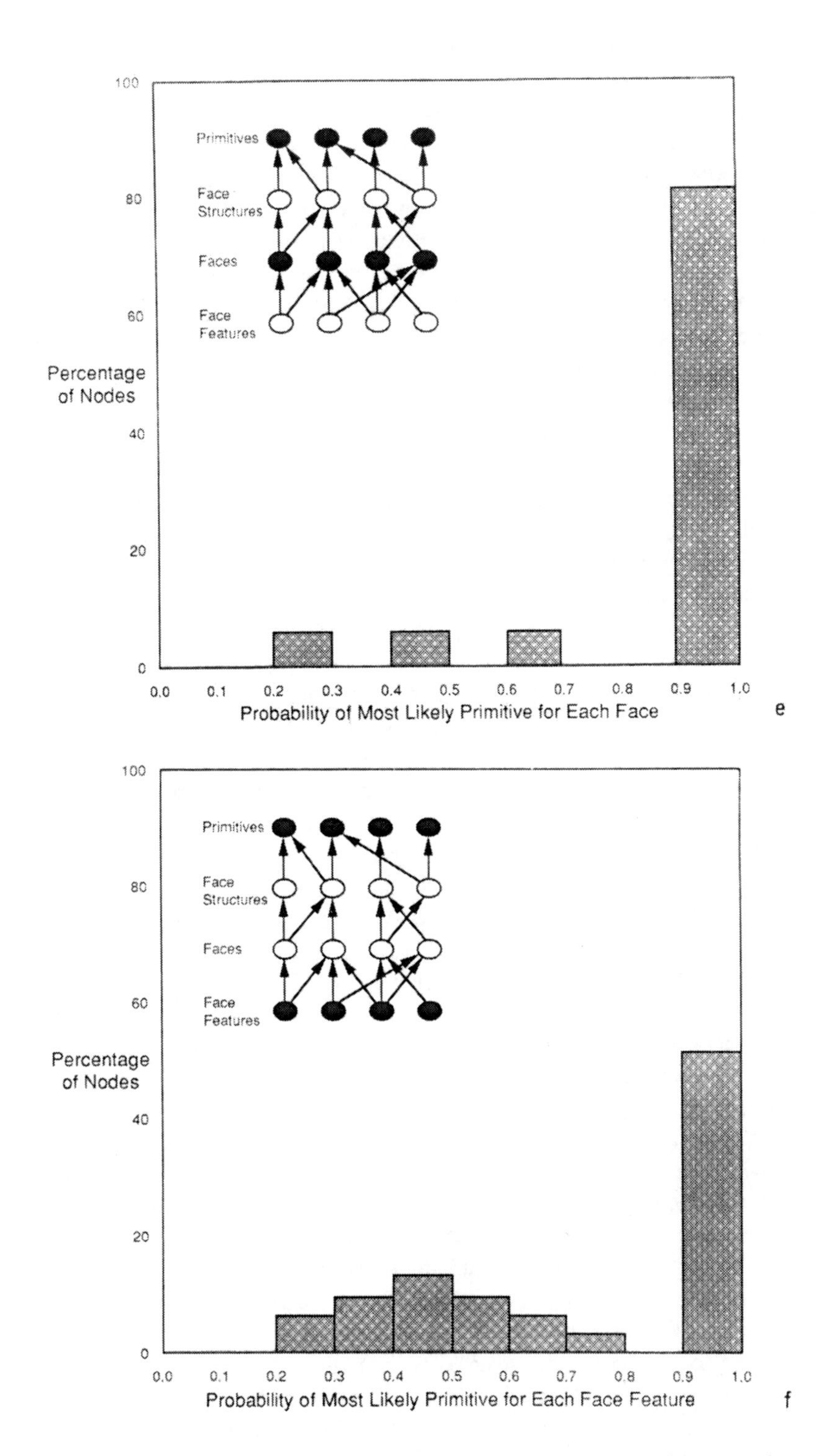

FIGURE 13.9. (e) Face to primitive mapping. (f) Face feature to primitive mapping.

rules, including parallelism, symmetry, and collinearity. From these groupings, simple 3-D inferences are made about the 3-D contours comprising the object; for example, parallel lines in the image imply parallel edges in the polyhedral object. The 3-D inferences are matched against manually identified instances of the properties in the model. Back-projected features are used to verify the object and constrain its position and orientation. Although SCERPO could be applied to unexpected-object recognition, the complexity of polyhedral models and the simplicity of the indexing features result in large indexing ambiguity. In addition, SCERPO's polyhedral models restrict its recognition domain to rigid objects. Our modeling scheme and indexing primitives, on the other hand, support the recognition of articulated objects.

Modeling objects using a set of qualitative primitives is not new. Mulgaonkar et al. (1984) describe a recognition system based on a set of generalized blob models including sticks, plates, and blobs. From the 2-D silhouette of an object, a graph-theoretic clustering technique yields a set of convex polygonal parts; internal image contours are ignored. The projected parts are then compared to 3-D part instances in the model database, subject to quantitative geometric and relational constraints. Like ACRONYM, the system is primarily top-down, starting with a model and matching image structures to the model-based predictions. Biederman (1985) proposed a set of primitives, called geons, based on the dichotomous and trichotomous properties of generalized cylinders. However, he failed to demonstrate how they may be extracted from the image, nor did he propose a control strategy for matching image features to models. Bergevin and Levine (1988a,b) have applied Biederman's geons to 3-D object recognition from 2-D images in a system called PARVO. Their approach to grouping lines consists of pairing segmentation points resulting from concave tangent discontinuities lying on the silhouette boundary of the object. From this pairing, line groups are formed, and internal contours are later assigned to the line groups on a second pass. The technique assumes that the segmentation points can be paired. In addition, PARVO assumes that a unique geon label can be assigned to each group of lines constituting a part. However, in the presence of occlusion or degenerate viewpoint, these assumptions may not be correct. Perhaps the greatest disadvantage of their approach is that it is dependent on their choice of geons as modeling primitives.

The viewer-centered representation of an object by a set of aspects was applied to 3-D object recognition by Chakravarty and Freeman (1982), and more recently by Ikeuchi and Kanade (1988). However, in these systems, the whole object is represented by the set of aspects. Thus, as the complexity of the object increases, so does the number of distinct aspects; automatically generating the distinguishable aspects is a difficult task. In our system, the aspects of a set of common parts or primitives have been generated and analyzed, and will be applicable to any objects constructed with these parts. Rather than matching against a large number of complex aspects, we plan to identify local instances of simple aspects. This allows us, like ACRONYM, to have articulated models, since we are matching aspects to primitives rather than to objects.

Conclusions

The inefficiency of most 3-D object recognition systems is reflected in the relatively small number of objects in their databases (on the order of 10); in many cases, algorithms are demonstrated on a single object model. The major problem is that these systems terminate the bottom-up primitive extraction phase very early, resulting in simple primitives such as lines, corners, or curvature points. Unfortunately, these primitives do not provide very discriminating indices into a large database. In fact, there may be many instances of such primitives in just one model, resulting in many hypothesized matches. The resulting systems are very top-down or model driven in nature. To achieve a more bottom-up recognition system requires that we index into the model database with more discriminating, higher order primitives. However, the more complex the indexing primitive, the more difficult the primitive extraction.

We propose a representation integrating constructions of 3-D volumetric modeling primitives at the database level with a set of aspects that describes the primitives at the image level. To reduce the number of aspects, our primitives are qualitative in nature, with the set of primitive aspects invariant to minor changes in primitive shape. The resulting integration of object-centered and viewer-centered models provides the foundation for a more bottom-up unexpected-object recognition system. The qualitative nature of the representation is ideal for qualitative recognition, and could provide a coarse front end for a more quantitative recognition system. Although we demonstrate our approach using a particular choice of 3-D primitives, the integration of object-centered and viewer-centered representations using a probabilistic aspect hierarchy is equally applicable to any representation scheme modeling objects as constructions of 3-D volumetric primitives.

Acknowledgments. The authors would like to thank Suzanne Stevenson, Larry Davis, and Peter Cucka for insightful discussions and for their comments on earlier drafts of this chapter.

References

Agin GJ, Binford TO (1976). Computer description of curved objects. IEEE Transact. Comp. **C-25**(4):439–449.

Bergevin R, Levine MD (1988a). Recognition of 3-D objects in 2-D line drawings: An approach based on geons. Technical Report TR-CIM-88-24, McGill Research Centre for Intelligent Machines, McGill University.

Bergevin R, Levine MD (1988b). Hierarchical decomposition of objects in line drawings. Technical Report TR-CIM-88-25, McGill Research Centre for Intelligent Machines, McGill University.

Besl PJ, Jain RC (1985). Three-dimensional object recognition. ACM Comp. Surveys **17**(1):75–145.

Biederman I (1985). Human image understanding: Recent research and a theory. Comp. Vision, Graphics, Image Process. **32**:29–73.

Binford TO (1971). Visual perception by computer. Proc. IEEE Conf. Systems Control, Miami, FL.

Binford TO (1982). Survey of model-based image analysis systems. Int. J. Robotics Res. **1**(1):18–64.

Bolles RC, Horaud P (1986). 3DPO: A three-dimensional part orientation system. Int. J. Robotics **5**(3):3–26.

Brooks RA (1983). Model-based 3-D interpretations of 2-D images. IEEE Transact. Pattern Anal. Machine Intell. **5**(2):140–150.

Chakravarty I, Freeman H (1982). Characteristic views as a basic for three-dimensional object recognition. Proc. SPIE Conf. Robot Vision, Arlington VA, 37–45.

Chin RT, Dyer CR (1986). Model-based recognition in robot vision. ACM Comp. Surveys **18**(1):67–108.

Dickinson SJ, Pentland AP, Rosenfeld A (1989). A representation for qualitative 3-D object recognition integrating object-centered and viewer-centered models. Technical Report CAR-TR-453, Computer Vision Laboratory, Center for Automation Research, University of Maryland.

Gardiner M (1965). The superellipse: A curve that lies between the ellipse and the rectangle. Sci. Am. **213**:222–234.

Grimson WEL, Lozano-Perez T (1984). Model-based recognition and localization from sparse range or tactile data. Int. J. Robotics Res. **3**(3):3–35.

Huttenlocher DP, Ullman S (1987). Object recognition using alignment. Proc. First Int. Conf. Comp. Vision, London, 102–111.

Ikeuchi K, Kanade T (1988). Automatic generation of object recognition programs. Proc. IEEE **76**(8):1016–1035.

Lamdan Y, Schwartz JT, Wolfson HJ (1988). On recognition of 3-D objects from 2-D images. Proc. IEEE Int. Conf. Robotics Automation, 1407–1413.

Lowe DG (1985). *Perceptual Organization and Visual Recognition*. Kluwer Academic Publishers. Norwell, Massachusetts.

Mulgaonkar PG, Shapiro LG, Haralick RM (1984). Matching 'sticks, plates and blobs' objects using geometric and relational constraints. Image Vision Comp. **2**(2):85–98.

Nevatia R, Binford TO (1977). Description and recognition of curved objects. Artif. Intell. **8**:77–98.

Pentland AP (1986). Perceptual organization and the representation of natural form. Artif. Intell. **28**:293–331.

Pentland AP (1987a). Recognition by parts. Proc. First Int. Conf. Comp. Vision, London, 612–620.

Pentland AP (1897b). Towards an ideal 3-D CAD system. SPIE Conf. Machine Vision and Man-Machine Interface, San Diego.

Requicha AAG (1980). Representations for rigid solids. ACM Comp. Surveys **12**(4): 437–464.

Rosenfeld A (1987). Recognizing unexpected objects: A proposed approach. Proc. 1987 DARPA Image Understanding Workshop, Los Angeles, 620–627.

Solina F (1987). Shape recovery and segmentation with deformable part models. Tech. Report MS-CIS-87-111, GRASP LAB 128, University of Pennsylvania, Philadelphia, PA.

Srihari SN (1981). Representation of three-dimensional digital images. ACM Comp. Surveys **13**(4):399–424.

Terzopoulos D, Witkin A, Kass M (1987). Symmetry-seeking models and 3D object reconstruction. Int. J. Comp. Vision **1**:211–221.

Terzopoulos D, Witkin A, Kass M (1988). Constraints on deformable models: Recovering 3D shape and nonrigid motion. Artif. Intell. **36**:91–123.

Thompson DW, Mundy JL (1987). Model-directed object recognition on the connection machine. Proc. 1987 DARPA Image Understanding Workshop, Los Angeles, 93–106.

Witkin AP, Tenenbaum JM (1983). On the role of structure in vision. In *Human and Machine Vision*, (J Beck, B Hope, A Rosenfeld, eds.), pp 481–543. Academic Press, New York.

14
From 2-D Images to 3-D Models

ALEX P. PENTLAND

Introduction

Any intelligent agent, be it robot or human, needs to continually update its knowledge about the surrounding environment, and images offer a rich source of relevant information. Unfortunately, however, in normal two-dimensional images the effects of viewing geometry, illumination, surface reflectance, and object shape are confounded together in a way that makes it very difficult to extract interesting information about the surrounding scene.

Nonetheless, early in evolution biological mechanisms developed to solve this difficult problem. The first solutions, however, must have involved relatively simple mechanisms if they were the results of blind evolutionary accident. Perhaps the best guess about what these first, simple vision systems were like comes from the observation that, following transduction of incident light by rods or cones, the first stage of almost all biological visual systems is to split up the incoming signal into a neural representation that localizes imaged activity in terms of retinal position, time, and spatial frequency (size) (Hubel and Wiesel, 1977). It seems likely, therefore, that it is possible to build simple, robust vision mechanisms based on filtering images in both space and time, and so our research group has devoted much effort toward searching for these mechanisms.

What we have found is quite interesting: it now appears to us that we may be able to obtain serviceable estimates of most of the interesting scene properties by examining the outputs of a few relatively simple filtering mechanisms. We have found, for instance, that we can extract good estimates of surface shape from image shading, texture, focus, and motion by use of one single type of image filtering process. Moreover, these same filters can be used to segment images into homogeneous texture regions and may be useful in stereo and color processing.

Space–Time Filtering and Spectral Power

It is widely accepted that in primates the initial cortical processing areas contain many cells that are tuned to orientation, spatial and temporal frequency, and to phase, although the tuning of these cells is relatively broad (Daugman, 1980).

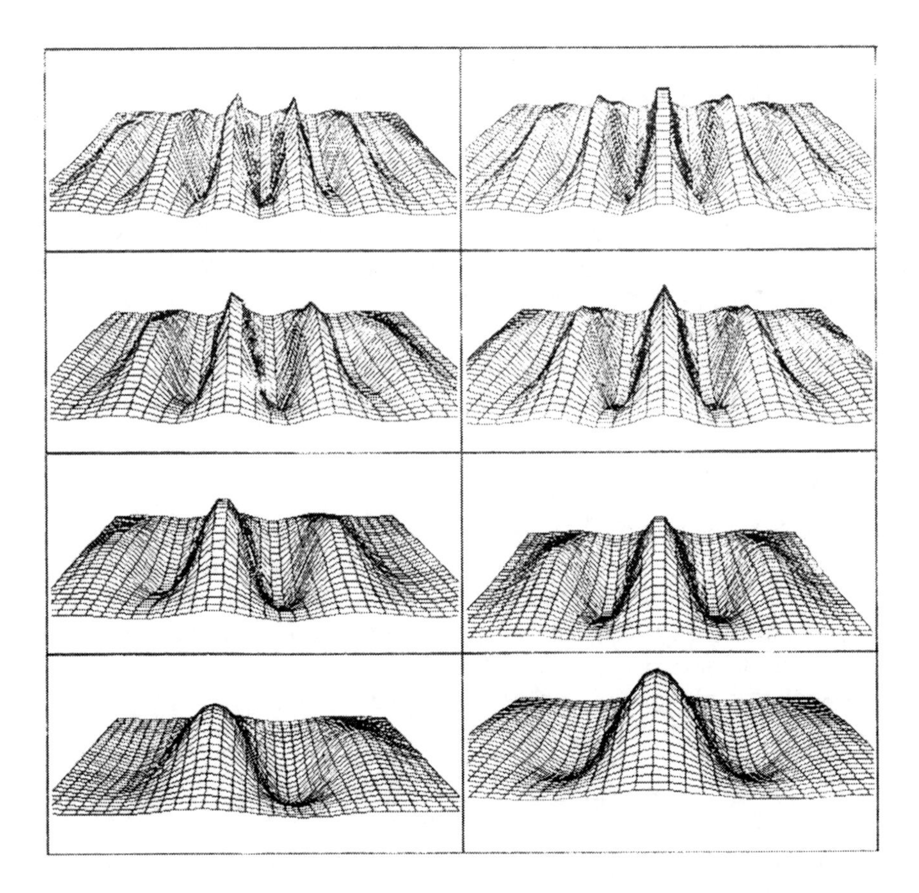

FIGURE 14.1. A set of filters that is selective in spatial frequency, orientation, and phase produces localized measurements of image frequency content. (From Pentland A [1989b], with permission.)

Figure 14.1 illustrates a set of filters that model the filtering properties thought to exist in cortex. Note that at each image location there are several sets of such filters, each with a different two-dimensional (2-D) image orientation.

The critical properties of these filters are that they are selective to orientation and spatial frequency (roughly, size) so that these filters decompose an image in much the same manner as the Fourier transform. Further, as in the Fourier transform, the filters come in quadrature pairs (roughly, sine and cosine phase pairs) so that the local phase information is available. The response of these filters can also depend on time, producing a set of filters that is selective in both space and time, so that they can fully characterize both the appearance and motion within a single patch of the image.

Although the details of exactly what these filters look like is unimportant for recovering shape estimates, I will for purposes of mathematical clarity assume that these filters are sine and cosine phase Gabor filters. The output of such a set of Gabor filters is exactly the Fourier transform of the image as seen through a Gaussian-shaped windowing function, so that we can easily characterize the output of these filters in terms of the spectral content of the image, and vice versa.

Perhaps the most important idea in using these filters is the notion that we can use them to measure the power (or energy) within some specific region of the Fourier domain (Pentland, 1984; Adelson and Bergen, 1985; Watson and Ahumada, 1985; Pentland, 1986a). Parseval's Theorem states that the integral of squared quadrature filter values over the spatial domain is equal to the integral of the squared Fourier components over the frequency domain (the Fourier power). Therefore, by squaring and adding our sine and cosine pair filters we can measure the Fourier power within a small range of spatial and temporal frequencies.

Image Formation

Being able to characterize filter output in terms of image spectral content, however, is only half the battle. We must also be able to characterize image spectral content in terms of scene properties, so that in the end we can relate filter response to scene characteristics. Luckily, the mathematics that accomplishes this is straightforward.

We will let $z = z(x,y)$ be a Lambertian surface illuminated by (possibly several) distant point sources, thus producing an image $I(x,y)$:

$$I(x,y) = \frac{p \cos \tau \sin \sigma + q \sin \tau \sin \sigma + \cos \sigma}{(p^2 + q^2 + 1)^{1/2}} \tag{1}$$

where p and q are the slope of the surface along the x and y image directions, respectively, e.g.,

$$p = \frac{\partial z(x,y)}{\partial x} \qquad q = \frac{\partial z(x,y)}{\partial y} \tag{2}$$

and $\mathbf{L} = (\cos \tau \sin \sigma, \sin \tau \sin \sigma, \cos \sigma)$ is a unit vector in the mean illuminant direction, where τ is the tilt of the illuminant (the angle the image plane component of the illuminant vector makes with the x axis) and σ is its slant (the angle the illuminant vector makes with the z axis).

Equation (1) can now be converted to a form that will allow us to relate image and three-dimensional (3-D) surface in terms of their Fourier transforms or filter responses. This is accomplished by taking the Taylor series expansion of $I(x,y)$ about $p,q = 0$ through the linear terms, to obtain

$$I(x,y) \approx \cos \sigma + p \cos \tau \sin \sigma + q \sin \tau \sin \sigma \tag{3}$$

This expansion produces a very good approximation whenever the illumination is oblique, or when the range of surface normals within the filtered area is small. In each of the following examples the quadratic terms of the expansion have also been considered. However, space prevents their treatment here. (See referenced articles for more details.)

Now let the complex Fourier spectrum $F_z(f,\theta)$ of $z(x,y)$ be

$$F_z(f,\theta) = m_z(f,\theta)e^{i\varphi_z(f,\theta)} \tag{4}$$

where f is the spatial frequency, θ is the orientation, $m_z(f,\theta)$ is the magnitude at position (f,θ) on the Fourier plane, and φ_z is the phase. Since p and q are partial derivatives of $z(x,y)$, their transforms F_p and F_q are simply related to F_z, e.g.,

$$F_p(f,\theta) = 2\pi f \cos(\theta) m_z(f,\theta)e^{i[\varphi_z(f,\theta)+\pi/2]} \tag{5}$$

$$F_q(f,\theta) = 2\pi f \sin(\theta) m_z(f,\theta)e^{i[\varphi_z(f,\theta)+\pi/2]} \tag{6}$$

Thus the Fourier transform of the image $I(x,y)$ is (ignoring the DC term)

$$F_I(f,\theta) = 2\pi f \sin(\sigma) m_z(f,\theta)e^{i[\varphi_z(f,\theta)+\pi/2]}[\cos\theta\cos\tau + \sin\theta\sin\tau] \tag{7}$$

This final equation is the desired result: it says that the Fourier transform of the image intensity surface $I(x,y)$ is a simple linear function of the Fourier transform of the height surface $z(x,y)$. As a consequence, we now know that the output of our filters is also a simple linear function of the surface shape. This fact allows us to look at the responses of our biological-like filters to the 2-D image data, and figure out what is happening in the surrounding three-dimensional world.

An Example: Extracting Surface Shape and Range

As an example of how simple filtering-based mechanisms can produce good estimates of scene properties, we will consider the problems of estimating surface shape and range by using image cues such as shading, texture, focus, and motion. In each case it will turn out that simple computations based on the output of our filters will produce a good estimates of shape.

Shape from Shading

Examining Eq. (7) shows that if given the illuminant direction then the Fourier transform of the surface can be recovered directly, except for an overall linear scaling and certain boundary conditions. Thus if the image Fourier transform is

$$F_I(f,\theta) = m_I(f,\theta)e^{i\varphi_I(f,\theta)} \tag{8}$$

then the Fourier transform of the z surface is simply

$$F_z(f,\theta) = \frac{m_I(f,\theta)e^{i[\varphi_I(f,\theta)-\pi/2]}}{2\pi f \sin\sigma[\cos\theta\cos\tau + \sin\theta\sin\tau]} \tag{9}$$

To recover surface shape using our set of filters, then, the transformations indicated in Eq. (9) must be performed. These transformations can be accomplished by (1) switching the role of the sine and cosine phase filters in the surface reconstruction, thus accomplishing the required $\pi/2$ phase shift, (2) reducing each filter's amplitude in proportion to its central frequency, thus accomplishing the $1/f$ frequency scaling, and (3) normalizing the average filter amplitude within

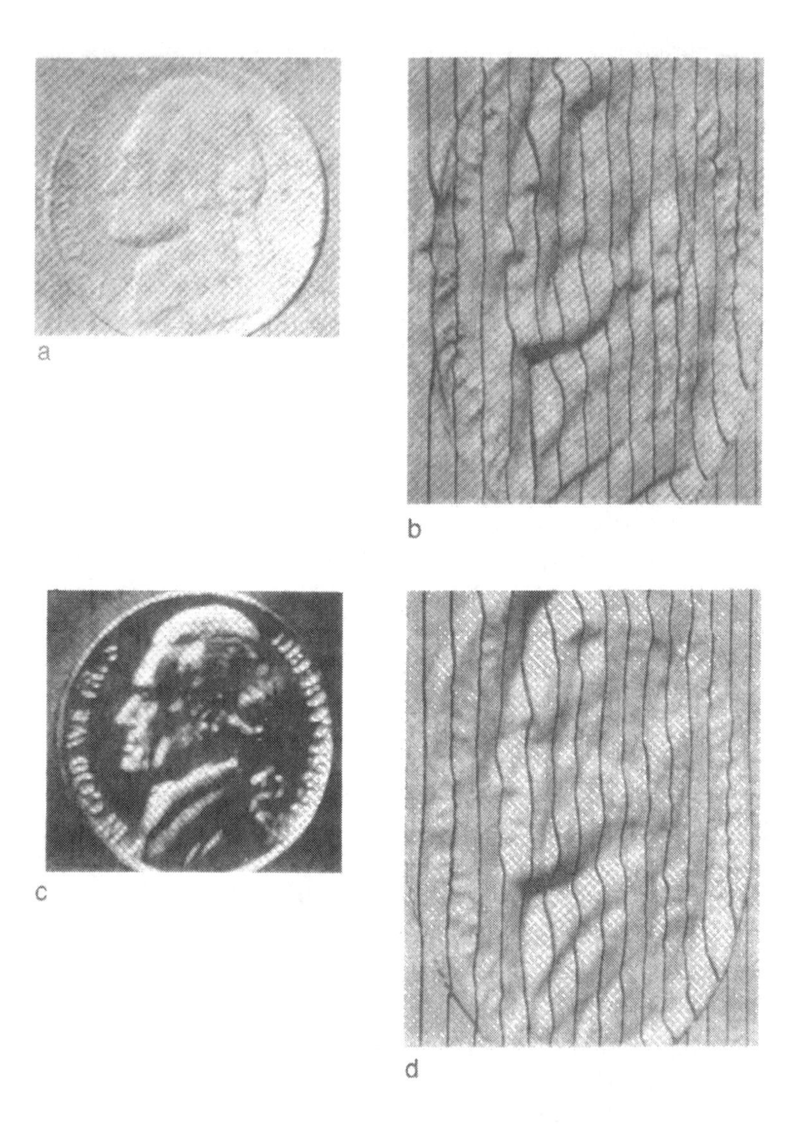

FIGURE 14.2. (a) A plaster cast of a nickel, (b) a shaded perspective view showing how the recovered 3-D surface would look from another point of view, with contours added to the face to enhance the viewer's perception of the recovered shape, (c) a shiny nickel, (d) a shaded perspective view of how the recovered 3-D surface would look from another point of view.

each orientation, thus removing the directional effects of the illuminant (as described in Pentland, 1982).

The final step, reconstruction of the height surface from these normalized filter responses, can be accomplished by a process nearly identical to the original filtering process, as when the filters form an orthonormal basis set they are exactly their own inverse. Thus to perform the z surface reconstruction via an

a

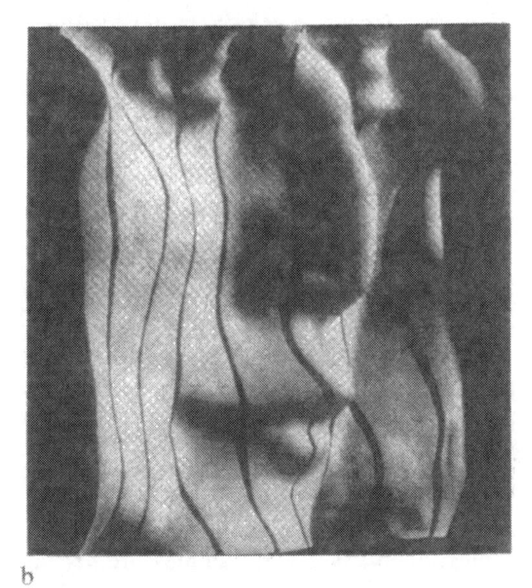
b

FIGURE 14.3. (a) An image, and (b) a shaded perspective view showing how the recovered 3-D surface would look from another point of view (only the face region is shown). (Parts a and b from Pentland A [1989b], with permission.)

inverse Fourier transform, one simply sums up all of the various filters proportional to their "activity" at each (x,y) position. The result of this summation will be the estimated surface shape within the windowed area of the image, the "receptive field" of the filters (for more details see Pentland, 1988a).

An example of recovering shape from image shading is shown in Figure 14.2. As a test of the accuracy of the shape recovery, I first used a plaster cast of a nickel, shown in Figure 14.2a, as such an object very closely approximates the diffusely reflecting surface assumed in the mathematical development above. To demonstrate to the reader the recovered 3-D shape, I have used computer graphics techniques to create a shaded, perspective view of how the surface would look from another point of view, as is shown in Figure 14.2b. I have also added straight "cuts" to the surface, which, when viewed from the side as in these images, provides contour information that aids the reader's perception of the recovered shape.

It can be seen that the shape of the nickel has been accurately recovered. To test the applicability of this approach to shiny surfaces, I then estimated surface shape for the shiny nickel shown in Figure 14.2c. The recovered shape is shown in Figure 14.2d; as can be seen, the technique works nearly as well for the shiny nickel as the diffusely reflecting plaster cast of Figure 14.2a.

A second example of such shape recovery is illustrated by Figure 14.3. Figure 14.3a shows an image widely used in data compression research. Figure 14.3b shows a shaded, perspective view of the recovered surface in the neighborhood of the face (again, lines have been added to help the viewer's perception of the recovered shape). The eyes, cheek, lips, nose arch, and nostrils can all be clearly seen in the recovered surface.

Shape from Texture

Extracting shape estimates from texture depends mostly on the foreshortening that occurs when we view a surface at an angle. When a patch of surface is slanted away from the viewer, foreshortening occurs along the direction in which the surface faces, causing an apparent compression of the surface's features along that direction. For most real-world textures this results in an increase in the "texture energy" (local Fourier power within a given frequency band) in the corresponding image direction.

Because of this foreshortening effect, then, the image orientation that has the most texture energy is exactly the angle τ in which the surface faces. (For simplicity we are assuming that the surface texture is distributed isotropically on the surface.) We can measure this oriented texture energy by squaring and then summing the sine and cosine phase filters of a particular orientation and size, as described above. Thus by finding the image orientation that has the largest texture energy we can estimate τ, the image-plane orientation of the viewed surface patch.

Once we know which way a patch of surface faces, we can then estimate the angle σ between the viewer and the surface normal of the patch. The amount of foreshortening, and thus the increase in texture energy along the tilt direction, is proportional to $1/\cos(\sigma)$. Thus if the texture energy along the tilt direction, is E_τ and the energy along the perpendicular direction is E_0, then the angle σ between viewer and surface normal is

$$\sigma = \cos^{-1}(E_0/E_\tau) \tag{10}$$

Finally, given τ and σ at each point, we can reconstruct the surface by use of a second set of filters as in the previous shading example, to obtain the surface shape (for more details see Witkin, 1981; Pentland, 1986a).

Figure 14.4 shows two examples of using a similar technique to estimate the shape of a textured log and a textured rock. Figure 14.4a shows the digitized image of a log and an isometric view of the recovered surface shape. Figure 14.4b shows a side view of a textured rock, and an isometric view of the recovered surface shape. As can be seen, the recovered surface shapes are approximately correct.

One of the more interesting points about this method of estimating surface shape is that it can be thought of either as a method of extracting shape from texture or as a method of extracting shape from shading. It turns out that both shading and texture behave very similarly for rough, textured surfaces, so that the texture energy measure defined above applies nearly as well to shading as to texture (see Pentland, 1986a, for additional details).

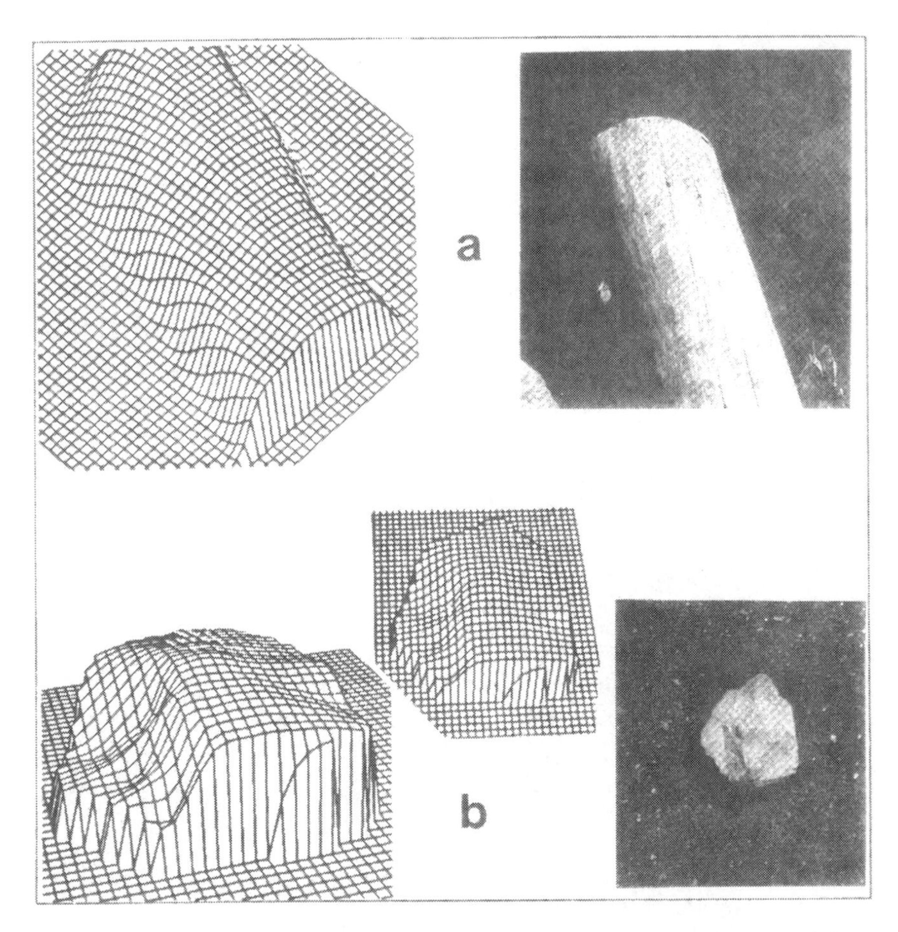

FIGURE 14.4. (a) An image of a textured log and the recovered surface shape, and (b) an image of a textured rock and the recovered surface shape.

Range from Focus and Depth of Field

Imagine that we have two images of a scene that are identical except for their depth of field, such as shown in Figure 14.5a and b. One way of obtaining such images in a biological system is to stimulate light receptors with different color sensitivities, as the depth of field in biological systems is quite different for different spectral bands. Other methods of obtaining two images with different depth of field are to vary aperture size or focal length within each saccade, a type of variation which actually occurs and can be detected experimentally in the human visual system (Crane, 1966).

In such image pairs the overall geometry is nearly identical, except that corresponding points are focused quite differently because of the difference in depth of field. For shape perception, given fixed imaging parameters (e.g., fixed

FIGURE 14.5. (a) An image taken with a wide aperture (small depth of field), (b) an image taken with a narrow aperture (large depth of field), (c) computed range; brighter areas are closer, darker areas are further away. (Parts a, b, and c all from Pentland A [1989b], with permission.)

accommodation), the critical fact is that the change in focus between the two images is a simple function of only one variable, the range, defined as distance between the viewer and the imaged point. To obtain an estimate of this distance, therefore, we need only compare corresponding points in the two images and measure the change in how sharply the image is focused by use of our spatiotemporal energy filters.

To extract range one starts just as with the shape-from-texture estimator, that is, by convolving data from the two images with quadrature pair filters and then squaring and summing the filter responses. The resulting values are then spatially averaged, resulting in a "power image" for each image, that is, in an estimate of the Fourier power within a particular spatial frequency band at each image location. These two power images (which differ only because of differences in focus) are then compared using a lookup table to produce an estimate of range, such as is shown in Figure 14.5c. In Figure 14.5c the recovered range information (distance between viewer and the surface at each point) is encoded by brightness; farther points are darker and nearer points are lighter (for more detail see Pentland, 1987; Bove, 1989).

Range from Motion

When a viewer moves through the world his changing viewpoint causes changes in the appearance and position of the surrounding surfaces. If these surfaces are textured then, just as in the shape-from-texture example described above, the apparent compression and stretching caused by viewer motion can be measured by use of the local texture energy, and from the measured amount of compression/stretching the distance to the viewed surface can be recovered.

However, because the compression and stretching occur over both time and space, we require energy filters oriented not only in space but also in time if we are to estimate the surface distance. Thus the filters shown in Figure 14.1 must be augmented by adding filters that respond to changes in spatial structure. The result is an expanded set of filters that has not only a spatial structure but also a temporal structure that is similar to that shown in Figure 14.1. Thus, for instance, the expanded set contains filters that are sensitive to leftward-moving high-frequency patterns or to flickering low-frequency patterns, as well as filters sensitive to static orientation and spatial frequency.

Figure 14.6a shows a single image from a fly-by sequence of Yosemite Valley in California. Spatiotemporal filters similar to those described above were applied to the image sequence, and the spatiotemporal energy measured at each point in both space and time. From this measured energy, the distance to the viewed surface was estimated as described in Heeger (1987).

The result is illustrated in Figure 14.6b and c. In Figure 14.6b the white area shows the true shape of the valley floor as determined from a digital terrain map. In Figure 14.6c the white area shows the estimated shape of the valley floor. As can be seen, the estimated shape is generally correct, especially in the

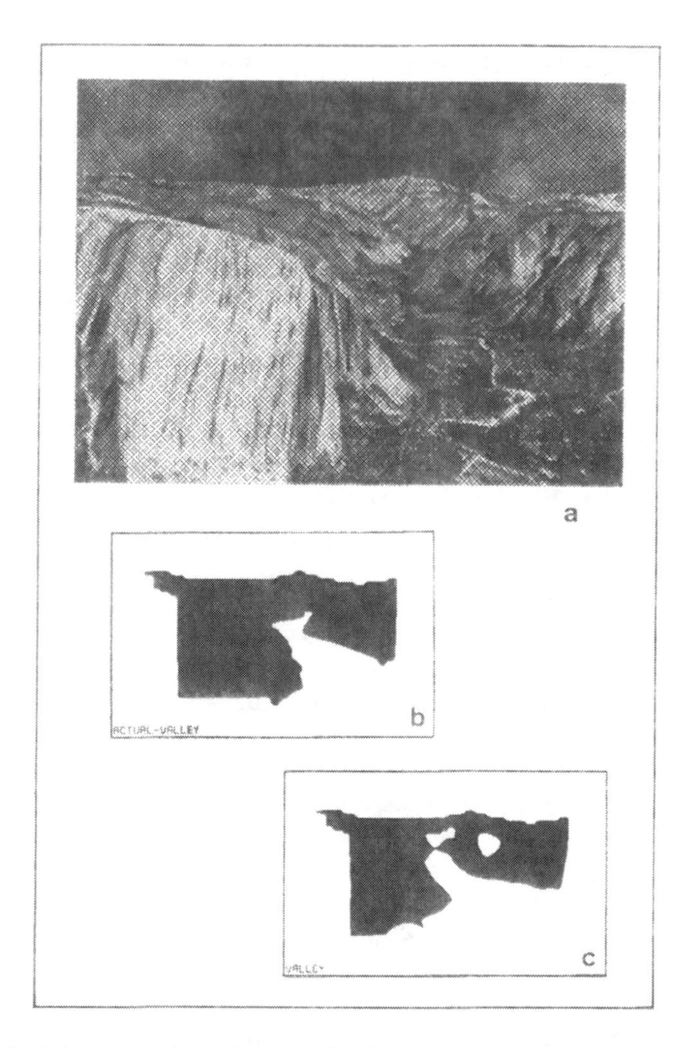

FIGURE 14.6. (a) A single image from a fly-by of Yosemite Valley, (b) the white area shows the true shape of the valley floor, (c) the white area shows the estimated shape of the valley floor. (Parts a, b, and c all from Pentland A [1989b], with permission.)

foreground. The major error is one roughly circular area that can be mostly attributed to "dropouts" in the original image data.

Fusing Information into a Single Percept

As the above examples make clear, it appears that simple, parallel filtering processes similar to those thought to exist in biological visual systems can produce useful estimates of many important scene characteristics. However, the output of these simple information extraction processes will not in general be a consistent, coherent description of the scene, because there are many special conditions under which one or more of them will fail.

To obtain a consistent and reliable percept, therefore, we need to be able to obtain a consensus among the information sources, so that we can explain as much of the image data as is possible. This suggests that robust statistical techniques, such as mode finding, will be necessary to extract a coherent estimate of scene structure from our noisy and error-prone information sources.

Eventually we must combine these point-by-point estimates of continuous scene properties (e.g., shape, reflectance) into more abstract, discrete symbolic descriptions that we can use to recognize objects despite variations in viewpoint, size, and details of shape. We believe that the fact that most objects have a clear part structure is critical in the transition from continuous, analog scene properties to discrete, symbolic descriptions (Pentland, 1986b). The idea, then, is that if we can group point-by-point estimates of scene properties into roughly convex, part-like "blobs," then we can obtain stable recognition by matching the part structure of objects in the scene to that of objects we have seen before.

As an example, a person can be described as having the following parts: head, body, upper and lower arms, upper and lower legs, hands, and feet. Each of these parts is connected to certain other parts, and each has a characteristic aspect ratio, shape, range of allowable positions, and so forth. It seems, therefore, that a good way to recognize when there are people in an image is to look for a set of part-like blobs that is arranged in a manner that is typical of people.

Because we must both find coherence within our point-by-point estimates of scene structure and break the image into part-like blobs, it seems reasonable to try to accomplish both tasks at the same time. These considerations provide the motivation for our approach to recognition, which is to find the simplest, most likely description of the point-by-point image evidence in terms of a set of part-like blobs.

By adopting this approach we have found that we can simultaneously combine conflicting, noisy scene estimates into a coherent estimate of scene structure, and at the same time obtain a maximum-likelihood decomposition of the scene into its component parts. Moreover, the mechanism by which we can accomplish this decomposition can be implemented as a simple, parallel neural network.

Finding Object's Parts

Many machine vision systems employ matched filters to find particular shapes in an image, typically using a multiresolution approach that allows efficient search over a wide range of scales. Thus in machine vision a natural way to locate the parts of objects within a scene is to make filter patterns that cover the spectrum of possible blob-like parts (such as is shown in Fig. 14.7a, match these part shapes against the image data (estimates of range, reflectance, etc.), and then pick the filter whose shape best describes the data. If the match between data and filter shape is sufficiently good, then we register the detection of a part whose shape is roughly that of the best matching filter.

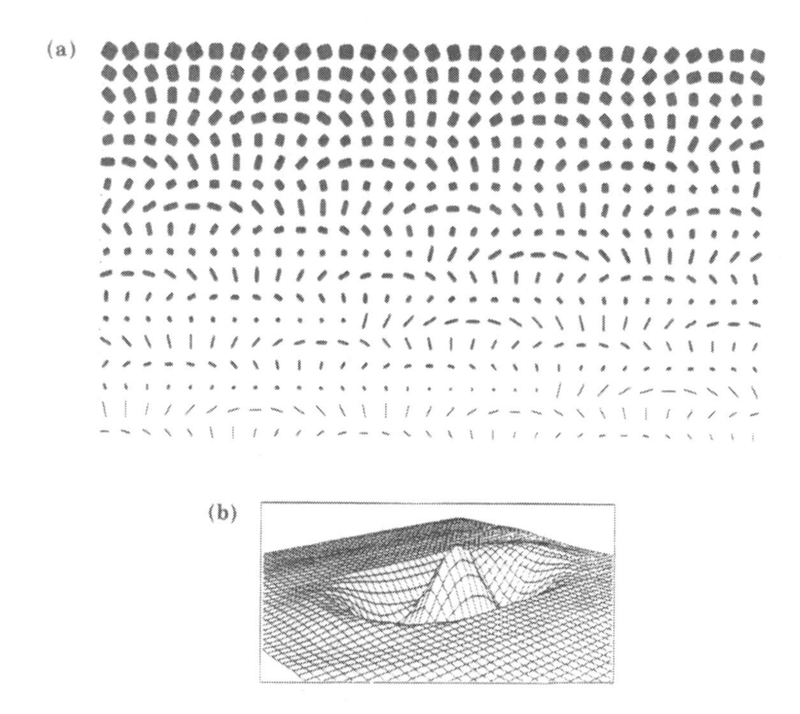

FIGURE 14.7. (a) Two-dimensional patterns used to segment image data into parts, and (b) spatial structure of a "receptive field" corresponding to one of these patterns. (Parts a and b from Pentland A [1989b], with permission.)

A biological version of this approach might use many hypercolumns each containing receptive fields with excitatory regions analogous to those shown in Figure 14.7a. The cell with the best matching excitatory field could then be selected by use of lateral inhibition. This arrangement of receptive fields and within-hypercolumn inhibition produces receptive fields with oriented, center-surround spatial structure, such as is shown in Figure 14.7b.

The major problem with such a filtering/receptive field approach is that all such techniques incorporate a noise threshold that balances the number of false detections against the number of missed targets. Thus we will either miss many of the object's parts because they do not quite fit any of our part shapes, or we will have a large number of false detections.

This false-alarm versus miss problem occurs in almost every image processing domain, and there are only two general approaches to overcoming the problem. The first is to improve the discriminating power of the filter so as to improve the false-alarm/miss tradeoff. The success of this approach depends on precise characterization of the target and so is not applicable to this problem.

In the second approach each nonzero response of a filter/receptive field is considered as an hypothesis about the object's part structure rather than being considered as a detection. One therefore uses a very low threshold to obtain a large number of hypotheses, and then searches through them to find the "real" detections. This approach depends on having some method of measuring the likelihood of a set of hypotheses, i.e., of measuring how good a particular segmentation into parts is as an explanation of the image data. It is this second, "best explanation" approach that we have pursued.

Global Optimization: The Likelihood Principle and Occam's Razor

The notion that vision problems can be solved by optimizing some "goodness of fit" measure is perhaps the most powerful paradigm found in current computational research (Hummel and Zucker, 1983; Hopfield and Tank, 1985; Ballard et al., 1983; Poggio et al., 1985). Although heuristic measures are sometimes employed, the most attractive schemes have been based on the likelihood principle (the scientific principle that the most likely hypothesis is the best one), i.e., they have posed the problem in terms of an a priori model with unknown parameter values, and then searched for the parameter settings that maximize the likelihood of the model given the image data.

Recently it has been proven (Rissanen, 1983) that one method of finding this maximum likelihood estimate is by use of the formal, information-theoretic version of Occam's Razor: the scientific principle that the simplest hypothesis is the best one. In information theory the simplicity or complexity of a description is measured by the number of bits (binary digits) needed to encode both the description and remaining residual noise; thus this new result tells us that both the Likelihood Principle and Occam's Razor agree that the best description of image data is the one that provides the bitwise shortest encoding.

This method of finding the maximum likelihood estimate is particularly useful in vision problems because it gives us a simple way to produce maximum likelihood estimates using image models that are too complex for direct optimization (Leclerc, 1988). In particular, to find the maximum likelihood estimate of an object's part structure one needs only to find the shortest description of the image data in terms of parts.

For the problem of decomposing an image into parts, it turns out that the task of finding the minimum length encoding can be reduced to a quadratic integer programming problem. To solve this problem we have developed a new numerical method that is both efficient and parallel; the technique can be described as a Hopfield–Tank neural network (Hopfield and Tank, 1985) placed in a feedback loop where the diagonal weights are initially quite large and decay over time until they finally reach the desired values (for more details see Pentland, 1988b, 1989a).

A biological equivalent of our solution method is to use a set of hypercolumns (each containing cells with the excitatory subfields illustrated in Fig. 14.7) that are tied together by a Hopfield–Tank network augmented by a time-decaying feedback loop. The action of this network is to suppress activity in all but a small subset of the hypercolumns. After this network has stabilized, each of the remaining active cells corresponds to exactly one part of the imaged objects. The characteristics of that cell's excitatory subfield correspond to the shape of the imaged part.

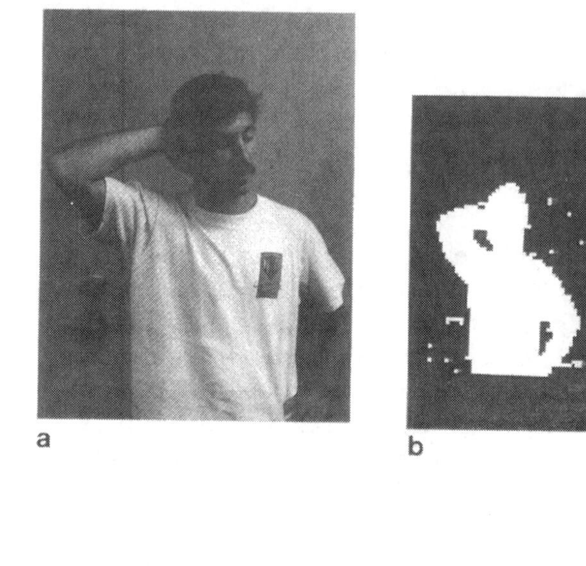

FIGURE 14.8. (a) Image of a person, (b) silhouette produced by thresholding a texture measure, (c) automatic segmentation into parts, (d) The Rites of Spring by Picasso, (e) a digitized version, (f) automatic segmentation into parts. (Parts a through f all from Pentland A [1989b], with permission.)

Some Examples: Breaking a Scene into Parts

The first example of segmenting an image into parts uses a real image of a person, shown in Figure 14.8a–c. The first step was to produce a silhouette, shown in Figure 14.8b, by automatic thresholding of a texture energy measure produced using the filters of Figure 14.1. The second step was to segment Figure 14.8b into parts, as shown in Figure 14.8c. Another example starts with Picasso's Rites of Spring, a black-and-white silhouette painting shown in Figure 14.8d. This image was digitized, as shown in Figure 14.8e, and segmented into parts, as shown in Figure 14.8f.

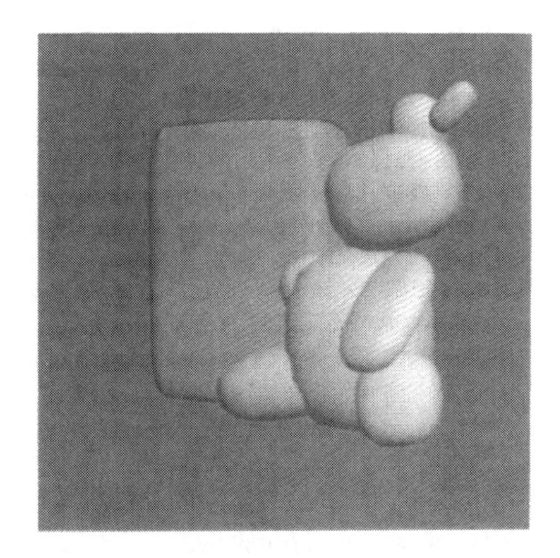

FIGURE 14.9. A three-dimensional solid model reconstructed from the range image in Figure 14.5(c).

A final example is the recovery of a volumetric model from the rabbit and book images shown in Figure 14.5a and b. From these two images a range image, shown in Figure 14.5c was created by the filtering method described above. This range image was then segmented into parts, and finally each part's position was adjusted to best fit the range data, as shown in Figure 14.9.

Summary

It seems likely that early in evolution visual systems were built from simple mechanisms that produced approximate information about the environment, and that during the course of evolution these simple mechanisms were progressively elaborated, eventually resulting in complex, accurate visual systems such as humans possess. Our research group has been trying to recreate this evolutionary progression, by asking if it is possible for simple, parallel mechanisms to produce interesting scene descriptions.

Perhaps surprisingly, we have found several simple mechanisms based on image filtering that can reliably extract point-by-point estimates of scene properties. We have also developed a simple, parallel grouping mechanism that can fuse these estimates into a high-level description of the scene's part structure. It is our hope that discovery of these simple parallel mechanisms will prove useful both for understanding biological vision and for building reliable, real-time machine vision systems.

Acknowledgment. This research was made possible by National Science Foundation Grant IRI-87-19920 and by ARO Grant DAAL03-87-K-0005

References

Adelson EA, Bergen JR (1985). Spatiotemporal energy models for the perception of motion. *J. Opt. Soc. Am.* **2**(2):284–299.

Ballard DH, Hinton GE, Sejnowki TJ (1983). Parallel visual computation. *Nature* (London) **306**:21–26.

Bove M (1989). Discrete Fourier transform based depth-from-focus. *Proc. Opt. Soc. Am. Conf. Image Understanding and Machine Vision*, N. Falmouth, MA, June 12–14, 115–119.

Crane H (1966). *A Theoretical Analysis of the Visual Accommodation System in Humans*. Final Report NAS 2-2760, NASA Ames Research Center.

Daugman J (1980). Two-dimensional analysis of cortical receptive field profiles. *Vision Res.* **20**:846–856.

Heeger D (1987). Optical flow using spatiotemporal filters. *Int. J. Comp. Vision* **1**(4):279–302.

Hopfield JJ, Tank DW (1985). Neural computation of decisions in optimization problems. *Biol. Cybernet.* **52**:141–152.

Hubel DH, Wiesel TN (1977). The Ferrier Lecture: Functional architecture of macaque monkey visual cortex. *Proc. R. Soc. London B* **198**:1–59.

Hummel RA, Zucker SW (1983). On the foundations of relaxation labeling processes. *IEEE Transact. Pattern Anal. Machine Intell.* **5**(3):267–287.

Leclerc Y (1988). Construction simple stable descriptions for image partitioning. *Proc. DARPA Image Understanding Workshop*, April 6–8, Boston, MA, 365–382.

Pentland AP (1982). Finding the illuminant direction. *Opt. Soc. Am.* **72**(4):448–455.

Pentland AP (1984). Fractal-based description of natural scenes. *IEEE Transact. Pattern Anal. Machine Intell.* **6**:661–675.

Pentland A (1986a). Shading into texture. *Artif. Intell. J. 29:147*–170.

Pentland A (1986b). Perceptual organization and the representation of natural form. *Artif. Intell. J.* **28**(2):1–38.

Pentland A (1987). A new sense for depth of field. *IEEE Proc. Pattern Anal. Machine Intell.* **9**(4):523–531.

Pentland A (1988a). Shape information from shading: A theory of human perception. *2d IEEE Int. Conf. Comp. Vision*, Tampa, FL, Dec. 3–8.

Pentland A (1988b). Automatic recovery of deformable part models. *M.I.T. Media Lab Vision Sciences Technical Report 104.*

Pentland A (1989a). Part segmentation for recognition. *Neural Comp.* **1**(1).

Pentland A (1989b). Progress toward a simple, parallel vision machine. *Optic News* **15**(5):9,26–32.

Poggio T, Torre V, Koch C (1985). Computational vision and regularization theory. *Nature* (London) **317**:314–319.

Rissanen J (1983). Minimum-length description principle. *Encyclopedia of Statistical Sciences*, Vol. 5, pp. 523–527. Wiley, New York.

Watson AB, Ahumada AJ (1985). Model of human visual-motion sensing. *J. Opt. Soc. Am.* **2**(2):322–342.

Witkin AP (1981). Recovering surface shape and orientation from texture. *Artif. Intell. J.* **17**:17–47.

15
Applied Machine Vision

PETER D. SCOTT

Introduction

Machine vision is the use of devices for non-contact sensing of x-ray, ultraviolet, visible light, infrared, ultrasonic energy, etc., to automatically receive and interpret an image of a real scene, to obtain information and/or control machines or processes.

Society of Mechanical Engineers, January 1985

Scope and Goals of This Chapter

We acquire information about the world around us through a rich complex of sensory channels. The texture of sensations that constitutes awareness of, say, a sunny clearing in the summer forest is our birthright, the product of evolutionary mechanisms refined through several billion years. In addition to the natural senses are those very much newer ones created by man for his own conscious purpose. We are driven to create artificial senses in order to extend the natural sensorium to phenomena and scales that evolution did not "instrument" for us, such as X-ray or magnetic resonance imaging, the delicate hearing-like sense of sonar, or the chemical smell sense of mass spectrometry. These nonbiological channels expose otherwise invisible domains to our systematic scrutiny. They allow us to experiment and understand, to craft new generations of technology from the previously unobserved. Most major innovations in science and technology have been preceded and stimulated by significant extensions to the natural sensorium by technical means.

But we do not create such mechanisms only to extend the reach of the human sensorium into new domains. We seek also to simulate natural sensory processes, endowing instruments with capacities that effectively mimic our own. Speech-input systems that can hear and interpret a set of voiced commands, some incorporating speaker identification, are already available. Robot gripper touch sensors with usable resolution rivaling the fingertips are under active development. And machines with capabilities that recall human vision, in some limited but useful form, have been common in industrial, scientific, and military applications for almost two decades.

Machine vision encompasses both of these aims: creation of systems that reach beyond natural biological visual limits of wavelength, sensitivity, speed, and size, and of systems that simulate specific functional aspects of human visual perception. Being free from evolutionary constraints, machine vision systems range freely beyond the visible portion of the electromagnetic spectrum to sense very much longer-wave energy, such as that used in radar imaging, and much shorter, such as ultraviolet and X ray.

Image acquisition for machine vision is not even confined to the electromagnetic spectrum; perfectly useful images are derived from ultrasonic and other compression waves, and from counting of energetic particles such as positrons and alpha particles. Clearly "vision" is understood in a broader context here: any energy source whose output can be focused onto a two-dimensional image plane, and whose spatial distribution in the image plane is determined by the objects it encounters along the way, will create an image useful for machine vision.

As indicated in the SME definition opening this chapter, the two key functions performed by every machine vision system are the acquisition of images (image capture) and the interpretation of images (image understanding). In this chapter the major approaches to these problems will be reviewed, with emphasis on methods most prominent in the current generation of machine vision systems and those on the immediate horizon. In an area as rich with divergent ideas and approaches as machine vision has become, the criterion of current usefulness is a severe test. Concepts of great promise that will surely find their place in future machine vision systems, such as neural networks, associative memories, and optical data processing, are not essential to current practice and will not be discussed. The central goal of this chapter is to present useful concepts in machine vision as a frame of reference yielding an alternative perspective on vision in general, and biological vision in particular.

Biological Vision and Machine Vision

Although the functions performed by natural and artificial vision systems have much in common, their similarities end short of a unified science of vision in man and in machine. The history of natural vision ranges over evolutionary time scales and has produced vision systems tuned to the evolutionary tasks of survival and reproduction. New generations of biological vision inherit parental structures and adapt them to changing circumstances for incremental evolutionary advantage. Redesign is gradual and "top-down" with basic molecules and primitive architectural structures preserved. In contrast, the history of machine vision is confined to a few decades, its basic building blocks are radically redesigned from the bottom-up every few years, and its goals shift rapidly as technological advances open new applications areas.

These different circumstances mitigate against full functional or structural commonality between biological and machine vision. The human retina does not operate on identical principles to the focal plane array of a solid-state camera, nor do the cells and neuronal pathways of the visual cortex closely resemble the transistor and conductor layouts in an integrated circuit. Thus even if the goals of a given vision task may be posed similarly for man and machine, it is unlikely that a successful implementation in one domain can be used as a blueprint for the design or understanding of the other. It is tempting for the machine vision designer to try to solve his problems by creating a close copy, a simulacrum of biological vision. But history and physics argue against it. Similarly, efforts to understand the flow of data and representations in biological vision structures using concepts borrowed from computer science have had limited success thus far.

Undoubtedly the gap between biological and machine vision is narrowing as life scientists unravel the complex multilayered functionality of natural vision, and simultaneously physical scientists and engineers work with neural network designs and artificial intelligence. But that gap is still there, and suggests that each be studied on its own terms and judged on its own merits. With this perspective it should come as no great surprise that what is readily available in one domain may be impossible in the other. Commercially available machine vision systems can accurately screen parts passing by on a conveyor belt at a rate at which the human eye perceives just a blur, yet no system anywhere can safely direct a vehicle down a well-marked roadway even at leisurely walking speed. There is no paradox here, only the difference in form and device between these two kinds of vision.

Although distinct, the disciplines of biological vision and machine vision still have much to contribute to one another. Though manifested quite differently, the goal of natural and artificial vision systems is after all the same: to extract useful information from images. Commonality of goal suggests that insights can be shared, if viewed from the right perspective. Machine vision design has in fact been continuously enriched by advances in mechanisms and models of natural vision [exemplified by the essential work of David Marr (Marr and Poggio, 1977) and colleagues elaborating a computational framework for natural vision]. As artificial intelligence and neural network methods move high level machine vision closer to the standard of cognitive judgment, we expect this favor to be returned by machine vision investigators. The discovery of those structures and transformations that work best in the artificial vision domain will provide important clues to the biological vision scientist searching for schema to understand high level vision processing. Just as the mathematical theory of automatic control systems has contributed significantly to understanding the dynamics of homeostasis and biorhythms, machine vision theory may well provide a framework for unraveling the abstract and very difficult problems of cognition and high level biological vision.

Origins and Current Applications

Comments on the Development of Machine Vision

Machine vision systems require both the acquisition of one or more images, and the automated extraction of useful information from them. Although the technology of image capture dates back almost four centuries to Galileo's telescope and the microscopes of Anton van Leeuwenhoek, the first applied machine vision systems did not appear until the late 1960s. What was unavailable in the interim were devices suited for automated interpretation of the image data. The electronic computer coupled a useful "brain" to the much earlier "eye" and made machine vision a reality.

Early machine vision applications reflected the much greater maturity of image capture technologies over the infant electronic computing technology. Little sophisticated data processing or decision analysis was evident, while the role of precision optics, image sensors, illumination, and careful scene framing was emphasized. Early capability was certainly concentrated on the capture side of the image capture–image understanding paradigm. A prominent first generation application was developed by Kodak Industries to inspect continuous sheets of newly manufactured photographic film for flaws (Hollingum, 1984). The film was linearly scanned across its feed direction with an infrared laser and the reflection captured and converted to a conventional electronic signal by a single infrared detector. Significant deviations from the expected signal level and expected bandwidth could then pinpoint manufacturing flaws and trigger relay circuits. By synchronizing the action of the relays with the feedstock motion, damaged sections of the continuous web could be identified and scrapped, saving the rest of the run. This system demonstrated careful attention to design of a sophisticated image capture module, and very limited decision logic applied to the image data, which was characteristic of the pioneering generation of machine vision. Tasks requiring a high level of image interpretation, such as the detection and classification of distinct parts that may appear in various orientations, were simply not feasible with computing equipment available at that time.

Although these early systems used general purpose computing devices and components, it was recognized from the beginning that effective number crunching to support image interpretation had its own unique characteristics and would work best with its own specialized hardware. Images have an inherent format unlike that of payroll records, linear programming tableaus, or other data structures for which early electronic digital computers were commonly optimized. There tends to be abundant parallelism in both image data and in the low level operations applied to image data, so memory and computing circuits taking effective advantage of this parallelism prove to be much more efficient. Another area that demanded unique hardware solutions was intraprocessor communications. The rapid pace at which even a single-camera system captures image data puts much greater demand on system bus bandwidth (the rate at which data can be moved into and out of memory) than typical nonimaging applications, stimulating the development of special purpose fast bus architectures for machine vision systems. In addition, standardized interfaces between inexpensive video cameras designed to television monitoring specifications, and the pixel formats needed for machine vision systems, were needed to drive down the price of image capture by permitting use of cameras manufactured for the much larger video market.

Growing availability of integrated VLSI (very large scale integration) design tools and fabrication services throughout the decade of the 1970s made complex custom microelectronic designs increasingly cost effective, and machine vision systems incorporating extensive specialized computing hardware in the areas of parallel processing, bus, and interface circuits soon appeared. The later generation of systems, beginning about 1980, has greatly improved data processing capability measured in speed or in memory size, most of which is directly attributable to special purpose hardware. With large dual-ported image memories and faster processing circuits comes the opportunity to view scenes at higher resolution, and make more subtle judgments, within time limits imposed by productivity requirements. Machine vision systems of this later generation are used to classify automotive part castings passing at several hundred per minute along a high speed conveyor belt, determine if the part is within specifications relative to multiple independent geometric parameters, and actuate a device to remove it from the line within seconds if it is defective. Such systems are flexible enough to rapidly "train" to new parts and new part descriptions, often by requiring only that an operator supply a clear view of the new part. Environments like this require an order of magnitude greater data flow and several orders of magnitude more computations than those of early generation vision systems offered just a few years previous.

The history of applied machine vision is rather short but has seen great advances in both power and flexibility, principally due to advances in the use of specialized computing hardware optimized for this purpose. The emphasis in early generation systems on sophisticated image capture subsystems led to relatively expensive, inflexible designs not easily reprogrammed or reconfigured to solve even closely related machine vision problems. Sharp improvements in the bus and processing speeds of the image analysis subsystems, together with the size of the image memories, are rapidly moving to bring the more mature "eyes" and the much more recent "brain" of this kind of vision system into closer balance.

Where Machine Vision Is Being Used

Machine vision has emerged from the research laboratory to the workplace, establishing firm niches in such diverse application areas as industry, medicine, and defense. The machine vision industry is currently in a phase of rapid growth, both outward to new applications areas and inward as machine vision presence in established applications areas becomes more dominant. From an installed base of zero in 1964, the market for complete vision systems rose to approximately $58 million by 1984, was an estimated $205 million in 1988, and has been projected to rise to $1.2 billion by 1994 (Ravich, 1987).

Sustaining this robust growth is the now well-documented cost-effective performance of machine vision technology when fairly simple vision-based judgments must be made rapidly and reliably at high repetition using images captured in well-controlled environments. Typical of such problems is the inspection of a flat metal part on a conveyor belt. A rigidly mounted camera triggered by timer or relay

TABLE 15.1. Current machine vision applications areas.

Application category	Specific examples
1. Industry	
1.1. Quality control and inspection	Inspect metal casting, verify fill level of bottle
1.2. Parts identification	High speed assembly line sorting of products
1.3. Assembly	Install surface mounted chips on pc boards
1.4. Robot guidance	High quality arc welding robot control
1.5. Product measurement	Report average grain size in steel ingot
1.6. Fringe measurement	Automated holographic interferometry
2. Medicine	
2.1. Automated laboratory tests	Blood counts, detection of abnormal cells
2.2. Computed diagnostic imaging	Computed tomography, magnetic resonance
2.3. Reading machines, vision aids	Page readers that output synthetic speech
3. Remote sensing	
3.1. Geophysical mapping and exploration	Identify likely drill sites for oil exploration
3.2. Weather prediction, air monitoring	Detect source sites of air pollution overload
3.3. Crop type and yield prediction	Guide futures market for specific crops
4. Defense	
4.1. Satellite surveillance	Detect ballistic launch, troop movements
4.2. Vision-based long-range guidance	Cruise missile guidance
4.3. Smart weapons and munitions	Air-to-air missiles, vision-guided bombs
4.4. Threat detection and management	Naval fleet defense systems
5. Other	
5.1. Scientific	Dynamics of turbulent flows, gait analysis
5.2. Security systems	Motion detection, intruder identification
5.3. Hazardous environments	Remote nuclear fuel rod inspection

need deliver only a single view of the well-illuminated, clutter-free scene. Interpretation may then require confirming only that the part is present, then computing geometric parameters such as its major and minor axis length, perimeter, and area for comparison with stored nominal values to determine part acceptability. Another example is the automated classification, based on simple shape and size parameters, and counting of two distinct populations of cells in a blood smear.

Table 15.1 lists the major current applications areas for machine vision technology. Excluded from this list are related applications in image processing and computer vision that share similarities to those listed, but that cannot be characterized as systems integrating image capture and automated image analysis subsystems. Also excluded are machine vision domains not yet proven outside the research laboratory, such as machine vision guided vehicular control.

Commercially available machine vision systems serving these applications may be grouped into three categories: board-level products, vision development systems, and turnkey systems. Compatible board sets, typically including frame grabbers, video cards, coprocessor cards, and coding/compression cards, permit the user great flexibility in configuring his application at minimum hardware cost. Development systems typically include a chassis containing camera and video interfaces, memory and image processing hardware made available to the user through a proprietary operating system designed to support vision applications programming. These systems require much less in-house engineering than board sets, and are better suited to fast prototyping. Turnkey systems are fully integrated "black boxes" designed to perform highly specific vision tasks with no programming in a user-friendly, typically menu-driven operator interface. These systems are ready to go right out of the packing crate, and require least in-house expertise, but are the most expensive and inflexible. Turnkey systems are often an order of magnitude more expensive than a set of board-level products with the same hardware capability. However, 6 months to a year is typically spent developing a board-level vision industrial application while the turnkey system goes to work immediately with no significant in-house development costs or manpower commitment.

The Functional Units of Machine Vision Systems

Machine vision begins with the capture of one or more images. An orderly sequence of transformations is then performed that abstracts the image information further and further from the raw pictorial domain, whose elements are the brightness levels at each point in each input image, while closing the distance to the desired symbolic domain, whose elements are classified objects and interpreted scenes. For instance a satellite image is just a pattern of intensity or brightness levels distributed on a focal plane. If the image is to be used to guide the satellite's attitude control system, several levels of abstraction from the raw brightness patterns are necessary. Stars must be discriminated from noise and labeled as stars. Confirmations of adjacent stars may be compared to stored constellation maps and the proper matches recognized. The resulting labeled sky map, together with the imaging model of the camera, may finally be used to identify the coordinates of the optical axis of the system, from which the attitude of the satellite follows. The original raw brightness levels were thus initially modified to reduce noise, then transformed to star/no-star labels, these aggregated into detected star clusters or constellations with computed location and orientation parameters, and these objects finally transformed to an optical axis with quantified orientation. At each step the elements of the corresponding data representation carry less of the original pictorial brightness information and summarize more concisely those features of the image needed to label the essential objects and their relationships.

The overall machine vision cycle may be conceptualized into six functional units, as shown in Figure 15.1. Each unit is described in a subsequent section of this chapter, with some important characteristics listed below:

- Image capture. The image may be formed from visible light or from a different region of the electromagnetic spectrum (such as infrared, X ray, or microwave). It may be of a different energy form entirely (positron emission, acoustic wave, etc.). The image may encode brightness (wave intensity or particle count),

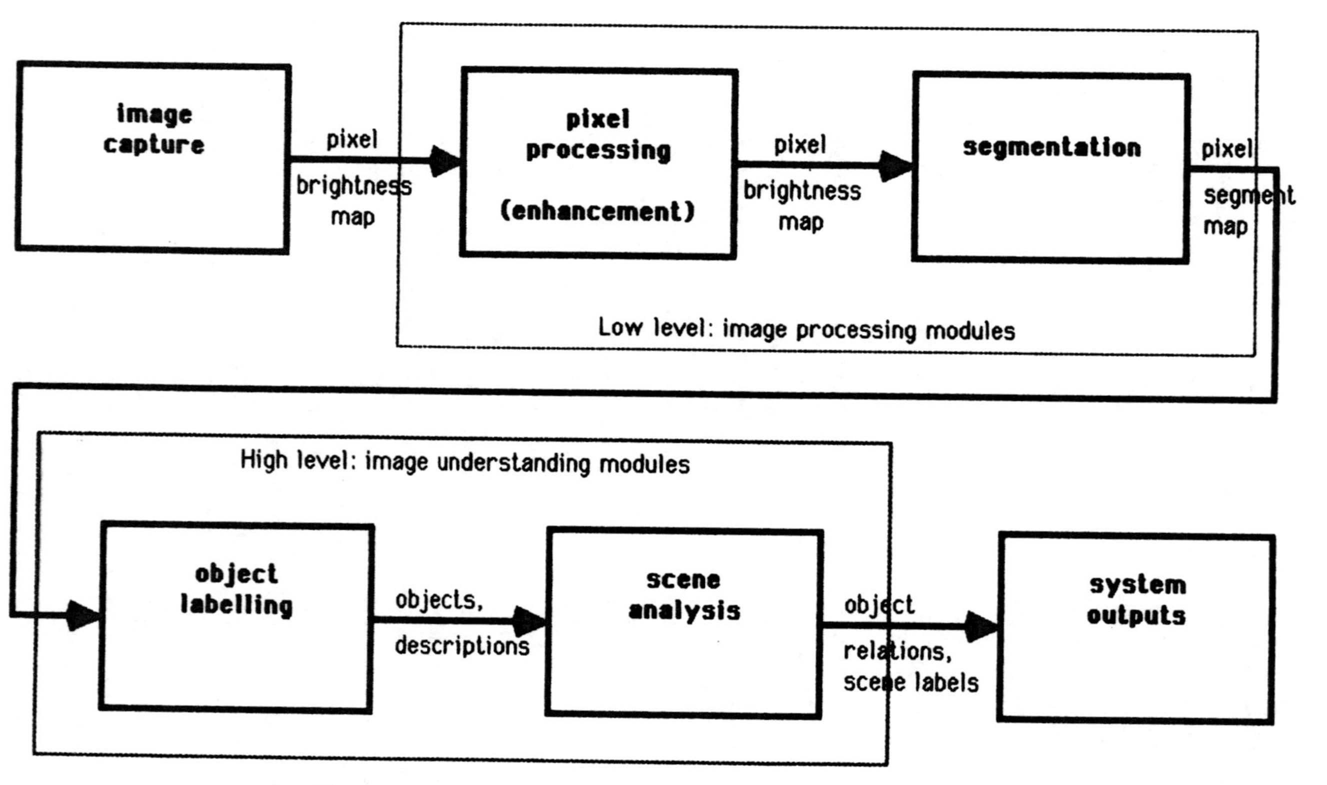

FIGURE 15.1. Machine vision system functional block diagram.

range, phase, or polarization. Each captured image is converted to a discrete set of quantized scalar values, or "pixels" (picture elements), for subsequent digital processing.

- Pixel processing. In preparation for interpreting the contents of the image, it is usually necessary to suppress noise and clutter, and enhance useful features of the image such as edges or holes in significant objects. The output of this process remains in the pixel brightness domain, and thus this image-to-image process is also referred to as image enhancement or filtering. Noise filters include low-pass, Wiener and median filters, feature filters include thresholding, edge enhancement operators, gray level rescaling, and distortion compensation resampling (interpolation of sensor data followed by nonuniform resampling to compensate for geometric distortion of the lens).
- Segmentation. Adjacent sets of pixels sharing common properties are identified and relabeled by a common segment number. Segments are commonly determined either by thresholding or by detecting the edges that form their boundaries. Segments are valuable abstractions from the pixel brightness domain to a new pixel association or "blob" domain, since blobs, bearing uniform properties in an image, frequently mark a single facet of a single object in the image. Segments form the raw material out of which objects and their features may be identified.
- Object labeling. From collections of segments, each of which is a region in pixel space, is abstracted an object, which projects its image into pixel space but is defined independently of its pixel representation. In a given image objects are revealed by their component segments, but are greater than these segments. A cubic block may present one, two, or three faces to the camera, and each face may be a rectangle or parallelogram. Regardless of its particular view-centered pixel attributes in a given "pose," it remains the same cube. Objects are typically identified by first extracting useful features from the segmented image, then matching these features against those of known, labeled object classes. Once identified, each object can be labeled by its name and then by its computed attributes. Frequently determination of the presence or the absence of an object with specific required features (e.g., length, area, number of holes) is the final required information sought by the machine vision process.
- Scene labeling. As objects can be abstracted from image segments, scenes can be abstracted from objects and their interrelationships. For instance, having identified objects of class "road," "dwelling," "field," and "vehicle" in close proximity we may classify an aerial photograph as a scene of the type "rural village." Scene labeling, by name and characteristics, is the final step in image understanding and is frequently used to establish the context for subsequent action. A robot must decide that it is a bin of parts that it is imaging before attempting to pick one of the parts out. Note that the representations useful to scene labeling are well abstracted from the earlier pixel domains. Thus hardware and software carefully tuned for scene labeling are likely to be quite different from those well suited to lower level image processing steps.

- System outputs. Vision system outputs vary widely, depending principally on the type and frequency of human supervisory intervention required in the overall task cycle. At one extreme, autonomous vision systems may output nothing more than a single relay signal to actuate the bumping of a defective part from a conveyor belt. Such unsupervised systems may additionally store and later report summary statistics. At the other extreme, systems functioning as "vision assistants" to screen image databases for specific features and then turn the images over to expert human judgment for final evaluation must output much more: a set of raw and a set of enhanced images, typically with graphic overlays showing what was found, as evidence for final human arbitration. For instance chest X rays may be screened by a machine vision system to detect small granularities or nodes that bear closer inspection by the radiologist, but that may otherwise be overlooked in the press of time or fatigue. Other classes of machine vision system outputs include detailed numerical control data to be downloaded to a robot or an assembly cell, part counts to inventory control, quality control data on the observed statistical properties of a production run to engineering, and location/orientation models to an onboard flight control system.

Image Capture

An image is a planar record of some energy field distributed in space. The energy source may be inherent within the objects being imaged, as in photos of stars or infrared mammography, or may be external to the imaging space. External illumination can reveal objects by reflection from their surfaces, by refraction as it passes through them, or by silhouetting as the illuminating field finds preferred paths around the objects. The energy field may be electromagnetic waves, acoustic waves, or populations of discrete energetic particles.

Regardless of its physical characterization, the usefulness of the image is measured by the clarity of the clues it contains, which reveals the presence and properties of significant objects in the imaging space. This may be enhanced by careful attention to the selection, placement, and intensity of the sources of illumination, the manner in which the energy field is mapped onto the imaging plane (lensing or focusing), and control of environmental factors such as competing energy sources and density of clutter or scatterers.

Each image, as it is captured, must be converted to a form compatible with analysis by digital computer, i.e., digitized. This is achieved by spatially sampling the image on a regular sampling lattice covering the image plane, and quantizing each sample (associating its value with one of a fixed finite number of distinct values), yielding an array of values called pixels. Rectangular sampling lattices are most often chosen, though hexagonal lattices are more efficient in sampling the plane and have other technical advantages as well. Beside its obvious merit of simplicity, which leads to inexpensive electronic circuit implementations and less complex computational algorithms, rectangular sampling has a major advantage over hexagonal sampling in the treatment of vertical and horizontal features. If hexagonal pixels are aligned so that horizontal edges look smooth, vertical edges will show a "sawtooth" profile. Rectangular pixels "staircase" oblique edges, but preserve smooth profiles simultaneously in the particularly important vertical and horizontal directions.

The number of discrete quantization levels available per pixel may be just two, in the binary or black-white image case, or range upward of 2^{36} (or roughly 68 billion) in some full color and multispectral systems. Since 2^B levels may be coded using B bits of memory, each doubling of the number of quantization levels requires one additional bit per pixel. The number of bits per pixel, or "depth" of the digitized image, is a central determinant of the complexity of the resulting machine vision system. Images eight bits deep (256 gray levels or colors) require eight times as much memory, and much more complex algorithms, compared to images one bit deep (two gray levels).

The most frequently selected imaging device for machine vision applications is a solid-state camera responding to visible light. Tube-type video cameras (vidicon, plumbicon) are also common (Spidell, 1987), as are cameras incorporating sensing arrays optimized for use in the infrared and ultraviolet bands. The active sensors in the camera may be deployed as a rectangular array, or gathered into a linear array. Linescan cameras are particularly cost effective when viewing a conveyor belt or other steadily moving target. X-Ray or ultrasonic imaging is frequently selected when the interior of opaque objects must be studied.

Although there has been great improvement on the image understanding side of the image capture–image understanding paradigm in recent years, it is still generally true that poor quality images will yield poor machine vision results. Noisy, distorted, or blurred images of low contrast and low resolution cannot be transformed by digital legerdemain to reliable scene descriptions. Careful attention to image capture hardware and environment is as important as the image understanding strategies in producing effective results.

Low Level Analysis: The Image Processing Modules

The mapping of one discrete image, or array of pixels, to another with more desirable features is called image processing. Image processing forms the low-level part of machine vision, that is the set of transformations whose input and output data structures adhere to the pixel structure of the image itself. The output of this set of modules is a segmented image from which symbolic rather than pixel-based representations may be abstracted.

Pixel Processing

Even with careful attention to image capture, the input image may differ in important ways from an ideal or desired image for the task at hand. Cues pointing to the location and characteristics of significant objects, while present, may not be sufficiently prominent. Other, undesirable image features (noise and artifact) may obscure these cues or make them difficult to quantify accurately. Blur, geometric distortion, and sensor nonlinearities create additional difficulties. A logical first step is to enhance the quality of the available image by a set of transformations designed to minimize the imperfections in the image capture process and maximize visibility of the useful cues. These transformations begin and end with images whose pixels take value in the original brightness domain, and are thus referred to as pixel processing (equivalently, image enhancement). Each transformation is a filter, creating a chain of increasingly enhanced images.

The specification and design of filters for pixel processing are better understood than perhaps any other aspect of machine vision. This is because pixel processing is an extension of signal processing, which has been an object of study since the birth of electronic communications over 100 years ago. A wide variety of filters, and filter design procedures, are available for the suppression of noise and clutter, the enhancement of desirable features, and compensation for nonlinearities and distortion (Wang et al., 1984).

Perhaps the simplest useful filtering operation is gray level rescaling, in which a transformation mapping input brightness values to output brightness values is defined and then applied independently to each pixel in the image. This filter is useful to correct for sensor nonlinearities, and to requantize (e.g., threshold) the image. By making the transformation space variant, differences in sensitivity between zones of the image sensor may be compensated as well. For instance the gray levels sensed on the periphery of the field may be scaled relative to the gray levels in the center if the lens design directs more light to the center of the image.

Although gray level rescaling is usually not linear, filters that do possess linear characteristics find broad application. Linear filters are commonly implemented as small kernel convolution filters or as Fourier domain filters (Mersereau and Dudgeon, 1984). In the former case each pixel is recomputed as some weighted average of its neighboring pixels, with weights and neighborhoods defining the "kernel" of the transformation. In the latter, specific bands of frequencies in the input image are identified and passed, while others are blocked. Linear filters are frequently used to reduce noise, since they possess desirable optimality properties in the RMS suppression of uncorrelated noise. They are also used extensively as feature filters, enhancing those features that can be closely linked to small kernel template shapes or specific frequency ranges. For instance linear filters that emphasize vertical, horizontal, and 45° diagonal edges can be used to synthesize an output image that at each pixel has brightness proportional to the maximum of this set. Such an image will emphasize edges with relative indifference to their orientation.

Any linear filter that reduces high-frequency noise will proportionately reduce desirable high-frequency features such as edges. To prevent loss of edge information while smoothing high-frequency noise, nonlinear small kernel filters based on replacement of each pixel by the median of pixels in its neighborhood have been developed. Median filters are effective at removing "salt-and-pepper" high-frequency noise without blurring edges of large (compared to neighborhood) sized, relatively smooth objects.

Imperfect optics can distort object shapes and sizes, particularly near the periphery of the image when a wide angle of view is being captured. By measuring the distortion as a function of position, a compensating inverse distortion function can be calculated. The sample image can then be interpolated and resampled at points computed to implement the inverse distortion.

If two images of the same image space are captured in close succession and then subtracted, stationary features will be canceled while the boundaries of moving objects and features will be emphasized. Time difference filtering is particularly useful where it is required to track a low contrast moving object in the presence of high contrast clutter, such as in cardiac angiography and radar imaging. It is vital that time difference filters operate on accurately registered images, i.e., images for which the optical axis has not shifted slightly between images due to small relative motion of the camera and image space. Misregistration leads to an undesired spatial-difference component in the time-difference filtered output, called "ghosting," which makes the desired moving features more difficult to characterize.

Pixel processing requires considerable memory and high computer instruction execution counts as full-resolution images are transformed into new images of equal size. The spatially invariant nature of many filters, however, makes these operations prime candidates for specialized hardware exploiting their single-instruction, multiple-data stream (SIMD) character. Using devices such as dual-ported memory, pipelining, and SIMD parallel processing it is not uncommon for relatively inexpensive machine vision systems to implement common imaging filters, such as small kernel convolution, at performance levels that would tax a mainframe computer costing an order of magnitude more.

Segmentation

Suppose we wish to determine which of several possible parts is currently passing under the camera on the production line. After pixel processing we still have only a regular array of brightness values, and need to determine which of these pixels belongs to the part being identified and which do not. Once this is decided, presumably the essential features of the revealed part outline can be matched against candidates to classify the part.

The process of marking each pixel not by its brightness but by a label shared with neighboring pixels that it resembles in some way is referred to as segmentation. Segmentation is a low-level task since both the input and output data structures are pixel maps, though the value of a pixel is abstracted during segmentation from its brightness to its attribute of association with specified neighboring pixels. The goal of segmentation is that each significant object in the image be concisely represented by a small number of segments, and that the background, noise, and clutter segments be as few and as distinct from object segments as possible. For instance in analyzing a LANDSAT satellite photograph for the purpose of predicting the yield of various crops, fields of uniform texture, corresponding to a single crop, would be useful segments, together with segments corresponding to water systems, rangeland, uncultivated land, etc. In this case the measure of similarity binding pixels into segments is their common texture. Other similarity measures include gray level or color, colocation inside detected boundaries, and segment template matching, as described below.

The simplest segmentation is into maximal connected sets of pixels of uniform gray level. Use of this strategy is usually limited to binarized images of flat objects originally captured under very controlled circumstances. With images greater than one bit deep, minor variations in illumination or surface finish tend to proliferate the number of segments unacceptably, as does the surface contouring visible with reflected illumination from truly three-dimensional objects. But where the conditions of binary data, flat objects, and controlled environment are met this simplest of segmentation strategies can be very effective (e.g., backlit, opaque, well-separated bottles on a conveyor belt).

Segmentation by the generation of maximally connected sets of equal gray level pixels may be generalized in several ways to accommodate more generally realistic assumptions (Haralick and Shapiro, 1985). The pixels bound into an individual segment need not be of identical gray level, but similar with respect to a suitable measure. The set may be nearly connected, useful when noise and thresholding disrupt the chain of pixels within a segment at a narrow bridge, or completely unconnected, as when detecting a "stars" segment in an astronomical photograph. These segmentation strategies are referred to as region growing. Once an initial partition of the pixel map into segments has been created, it may be recursively refined by a process known as split-and-merge. Adjacent segments may be merged together if they are sufficiently similar, while individual segments may be split if inadequately homogeneous, or if they contain subsegments more similar to adjacent segments than to each other. Viewed as a process of successive refinement, split-and-merge may be initiated from any convenient initial segmentation, such as the maximal connected set of uniform gray level pixels described above, or even by considering each pixel as an individual segment. In designing split-and-merge algorithms, care must be taken to define the split criterion in such a way that the number of segments does not grow very large (fragmentation) or the partitioning begins to cycle periodically.

An alternative to region growing is to define segments by detecting their bounding edges. Although there may be considerable gray level variability within a single segment, making the proper region-growing similarity measure difficult to construct, edge strategies can be very powerful since edges are high contrast features and segments may be simply defined as those connected sets enclosed by edges. Consider the image of a rubber ball taken in sunlight. For machine vision purposes the "ball" pixels in the image should ideally be abstracted to a single segment since there are no useful features on the surface of a sphere. But the constantly changing relative orientation between the outward pointing normal of the ball's surface and the line of sight to the sun will produce a wide variety of gray levels in the resulting image, brightest where the inner product is largest. This gray level variation may contain useful cues for certain tasks, such as recovering surface orientation and reconstructing the three-dimensional contour of the ball from this one photograph ("shape-from-shading"), but is not helpful in segmentation. Region-growing techniques, looking for close similarity between adjacent pixels, will tend to settle on several roughly annular segments to represent the ball. If however the image is initially screened for edges, the relatively smooth internal variation due to shading will fail any reasonable edge criterion while the actual bounding outline of the ball will be so labeled (assuming adequate contrast between the ball and its background segment). Once this bounding edge has been determined to enclose a single connected set of pixels, this can be marked as a single segment.

Segmentation by the detection of bounding edges usually operates on an image that has been pixel processed with a feature filter designed to enhance edges and suppress the smooth gray level variation internal to segments. Edges in these images, while more prominent, tend to be broken into disconnected chunks of variable thickness, accompanied by many small false edge artifacts. While the human eye can often easily visualize out of these data the true edges of the significant object segments, there is no general agreement on how this is done by the human observer or how it should be done by a machine. Useful techniques include edge following, in which an edge is grown outward in its most likely direction while keeping its width constant, edge segment chaining, in which regions marked as edge segments are connected and merged until additional merging results in unacceptable modification to the image, and Hough transform. In the latter case the image is mapped from pixel space to parameter space, whose coordinates correspond to the model parameters of the set of bounding edges being sought. For instance, if linear edges are being sought, the parameters would consist of the angle of the line and its separation from the origin. High values in a given cell in parameter space then indicate the presence of the corresponding edge in the image (Duda and Hart, 1972).

As suggested in an earlier example, texture can also be used as the basis for segmentation. Given an aerial view where a forest meets a field planted in mature corn, there may be considerable difficulty using edge-based segmentation. The local contrast across such a boundary is not necessarily greater than average local contrast within either of the segments. Edge enhancement filters are liable to find edge elements everywhere. What changes as the segment boundary is crossed is not local contrast or average gray level, but texture. Textures may be classed as deterministic, such as the periodically repeating fabric pattern of a hound's-tooth jacket, or stochastic, such as our wood lot and corn field. Once the textures have been identified and modeled, their relative fit to a given region in the image may be computed and used as the basis for segmentation.

Most of these segmentation strategies result in algorithms with more branching and less regular, parallel structure than the pixel processing filters described previously. Less has been achieved in the effort to reduce segmentation to highly efficient special purpose hardware structures driven by stable algorithms. For this reason, those segmentation strategies with maximum inherent parallelism and least branching, most notably threshold-based segmentation and the Hough transform, are favored by systems designers looking for opportunities to maximize throughput. Perhaps in a forthcoming generation a full spectrum of "segmentation chips" will be available, but for now machine vision designers must work around the lack of segmentation standards and hardware.

High Level Analysis: The Image Understanding Modules

Vision is the process of capturing images and interpreting their contents. Each image contains cues bearing on the identity and properties of objects lying within the field of view, and how these objects are juxtaposed. Recovering a scene description from the image database, that is the location, label, and description of each significant object, is called the "inverse imaging problem." Given an explicit imaging model (which specifies the geometric and optical parameters of the image capture subsystem, the illumination and light scattering properties of the medium as required to predict the pixel brightness map for any scene) the inverse imaging problem requires recovery of the object labels, properties, and interrelationships from the image database. Every vision task requires at least a partial solution, to varying degree, of the inverse imaging problem. Whereas low level processing prepares the way by enhancing the cues pointing to significant objects while suppressing irrelevant or misleading cues (due to distortion, noise and clutter), completing this task requires moving from the tangible pixel domain of the image processing modules to the abstract labeling domains of the high level image understanding modules: object labeling and scene analysis.

The task of recognizing objects and their interrelationships is different in kind from the low level processing tasks discussed previously, and more difficult. The input and output of low level processes are both pixel maps, usually related by simple local transformations. Thus the basic data structures required to perform these operations are of the same size and layout as the image itself, and together with the usually straightforward, densely repeated operations lend themselves to clear-cut programming logic. In contrast high level operations require data structures representing the objects that lie beyond the pixel maps, and there is no general agreement how objects in the real world may be best represented in a digital computer. Objects may be described by their topological, their geometric, and/or their gray level (metric) attributes. They may also be characterized using surface patch models that describe their bounding surfaces, by space-filling models consisting of primitive volume elements or "voxels," prisms, or generalized cylinders. The features and models of these representations range from "flat" or two-dimensional, to characterizing the three-dimensional orientations of those visible surfaces only within a specific view, to being fully three-dimensional (describing each object in the scene in three-dimensional object-centered coordinates) (Ballard and Brown, 1982). Even once a specific representation scheme is chosen, it is still not clear how to perform basic understanding tasks such as the selection and estimation of key features or the detection and characterization of significant objects (see also Chapter 13).

As described in the previous section, low level entities such as image segments, defined in terms of their visible pixel space properties, can be unambiguously computed directly from image gray level data. For instance, segments may be derived from a region growing criterion, then improved by split-and-merge techniques. Unlike segments, objects are not defined directly by their pixel space properties but as entities in the real world that have been ambiguously projected onto the pixel domain to be registered in the image. Clearly labeling of objects is a much more difficult task than the labeling of segments. Pixel maps contain segments, but they do not contain, only suggest, objects. The ambiguity inherent in object labeling may derive from the inability to see around to the back side of visible objects or from the occlusion of part of one object by another. It may additionally be caused by resolution limitations or blur that reduce visibility of small but significant features. The lack of adequate illumination or of shadowing of key surfaces may also make object recognition problematic.

Resolving the ambiguities inherent in inverse imaging is equivalent to solving an ill-posed problem (Poggio and Torre, 1984). This difficulty arises in many inverse problems for which the corresponding forward problem fails to be one-to-one, or fails to be onto. The forward imaging problem is not in general one-to-one, since several distinct object distributions may result in the same image due to occlusion etc., nor is it onto, since certain incongruous images cannot be derived from any possible object distribution. Some well-known optical illusions stand as clear demonstration that the interpretation of images is neither one-to-one nor onto. Line drawings that seem to flip spontaneously between two different underlying interpretations demonstrate failure of one-to-oneness, and impossible objects such as those conjured up in the woodcuts of Escher vividly portray failure of onto-ness. The defining characteristic of an ill-posed problem is that its solution does not depend continuously on the input data. Indeed, even the basic properties of existence and uniqueness of the solution of an ill-posed problem may change with small changes in the data. This is a highly undesirable characteristic for machine vision, since minor fluctuations in the input data are inevitable and the system output should not fluctuate abruptly, changing its scene interpretation radically in response to low level noise or calibration drift. Ill-posed problems may be regularized by choosing an auxiliary criterion by which a set of consistent candidate solutions may be ranked, such as a minimum-norm criterion. They may also be approached by adding constraints to the range of the inverse problem that limit the number of possible interpretations available for a given noisy image, ideally to a single feasible object distribution for a single noisy image. Both these approaches are important in image understanding.

Object Labeling

Object labeling is the process of classification and characterization of significant discrete entities within the field of view. The object labeling module, present in all machine vision systems, operates on pixel maps imported from low level processing steps such as enhancement and segmentation. At one extreme, object labeling may require nothing more than passing on segment labels as object labels, along with calculation of a few simple geometric characteristics such as the centroid or area of the object. At the other extreme the gray level, surface orientation, and segment pixel maps may all be used to derive high-dimensional model-based feature sets whose interrelationships are encoded in a graph structure that is then compared against stored subgraphs for object location and labeling.

The first step in object labeling is classification. For instance, if several different kinds of machine parts may be passing on a conveyor belt, we need first classify a part as a gear before further characterizing it by counting its teeth or measuring its pitch. Viewed as a problem in pattern recognition, the essential steps in object classification may be identified as the selection and extraction of features, followed by the classification of the resulting feature vector according to statistical or structural criteria. The features may be computed directly, such as the perimeter or the number of connected segments in an object, or by first fitting the data to models, such as modeling a curved edge by its best polynomial approximant of given order and then characterizing it by the corresponding coefficients. Once an object is classified, any additional characterizations required for objects of that class may subsequently be computed using methods appropriate to that object class. For instance, a circle may be characterized by its diameter computed as four times its area-to-perimeter ratio, and its eccentricity, computed as the ratio of lengths of its major and minor axes.

Object labeling strategies are perhaps most naturally grouped according to the required "depth of interpretation" measured by the dimensionality of the features and models used to support the classification. As depth of interpretation increases, the system must detect and make judgments on features that relate more closely to the complete three-dimensional nature of the object and less on the features immediately apparent in the raw image. With increasing depth of interpretation comes the ability to perform more robust classification, for instance, discriminating a rounded dome from a flat disc even though each has the same circular silhouette. On the other hand the price paid for increased depth of interpretation is severe: more complex algorithms operating on higher dimensional data structures, longer execution time, and less well-understood performance characteristics.

- One-dimensional labeling is useful when each object class is distinguished simply by a length feature in a single scan line. After perhaps thresholding, only one-dimensional run lengths need be computed, presenting a very simple and rapid computational cycle. Consider, for instance, a binary line scan camera looking straight down at an assembly line with scan direction across the assembly line. Suppose the linear field of view of this camera on the surface of the line is illuminated at an oblique angle by a single "light stripe" (like that between two venetian blind slats in the late afternoon). When an object comes along the line, the camera sees a run of dark pixels corresponding to the width of the object, since the light stripe follows the surface of the object away from the linear field of view of the camera. Each scanline can be thought of as a one-dimensional image, with features corresponding to runs of black pixels and white pixels. If the cross-sectional width of each object at each point along its length characterizes its object class then one-dimensional classification is adequate. Since the width of most realistic objects typically tends to vary considerably along its length, and this measure is also sensitive to angular alignment of the object on the line, one-dimensional classification has limited application. Where applicable, however, such as in the determination of the presence or absence of a part at a given time, its simplicity and speed often make it the method of choice.

- Two-dimensional labeling is based on features and models that can be represented in a plane. Since the images from which these structures must be derived are also described in a plane, compatibility between input and derived data structures frequently leads to efficient algorithms for this class of methods. Two-dimensional features may be of several types: topological features such as the number of holes and connected segments, geometric features such as positions of points of maximum curvature along the bounding segment edges, metric features such as area, perimeter, or aspect ratio, and *xy* moments such as the centroid. Features may also be derived from local properties of segments of the object, like the distance between centers of a pair of holes drilled through a metal plate. The ideal circumstance for use of two-dimensional classification is the case of well-separated (nonoccluding) diffusely backlit objects distinguishable by their silhouettes and presented to the camera in one of a limited number of stable "poses." Examples include the high-speed identification of bottled products by their silhouettes, and quality control inspection of circuit features such as conducting traces and drilled holes in printed circuit board artwork. Reflected (frontlit) rather than backlit illumination is often necessary due to an opaque or cluttered background surrounding the objects to be classified. Reflected illumination can be accommodated as long as adequate contrast between object and background is available, and the surface finish characteristics of the object are taken carefully into account in the design of the image capture environment and algorithms. For instance, an image of a lake taken from a lowflying aircraft has a very different distribution of gray levels on a still bright day than on an overcast windy day producing some whitecaps. Highly polished metal parts will exhibit striking gray level changes, which make two-dimensional segmentation difficult, since physical features will not tend to correspond even approximately to segments exhibiting uniform gray level.

Models for two-dimensional classification are based either on representing the boundary of the object or characterizing the way it fills the plane. Exact yet efficient boundary coding for two-dimensional objects may be accomplished using a chain code, in which a connected and thinned boundary is described by a series of counterclockwise movements in the plane from (say) the right-most pixel within the boundary. Each movement along the boundary can be coded by three bits of data, since there are only eight directions to the nearest neighbor on the boundary (or two bits if four-nearest neighbor connectivity is used). The sequence of directions of movement then encodes the boundary. Boundary models based on Fourier descriptors, splines, and Bezier curves (Chin and Dyer, 1986) have also been extensively used. Exact space-occupancy coding for two-dimensional objects may be efficiently achieved through the use of quadtrees. Suppose an irregularly shaped object is contained in a square region of the plane. Quartering this region, denote each subregion by the symbolic value "white" if there are no object pixels in that quarter, "black" if all the pixels in that quarter are object pixels, or else "gray." Further quarter all gray regions, continuing the labeling recursively until no deeper gray regions remain (at full, single pixel resolution all quarters are either black or white). Where the exact space-filling description of quadtrees is not needed, model-based shapes generated from unions and intersections of simple geometric constructs such as rectangles and figures of revolution generate very efficient space-filling representations, particularly of smooth objects.

Once the basis for object representation has been chosen and features computed, matching feature sets to a "vocabulary" of reference objects remains to complete the object classification task. For two-dimensional labeling, template matching is particularly attractive. Images of each reference object are used as convolution kernels, with object classification and localization based on peaks detected in the resulting convolution image. Template matching is equivalent to correlating each reference object to the object being classified, and at each possible location in the plane.

A limitation in template matching is that the correlation between object and template is highly sensitive to both scale and orientation. Thus, for instance, as a target grows larger in approaching the camera, the template match with its less magnified reference image will become less distinct. A simple technique for reducing scale and orientation sensitivity is to normalize the object representation with respect to these parameters before matching. The object can be scaled to unit size (fixed area measured in object pixels or fixed bounding box size) and rotated so that its main axis has a standard alignment. Matching then proceeds against similarly normalized templates. A potentially less manageable problem is that a given real-world object will map into many different two-dimensional images depending on what facet (planar projection) of the object will be recorded from the selected camera location. If the object always presents one of a very limited number of facets, such as a rotationally symmetric bottle that will always present the same pose to the camera, or a machine part that has only a few stable configurations on the conveyor line, then template matching may proceed as previously described, with each facet included separately in the vocabulary of reference images.

Other matching techniques useful for two-dimensional object classification operate in more abstract matching domains. Statistical pattern matching is commonly framed in "feature space," the n-dimensional space whose coordinates are the values of n scalar features chosen to represent the object. Clustering and discriminant function analysis may be employed to discover the best feature space match within a set of reference objects (Mantas, 1987). Graph matching employs a representation of the object as a connected graph whose nodes represent features and arcs represent association between features. Although these methods are more tolerant of minor variations in lighting and shape they are more computationally demanding. Certainly the simplicity and effectiveness of template matching in appropriately controlled conditions recommend it for two-dimensional object classification, with statistical pattern analysis and graph-matching useful where pixel-based template matching is inadequate.

- Two and one-half-dimensional labeling methods employ features and models that interpret objects as collections of visible surfaces. Features defined here that are without equivalent in lower dimensional methods include range, surface orientation, and surface relief texture. Since lower dimensional models are flat and do not support the concept of one object being located "behind" another object, the present case is the lowest dimensional setting in which occlusion may be systematically handled. Other useful capabilities of these methods include the classification of objects that may present many possible facets to the camera, and those objects whose illumination, surface relief, and surface finish makes flat two-dimensional segmentation into useful segments impossible.

The clearest data from which to extract two and one-half-dimensional (visible surface) structures are range images. In such images range to the nearest opaque object along the line of sight is sensed as a function of pixel location in the image plane. This can be done using time-of-flight measurements based on laser optics or on ultrasound. Range can also be determined directly from holographic images or by the use of structured light techniques in which a grid pattern is projected onto the object and its range image determined from the deformation of the grid (Mantas, 1987).

Where only the usual brightness images using unstructured light are available, surface orientation and range may be still inferred from various cues using methods known collectively as "shape-from-X" (for instance, shape-from-stereo). Taking its cue from biological vision, shape-from-stereo algorithms operate on simultaneous stereo pairs of images, first matching to get corresponding points in the image pair and then computing range by triangulation. A related technique, shape-from-photometric-stereo, uses differences in appearance created by changing the direction of illumination, rather than spatial triangulation, to derive surface information. Shape-from-shading uses illumination and reflection models to predict how the brightness of a surface patch will vary with its surface orientation, and then computes a best fit to the actual brightness data, thus inferring orientation from a single image. The distribution of orientation vectors over the object may then be used directly as features, or integrated to compute distributed range estimates. For instance, it is frequently difficult to discriminate "orientation edges," which represent boundaries of three-dimensional objects, from "brightness edges," which may be formed by a shadow falling across a flat segment. Using surface orientation, a brightness edge in the image along which the computed orientation is perpendicular to the camera line of sight specifies an object boundary, not just a shadow or internal edge. Other shape-from-X methods include shape-from-motion, shape-from-contour (deriving the surface shape of an object from the shape of its silhouette), and shape-from-texture (Brady, 1982; see also Chapter 14).

Whether the resulting data are represented by a surface orientation ("needle") map or by a range map, surface features must be extracted before matching and object labeling may be accomplished. As in the two-dimensional case, these features may be computed directly from the data or may be based on models. Direct features include surface area, average surface curvature, and surface smoothness (variability of surface curvature). Model features are computed by fitting the object using a selected model set, then characterizing the object by the resulting best-fit model parameters. Surface patch models use geometric surface constructs, such as ellipsoids, paraboloids, or multinomials to describe each region of the surface. Ribbon models describe objects not by their surface "space-filling" or patch properties but by the boundaries between patches. Consider describing a surface by making a wire frame model of the surface and then characterizing the shape of the segments of wire as parameterized curves in space.

The efficiency of template matching is a major incentive to the use of two-dimensional methods. The greater complexity of view-centered collections of surfaces makes template matching generally ineffective as a matching strategy in the two and one-half-dimensional case. For a given object in the vocabulary, a great number of view-dependent templates would typically be needed, and the template match would proceed over three-dimensional space, requiring many more calculations. Feature space methods and graph matching may both be employed as described previously, though the feature sets and the directed graphs describing the objects tend to be more complex in the present case.

- Three dimensional labeling methods use object rather than view centered descriptions and thus result in the most complete and "objective" descriptions of the region of space imaged by the camera. While a visitor to a museum can get a sense of a sculpture from a single perspective, it is only by walking around the object and permitting an integrated, view-independent perception of the statue that the fullest impression is gained. This is particularly important when studying groups of objects which may obscure one another in certain views but each be in turn revealed as we walk around or through the object collection. Three dimensional methods are also in a sense the most intuitively appealing basis for object description, since rigid objects are indeed three dimensional and permit the most natural model-object correspondence.

While attractive from completeness and correspondence arguments, three-dimensional labeling methods are based on models and features that are furthest removed from the primal data domain of pixel brightness values in single images. These methods are predictably most demanding of imaging and computing resources (multiple images, large data structures, complex algorithms) and least well developed for machine vision application at the present time. Much of the early research in three-dimensional labeling was devoted to "microworld" environments, in which highly controlled laboratory conditions and primitive block-like objects greatly simplified the labeling problem. Although building a necessary foundation, results based on microworlds constraints do not well describe most applications environments and the results have been of little immediate usefulness.

Robust sets of three-dimensional objects may be represented by space-filling or boundary models (Srihari, 1981). Exact space filling may be efficiently encoded using octrees, the three-dimensional extension of quadtrees in which eight three-dimensional octants are labeled white, black, or gray at each level of spatial discrimination. Space-filling models include those built from union and intersection of volume primitives (cubes, spheres, prisms, etc.) and those based on the generalized cylinder. A generalized cylinder is built by first selecting an arc to serve as its spine. At each point along the spine a cross-sectional shape is defined. As the sequence of cross-sectional shapes is wept along the spine of the generalized cylinder its volume is defined. Boundary models, on the other hand, describe a three-dimensional object by its set of bounding surfaces and their interrelationships. Perhaps the simplest boundary model consists of purely planar surfaces, which can then be parameterized by their linear bounding edges. Other boundary models based on surface patch descriptions, as described in the previous section, have proved useful.

The principal difficulty in three-dimensional object labeling is not the lack of suitable representations and models, but rather the effective matching of these structures against the set of reference objects. Either the set of data images must be combined and abstracted upward into three-dimensional object-centered coordinates, with matching done in object space, or candidate objects must be projected down into two-dimensional image space and the matching done there (cf. Chapter 13). Since real world objects cannot typically be characterized by a small set of simply extracted features, matching in object space normally involves a complex and resource-consuming graph search. Features such as bounding arcs, faces, and corners are computed from the image-based representations and linked into a directed object graph (Besl and Jain, 1985). Similarities between subgraphs of this graph and subgraphs of the stored set of reference graphs indicate tentative matches that can then be further refined. This process is significantly simplified if images are compared to images, rather than objects to objects, since images may be matched using more direct "flat" techniques described earlier. The price paid for this simpler matching downward in the image space is the proliferation of images that may be needed to describe a single object. For instance there are approximately 100 views of an object that are displaced at least 30° apart. A promising compromise is to use the computed three-dimensional object features to constrain the set of candidate reference objects and points of view from which these reference objects may be being viewed, and then project only this limited number of views down into image space for subsequent matching. But however done, three-dimensional matching methods are complex and costly. Those applications in which three-dimensional methods are most necessary, such as robot vision and autonomous vehicle guidance, have been among the last to mature.

Scene Analysis

All preceding steps in the machine vision paradigm, both low and high level, have been directed toward the systematic discovery and characterization of those objects evident in the set of images. With object labeling complete, the contents of the images have been catalogued and interest may shift from content to meaning. In the most straightforward case the significance of a scene (labeled collection of objects) is immediately apparent from the object labels themselves, such as an automated part classifier on a high-speed conveyor belt. Having identified a part as a gear, and characterized the number of teeth on the gear, the part is classified and the scene immediately "understood" without further analysis. Whatever system outputs are needed, such as control signals to be sent to the belt drive to route the part properly and an updated count to the inventory database management system, can now be computed. However, in other applications the meaning of the scene is derived from the output of the object labeling module only after an additional, extensive process of abstraction and classification. A satellite photograph revealing geological objects including specific rock formations and drainage basins might suggest the presence of a valuable ore deposit, but the same objects in slightly different geographic relationship might not have this significance at all.

The process of determining a label (classification and numerical characterization) for the entire scene is called scene analysis. Note that labels may also be associated with objects, as described in the previous section. Here we consider a label describing the entire scene. This label is the broadest abstraction of the machine-visual process begun with the acquisition of one or more images, and serves as the final "action summary" used to drive actuators and other system outputs. Overall, a process of data compression begun with low-level processing of large volumes of pixel data into many fewer features and segments, and these into even more compact object labels, is completed as the objects and their relationships are summarized into a single scene label sufficient to drive the required outputs.

At the furthest remove from standardized pixel domains, scene analysis in its methods and structures is the most abstract, the most application-specific of the machine vision modules, and the least well understood. It is here that the highest "cognitive" functionality of the system is found, and as in biological vision, these functions are the most elusive to understand and difficult to implement. Applications demanding high levels of scene analysis have proved most refractory and still largely remain in research laboratories. Most successful machine vision applications rely on simple scene analysis based either on logic tables or scene graph matching. In the former case the scene label is decided as a directly computable logic function of the object labels. For instance, if objects classified by the labels subassembly 1 and subassembly 2 are both present and in their expected orientations, then send an enable signal to the assembly robot giving the subassembly location coordinates. The scene label in this case would be something like "OK for assembly processing," and include some location descriptors taken directly from the object labels. A scene graph, on the other hand, is a directed graph whose nodes are the objects (including their labels) and whose arcs are the relations between objects. For instance object A may lie on object B, or objects A–F may be connected together to form a rectangular frame. The scene graph may be matched against stored reference scene graphs to classify the scene and extract its complete characterization (Duda and Hart, 1973). Much has yet to be discovered concerning efficient ways to encode scenes, particularly on the incorporation of prior knowledge, constraints, and learning. Scenes approaching the diversity and complexity of "natural scenes" will continue to present great challenges in the years ahead.

Trends and Conclusions

Machine vision consists of a family of modules and techniques of proven value in applications such as quality control inspection, mensuration, remote guidance, and automated manufacture. Continued progress in image acquisition subsystems, in computer hardware and software, and in the availability of increasingly effective integrated vision application systems will enhance its attractiveness in new application areas and further secure its position in current ones. We conclude by mentioning some trends in these areas with potential significance for coming generations of machine vision equipment.

Image acquisition systems for machine vision are becoming larger, faster, and more versatile, while a continuing trend of price reduction makes them more cost effective at a given performance level. Solid-state focal plane array cameras based on charge transfer devices (CCD, CID) have largely superseded tube type cameras for industrial applications in the last decade, due to advantages of size, weight, power, ruggedness, low bloom, and high sensitivity. Megapixel cameras (1024×1024 pixels and beyond) that operate at video rate (at least 30 frames per second) with image depth of at least 256 gray levels or colors are expected in the next 2–5 years. Active range imagers, particularly those based on scanning laser rangefinders, may be expected to become faster and more highly resolved in the coming decade as they increase in acceptance (Nitzan, 1988). For high precision measurement and inspection of small structures (such as integrated circuit masks and chips), holographic and interferometric ranging using laser illumination is a promising new route into the domain of three-dimensional measurement and representations.

Vision system computer-related hardware may be expected to continue the trend toward specialization and high performance already apparent in the current generation of products. On the chip level, continued improvement in frame memory design will be needed to support mass data transfers and high-speed buses. For instance, a megapixel camera with 30 frames per second and one byte per color primary will have to move at least 90 megabytes of data per second, stretching both bus and memory technology. Specific signal and image processing functions, such as the Fourier Transform and "morphology filters," until recently limited to software implementation are now available as chip sets with greatly improved response times. For instance, the Zoran 34161 coprocessor will compute a 1024 point complex Fast Fourier Transform in 2.4 msec, permitting frequency domain pixel processing in many realtime applications that could not utilize this technique at slower software response times. On the system level vision-specific architectures that offer extensive parallelism, vectorization, and pipelining (simultaneous calculation of multiple steps in a problem) (Maresca et al., 1988) while retaining the flexibility needed for imaging applications are coming to dominate all but entry level designs. Inexpensive read–write optical disk systems with capacity measured in the thousands of megabytes, and optimized for image-type direct memory access mass data transfers, have recently become commercially available. These devices may be expected to have an immediate impact on the size and complexity of the "knowledge base" that will be made available to machine vision systems, particularly extended libraries of reference objects for matching, and more detailed and realistic constraints for scene analysis.

In the domain of software several current trends may be expected to develop momentum in the next few years. The depth of interpretation, frequently centered at two dimensions in current industrial applications, will more commonly operate at two and one half dimensions and in specific areas such as robot vision and automated vehicular control, reach to three dimensions (and beyond, to four-dimensional space–time sequences) in the coming decade. "Flat" vision problems have by and large been solved, while many challenges remain for view-surface and object-volume methods. Increased depth of interpretation frees the designer to operate with richer vision data and less stringently constrained environments, moving applications from the highly controlled environment of (say) the conveyor belt, to the cluttered and dusty factory floor. Artificial intelligence, neural networks, and expert systems will enhance the cognitive or high-level layer of machine vision as "knowledge-based vision" gradually moves from the laboratory into systems and applications (Rao and Jain, 1988). Improved understanding of the intimate relationship between measurement and model is permitting resolution of object features well below apparent physical resolution limits, and continued development of these superresolution and subpixel algorithms will be important for many applications involving precision gauging. "Data fusion" concepts will become increasingly important as multiple images (perhaps from several distinct cameras and locations) must be combined with data from disparate sensors such as tactile and point-distance sensors to create a large heterogeneous database from which vision-based interpretations are to be extracted.

Finally, it may be appropriate to close on a cautionary note: what not to expect. Cameras with focal plane arrays approaching in either density or number the photosensitive cells of the human retina cannot be designed using current technologies. The human observer will continue to sense, and act on, richer detail than the camera can provide the computer. Also, we cannot expect difficult open questions concerning the representation, interpretation, and updating of knowledge to be settled soon. Thus, higher level functionality such as scene analysis for natural scenes, which depends on efficient use of a deep and general "world" knowledge base, will continue to be sharply limited for the foreseeable future. In sum, although machine vision is currently producing effective solutions and continues to develop rapidly, a general-purpose simulacrum of the human visual capability should not be expected anytime soon.

References

Ballard DH, Brown CM. (1982). *Computer Vision*. Prentice-Hall, Englewood Cliffs, NJ.

Besl PJ, Jain RC. (1985). Three-dimensional object recognition. Assoc. Comp. Machinery Surveys **17**:77–145.

Brady M. (1982). Computational approaches to image understanding. Assoc. Comp. Machinery Surveys **14**:3–71.

Chin RT, Dyer CR. (1986). Model-based recognition in robot vision. Assoc. Comp. Machinery Surveys **18**:67–108.

Duda RO, Hart PE. (1972). Use of the Hough transform to detect lines and curves in pictures. Commun. Assoc. Comp. Machinery **15**:11–15.

Duda RO, Hart PE. (1973). *Pattern recognition and Scene Analysis*. Wiley, New York.

Haralick RM, Shapiro LG. (1985). Image segmentation techniques. Comp. Vision, Graphics Image Process. **35**:100–132.

Hollingum J. (1984). *Machine Vision: The Eyes of Automation*. IFS Publications. Springer-Verlag, Berlin.

Mantas J. (1987). Methodologies in pattern recognition and image analysis—a brief survey. Pattern Recog. **20**:1–6.

Maresca M, Lavin MA, Li H. (1988). Parallel processing for vision. Proc. IEEE **76**:970–981.

Marr D, Poggio T. (1977). A theory of human stereo vision. AI Memo No. 451, AI Laboratory, MIT.

Mersereau A, Dudgeon D. (1984). *Multidimensional Digital Signal Processing*. Prentice-Hall, Englewood Cliffs, NJ.

Nitzan D. (1988). Three dimensional vision structure for robot applications. IEEE Transact. Pattern Anal. Machine Intell. **10**:291–309.

Poggio T, Torre V. (1984). Ill-posed problems and regularization analysis in early vision. AI Memo No. 773, AI Laboratory, MIT.

Rao AR, Jain R. (1988). Knowledge representation and control in computer vision systems. IEEE Expert Systems 98–113.

Ravich LE. (1987). Technological and cost improvements spur growth of image processing market. Laser Focus/Electro-Opt. Dec.: 148–155.

Spidell E. (1987). Selecting the optimum image sensor: Camera tubes vs. solid state arrays. Laser Focus/Electro-Opt. Jan.: 158–163.

Srihari SN. (1981). Representation of three-dimensional digital images. Assoc. Comp. Machinery Surveys **13**:399–422.

Wang DCC, Vagnucci AH, Li CC. (1984). Digital image enhancement: A survey. Comp. Vision, Graphics Image Process. **33**:363–381.

Glossary

[This Glossary has been compiled by the editor from definitions submitted by the coauthors.]

2-Deoxyglucose: A glucose analogue used in experiments that provides radioactive labeling of physiologically activated brain regions (also 2DG).

2-D: Two-dimensional.

3-D: Three-dimensional.

Achromatic Color Difference: The color difference of lightness (q.v.) that remains, for example, between two parts of a black-and-white reproduction, where the relative spectral distributions of the light reflected or emitted are identical.

Algorithm: The precise, formal statement of a computational procedure for solving a specified problem.

Analyzers: Psychophysically identified mechanisms that are sensitive to a limited range of values on a specific stimulus dimension (q.v.), and that are sensitive to that range of values in a limited region of the retinal image. For example, a mechanism sensitive to low spatial frequencies and localized to the central fovea would be such an analyzer.

Analyzer Tuning Width: The difference in the measures of two stimuli that are perceived to be independent on the same stimulus dimension (q.v.). For example, the difference between a threshold stimulus and the closest subthreshold stimulus that fails to affect it is a tuning width.

Anterograde: From cell body toward the axon terminals.

Aperture Colors: Colors seen one at a time through an "aperture" against a black background.

Attachment Surface: A 3-D surface of a primitive (q.v.) onto which another primitive can be attached.

Axon: The process of a neuron that is specialized to carry action potentials over long distances.

Backward Masking: The influence of a subsequently presented adapting or "masking" stimulus on the threshold of an earlier presented test stimulus. As the word "masking" implies, masking stimulus presentation generally increases the threshold of the test stimulus.

Band of Baillarger: Band of layer 4 in cortex containing terminals of thalamic fibers.

Bandwidth: A measure of the range of values on a stimulus dimension (q.v.) to which a detection mechanism is sensitive. It is usually expressed in octaves (q.v.). The bandwidth is often expressed at half amplitude, covering the range of values over which the mechanism gives a response at least half its maximum response.

Basic Color Terms: White, gray, black, red, green, yellow, blue, orange, purple, pink, and brown—names for the fundamental color categories first identified by Berlin and Kay.

Binder: The substance or mechanism that allows a colorant, such as a dye or pigment, to be bound to its substrate (textile, paper, etc.).

Blackbody Radiator: A radiator of uniform temperature whose radiance in all parts of the spectrum is the maximum obtainable from any radiator at a given temperature. It absorbs all radiant energy that falls on it.

Bleaching: The change in color from the red of a dark adapted retina to white after it has been exposed to light. During this process rhodopsin (q.v.) loses its chromophore (q.v.), and if it is not regenerated, another photon response cannot occur.

Blob: A local zone of high cytochrome oxidase concentration within the upper layers of striate cortex (also patch, puff, dot, spot).

Bloch's Law: Up to some limiting critical duration, psychophysical thresholds for short duration visual stimuli are determined by the product of the intensity and the duration of the stimulus. For typical experimental conditions, the critical duration is usually under 100 msec.

Bottom-up Recognition: Interpreting image data by proceeding from less to more complex structures without the aid of specific a priori models of the scene.

Brightness: The subjective, intensive attribute of a light sensation provided by an aperture color (q.v.), uncontaminated by context.

Brightness Constancy: The tendency for an object to have the same apparent brightness, regardless of the ambient level of illumination (e.g., moonlight vs. sunlight).

Bunsen–Roscoe Photochemical Law: The amount of photoproduct in a reaction depends on the integral of the amount of light presented with respect to time. For a constant intensity photostimulus, the amount of photoproduct is proportional to the product of the intensity and the duration of the photostimulus.

CAT-301: Antigen found in cat spinal cord and in portions of the magnocellular pathway in the primate visual system.

Caudal: Toward the tail.

Cell Bodies (Somas): The term refers to the central part of a neuron from which the dendrites and axons emerge. The soma contains the nucleus and most of the metabolic machinery required for the maintenance of cell function.

Center-Surround Cell: Gives stronger responses to small spots than to large spots (may be color opponent).

Centric: Radially symmetrical orientation column array in cortex.

CFF (or CFFF): See Critical Flicker Frequency or Critical Flicker Fusion.

Channel (Molecular): Membrane-bound molecules that span the membrane and allow ions or other substances to pass through. Often these channels are linked to receptors for neurotransmitters in which case they are called ligand-gated channels. Channels that open or close depending on the voltage across the cell membrane are termed voltage-gated channels.

Channel (Psychophysical): A set of analyzers (q.v.) that responds (or is sensitive) to the same stimulus dimensions, e.g., the analyzers that are sensitive to low spatial and high temporal frequencies.

Chromatic Adaptation: Modification of visual response to a test color brought about by chromatically non-neutral stimuli, usually pre-exposed.

Chromatic Color Difference: The color difference that remains when the relative spectral distributions of two stimuli differ at equal luminance.

Chromatic Discrimination: Perception of a difference between chromatic colors seen at equal luminance.

Chromaticity: Specification of a color in terms of coordinates that are the ratios of each tristimulus value (q.v.) to their sum.

Chromaticity Diagram: A two-dimensional chart in which colors are represented in terms of their chromaticity coordinates.

Chromophore: The light-absorbing retinoid group (a vitamin A derivative) attached to opsin in rhodopsin in rods and cones.

CIE: Commission International de l'Eclairage (French for International Commission on Illumination), an international standards body.

CO: Cytochrome oxidase.

Color Categorization: Tendency to group colors that differ, sometimes very significantly, but which are nevertheless similar in hue, and which tend to be identified with the same basic color term.

Color Column: Region of striate (or prestriate) cortex within which the cells respond preferentially to the same color.

Color Constancy: The tendency of the perceived chromatic color of a surface to remain constant despite changes in the spectral distribution of the illuminant.

Color Context: Colors in the peripheral visual field which affect the appearance of a color being fixated and attended to.

Color Induction: The capacity of two or more nearby fields to alter one another's color appearance relative to that of any one of them seen in isolation.

Color Order: The arrangement of color samples in a continuously varying, three-dimensional space, usually with two chromatic dimensions and one achromatic one.

Colorant: A material, usually a dye or pigment, bound to a material, which largely determines its diffuse spectral reflectance and thereby its perceived color.

Colorimetry: A technique of measuring color stimuli to provide a system of color-stimulus specification. The procedure begins with radiometric measurement, from which tristimulus values (q.v.) and chromaticity coordinates are calculated.

Complex Cell: Orientation-selective cell in visual cortex, without separate ON and OFF regions in its receptive fields.

Computational Vision: A branch of computer science dealing with the precise formulation and solution of problems in visual information processing.

Conductance, cGMP Sensitive: The membrane conductance that is affected by cyclic guanisine monophosphate (cGMP); it is synonymous with "light-sensitive conductance."

Conductance; Light Sensitive: See cGMP sensitive.

Conductance; Inner/Outer Segment: The membrane conductance of the inner/outer segment of a photoreceptor.

Cones (L, M, and S): The photopic (q.v.) photoreceptors for daylight color vision, which are found in three types with overlapping spectral sensitivities. L, M, and S refer to long, middle, and short-wavelength peak sensitivities (sometimes called R, G, and B) of the three types of cones.

Contralateral: The opposite side of the brain.

Crawford Adaptation: Measurement of visual sensitivity at many fine ($\sim$0.1 sec) intervals with respect to the time of onset of an adapting flash. With Crawford's procedure, measurement is made both at negative (where the test flash precedes the onset of the adapting flash) as well as positive SOA (q.v.) values.

Critical Duration: See Bloch's law.

Critical Flicker Frequency: The maximal frequency of a flickering light at which it can be detected as flickering (i.e., does not appear as a steady light).

Critical Flicker Fusion: This refers to the highest flicker rate at which a pattern is no longer perceived as flickering, the rate of temporal change at which the visual system can no longer follow the changes.

Critical Period: A time during either pre- or postnatal brain development when information from the outside world is more or less permanently encoded within the brain.

Dark Adaptation: The visual changes resulting from lack of exposure to light. Classically, dark adaptation is considered to be the characteristic decrease in detection threshold (i.e., the increase in sensitivity) resulting from exposure to darkness. In psychophysics, depending on the task, dark adaptation can involve either an increase and/or a decrease in sensitivity.

Dark Current/Potential: The membrane current/potential of a photoreceptor in the dark.

Dark Threshold: The smallest stimulus that will produce a just detectable response in the dark.

Degenerate View: An image of an object taken from a viewpoint such that one or more distinguishing features of the object are not visible.

Dendrite: A process that emerges from the soma of a neuron, conventionally thought to receive inputs from other cells, but they also have output synapses.

Dendritic Field: The area or volume of neural tissue that is covered by the dendritic processes of a particular cell.

Depolarization: An increase in relative potential inside the cell membrane as compared to the extracellular field.

Deuteranopia: A type of dichromacy (or "color blindness") that is inherited in an X-linked recessive fashion. Deuteranopes lack the normal middle wavelength (green-cone) photopigment, but this condition is also associated with a number of known and unknown visual abnormalities.

Diffuse Reflection: Light that is selectively reflected (reradiated) from a surface in many directions, whose spectral distribution carries the information for perceiving the color of the surface.

Dimension of Image or Stimulus: Some characteristic quality of images or stimuli that can be scaled as a continuum of values, e.g., spatial frequency is such a dimension.

Disparity: Noncorrespondence between the retinal locations in the right and left eyes (for a point in space, or for the binocular receptive field of a cortical cell).

Dispersion: The variable refraction of light depending on its wavelength.

Dowling–Rushton Relationship: Visual threshold is directly proportional to the logarithm of the amount of bleached photopigment.

Driving Force: Refers to the potential energy available to move an ion across the membrane. The ion's concentration gradient across the membrane and the voltage across the membrane are two factors that determine this driving force.

Duplicity Theory: Vertebrate vision is mediated by two different types of photoreceptors, the rods and cones, which function under different levels of illumination.

Dye: One of the two principal forms of colorants, along with pigments, which are deliberately used to change the original color of a surface.

Ectopic: Mispositioned.

Endspectral: 400–475 nm and 625–700 nm wavelengths.

Face: A closed cycle of 2-D image contours.

Face Feature: A subset of the contours comprising a face.

Face Structure: A connected set of faces representing an aspect of a primitive (q.v.).

Ferry–Porter Law: Critical flicker frequency (or CFF) increases with the logarithm of stimulus intensity.

Fluorescence: The ability of certain substances to absorb energy and emit it at a different wavelength, usually a longer one.

Focal Colors: Red, yellow, green, blue, black, white, orange, purple, pink, brown, and gray (see Berlin and Kay, 1969).

Field Potential: An extracellular voltage which arises from the change of potential of a large number of cells. This change in transmembrane potential produces a current outside the cells which can be detected as a voltage change across the surface of a piece of tissue.

Gabor Filters: Mathematically, the operation applied by a Gabor filter is characterized by a sinusoid multiplied by a Gaussian. The resulting function is an oscillation that decreases from a central peak to very small amplitudes going away from the peak. It was used by Gabor in developing his information theory and more recently has been applied in filtering operations analogous to biological receptive fields.

Ganglion Cell: Retinal cells whose axons form the optic nerve.

Ganglion Cells (Primate): ON center: Depolarizes when light strikes its receptive field center. May receive input from L, M, or S cones. OFF center: Hyperpolarizes when light strikes its receptive field center (depolarized at light offset). May receive input from L, M, or S cones.

Gas Discharge Lamps: Lamps that emit light as the result of gas atoms that are excited by an electric current.

Generalized Cylinder: A 3-D volumetric primitive (q.v.) defined by sweeping an arbitrary cross-section shape along an arbitrary curve while allowing the size of the shape to vary according to an arbitrary sweep function.

Glia: Nonneuronal cells found associated with neurons. They are thought to provide nutritive support for neurons, but recently many more functions for glia have been discovered including a role in release of neurotransmitters. The principal glial cell in the retina is called the Müller cell.

Glutamate: An amino acid that is probably the transmitter released by optic nerve terminals, as well as at other neuronal synapses.

Harmonic: The term harmonic refers to any sine wave grating of frequency $k \times$ f relative to a fundamental frequency *f*, where *k* is a positive integer. The third harmonic is approximately 1.6 octaves above the fundamental frequency.

High Pass Filter: A system or structure, e.g., of cells or synapses, that faithfully transmits fast, high-frequency signals and attenuates or suppresses slow, low-frequency signals.

Horopter: An imaginary line through the point of fixation and forming an arc around one's head, which contains all points in focus at that accommodation setting of the eyes.

HRP: Horseradish peroxidase. It is used to anatomically visualize neurons.

Hypercolumn: Global array of visual cortical columns for a given locus of visual space.

Hypercomplex Cell: Orientation- and length-selective cell in visual cortex.

Impulse Response: This is the response to a high-energy stimulus, either very brief (temporal impulse response) or spatially very localized (a point of light, produces the spatial impulse response). If a system is linear, the impulse response describes the system completely, and the response to any other input can be determined.

Incandescence: The self-emission of radiant energy in the visible spectrum caused by the thermal excitation of atoms or molecules, most often by heating a tungsten filament electrically.

Increment Threshold Function: A plot of the threshold intensity of a light flash as a function of the intensity of the background on which it is superimposed. In general, increment threshold functions are plotted on logarithmic coordinates.

Inverse Imaging Problem: Recovering a scene description from the computed image data base.

Ion Exchanger: A membrane-bound molecule or structure which facilitates the exchange of ions across the membrane.

Ion Pump: A membrane-bound molecule or structure that moves ions across the membrane at the expense of energy stored in the cell. An example is the NA^+–K^+ pump, an ATPase or enzyme that hydrolyzes ATP and "pumps" NA^+ out of and K^+ into the cell.

Ipsilateral: Same side of the brain.

Isoluminant Colors: Two or more colors that differ only in their chromatic color (q.v.), with luminance and achromatic color (q.v.) held constant.

Isomerization of Rhodopsin: This refers to the change of conformational state of the molecule as it occurs after absorbing a photon or due to thermal energy fluctuations.

Kanisza Triangle: A well-known and compelling display of illusory contours, first created by Gaetano Kanisza (see Chapter 9, Fig. 9.6).

Labeled Analyzer: Analyzers are selective in their responding to the dimensions to which they are tuned. If information about which stimuli they are responsive to is available to processes that underlie discrimination responses, then the analyzer is labeled.

Lambertian Surface: A pure matte surface that looks equally bright in all directions. The light reflected at an angle i to the incident ray has an intensity proportional to cos i.

Lateral Inhibition: A common form of inhibition where cells inhibit neighboring cells which are on the same synaptic level. Examples are the horizontal cells and amacrine cells in the retina.

Layers of Striate Cortex: Upper layers: 1, 2, and 3. Middle layers: subdivisions of 4 (4A, 4B, 4Cα, 4Cβ). Lower layers: 5 and 6.

LGN: Lateral geniculate nucleus of the thalamus.

Ligand: A molecule that binds to a receptor molecule.

Light Adaptation: The visual changes resulting from exposure to increased illumination. Classically, light adaptation is considered to be the characteristic increase in detection threshold (i.e., the decrease in sensitivity) that results from exposure to illumination. In psychophysics, depending on the task, light adaptation can involve either a decrease and/or an increase in sensitivity.

Lightness: The subjective intensive attribute of a light sensation, provided by a surface color seen in a complex environment.

Linear System: A system defined by the following relationship of input and output: If $R(x)$ is the response to input x, and $R(y)$ is the response to input y, then for any x, y, input $x + y$ will produce response $R(x) + R(y)$.

Low Pass Filter: A system or structure such as of cells or synapses in which slow, low-frequency signals are carried faithfully but fast, high-frequency signals are largely attenuated or reduced.

Luminance: Luminous flux of an extended surface as measured at a point at which a unit of the projected area of the surface subtends a unit solid angle. Related to radiance by the photopic luminous efficiency function which takes into account the spectral sensitivity of the eye.

Luminous Efficiency Function [$V(\lambda)$]: An empirically determined and standardized function which weights radiant energy as a function of wavelength according to the sensitivity of the eye.

Machine Vision: The mechanization of visual tasks and processes and their application to industrial and information processing operations.

Magnification: The relationship between the point-to-point distance on a body surface (e.g., the retina) and the equivalent point-to-point distance on a map of that body surface in the brain (e.g., the visual cortex).

Magnocellular: Inner large-cell layers of the LGN, numbered 2 and 1.

Matched Filter: A filter which maximizes signal-to-noise ratio so that a waveform of known shape can be separated from random noise.

McCullough Effect: Exposure to a colored grating in one orientation alters the apparent color of a grating presented in other orientations. For example, prior exposure to a red against white vertical grating causes an achromatic horizontal grating to be tinged green.

Membrane Conductance: The electrical conductance of the membrane. It can consist of several components, subserved by different channels and affected by different stimulating agents.

Membrane Current: The electric current flowing through a membrane.

Membrane Potential: The difference in voltage on the two sides of a membrane.

Mesopic: Levels of illumination over which both rod and cone photoreceptors operate. See also Scotopic and Photopic.

Metamerism: (1) Condition in which two stimuli match visually although differing physically. (2) The degree to which a material appears to change color when viewed successively under different illuminants.

Midspectral: 475–625 nm wavelengths.

MT: Middle temporal visual area (V5) of cortex.

Nasal: The part of the eye near the midline (or nose); opposite of temporal.

Network Adaptation: Visual adaptation that is attributed to retinal neurons postsynaptic to the photoreceptors.

Neural Adaptation: Properties of visual adaptation that are not related to the concentration of photopigment within the photoreceptors, but to changes in neuronal membrane properties of the photoreceptors and/or postreceptor visual neurons.

NMDA: N-Methyl-*D*-aspartate, a molecule that binds tightly to one of the types of glutamate receptor.

Octave: An octave corresponds to a change in frequency by a factor of 2.

Ocular Dominance Column: Region of striate cortex within which the cells respond preferentially to input from either the right or the left eye.

OFF Region: Receptive field subregion associated with cell responses to darkness (or removal of light).

ON Region: Receptive field subregion associated with cell responses to light.

Opponent Colors: Colors whose hues, when additively mixed, cancel one another and disappear.

Orientation Shift: An abrupt change in the orientation tuning of cells encountered in a smooth trajectory through striate cortex (most typically, from upper to lower layers).

Orientation Column: Region of striate cortex within which the cells respond preferentially to the same line orientation.

Orthographic Projection: A parallel projection onto the image plane.

Parvocellular: Outer small-cell layers of the lateral geniculate nucleus (LGN), numbered 6, 5, 4, and 3.

Patch Clamp: With this technique a tight seal is formed between the mouth of a glass capillary and a small patch of membrane. Voltage (or current) can be clamped to a constant value and the current (or voltage) across the patch can be measured.

Pathway (Psychophysical): A set of channels which codes a set of stimulus dimensions, such as contrast, in relative isolation from the coding of other stimulus dimensions, such as color.

Photochemical Adaptation: Changes in visual sensitivity resulting from bleaching photopigment.

Photochemical Theory: An attempt to explain all known properties of human visual adaptation in terms of hypothetical photochemical processes. Today, photochemical theory is remembered primarily in a historical context and is usually associated with the ideas of Selig Hecht.

Photopic: Brighter levels of illumination over which only cone photoreceptors operate. See also Mesopic and Scotopic.

Pigment: One of the two principal forms of colorants, along with dyes, that are deliberately used to change the original color of a surface.

Pixel: An acronym for "picture element." In computer vision a picture is composed of square pixels (see, e.g., Chapter 12, Fig. 12.1).

Plasma Membrane: Synonymous with the membrane separating the outside from the inside of a cell.

Prestriate Cortex: Areas V2, V3, V3A, V4, MT(V5).

Primary Colors: Red, yellow, green, and blue [see Hering, 1878 (1964), Chapter 8].

Primitive: An elementary object used to construct object models. For example, the primitives can be a restricted class of generalized cylinders.

Principle of Univariance: The concept that a single type of cone (or rod) photoreceptor responds to light more or less strongly, with information about the wavelength of the absorbed light being lost.

Protanopia: A type of dichromacy (or "color blindness") that is inherited in an X-linked recessive fashion. Protanopes lack the normal long wavelength (red cone) photopigment, but this condition is also associated with a number of other known and unknown visual abnormalities.

Radiance: Radiant intensity per unit projected area. Related to luminance by the photopic luminous efficiency function which takes into account the spectral sensitivity of the eye.

Receptive Field: This consists of that set of cells from which a target cell (q.v.) receives input and which produces a response of the target cell. Referred to the retina, the receptive field of a target cell, such as in LGN or V1, consists of a retinal neighborhood that contributes to the response of the target cell.

Receptor (Molecular): A molecule that is activated by a chemical such as a neurotransmitter and that can alter the state of the cell in which it is found. Many types of receptors contain channels that can open up to provide a pathway for ions to move across the cell membrane.

Reflection: Process whereby light incident on a surface is absorbed and reradiated at the same wavelength. If the wavelength changes, the process is called fluorescence instead.

Response Latency: The time interval between stimulus and response.

Resting Potential: The membrane potential in the resting state, in the absence of activity.

Retinal Coordinates: A coordinate system to describe image size or location that is fixed with respect to the retina. Usually distance is expressed in units of visual angle.

Retinal Densitometry (also called "Fundus Reflectometry"): A technique that permits measurement of photopigment density in the intact living eye of humans or other vertebrates.

Retinex: A scheme that attempts to predict color appearance of test colors in a complex situation where color context has a significant influence (see Chapter 8 and, e.g., Land, 1986).

Retinotectal: Referring to the projection from the retina to the tectum of cold-blooded vertebrates or to the superior colliculus of mammals.

Retrograde: From axon terminal toward cell body.

Reversal of Mapping: The location in a brain map at which a particular mapped parameter (e.g., color, orientation) shows an abrupt change in mapping direction.

Reversal Potential: The voltage at which no current will flow through the membrane or through a particular type of channel. At this voltage the concentration gradient and electrical gradient produce equal and opposite forces on ions which carry current through the membrane or channel.

Rhodopsin: The type of light-absorbing photopigment found in rod photoreceptors of all land vertebrates including humans.

Ricco's Law: Up to some critical limiting area, psychophysical thresholds for spatially small stimuli are determined by the product of their area and their intensity. Under dark adapted conditions in the peripheral retina, Ricco's law holds for circular stimuli up to about 1° in diameter. For central vision and under more light adapted conditions, Ricco's law holds only for considerably smaller stimuli.

Rod–Cone Interaction: An interaction between signals stemming from the activity of the rod- and cone-related visual system. SRCI (q.v.) is one of many forms of rod–cone interaction.

Rod Intrusion: Complications in colorimetry and color appearance that are caused by signals from rods that interact with those from the three classes of cones.

Rod Monochromia: An extremely rare form of complete color blindness in which the affected individual has no functioning cone photoreceptors but has apparently normal rod vision.

Rostral: Toward the head.

Scatter: Process causing light, as it passes through a medium, to interact with the particles of the medium so as to change its direction of travel and, in some cases, its wavelength.

Scotopic: Dimmer levels of illumination over which only rod photoreceptors operate (levels of illumination that are above absolute threshold but below cone threshold).

Simple Cell: A cortical, orientation-selective cell, with separate ON and OFF regions in its receptive field.

SOA: See Stimulus Onset Asynchrony.

Spatial and Temporal Frequency Tuning: Observers, analyzers, and neurons have been shown to be differentially sensitive to spatial and temporal modulation of patterns. A function that describes sensitivity on a spatial or temporal frequency axis is said to measure the tuning.

Spatial Frequency: A measure of the fineness of a repeating visual pattern in terms of the number of points or lines per unit distance.

Specific Threshold: The minimal intensity of a flash necessary for detecting its color. Under dark adapted conditions, specific threshold generally has a much higher intensity value than detection threshold (i.e., the intensity necessary for detecting the presence of a flash but an intensity that does not necessarily permit correct identification of the stimulus size, color, or shape).

Spectral Distribution: A function that describes the relative amount of light which is emitted, transmitted, or reflected as a function of wavelength.

SRCI: See Suppressive Rod–**Cone Interaction.**

Stereoscopic Vision: The aspect of vision concerned with three-dimensional spatial relationships.

Stimulus Dimension: See Dimension.

Stimulus Onset Asynchrony: This term applies in psychophysical experiments in which two different stimuli are pulsed and presented at various intervals in respect to one another. The stimulus onset asynchrony (SOA) value is the time between the onset of the adapting or masking stimulus (which is usually more intense and of longer duration) and the onset of the test flash. In general, the observer adjusts the intensity of the test flash for threshold. The SOA is referred to by a variety of other terms in the literature, most commonly the delay interval.

Stripe of Gennari: A prominent stripe in layer 4 of V1 where thalamic afferents terminate.

Suppressive Rod–Cone Interaction: A tonic inhibitory influence of dark adapted rods on cone pathways. It occurs in the distal retina.

Surface Colors: The colors seen on actual surfaces in the visual world.

Synapse: The connection between cells that can either be electrical, referred to as gap junction, or can be chemical, in which case it uses a neurotransmitter.

Synapse, Sign Conserving: A synapse in which the postsynaptic cell changes potential in the same direction as the presynaptic cell, e.g., when the presynaptic cell depolarizes, so does the postsynaptic cell.

Synapse, Sign Inverting: A synapse in which the postsynaptic cell changes potential in the opposite direction from the presynaptic cell, e.g., when the presynaptic cell depolarizes, the postsynaptic cell hyperpolarizes.

Target Cell: This is a cell which is the target of inputs from a set of cells forming the receptive field (q.v.) of the target cell.

Temporal: Toward the temple, toward the side; opposite of nasal.

Temporal Frequency: It is possible to temporally modulate spatial frequency and other stimuli, producing a flickering appearance and sometimes apparent movement. This temporal modulation can be measured in terms of the number of sinusoidal modulations of the pattern elements from light to dark to light again that occur in 1 sec. This is the temporal frequency of the flickering pattern measured in cycles per second, or Hz.

Tetrodotoxin: A drug that blocks voltage-sensitive sodium channels and thus blocks the initiation and propagation of action potentials.

Threshold: The smallest stimulus that produces a just detectable response.

Top-Down Recognition: Using specific a priori models to constrain the interpretation of image data.

Topography: The orderly relationship between the positions of one group of cells, such as ganglion cell bodies in the retina, and the positions of their target nuclei, such as the optic tectum.

Transient Tritanopia: A transient (~1 sec) tritanopia (q.v.) in normal human subjects induced by prior exposure to intense yellow stimuli.

Trichromacy: The nature of human color vision that, in its initial stage, consists of three different classes of signals that originate in three different classes of cone photoreceptors.

Tristimulus Values: Relative excitations of the three kinds of cone photoreceptors, or other functions such as the x, y, z functions of the CIE system of colorimetry which predict the same color matches.

Tritanopia: An exceedingly rare type of dichromacy (or "color-blindness") that is inherited. Tritanopes lack the normal short wavelength (blue-cone) photopigment.

TTX: Abbreviation for tetrodotoxin (q.v.).

TVI Curves (threshold vs. illuminance or threshold vs. intensity functions): The name for increment threshold curves used by W. S. Stiles and other investigators interested in human color vision. The term is virtually synonymous with the terms "light adaptation function" or "increment threshold function."

V1: Striate cortex, area 17.

Visual Adaptation: The change in visual processing resulting from exposure to darkness or light.

Visual Aftereffects: Changes in vision following prolonged exposure to specific types of stimuli. The prior adaptation condition (which induces the aftereffect) and subsequent test conditions (under which the aftereffect is observed) do not usually involve changes in overall illumination (e.g., the tilt aftereffect, the waterfall illusion) and, hence, involves mechanisms different from those underlying light and dark adaptation.

Voltage-Sensitive Membrane Conductance: That membrane conductance which is affected by the membrane potential.

Weber's Law: The just noticeable difference (i.e., *jnd* or ΔI) in intensity of a stimulus is a constant percentage of the initial value (I) of a stimulus. $\Delta I/I$ is equal to the Weber constant.

Index

CPSIA information can be obtained at www.ICGtesting.com
Printed in the USA
LVOW12s1609260514

387280LV00007B/140/P